V International Conference on Photonics and Information Optics (PhIO 2016)

Journal of Physics: Conference Series Volume 737

Moscow, Russia
3 – 5 February 2016

Editor:

N.N. Evtikhiev

ISBN: 978-1-5108-2993-0
ISSN: 1742-6588

Printed from e-media with permission by:

Curran Associates, Inc.
57 Morehouse Lane
Red Hook, NY 12571

Printed by Curran Associates, Inc. (2016)

For permission requests, please contact the Institute of Physics
at the address below.

Institute of Physics
Dirac House, Temple Back
Bristol BS1 6BE UK

Phone: 44 1 17 929 7481
Fax: 44 1 17 920 0979

techtracking@iop.org

Additional copies of this publication are available from:

Curran Associates, Inc.
57 Morehouse Lane
Red Hook, NY 12571 USA
Phone: 845-758-0400
Fax: 845-758-2633
Email: curran@proceedings.com
Web: www.proceedings.com

Table of contents

Volume 737

V International Conference of Photonics and Information Optics
3–5 February 2016, Moscow, Russia

Accepted papers received: 26 July 2016
Published online: 13 September 2016

Preface

Papers

Coherent and nonlinear optics

Optical materials, the interaction of radiation with matter

Optical measurements, biophotonics

Optical information processing and holography

Preface

This special volume of JPCS contains the "Proceedings of the V International conference on photonics and information optics (PhIO 2016)". The conference was held in Moscow, Russian Federation in February 3-5, 2016.

Organizers of the conference:

- The Russian Academy of Sciences
- National Research Nuclear University «MEPhI» (Moscow Engineering Physics Institute).

The conference included 10 meetings in total: two plenary sessions and eight section meetings, also poster sessions took place. 100 oral and more than 120 poster reports were presented by scientists from 130 organizations of more than 10 countries.

The V International conference on photonics and information optics carried on traditions of previous forums held in National Research Nuclear University «MEPhI» (Moscow Engineering Physics Institute): wide range of topics, broad scope of participants and the involvement of young scientists and students. Relevance and importance of researches on photonics and information optics, need of an exchange of new ideas and methods between experts both in the Russian Federation, and at the international level, cause noticeable interest in this conference not only among scientists of the CIS, but also in Europe and Asia.

Chairmen of the program committee are members of the Presidium of the Russian Academy of Sciences, president of Kotelnikov Institute of Radio-engineering and Electronics of the Russian Academy of Sciences, academician Yu.V. Gulyaev and head of the Laser Physics Department of National Research Nuclear University «MEPhI» (Moscow Engineering Physics Institute), professor N.N. Evtikhiev Conference themes: coherent and nonlinear optics, fiber and integrated optics, optics of crystals and acousto-optics, interaction of radiation with matter and optical materials, optical measurements and biophotonics, holography and digital optics, optical communication and optoelectronic data processing.

The articles based on materials, selected by the committee, presented at the V International conference on photonics and information optics, are included in this special issue.

The next conference is the VI International conference on photonics and information optics. The conference will be held in the National Research Nuclear University «MEPhI» (Moscow Engineering Physics Institute) in Moscow, Russian Federation under the auspices of the Russian Academy of Sciences and the Ministry of Education of the Russian Federation in February 1-3, 2017. The web site of the conference is http://fioconf.mephi.ru. We look forward to seeing all of you at the VI International conference on photonics and information optics.

Co-chairmen of program committee of PhIO 2016, professor N N Evtikhiev

Journal of Physics: Conference Series **737** (2016) 011002 doi:10.1088/1742-6596/737/1/011002

Peer review statement

All papers published in this volume of *Journal of Physics: Conference Series* have been peer reviewed through processes administered by the proceedings Editors. Reviews were conducted by expert referees to the professional and scientific standards expected of a proceedings journal published by IOP Publishing.

Journal of Physics: Conference Series **737** (2016) 012001 doi:10.1088/1742-6596/737/1/012001

Mathematical modelling of microwave photonic time-stretch system

E Yu Zlokazov, R S Starikov, V A Nebavskiy

National Research Nuclear University MEPhI (Moscow Engineering Physics Institute), 31 Kashirskoe shosse, , 115409 Moscow, Russia

E-mail: ezlokazov@gmail.com

Abstract. The paper highlights the mathematical modeling results of optical pulse stretching process in fiber due to group velocity dispersion effect.

1. Introduction

Investigations described in the present article are aimed to mathematical modeling of broadband optical pulses propagation in single mode fiber and explore the temporal stretching of pulses due to GVD (group velocity dispersion) effect. GVD induced stretching of broadband optical pulses allows realization of bandwidth compression method that can be used in different information processing systems such as wideband PhADC (photonic analogue-to-digital converters) [1-4]. Advantage of this method is the possibility to construct a system capable to capture the wideband (up to 1THz) radiofrequency (RF) signal using limited bandwidth ADCs (analogue-to-digital converters) with record resolution [3].

Figure 1 illustrates the principle of PhADC based on optical pulses time stretching. Super-continuum (SC) source generates wideband optical pulse; firstly a pulse propagates through the first dispersion module D1 with fiber length L1. GVD causes the temporal shift of pulse spectral components such that each lower wavelength component goes after higher wavelength components. Electro-optic light modulator modulates the chirped super-continuum pulse by input analogue RF signal. Second dispersion module D2 with fiber length L2 provides next temporal shift of modulated pulse optical components resulting the widening of pulse RF modulated envelope. This operation provides the compression of encoded RF signal bandwidth, thus photo diode and low-bandwidth electronic ADC can be used to capture and digitize the stretched sample.

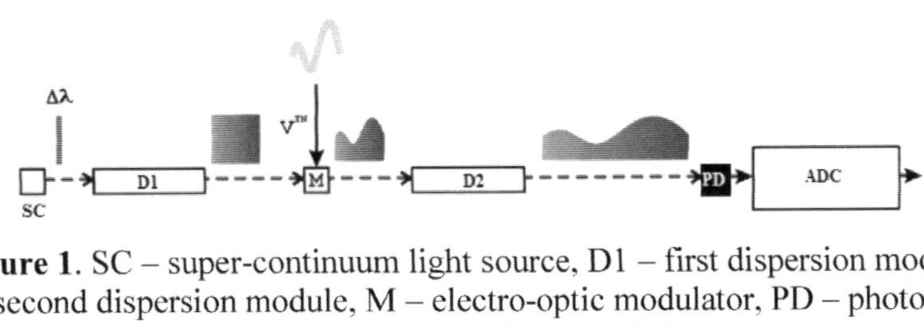

Figure 1. SC – super-continuum light source, D1 – first dispersion module, D2 – second dispersion module, M – electro-optic modulator, PD – photo-diode, ADC – electronic "slow" analogue to digital converter

Input RF sample stretch factor can be determined as

$$N = \frac{\tau_2}{\tau_1} = \frac{L_1 + L_1}{L_1} \quad (1)$$

where τ_1, τ_2 are the lengths of an optical pulse after first and second dispersion modules correspondingly. This determines the length of radio signal to be processed according to

$$\tau_1 > \tau_{RF} \text{ or } \tau_1 \gg \frac{1}{\omega_{minRF}} \quad (2)$$

where $\tau_{RF}, \omega_{minRF}$ are the temporal length and the lowest frequency of RF signal. Another one limitation of this system is that optical bandwidth must be wider than input RF bandwidth:

$$\Delta\omega_{opt} \gg \omega_{maxRF} \quad (3)$$

In the case of continuous time signal processing a multichannel version of the described optical scheme can be used [2-4].

2. Mathematical models

2.1. Super-continuum light source and fiber

The theory of electro-magnetic wave propagation in dispersive medium was used to provide the analytical model of proposed system. Using the generalized nonlinear Schrödinger equation, slowly varying amplitude approximation (SVAA), and single-mode fiber model in coordinate system moving with optical field group velocity the variation of optical amplitude envelope [5] A can be described by:

$$\frac{\partial A}{\partial z} + \frac{\alpha}{2}A + \frac{i}{2}\beta_2\frac{\partial^2 A}{\partial T^2} - \frac{1}{6}\beta_3\frac{\partial^3 A}{\partial T^3} = i\gamma\left[|A|^2 A + \frac{2i\,\partial}{\omega_0\,\partial T}(|A|^2 A) - T_R A\frac{\partial|A|^2}{\partial T}\right] \quad (4)$$

For successful performance the parameters of a system must be chosen such that the nonlinear effects will be negligible comparing to dispersion effects. Also it is useful to introduce the normalized amplitude in order to exclude the signal decay from equation (4):

$$A(z,T) = \sqrt{P_0}e^{-\frac{\alpha z}{2}}U(z,T) \quad (5)$$

Where P_0 – peak impulse power. The propagation equation can be written as:

$$\frac{\partial U}{\partial z} = -i\frac{1}{2}\beta_2\frac{\partial^2 U}{\partial T^2} + \frac{1}{6}\beta_3\frac{\partial^3 U}{\partial T^3} \quad (6)$$

Where β_2, β_3 – 2nd and 3rd order propagation constants (DVG-2, 3).
Using Fourier method one can obtain:

$$U(T,z) = \left(\frac{1}{2\pi}\right)\int_{-\infty}^{\infty}\hat{U}(\omega,0)K(\omega,z)e^{-i\omega T}d\omega \quad (7)$$

$$K'(\omega,z) = \exp\left(\frac{j\beta_2\omega^2 z}{2} + \frac{j\beta_3\omega^3 z}{6}\right) \quad (8)$$

$K'(\omega,z)$ – fiber impulse response function depending on z, $\hat{U}(\omega,0)$ – Fourier transformation of a signal in fiber input, $U(T,z)$ – signal function at the fiber point z.

2.2. Modulator

We used a Mach–Zehnder modulator in analytical model. Suppose it biased in the central point of operating performance curve linear region, in this case the envelope function of an optical pulse passed through the modulator can be written as:

$$U_{out}(t) = U_{in}(t)\cos\left(\frac{\pi}{4} + \frac{m}{2}Y(t)\right), \quad (9)$$

where m – is a modulation depth, $Y(t) = \frac{U_{max}}{U_\pi}s(t)$ – RF signal function, $\frac{U}{U_\pi} = \frac{m}{2}$, U_π – half-wave voltage, U_{max} – is a maximum value of Y(t).

2.3. Photo diode

The current of a photo diode illuminated by optical field $E_{in}(T)$ can be described as

$$I(t) = KE_{in}^{*}(T)E_{in}(T), \quad (10)$$

where

$$K = \frac{c\varepsilon_0 n}{2} R_{pd} A_{ef}, \quad (11)$$

A_{ef} – fiber modal efficiency, R_{pd} – photo diode sensitivity, ε_0 – dielectric constant, n – light permittivity, c – light speed in vacuum. Factually a photo diode performs as quadratic detector. The equations introduced in this section describe all the modules of optical system presented in figure 1.

3. Modeling examples

We used a Gaussian shaped super-continuum pulse with temporal length of 833fs in the model of optical signal stretch system (figure 2).

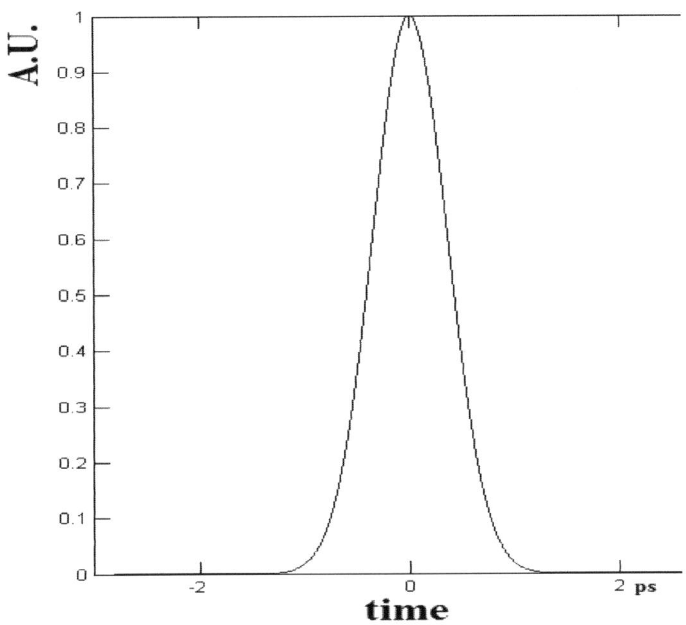

Figure 2. SC pulse intensity envelope

The first dispersion module D1 has the fiber length $L_1 = 2\text{km}$ and 2nd and 3rd order dispersion constants $\beta_2 = 20\frac{ps^2}{km}$, $\beta_3 = 0.2\frac{ps^3}{km}$. Figure 3 represents the envelope of source pulse after propagation through D1.

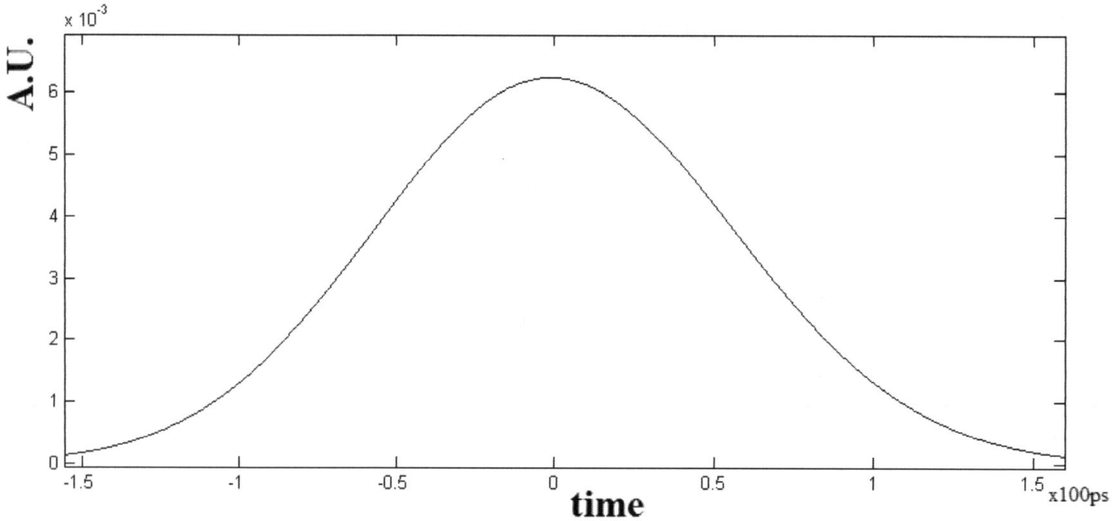

Figure 3. SC pulse envelope after propagation through D1

After D1 the stretched SC pulse is modulated by sinusoidal RF signal with the frequency of 30GHz and modulation depth m = 20% (see figure 4a).

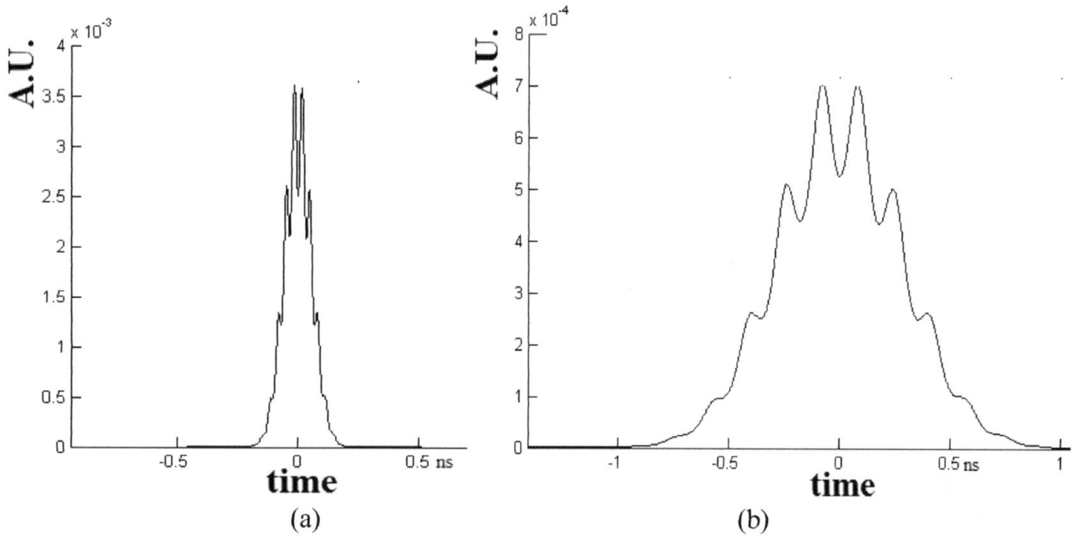

Figure 4. RF modulated pulse envelope (a); pulse envelope after propagation through D2 (b)

Being modulated the optical pulse is sent to second dispersion module D2 with fiber length L_2 = 8km, dispersion parameters are chosen the same as for D1 module. The modulated and stretched in D2 optical pulse envelope is demonstrated in figure 4b. Figures 5a and 5b represent the normalized amplitude function and spectrum of a signal in the output of PD after subtraction of Gaussian pulse envelope function.

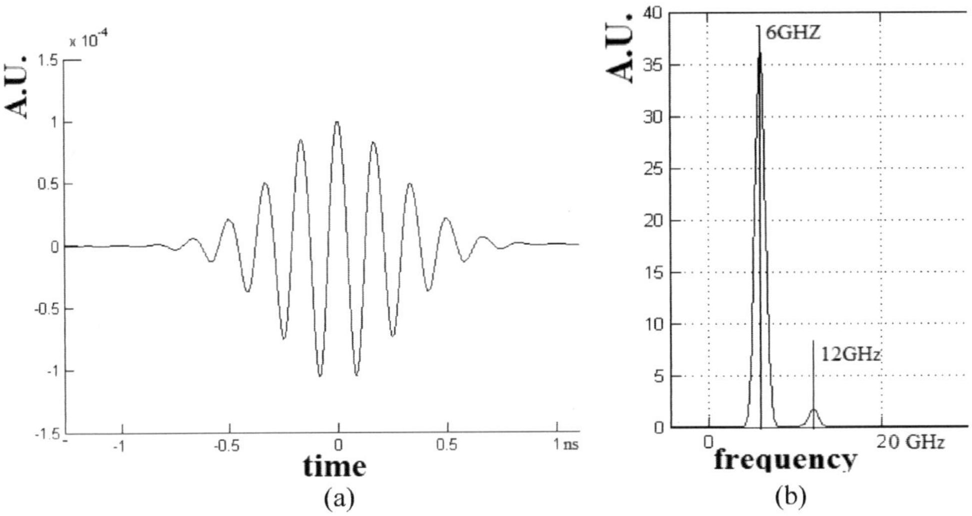

Figure 5. Pulse amplitude function at PD output after subtraction of Gaussian envelope term (a) RF spectrum of this signal (b)

The view of a spectral distribution of RF signal at PD output demonstrates N times stretch of the signal corresponding to (1), also it shows the appearance of spur 2nd harmonic.

4. Conclusion

The reviewed analytical model was used in development of a special software capable the simulation of time-stretch PhADC system. Modeling results demonstrate the stretch of RF signal using the system under research and appearance of spur harmonic caused by electro-optic modulator nonlinearity.

Acknowledgments
This work was supported by the Russian foundation of basic researches, project number 14-29-08174.

References

[1] Caputi W 1971 *IEEE Trans. Aerosp. Electron. Syst.* pp 269-278
[2] Valley G 2007 *Optics Express* **15** pp 1955-1982
[3] Fard A, Gupta S, Jalali B 2013 *Laser Photonics Review* **7** pp 207-263
[4] Starikov R S 2015 *Achievements of Modern Radioelectronics* **2** pp 3-39
[5] Agrawal G P 2001 *Nonlinear Fiber Optics (3rd ed)* (San Diego, CA: Academic) ch 3

Journal of Physics: Conference Series **737** (2016) 012002 doi:10.1088/1742-6596/737/1/012002

Photonic Beamformer Model Based on Analog Fiber-Optic Links' Components

V A Volkov[1], D A Gordeev[1], S I Ivanov[2], A P Lavrov[2], I I Saenko[2]

[1] Open Joint Stock Company "Research Institute "Vektor", Academician Pavlov str. 14a, Saint Petersburg 197376, Russia

[2] Peter the Great St.Petersburg Polytechnic University, Polytechnicheskaya str. 29, Saint Petersburg 195251, Russia

e-mail: lavrov@ice.spbstu.ru

Abstract. The model of photonic beamformer for wideband microwave phased array antenna is investigated. The main features of the photonic beamformer model based on true-time-delay technique, DWDM technology and fiber chromatic dispersion are briefly analyzed. The performance characteristics of the key components of photonic beamformer for phased array antenna in the receive mode are examined. The beamformer model composed of the components available on the market of fiber-optic analog communication links is designed and tentatively investigated. Experimental demonstration of the designed model beamforming features includes actual measurement of 5-element microwave linear array antenna far-field patterns in 6-16 GHz frequency range for antenna pattern steering up to 40°. The results of experimental testing show good accordance with the calculation estimates.

1. Introduction

Operation in an ultrawide frequency band (over several octaves) is highly desirable for some up-to-date types of phased array antennas (PAA). Beamformers based on true-time-delay (TTD) technique permit to form scanning PAA beam or a set of switched beams free of squint effect and achieve ultrawideband array antennas operation. For the last decades, the TTD beamforming has been often associated with photonic technologies, in view of their advantages compared with traditional electronic technique, such as wide instantaneous bandwidth, low loss in signal transmitting, small size and weight, high resistance to electromagnetic interference and flexible designing [1 - 3]. So, applying the microwave photonic devices for development of PAA beamforming systems affords the means to overcome conventional radio electronic beamformer limitations. Therefore, many beamforming architectures based on microwave photonics have been proposed and evaluated in the past years, for PAA both in transmission and reception modes. Most of the proposed architectures using fiber optic components to provide control of one- or multibeam antenna array pattern require specially developed units, such as chirped fiber Bragg gratings, fast tunable lasers with narrow spectral linewidths, reflection fiber segments, optical filter arrays and others. Meanwhile, up to date some manufactures have developed a family of microwave photonic links operating up to 40 GHz, offering various levels of bandwidth, noise figure and dynamic range. The performance characteristics of these modern analog photonic data communication links' components have attained the level, which allows their using to construct experimental prototypes of PAA beamforming systems and further designing of advanced beam forming network. Implementation of widely developed analog fiber-optic link components to compose the antenna array beamformer (BF) arrangement has been extensively investigated [4 - 6]. Some of proposed architectures were found to be the most suitable for this way of BF prototype designing; however, there are special requirements (namely, phase-frequency dependence, intrinsic time delays in transmitter and receiver modules) which the BF photonic components have to meet anyway while the fiber-optic communication links manufacturers do not take them into account. Here we consider TTD photonic BF for PAA in receive mode based on dense wavelength division multiplexing (DWDM) components and chromatic dispersion of a single mode optical fiber and implemented with units and components available at the market of modern fiber-optic communication systems [6]. Also we present some results of a kind of BF experimental model assembling and adjustment and its performance investigation including the microwave receiving linear PAA far-field pattern measurement in the 6-15 GHz frequency band. Experimental results show squint-free beamforming achieved with the proposed beamformer model.

2. Photonic Beamformer Model Architecture and Principle of Operation

The developed beamformer architecture based on the components implementing a DWDM technology approach and optical fiber chromatic dispersion is shown in Figure 1 [4, 6, 7]. This scheme is designed for the N element's linear PAA and uses optical comb – a set of N lasers with different but uniformly spaced (step $\Delta\lambda$) wavelengths, the total wavelength band is $(N - 1)\cdot\Delta\lambda$. The RF signals from the antenna elements A_1 .. A_N modulate a set of laser diodes. Electrooptic conversion is achieved either by

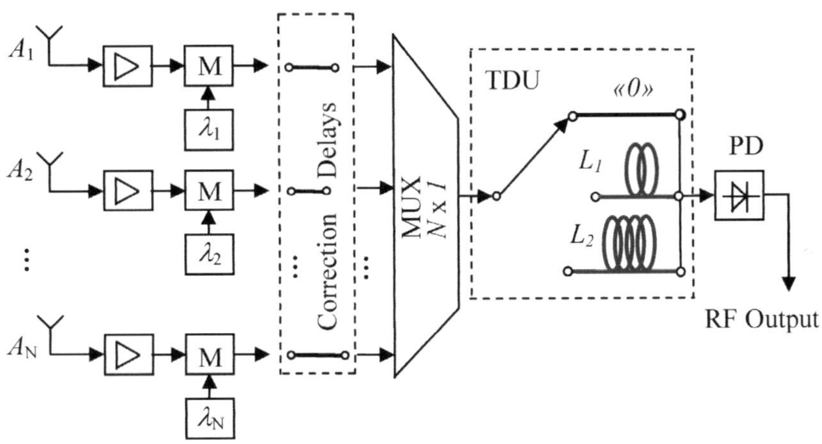

Figure 1. Architecture of photonic beamformer with fiber chromatic dispersion based interchannel delays.

direct modulation of a lasers or by using an external modulators as represented in Figure 1 (scheme elements M). Further, the intensity-modulated optical carriers are combined into a single fiber by a multiplexer (MUX N x 1 unit) and fed into a time delay unit (TDU). For practical application TDU may include one optical switch n x n and $(n - 1)$ fiber segments whose lengths vary exponentially: ΔL, 2 ΔL, 4 ΔL, etc. as in [7]. Switch control allows to change the total length of the fibers from ΔL to $L = (2^{n-1} - 1)\,\Delta L$ with increment ΔL. As all the optical carriers share with the same light paths, the time delay differences between adjacent channels are produced by the fiber chromatic dispersion D measured in ps/(nm·km). Time delay $\tau = \Delta\lambda{\cdot}L{\cdot}D$ introduced between adjacent channels results in the respective tilting of PAA beam. Photodiode (PD) at TDU output converts sum of delayed intensity-modulated optical carriers back to microwave domain. In Figure 1 instead of optical switch n x n and $(n - 1)$ fiber segments one can see 2 fiber lengths L_1 and L_2 successively inserted between the multiplexer and the photodiode thus giving (together with "*0*" - length position) 3 interchannel delay values: 0, τ_1 and τ_2 corresponding to 0, θ_1 and θ_2 beam pointing angles. The "Correction Delays" unit contains fiber segments with strictly sized up lengths equalizing the interchannel delays from RF sources (scheme elements A_1 .. A_N) to the MUX output. Hence, it corresponds to a flat phase law tilted 0 to PAA base line – the PAA beam directed normal to PAA base.

In the considered architecture of 1-D beamformer the number of PAA beams can be increased (as for example in [4]) that only requires a higher splitting ratio and another set of dispersive fibers and photodiodes.

3. Research of Photonic Beamformer Key Components

Among the photonic BF model's key components depicted in Figure 1 the units converting radio signal into optical one (laser jointly with modulator) and backward (photo receiver) are crucial for BF correct performance. In our BF model arrangement the transmitters and receiver modules from analog fiber-optic links OTS-2-18 (Emcore [8]) were used which can operate at 0.1..18 GHz bandwidth. In-built microprocessor-based transmitter control for laser bias and temperature as well as Mach-Zehnder modulator bias provides better consistent performance operation and allows for appreciably reduce the measurement results variations. The receiver module has an additional RF amplifier with 15 dB gain after photodetector. Transmitter unit includes 3 separate optically combined elements: DFB laser with wavelength corresponding to ITU grid with 100 GHz step, $LiNbO_3$ Mach-Zander modulator (MZM) with half-wave voltage approx. 5.8 V, optical 10 dB directional coupler and photodiode for MZM quadrature working point stabilization. Transmitter optical power in the fiber is about 10 dBm.

Figure 2 shows the transfer function P_{OUT} (P_{IN}) for one of OTS-2-18 links. It was measured with two input RF signals with frequencies 3.99 and 4.00 GHz and equaled powers: the principal harmonic power (line 1), the third order intermodulation response (line 2), and the receiver output noise P_{min} = – 155 dBm, averaged through the total frequency bandwidth 18 GHz and related to the 1 Hz bandwidth. More detail performance characteristics estimations are given in [9].

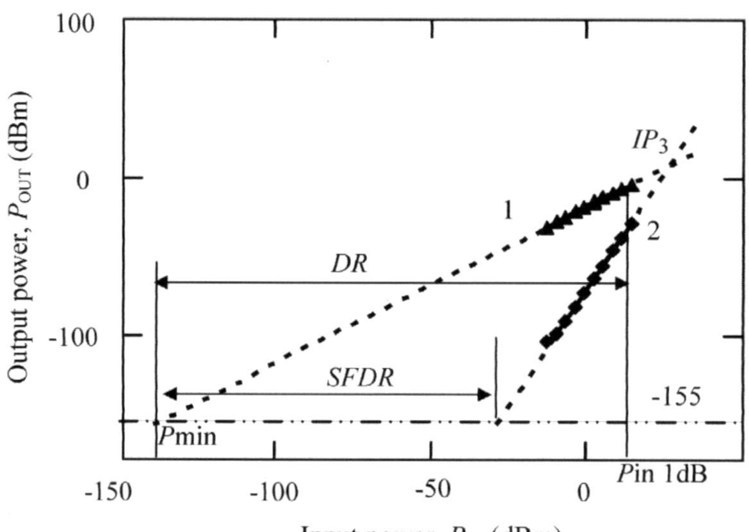

Figure 2. Fiber-optic OTS-2-18 link transfer function.

The noise power was measured with unmodulated receiver optical input (approx. 10 dBm). The measured link gain ranges from –3 dB at 1 GHz to –7 dB at 18 GHz. The 1 dB compression point $P_{IN\ 1dB}$ = 16.5 dBm, and the minimum input level is –144 dBm (when line 1 crosses the noise level P_{min}). Thus, the 1 dB compression dynamic range comes to 160.5 dB and the spurious free dynamic range is about 113 dB. The input signal level *IP*3 corresponding to the third order intermodulation products amounts to 25 dBm (P_{IN} when lines 1 and 2 are intersected).

The channels' phase-frequency characteristics are highly significant for BF performance. Figure 3 shows both of the total phase-frequency characteristic (argument of $S_{21}(f)$) and its deviation $\Delta\varphi(f)$ from a linear function for the frequency band of up to 20 GHz. It was measured for direct connection of transmitter and receiver units. One can see significant link time delay. Intrinsic time delays brought by link modules were measured separately and resulted in amount up to 17.4 ns for transmitter and 15.6 ns for receiver modules.

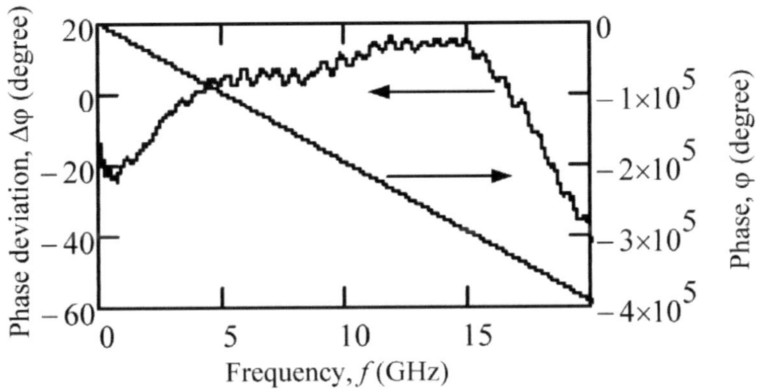

Figure 3. Phase-frequency response for OTS-2-18 link.

Summation of optical signal from transmitters is fulfilled by 8-channel multiplexer whose channels are centered in accordance with ITU standard channels from 27 to 34 with 100 GHz step. Its insertion loss is approx. 1.5 dB, –1 dB bandwidth approx. 65 GHz, that is sufficient for transmission of intensity modulated optical carrier with RF frequencies up to 20 GHz and more. Nevertheless, it should be mentioned that all of the transmitters' and corresponding multiplexer channels' midmost wavelengths have to be accurately alignment according to ITU channel plan and avoid mismatching during operation. We have measured passband characteristics for all multiplexer channels with optical spectrum analyzer and found good wavelengths setting and flat-shape channel passband curves behavior [10].

As it was pointed earlier one need to align the delays of all BF channels from MZM inputs to common MUX output. This can be done by insertion of equalizing fiber patch cord in each BF channel (see Figure 1: Correction Delays unit). We measured channels time delays τ_{CH} by Vector Network Analyzer (R&S, ZVA 40) using the relation: $\arg[S_{21\,k}(f)] = 2\pi f \cdot \tau_{CH}$. Further one can calculate necessary lengths of corrective patch cords for all BF channels to provide their identical delays at the TDU input. TDU in BF model consists of SM fibers with lengths labeled 0, L_1, L_2 corresponding to PAA beam pointing angles θ_0, θ_1, θ_2 in accordance with relation $L = d{\cdot}\sin(\theta)/(c{\cdot}D{\cdot}\Delta\lambda)$ where d is the distance between neighboring antenna elements and c is the speed of light in a vacuum.

4. Beamformer Model Design and Setup

Initially BF model was intended for measurements in conjunction with PAA with element spacing $d = 25$ mm. However actual measurements with 1-D PAA for demonstration of the designed BF model beam steering performance were accomplished with wideband 1-D PAA composed of $d = 10$ mm spaced Vivaldi patch radiating elements. To approximate the initially chosen range of delays between BF channels (and dispersion fiber lengths L_1, L_2) with the distance between radiating elements the last ones were connected to five BF inputs selectively through one thus giving radiating elements actual spacing $d = 20$ mm. This spacing jointly with specified BF dispersion fiber lengths $L_1 = 1600$ m, $L_2 = (L_1 + 1500)$ m and 0.8 nm wavelength spacing (100 GHz ITU grid) results in beam pointing angles $\theta_0 = 0°$, $\theta_1 = 18.9°$, $\theta_2 = 38.7°$. The dispersion fiber lengths were made as two coils of standard SMF-28 fiber with lengths 1600 m and 1500 m and chromatic dispersion about $D = 17$ ps/nm·km.

The results of time delay measurements of used components have been converted to their tantamount electrical lengths. Because of large difference in electrical length of our MUX channels (up to 6 m, as 8 x 1 MUX from ATW has inner structure as a chain) we have replaced the MUX by 8 x 1 splitter/combiner ($LiNbO_3$ Planar Light Circuit, PLC). Optical losses in 8 x1 combiner channels were 11.5±0.35 dB, channels electrical lengths (relying on 1 m patch-cords) were in range 1558.1±1.2 mm. Hereafter five corrective patch-cords with the calculated lengths: 200, 226, 254, 550, 624 mm were made and inserted in BF model channels at the combiner inputs. Table 1 shows the results of time delays alignment in designed BF model (with corrective patch-cords) for 3 beam positions: $\theta_0 = 0°$, $\theta_1 = 15°$, $\theta_2 = 30°$. 'Exp' means experimental value, 'Calc' means expected (calculated) value. PAA channel numbers (as to its central element) are +2, +1, 0, –1, –2 and they are connected with transmitter units OTS-2T with ITU channel numbers 27, 28, 29, 30, 31 respectively. Angles $\theta_0 = 0°$, $\theta_1 = 15°$, $\theta_2 = 30°$ are given for case "initial" PAA with $d = 25$ мм.

Table 1. Time delays adjustment in BF model for 3 beam positions.

Channel # in PAA	ΔS_{21}, dB	$\Delta\tau_{CH}$, ps, Exp / Calc		
		$\theta_0 = 0°$, $L_0 = 0$ m	$\theta_1 = 15°$, $L_1 = 1600$ m	$\theta_2 = 30°$, $L_2 = 3100$ m
+ 2	–1.4	0.05 / 0.0	43.03 / 43.14	83.48 / 83.33
+ 1	1.2	0.003 / 0.0	21.90 / 21.57	42.68 / 41.67
0	0.1	0.44 / 0.0	0.0	0.0
– 1	1.4	–1.89 / 0.0	–22.56 / –21.57	–43.12 / –41.67
– 2	–1.3	1.41 / 0.0	–42.64 /–43.14	–82.75 / –83.33

The values in columns 2 and 3 show results of BF model alignment in its initial state $\theta_0 = 0°$, they are given relatively to mean values over five channels. S_{21} mean value was –25.0 dB, mean value of electrical length over 5 BF channels was 11.708 m. Time delays in columns 4 and 5 are given relatively to delays for central PAA element #0, serviced by the laser with a wavelength at ITU channel #29. These delays are measured at RF frequencies near 4 GHz. Because of small differences between 'Exp' and 'Calc' time delay values in columns 4, 5 one can expect BF model functioning in wide frequency range with no squint effect.

5. Photonic BF Model Test Results

An experimental demonstration of the BF prototype performance consisted in far-field pattern measurements of the 1-D antenna array assembled with the photonic BF model in anechoic chamber. In our pattern measurements a wideband reference horn fed by a vector network analyzer (with additional RF power amplifier) radiates a microwave signal. The PAA under test located in front of the reference horn is fixed on an azimuth scan drive. The distance between the horn and array is about 6 m to fulfill far-field condition requirement. Microwave signal from BF (RF output of its optical receiver unit) is returned to vector network analyzer and compared with signal that feeds through RF power amplifier reference horn during azimuth scan. For each azimuth position – 1° step – network analyzer scans wide frequency range from 6 to 15 GHz. These measurements have been done for 3 different lengths of dispersive fiber in our TDU.

Figure 4 represents the far-field patterns measured with the BF for 3 optical fiber lengths at the middle frequency 9 GHz. Insertion of 0, 1600 and 3100 m optical fibers shifts the antenna pattern to $\theta_0 = 1.3°$ (red), $\theta_1 = 18.8°$ (blue), and $\theta_2 = 36.8°$ (green), respectively. The measured shifts are close to the calculated ones when time delays between two neighbouring radiating elements are 0, 21.57, 43.14 ps, see Table 1. Small "0"- beam deviation from the normal position (0°) was generated by residual differences about a few tenth of a millimeter in total channel electrical lengths including RF cables between PAA and the BF model inputs. The sidelobe level for $\theta_0 = 1.3°$ is approx. –13 dB in accordance with the uniform amplitude distribution. Figure 5 shows the far-field patterns in polar form for 3 optical fiber lengths at frequencies 6 (red), 9 (blue), and 12 GHz (green), respectively. One can see no squint effect in wide frequency range for PAA with designed photonic BF model.

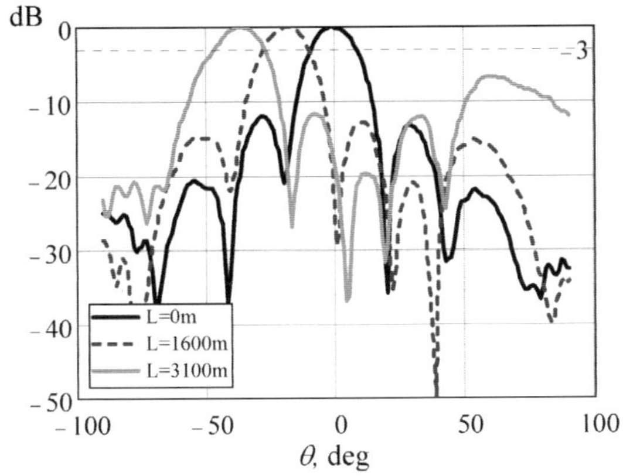

Figure 4. The PAA far-field patterns at 9 GHz: control of beam position.

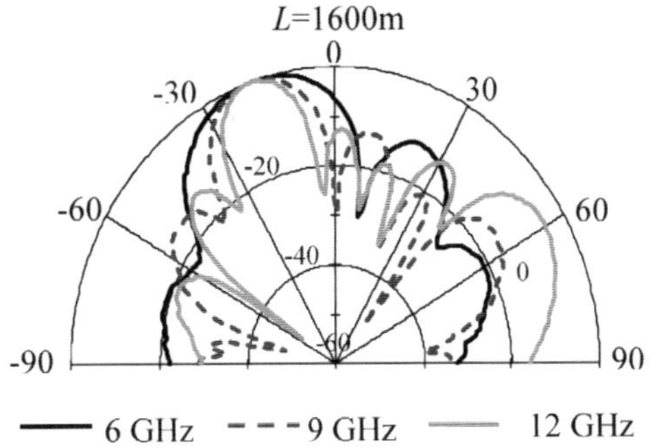

Figure 5. The PAA far-field patterns for 6, 9, and 12 GHz at 18.8°, no squint effect.

6. Conclusion

The model of photonic beamformer for ultrawideband microwave phased antenna array has been developed and tested. We have demonstrated the ability of 1-D BF based on DWDM multiplexing and optical fiber chromatic dispersion to steer the beam of PAA in the receive mode. This BF architecture realizes true-time-delay approach to beam steering where different time-delay distributions are provided with a set of dispersive fiber lengths. The beamformer model containes the components available on the market of fiber-optic analog communication links. Experimental demonstration of the designed model beamforming features includes actual measurement of 5-element microwave linear array antenna far-field patterns for 3 beam directions up to 36° at frequencies 6..12 GHz. The results of BF model experimental testing show no squint effect and good accordance with the calculation results.

References

[1] 2009 *Microwave Photonics: Devices And Applications ed S Iezekiel* (Chippenham: Wiley-IEEE Press)
[2] Tur M, Yaron L, Rotman R and Raz O 2011 *Proc. OSA/OFC/NFOEC 2011 (Los Angeles)* Paper OThA6
[3] Yao J P 2012 *IEEE Photonics Society Newsletter* **26**(3) 5-12
[4] Blanc S, Alouini M, Garenaux R, Queguiner M and Merlet T 2006 *IEEE MTT Trans.* **54** 402-11
[5] Minasian R A, Chan E H W and Yi X 2013 *Optics Express* **22**(19) 22918-36
[6] Lavrov A P, Ivanov S I and Saenko I I 2014 *Internet of Things, Smart Spaces, and Next Generation Networks and Systems ed S Balandin et al* ser. *LNCS* 8638 (Springer Int. Publ. Switzerland) pp 647-55
[7] Yang Y, Dong Y, Liu D, He H, Jin Y and Hu W 2009 *Chinese Optics Lett.* **7** 118-20
[8] http://emcore.com/wp-content/uploads/2016/03/Optiva-OTS-2-18GHz-Unamplified.pdf
[9] Ivanov S I, Lavrov A P and Saenko I I 2015 *J. Opt. Technol.* **82**(3) 139-46
[10] Ivanov S I, Lavrov A P and Saenko I I 2015 *Proc. 2015 Int. Siberian Conf. on Control and Communications (SIBCON) (Omsk, Russia)* (New York: IEEE) pp 1-4

Multiplexing and demultiplexing of the complex signal in the singular beams propagating in a few-mode optical fibers: an experiment

S Halilov, A Ilyasova, A Rubass and A Pogrebnaya

V.I. Vernadsky Crimean Federal University, Vernadsky Prospekt, 4, Simferopol, 295007, Russian Federation.

E-mail: kurioza.9@gmail.com

Abstract. We have experimentally demonstrated the possibility of multiplexing and demultiplexing the complex signal in the singular beam propagating in a few -mode fiber. Stokes polarimetry method conducted field analysis of the structure at the outlet of a few-mode fiber. The possibility of identifying each of the input signals of the wavelengths and orbital angular momentum. It is found that in the radiation field at the exit of a few-mode fiber, C-points each of which corresponds to the input beam polarization. The topological index of these C-points corresponding to the initial charge of the beams.

1. Introduction

The main tasks of modern optics include: - the creation of instruments and devices for high-speed data transmission over long distances; - increasing data capacity of communication channels; - encoding and decoding transmitted signals [1-3]. Recently, as a communication channel commonly used optical fibers. In [4-6] were shown the processes of propagation of singular beams with unique properties in optical fibers. This propagation is accompanied by screwing a helical beam wave front in the fiber environment, and between fields of their own beam mode energy is exchanged. Singular beams are of fundamental interest in finding methods of increasing the transmittance of optical fibers. By the process of applying the transmitted signals include multiplexing, and the separation of signals – demultiplexing. A number of studies [7-10] have been highlighted principles of multiplexing and demultiplexing of signals in optical fibers. The proposed experimental design allows for: multiplex and demultiplex composite signal in the optical-mode fiber; analyze the field at the exit of a few-mode fiber by Stokes-polarimetry for different combinations of folded signals, depending on the signs and magnitudes of topological charges of optical vortices and direction of the circularly polarized components.

2. Multiplexing signal

To investigate the signal multiplex in the few-mode fiber used in the experimental setup shown in Figure 1.

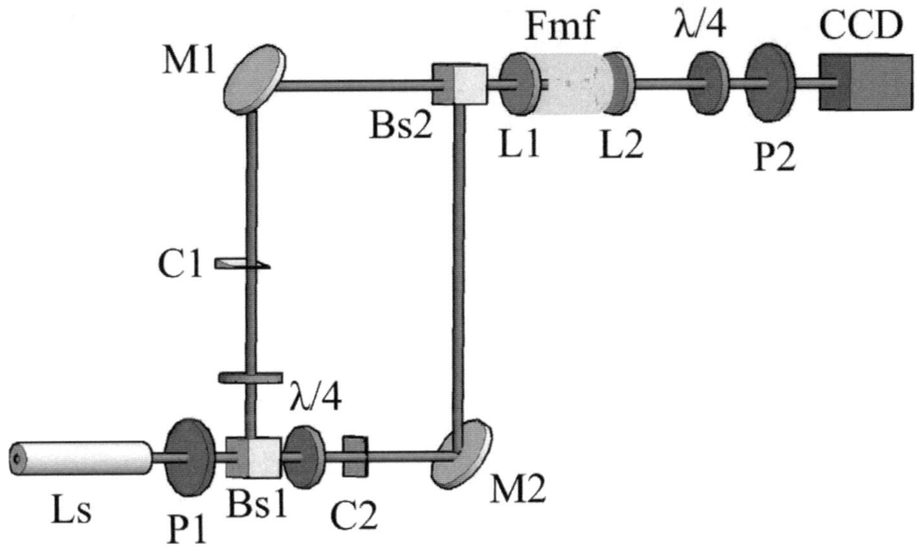

Figure 1. Schematic of the experimental setup: (Ls) Laser, (P) polarizer, (λ/4) quarter-wave plate, (Bs) beam splitter, (L) lens, (M) mirror, (C) dielectric wedge, (Fmf) few-mode fiber (V = 3.4), (CCD) camera.

The linearly polarized Gaussian beam from the light source Ls with a wavelength $\lambda = 0{,}63 mkm$, the beam is split into two with a beam splitter Bs_1. Using records $\lambda/4_{1,2}$ from the resulting beams distinguish orthogonally directed circularly polarized components - right and left. Then, placing the optical path of the beams, dielectric wedges $W_{1,2}$ generate singular beams with given values of topological charges $l = \pm 1$ [11, 12]. The second beam splitter Bs_2 collect beams, thereby setting the complex signal. Further, already multiplexed signal is transmitted is wound few-mode fiber Fmf. Analysis of the multiplexed signal is produced at the output of the fiber mode field for the state. The identification of each of the superimposed beams can be made from the values of the orbital angular momentum. Also, the state received the total field at the output of the fiber, the method of Stokes-polarimetry, we can identify each of the input signals.

We have a number of possible combinations of signal multiplexing have been implemented in the inlet mode fibers:

1) Multiplexing of the beams with the same values and different signs of topological charges of optical vortices;

2) Multiplexing beams with identical and orthogonal directions of circular polarization.

The resulting intensity distribution maps and polarization multiplexing in the process of the composite signal at the output of a few-mode fiber, shown in Figure 2.1-3.

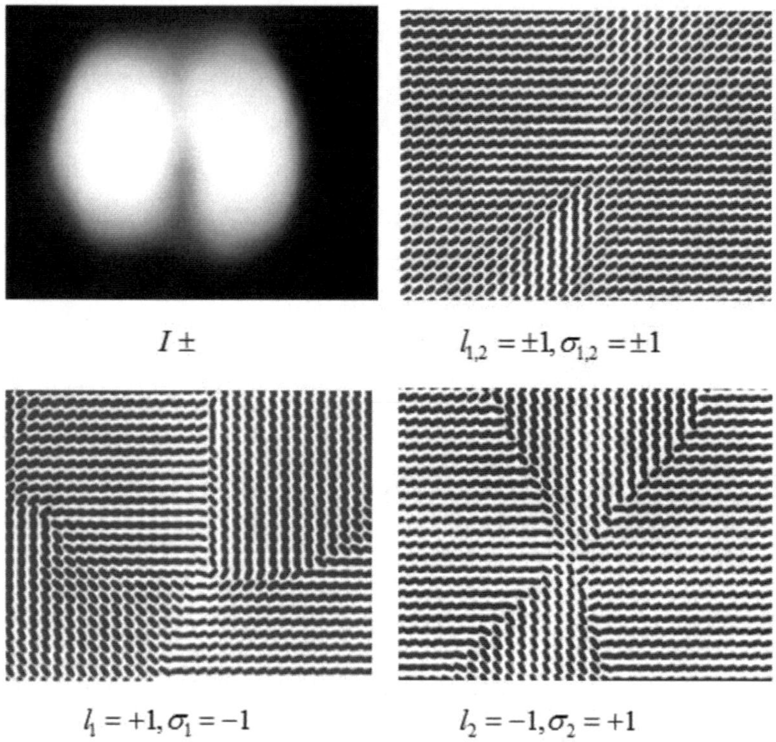

Figure 2.1.

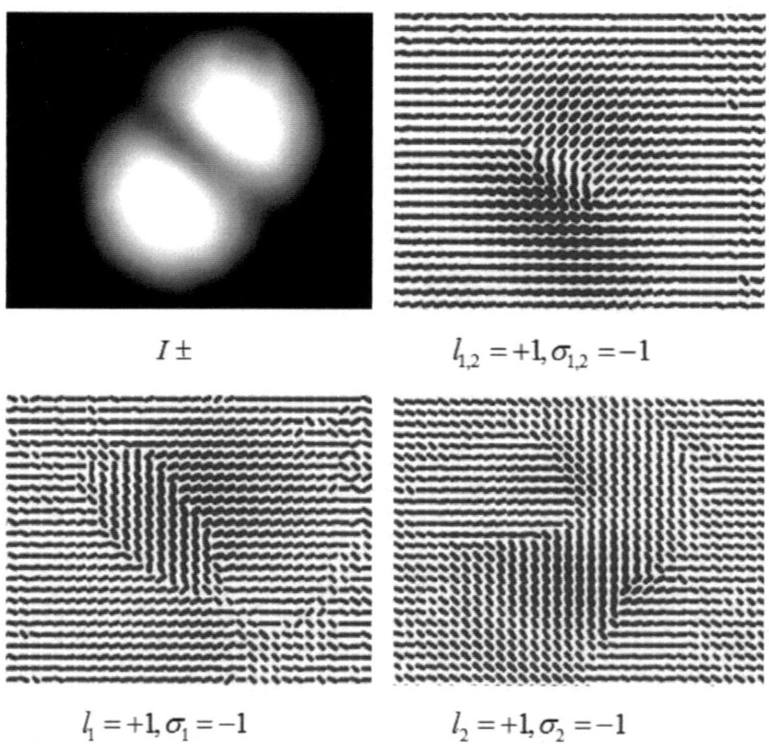

Figure 2.2.

Journal of Physics: Conference Series **737** (2016) 012003 doi:10.1088/1742-6596/737/1/012003

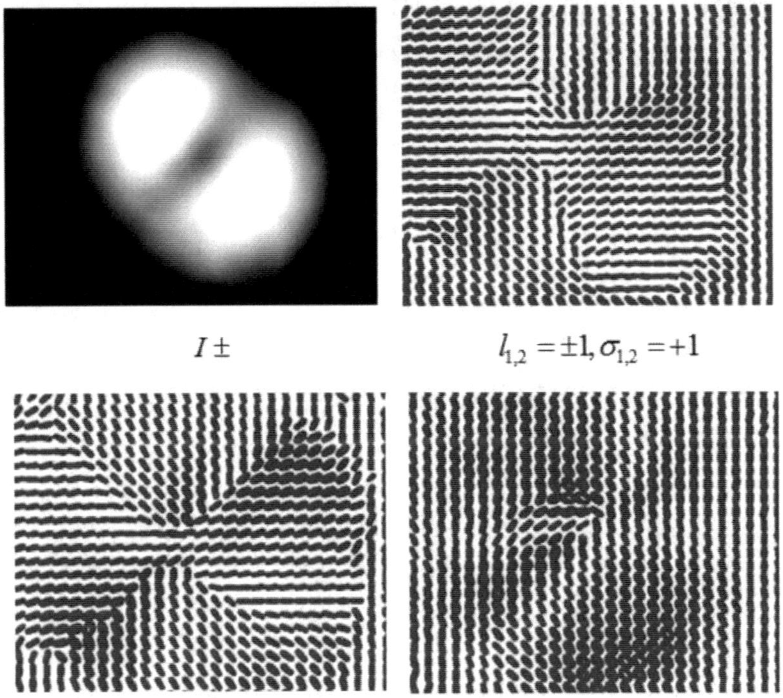

$I\pm$ $\quad$ $l_{1,2}=\pm 1, \sigma_{1,2}=+1$

$l_1=-1, \sigma_1=+1$ $\quad$ $l_2=+1, \sigma_2=+1$

Figure 2.3.

Figure 2.1-3. Maps of the intensity distribution and polarization multiplexing in the composite signal at the output of a few-mode fibers: (1) the distribution of the intensity and polarization of the beam with a wavelength $\lambda = 0,63 mkm$, topological charges, $l_1 = +1$ and $l_2 = -1$, left $\sigma_1 = -1$ and right-circular $\sigma_2 = +1$ polarisations; (2) the distribution of the intensity and polarization of the beam with a wavelength $\lambda = 0,63 mkm$, topological charges $l_1 = +1$ and $l_2 = +1$, left-circular $\sigma_{1,2} = -1$ polarisations; (3) the distribution of the intensity and polarization of the beam with a wavelength $\lambda = 0,63 mkm$, topological charges $l_1 = -1$ and $l_2 = +1$, right-circular $\sigma_{1,2} = +1$ polarisations.

3. The signal demultiplexing

Investigation signal demultiplexing process was carried out on the experimental setup, shown in Figure 3.

Gaussian beam emitted from the light source Ls_1 with a wavelength $\lambda_1 = 0,63 mkm$ using a polarizer P_1 and a quarter wave plate $\lambda/4_1$ is converted into right-circularly polarized beam. Further, during the passage of the dielectric wedge W_1 beam is converted into a singular, carrying an optical vortex with a given value of the topological charge. The second Gaussian beam emitted from the light source Ls_2 with a wavelength $\lambda_2 = 0,53 mkm$ of passing through a polarizer P_2 and a quarter wave plate $\lambda/4_2$ is converted into left-circularly polarized beam. Two beam from the light source Ls_1 and Ls_2 going to fission Bs and dice fall to the end of the wound-mode fiber Fmf. The output from the fiber by means of beams share the prism space. We perform an analysis of states separated beams by Stokes-polarimetry. In the process of demultiplexing identify each of the original beams and the

Journal of Physics: Conference Series **737** (2016) 012003 doi:10.1088/1742-6596/737/1/012003

wavelength λ_1 and λ_2 mode composition, in a complex signal. Research demultiplexing process is also performed for a variety of signal multiplexing combinations (for the same values and different signs of topological charges of optical vortices with identical and orthogonal directions of circular polarization).

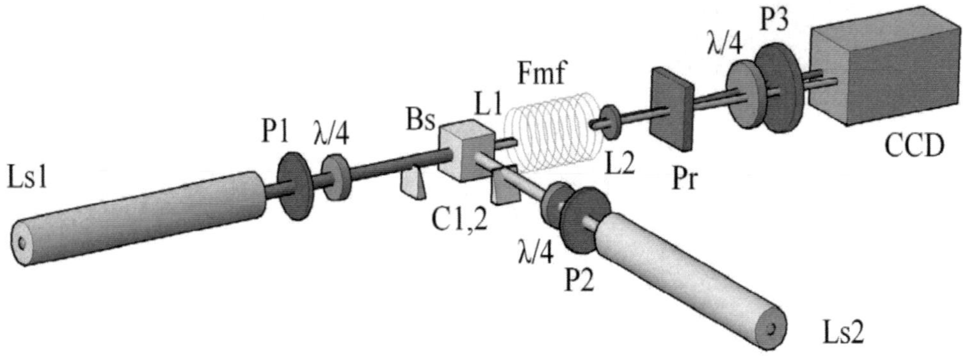

Figure 3. Schematic of the experimental setup: (Ls) Laser, (P) polarizer, (λ / 4) quarter-wave plate, (C) a dielectric wedge, (Bs) beam splitter, (L) lens, (Fmf) few-mode fibers (V = 3.4), (Pr) prism, (CCD) camera.

The resulting polarization distribution maps in the process of demultiplexing the composite signal at the output of a few-mode fibers, are shown in Figure 4.1-3.

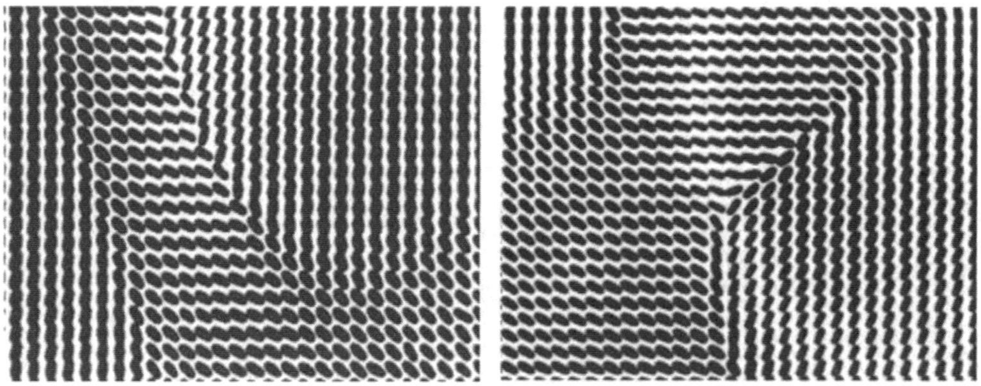

$l_1 = +1, \sigma_1 = -1$ $l_2 = -1, \sigma_2 = +1$

Figure 4.1.

Journal of Physics: Conference Series **737** (2016) 012003 doi:10.1088/1742-6596/737/1/012003

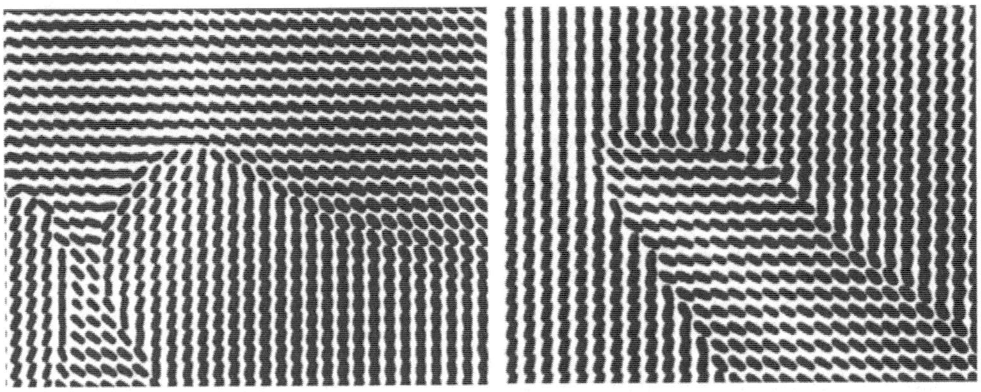

$l_1 = -1, \sigma_1 = -1$ $l_2 = -1, \sigma_2 = -1$

Figure 4.2.

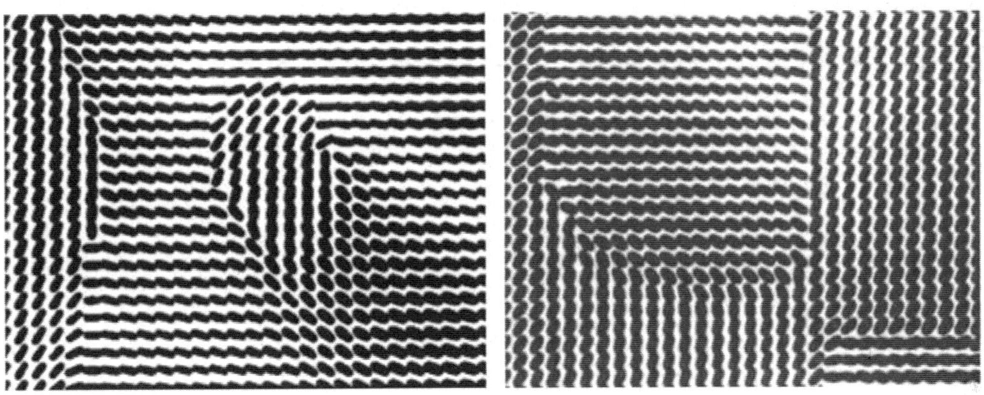

$l_1 = +1, \sigma_1 = -1$ $l_2 = +1, \sigma_2 = -1$

Figure 4.3.

Figure 4.1-3. Maps polarization distribution in the process of demultiplexing the composite signal at the output of a few-mode fiber, green - light beam with a wavelength of $\lambda_1 = 0,53mkm$, red - light beam with a wavelength $\lambda_2 = 0,63mkm$: (1) the polarization distribution for beams with topological charges $l_1 = +1$ and $l_2 = -1$, left $\sigma_1 = -1$ and right-circular $\sigma_2 = +1$ polarisations; (2) the polarization distribution for beams with topological charges $l_1 = -1$ and $l_2 = -1$, left-circular $\sigma_{1,2} = -1$ polarisations; (3) the polarization distribution for beams with topological charges $l_1 = +1$ and $l_2 = +1$ left-circular $\sigma_{1,2} = -1$ polarisations.

4. Conclusion

We have experimentally demonstrated the possibility of multiplexing and demultiplexing the complex signal in the singular beam propagating in a few -mode fiber. Stokes polarimetry method conducted field analysis of the structure at the outlet of a few-mode fiber. The possibility of identifying each of the input signals of the wavelengths and orbital angular momentum. It is found that in the radiation field at the exit of a few-mode fiber, C-points each of which corresponds to the input beam polarization. The topological index of these C-points corresponding to the initial charge of the beams.

References

[1] Shapiro J, Guha S and Erkmen B 2005 *J. Opt. Netw.* **4** pp.501–516
[2] Willner A, Wang J and Huang H 2012 *Applied physics. Science* **337**, pp.655–656
[3] Matsuo S, Sasaki Y, Akamatsu T, Ishida I, Takenaga K, Okuyama K, Saitoh K and Kosihba M 2012 *Opt. Express* **20**, 28398–28408
[4] Fadeyeva T, Volyar A, Zhilaitis V, Soskin M *Proc of SPIE* **3904** pp.112-123
[5] Fadeyeva T, Volyar A, Zhilaitis V, Soskin M *Proc of SPIE* **3904** pp.124-130
[6] Volyar A, Fadeyeva T, Shvedov V *Proceedings of SPIE* **4403** pp. 153-161
[7] Gibson G, Courtial J, Padgett M, Vasnetsov M, Pasko V, Barnett S and Franke-Arnold S 2004 *Opt. Express* **12** pp.5448–5456
[8] Richardson D, Fini J and Nelson L 2013 *Nat. Photonics* **7** pp.354–362
[9] Bozinovic N, Yue Y, Ren Y, Tur M, Kristensen P, Huang H, Willner A and Ramachandran S 2013 *Science* **340** pp 1545–1548
[10] Wang J, Yang J, Fazal I, Ahmed N, Yan Y, Huang H, Ren Y, Yue Y, Dolinar S, Tur M and Willner A 2012 *Nat. Photonics* **6** pp.488–496
[11] Shvedov V, Izdebskaya Ya, Alekseev A and Volyar A 2002 *Technical Physics Letters* **28** **(3)** pp 256–259
[12] Izdebskaya Ya, Shvedov V, Volyar A 2004 *Ukr. J. Phys. Opt.* **5(3)** pp 96-99

Solitary wave generation from continuum in asymmetric oppositely directed nonlinear waveguide coupler

E V Kazantseva[1], A I Maimistov[1,2]

[1]National Research Nuclear University MEPhI (Moscow Engineering Physics Institute), 31 Kashirskoe shosse, 115409 Moscow, Russia
[2]Moscow Institute of Physics and Technology, 9 Institutskiy per., Dolgoprudny, Moscow Region, 141700, Russia

E-mail: elena.kazantseva@gmail.com

Abstract. In a model which describes asymmetric oppositely directed nonlinear waveguide coupler it was observed in the numerical simulation a phenomenon of solitary wave formation from the input constant continuous wave set at the entrance of a waveguide with negative index of refraction. Threshold value of the amplitude of the constant continuous wave, which defines the condition of appearance of the first solitary wave, decreases with increasing of the parameter of nonlinearity. The period of solitary wave formation decreases with increasing of the continuum wave amplitude.

1. Solitary wave solution

Propagation and interaction of coupled waves in asymmetric oppositely directed waveguide coupler is described by the system of equations for the normalized amplitudes e_1 and e_2 of electric field of the light waves propagating in the waveguides with positive and negative refraction indices correspondingly [1-3]:

$$i\left(\frac{\partial}{\partial\zeta}+\frac{\partial}{\partial\tau}\right)e_1+e_2+r|e_1|^2e_1=0,\ i\left(\frac{\partial}{\partial\zeta}-\frac{\partial}{\partial\tau}\right)e_2-e_1=0, \qquad (1)$$

Nonlinearity of a waveguide with positive index of refraction is characterized by a parameter r. A waveguide with negative index of refraction is linear. That is the reason to consider a waveguide coupler consisting of two waveguides one of which possesses positive and another one – negative index of refraction, as asymmetric nonlinear waveguide coupler. The nonlinearity of the whole device, a waveguide coupler, results from nonlinearity of the first waveguide with positive index of refraction due to interaction of the waves in both channels of a coupler. The interaction between the waves in different waveguides of a coupler occurs due to a term with e_2 in the wave equation for the normalized amplitude e_1 of the electric field in a waveguide with positive index of refraction, and a term with e_1 in the wave equation for the normalized amplitude e_2 of the electric field in a waveguide with negative index of refraction.

A solitary wave solution of the system of the equations (1) could be found, if we consider the normalized amplitudes of the electric fields as variables with real amplitudes $a_{1,2}$ and phases $\phi_{1,2}$: $e_1 = a_1 e^{i\phi_1}$, $e_2 = a_2 e^{i\phi_2}$. Also we assume that a solitary wave solution depends on a single variable $\eta = (\zeta + \beta\tau)(1-\beta^2)^{-1/2}$. A parameter β determines the velocity of solitary wave propagation in nonlinear waveguide coupler. The positive values of the parameter β correspond to the solitary waves propagating in the direction opposite to the axis ζ, while negative values of the parameter β correspond to the solitary waves propagating in the direction of the axis ζ. The real amplitudes $a_{1,2}$ of the components of the solitary wave (one of the components propagates in the waveguide channel with positive refraction index, another component propagates in the waveguide channel with negative index of refraction) are depend on the parameter β. These amplitudes are defined by the following expressions [1,2]:

$$a_1^2(\eta) = \frac{4}{\theta(1+\beta)\cosh 2(\eta-\eta_0)}, \quad a_2^2(\eta) = \frac{4}{\theta(1-\beta)\cosh 2(\eta-\eta_0)}, \tag{2}$$

a parameter η_0 defines the position of the center of the solitary wave, and the parameter θ is defined as following:

$$\theta = \frac{r}{1+\beta}\sqrt{\frac{1-\beta}{1+\beta}}.$$

The phases of the solitary waves are:

$$\phi_1(\eta) = \phi_1(-\infty) + 3arctg\left(e^{2(\eta-\eta_0)}\right), \ \phi_2(\eta) = \phi_2(-\infty) + arctg\left(e^{2(\eta-\eta_0)}\right), \tag{3}$$

with the condition

$$\cos(\phi_1(-\infty) - \phi_2(-\infty)) = 0.$$

This condition fulfills at $\phi_1(-\infty) = 0$, $\phi_2(-\infty) = -\pi/2$.

Interrelation between the amplitude and propagation velocity of the solitary wave is typical for the solitons – nonlinear solitary waves which propagate without change of the form, with constant amplitude and velocity. The solitons are quite robust against perturbations and the collisions between themselves, revealing the particle-like properties under interactions.

2. Numerical simulation of solitary wave generation in asymmetric oppositely directed nonlinear waveguide coupler

In the numerical simulation the normalized amplitudes of the electric fields $e_1(\tau, \zeta = 0)$ and $e_2(\tau, \zeta = \zeta_L)$ at the inputs (at $\zeta = 0$ and at $\zeta = \zeta_L$, where ζ_L is a length of a waveguide coupler) of the channels of the two-channel nonlinear waveguide coupler, where one channel has positive, and another one has negative refractive index, were set as following:

$$e_1(\tau, \zeta = 0) = 0, \ e_2(\tau, \zeta = \zeta_L) = a. \tag{4}$$

Parameter a defines the constant amplitude of the continuum wave set at $\zeta = \zeta_L$ in the waveguide channel with negative index of refraction. In the numerical simulation the length of a waveguide coupler is set as $\zeta_L = 40$. In the numerical simulation it was found that if the constant amplitude a of continuous wave set at the input of a waveguide with negative index of refraction, i.e. at the end side of the coupler, as in a channel with negative index of refraction the waves propagate in the direction opposite to ζ axis, exceeds a threshold value, this constant continuous wave transforms due to nonlinear phase modulation into solitary wave propagating in asymmetric oppositely directed nonlinear waveguide coupler.

For example, in the figures 1-3 is illustrated the process of solitary wave formation in the oppositely directed nonlinear waveguide coupler, nonlinearity of the coupler is characterized by the parameter of nonlinearity equal to 1 ($r = 1$). The amplitude of continuum wave at the input of waveguide with negative index of refraction was set to different values. If the value of the amplitude a is below the threshold value, the solitary wave does not form in the oppositely directed nonlinear waveguide coupler. At the amplitude of constant continuous wave equal to 2, the solitary wave appears from the continuum radiation and propagates in a waveguide coupler (figure 1). In the figure 1 (a) is shown the absolute value of the amplitude of the normalized electric field $|e_1|$ in a waveguide with positive index of refraction, in the figure 1 (b) is shown the absolute value of the amplitude of the normalized electric field $|e_2|$ in a waveguide with negative index of refraction. The amplitudes of the components of the solitary wave in the waveguides with positive and negative indices of refraction correspond to the relation (2) which defines the dependence of the amplitude of the solitary wave on velocity parameter β and nonlinearity coefficient r. The velocity of solitary wave is defined as $v_s = -\beta$.

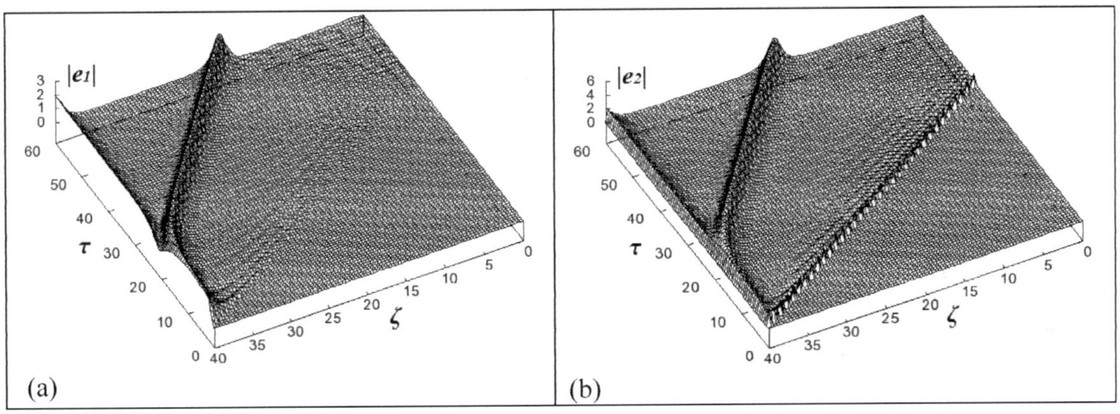

Figure 1. Formation of a solitary wave from constant continuous wave with amplitude $a = 2$, which is set at the entrance of the waveguide with negative index of refraction (at the end side of the asymmetric oppositely directed nonlinear waveguide coupler at $\zeta = \zeta_L = 40$). The nonlinearity parameter $r = 1$.

At increase of the amplitude of the constant wave the amount of solitary waves appearing at the same time interval is increases. The propagation velocity of the solitary waves here is a negative value, as these pulses propagate from the end side of the two-channel waveguide coupler in a direction opposite to ζ axis. The absolute value of the velocity of formed solitary waves (as well as a value of the parameter β) and their amplitude increase with increasing of the amplitude of the input constant continuous wave.

At the amplitude of constant continuous wave set in the entrance of the waveguide with negative index of refraction (at $\zeta = \zeta_L$) equal to 2.5, there are five equal solitary waves appear in the asymmetric oppositely directed nonlinear coupler (figure 2). After their formation from the continuum wave these solitary waves propagate with the same velocities and they have the identical amplitudes.

At the amplitude of constant continuous wave equal to 4, there are eleven solitary waves appear at the considered time interval (figure 3).

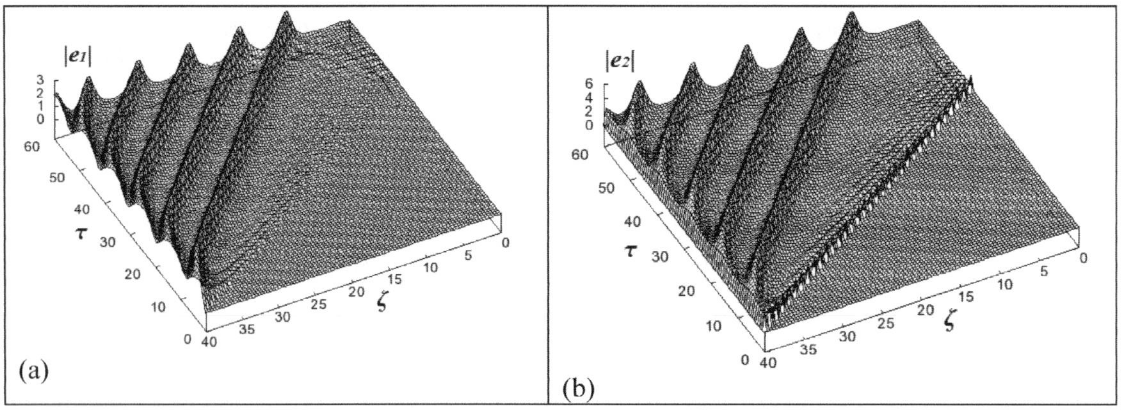

Figure 2. Formation of solitary waves from constant continuous wave with amplitude $a = 2.5$, which is set at the entrance of the waveguide with negative index of refraction (an the end side of the asymmetric oppositely directed nonlinear waveguide coupler at $\zeta = \zeta_L = 40$). The nonlinearity parameter $r = 1$.

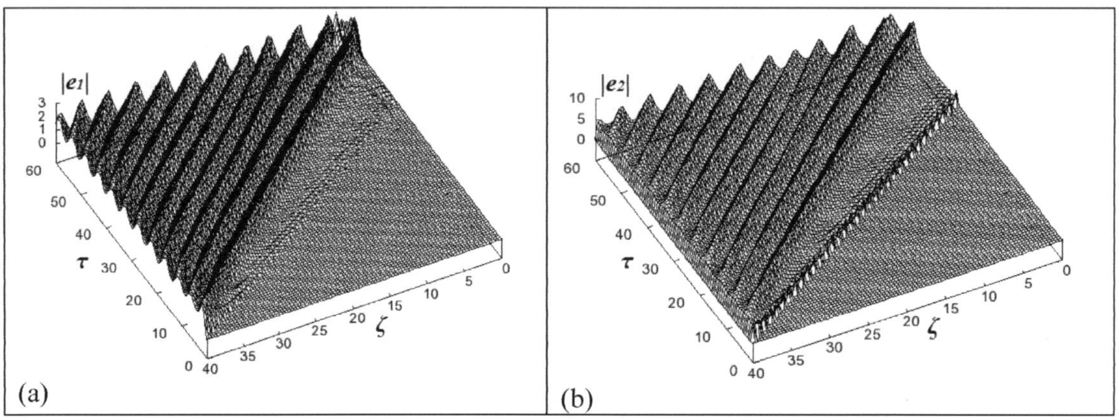

Figure 3. Formation of solitary waves from constant continuous wave with amplitude $a = 4$, which is set at the entrance of the waveguide with negative index of refraction (at the end side of the asymmetric oppositely directed nonlinear waveguide coupler at $\zeta = \zeta_L = 40$). The nonlinearity parameter $r = 1$.

3. The influence of nonlinearity of the waveguide coupler on the threshold value of the amplitude of the input wave at which the first solitary wave forms, on the period of the solitary wave formation and on the solitary wave parameters

The numerical simulation performed for several values of the parameter of nonlinearity $r = 1, 0.5, 0.25, 0.1$, revealed that the threshold value of the amplitude a of the constant continuous wave, which defines the condition of appearance of the first solitary wave, increases with decreasing of the parameter of nonlinearity.

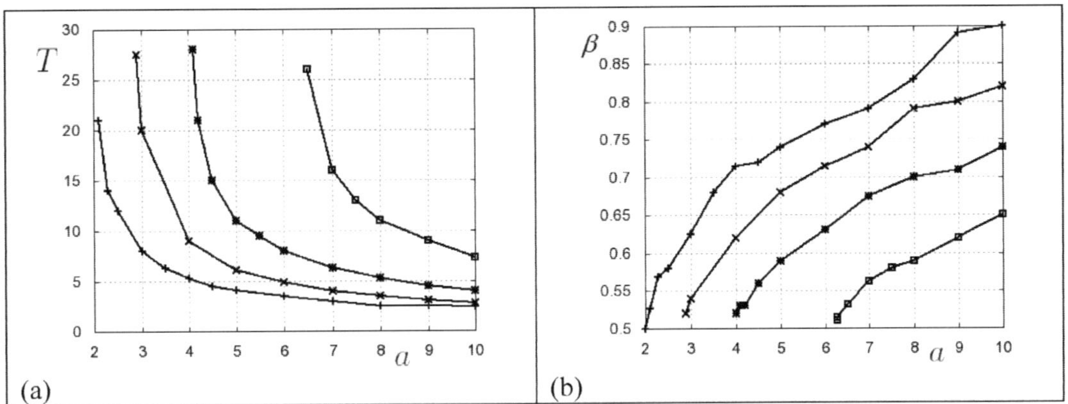

Figure 4. The plots illustrating the dependence of the period T (figure 4, a) of the solitary wave formation along τ axis, and the velocity parameter of solitary waves β (figure 4, b) on the constant amplitude a of the input continuum wave. The plots are shown for the values of nonlinearity parameter $r = 1$ (solid line with crosses), $r = 0.5$ (solid line with inclined crosses), $r = 0.25$ (solid line with stars), $r = 0.1$ (solid line with squares).

In the figure 4 (a) the plots are shown for the different values of the nonlinearity parameter $r = 1, 0.5, 0.25, 0.1$, illustrating the dependence of the time interval T (along τ axis), defining the period of formation of the solitary wave, on the amplitude parameter a of a constant continuous wave. At the figure 4 (b) for the same values of the parameter of nonlinearity the plots are shown, they are illustrate the dependence of the parameter β, characterizing the propagation velocity of the formed solitary waves as a function of the amplitude parameter a of an input constant continuous wave, set at $\zeta_L = 40$ in a channel with negative index of refraction.

Conclusion

A constant continuous wave of certain amplitude which is set at the entrance of the waveguide with negative index of refraction in the asymmetric oppositely directed nonlinear waveguide coupler (from the output of the waveguide coupler) could transform (due to nonlinear phase modulation) into solitary wave propagating in a two-channel nonlinear waveguide coupler. It is found that the threshold value of the amplitude a of the constant continuous wave, which defines the condition of appearance of the first solitary wave, decreases with the increase of the parameter of nonlinearity r.

Increase of the amplitude of the input constant continuous wave set at $\zeta = \zeta_L$ in the waveguide with negative index of refraction of the asymmetric oppositely directed nonlinear waveguide coupler, results in increasing of the amount of solitary waves, formed at the considered time interval. The absolute value of the solitary wave velocity and the amplitude of the formed solitary waves increase with increasing of the amplitude of the input constant continuous wave.

Acknowledgments

The research is supported by the Russian Foundation for Basic Research (grant № 15-02-02764) and Russian Scientific Foundation (grant № 14-22-00098).

References

[1] Kazantseva E V, Maimistov A I and Ozhenko S S 2009 *Phys. Rev.* A **80** 043833
[2] Maimistov A I, Kazantseva E V and Ozhenko S S 2011 *Bulletin of the Russian Academy of Sciences, Physics* **75 (2)** 165-167
[3] Kazantseva E V and Maimistov A I 2015 *Quantum Electronics* **45 (12)** 1137-1142
[4] Kazantseva E V and Maimistov A I 2015 *EPJ Web of Conferences* **103** 04005

Adiabatic modulation of cnoidal wave by Kuznetsov – Ma soliton

V A Makarov, V M Petnikova, and V V Shuvalov
Faculty of Physics and International Laser Center of M.V. Lomonosov Moscow State University, Russia

E-mail: v_petnikova@yahoo.com

Abstract. The problem of nonlinear interaction of a cnoidal wave (a “fast” component of vector light field) with localized in time and periodic in space control signal in the form of Kuznetsov-Ma soliton (a "slow" component of the same field) is analytically solved in the adiabatic approximation. The conditions which must be fulfilled for stable propagation of the obtained solution with amplitude and frequency modulation are determined.

1. Introduction

Self-consistent analytical solutions of nonintegrable nonlinear problems of wave physics [1] (breathers, solitons, cnoidal waves, and others) are interesting from the viewpoint of urgent problems of ultrafast data transmission [2-8]. These solutions are either partial [9,10] or approximate, obtained, e.g., by perturbation theory methods [11,12] or after linearization [13,14] of nonlinear equations. However, the latter not always adequately describe nonlinear wave propagation. To find approximate analytical solutions of the aforementioned problems, the adiabatic approximation [15-18] was used in [19-23], which appeared very efficient to describe not only interactions of cnoidal waves (solutions with separable space-time variables [19-22]), but also the interaction of the cnoidal wave with the rational soliton [23] for which space-time variables are not separable. It was shown that the slow control signal (rational soliton) causes amplitude and frequency modulation of the rapidly varying information signal (cnoidal wave) localized in the region of the signal intensity variation.

This paper is devoted to the study of the adiabatic interaction of the time-limited and periodically varying in space Kuznetsov-Ma (KM) soliton [24] with the fast cnoidal wave. Periodic in coordinate space-time amplitude and frequency modulation of the cnoidal wave, much more complex than in the case of the rational soliton [23], should be expected.

2. Adiabatic interaction of electric field components

Nonlinear propagation of the plane elliptically polarized electromagnetic wave through an isotropic nonlinear gyrotropic medium is described by the nonintegrable system of nonlinear Schrödinger equations [13,14,19-23] for slowly varying orthogonal circularly polarized electric field strength vector components $A_\pm(z,t)$:

$$\frac{\partial A_\pm}{\partial z} - i\frac{k_2}{2}\frac{\partial^2 A_\pm}{\partial t^2} + i\left[\mp\rho_0 + \left(\frac{\sigma_1}{2}\mp\rho_1\right)|A_\pm|^2 + \left(\frac{\sigma_1}{2}+\sigma_2\right)|A_\mp|^2\right]A_\pm = 0. \quad (1)$$

where z – is the propagation coordinate, t is the time in the travelling coordinate system, the parameter $k_2 = \partial^2 k/\partial^2\omega \neq 0$ accounts for the second-order frequency dispersion; $k(\omega)$ is the wavenumber; ω

is the carrier frequency; $\sigma_1 = 4\pi\omega^2\chi^{(3)}_{xyxy}/kc^2$ and $\sigma_2 = 2\pi\omega^2\chi^{(3)}_{xxyy}/kc^2$ are defined by independent components of the local cubic nonlinearity tensor $\hat{\chi}^{(3)}(\omega; -\omega, \omega, \omega)$, $\rho_{0,1} = 2\pi\omega^2\gamma_{0,1}/c^2$ are defined in terms of pseudo-scalar constants $\gamma_{0,1}$ of linear and nonlinear gyration.

The adiabatic approximation [19-23] presupposes different time scales of amplitude variations in system (1). This makes it possible, e.g., to consider $|A_+(z,t)|^2$ in the equation for the fast field component $A_-(z,t) = r_-(z,t)\exp(i\kappa_- z)$ to be almost unchanged, and in the approximation

$$\partial r_-(z,t)/\partial z \ll i\kappa_- r_-(z,t) \tag{2}$$

(small spatial modulation), to obtain from Eq. (1) the ordinary differential equation for r_- [23]:

$$\frac{d^2 r_-}{dt^2} = k_2^{-1}\{[2(\kappa_- + \rho_0) + (\sigma_1 + 2\sigma_2)|A_+|^2] + (\sigma_1 + 2\rho_1)r_-^2\}r_- \tag{3}$$

Here κ_- is the free problem parameter (separation constant). One of the solutions to (3) can be written in the form

$$r_-(z,t) = B_-(z,t)\mathrm{cn}[\varphi_-(z,t), \mu_-], \tag{4}$$

containing the elliptic Jacobi function [25] with the free parameter μ_-. The slowly varying amplitude $B_-(z,t)$ and argument $\varphi_-(z,t)$ are given by the formulas:

$$B_-(z,t) = 2\mu_- \left[\frac{\kappa_- + \rho_0}{k_2(2\mu_-^2 - 1)(\sigma_1 + 2\rho_1)}\left(1 + \frac{\sigma_1 + 2\sigma_2}{2(\kappa_- + \rho_0)}\lfloor A_+(z,t)\rfloor^2\right)\right]^{1/2}, \tag{5}$$

$$\varphi_-(z,t) = \left[\frac{2(\kappa_- + \rho_0)}{k_2(2\mu_-^2 - 1)}\right]^{1/2} \int_0^t dt' \left[\left(1 + \frac{\sigma_1 + 2\sigma_2}{2(\kappa_- + \rho_0)}\lfloor A_+(z,t)\rfloor^2\right)\right]^{1/2}, \tag{6}$$

found neglecting the time derivatives of B_- and frequency $\nu(z,t) = \partial\varphi_-(z,t)/\partial t$. The slow electric field component $A_+(z,t)$ of the propagating wave is found similarly to [19], substituting the time-averaged value $r_-^2(z,t)$

$$\langle r_-^2(z,t)\rangle_t \approx \frac{4\mu_-^2(\kappa_- + \rho_0)}{k_2(2\mu_-^2 - 1)(\sigma_1 + 2\rho_1)}\langle \mathrm{cn}^2[\varphi_-(z,t), \mu_-]\rangle_t = \frac{4(\kappa_- + \rho_0)[\mathrm{E}(\mu_-)/\mathrm{K}(\mu_-) + \mu_-^2 - 1]}{k_2(2\mu_-^2 - 1)(\sigma_1 + 2\rho_1)} \tag{7}$$

to the first equation of system (1), which as a result takes the form

$$\frac{\partial A_+}{\partial z} - \frac{ik_2}{2}\frac{\partial^2 A_+}{\partial t^2} + i(-\tilde{\rho} + \tilde{s}|A_+|^2)A_+ = 0. \tag{8}$$

In Eq. (7), $\mathrm{K}(\mu_-)$ и $\mathrm{E}(\mu_-)$ are complete elliptic integrals of the first and second kind [25]; in Eq. (8), $\tilde{\rho} = \rho_0 + 2(\kappa_- + \rho_0)q$, $\tilde{s} = \sigma_1/2 - \rho_1 - (\sigma_1 + 2\sigma_2)q$, where

$$q = \frac{(\sigma_1 + 2\sigma_2)}{(2\mu_-^2 - 1)(\sigma_1 + 2\rho_1)}\left[\frac{\mathrm{E}(\mu_-)}{\mathrm{K}(\mu_-)} + \mu_-^2 - 1\right]. \tag{9}$$

Equation (7) has the solution in the form of the KM soliton if $\tilde{s} > 0, k_2 < 0$

$$A_+(z,t) = \exp[iz(\tilde{\rho} - P\tilde{s})]\left[1 + \frac{2(1-2a)\cos(zbP\tilde{s}) + ib\sin(zbP\tilde{s})}{\sqrt{2a}\cosh[bt(0.5P\tilde{s}/|k_2|)^{1/2}] - \cos(zbP\tilde{s})}\right], \tag{10}$$

where $b = 2[2(2a-1)]^{1/2}$, $a > 0.5$ and P are the free parameters. Substitution of Eq. (10) into (5), (6) defines the function $m(z,t) = [1 + (\sigma_1/2 + \sigma_2)\lfloor A_+(z,t)\rfloor^2/(\kappa_- + \rho_0)]^{1/2}$ of the space-time modulation of the information signal (4). Formulas (4-10) specify a new set of approximate analytical solutions of the nonlinear system (1) with the free parameters κ_-, μ_-, a and P. They are limited only by the range of application of the adiabatic approximation.

3. Range of application of the adiabatic approximation

The applicability condition of the adiabatic approximation is reduced, first, to satisfying the assumption about the slow variation of the KM soliton in time in comparison with the cnoidal wave having the period $T_- = 4\mathrm{K}(\mu_-)/(d\varphi_-/dt)$. The KM soliton width $2\Delta t_+$ is estimated by the interval between the points at which $A_+(z=0)$ vanishes. In this interval, a sufficient number of cnoidal wave periods should be settled. A consequence of this is the inequality

$$\left[\frac{|k_2|}{(2a-1)\tilde{s}}\right]^{1/2} \mathrm{Arch}\left(\frac{4a-1}{\sqrt{2a}}\right) \gg 2\mathrm{K}(\mu_-)/(d\varphi_-/dt). \tag{11}$$

Second, the contribution of spatial modulation of the information signal to $\partial A_-/\partial z$ should be small (see (2)). Assuming that a is close to unity, from (11) and (2), we find the following inequalities

$$\frac{P\tilde{s}}{|\kappa_- + \rho_0|} \ll \frac{1}{4(2\mu_-^2 - 1)\mathrm{K}^2(\mu_-)}, \quad (12)$$

$$\frac{P|\tilde{s}(\sigma_1 + 2\sigma_2)|}{|(\kappa_-(\kappa_- + \rho_0)|} \ll \frac{10T_-}{\Delta t_+}, \quad (13)$$

from which it follows that $P\tilde{s}$ should be small. Both these requirements can be satisfied either by limiting the KM soliton pedestal level, or by lowering the medium nonlinearity.

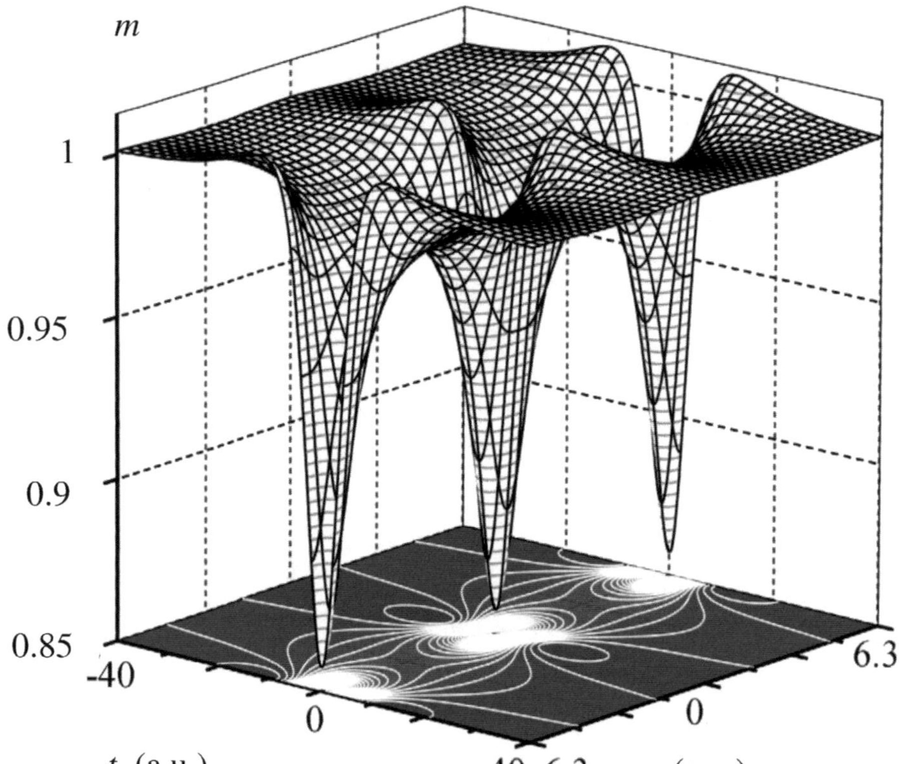

Figure 1. The function $m = [1 + (\sigma_1/2 + \sigma_2)|A_+(z,t)|^2/(\kappa_- + \rho_0)]^{1/2}$ of stable space-time modulation of the “fast” cnoidal wave by the "slow" Kuznetsov-Ma soliton.

Figure 1 illustrates the stable in time and space amplitude and frequency modulation of the approximate solution for the fast field component by the slow signal in the form of the KM soliton. The modulation function $m(z,t)$ is constructed at $a = 1$, $P = 0.2$, $k_2 = -1$, $\kappa_- = -6$, $\sigma_1 = 1$, $\sigma_2 = 0$, $\rho_0 = 1$ and $\rho_1 = 0.013$ satisfying (12,13).

4. Conclusion

In this study, the adiabatic approximation was applied to the approximate analytical solution of the nonintegrable problem of the interaction of the KM soliton ("control" signal) with the cnoidal wave ("information" signal). The space-time amplitude and frequency modulation of the cnoidal wave by the KM soliton was obtained. The applicability conditions of the adiabatic approximation, limiting the KM soliton power and the nonlinearity value, were found. These conclusions are in qualitative agreement with the results of recent studies of the interactions of nonlinear waves of different types [4,6,26,27].

References

[1] Christiansen P L et al 2000 *Proc. Royal Soc. London* A **456** 2263
[2] Chen Zhigang et al 2012 *Reports on Progress in Physics* **75** 086401
[3] Petnikova V M and Shuvalov V V 2007 *Phys. Rev.* E **76** 046611
[4] Bludov Yu V et al 2010 *Eur. Phys. J. Special Topics* **185** 169
[5] Kalla C 2011 *J. Phys. A: Math. Theor.* **44** 335210
[6] Zhao L and Liu J 2013 *Phys. Rev.* E **87** 013201
[7] Cheng Xue-Ping et al 2014 *Phys. Rev.* E **89** 043202
[8] Yushko O V and Redyuk A A 2014 *Quantum Electronics* **44** 606
[9] Makarov V A et al 2012 *Quantum Electronics* **42** 117
[10] Makarov V A et al 2012 *Quantum Electronics* **42** 1118
[11] Ma, Y C. 1979 *Studies in Applied Mathematics* **60** 43
[12] Boyd J P 1984 *J. Math. Phys.* **25** 3402
[13] Makarov V A et al 2013 *Laser Phys. Lett.* **10** 075404
[14] Makarov V A et al 2014 *Quantum Electronics* **44** 130
[15] Davydov A S 1965 *Quantum mechanics* (Oxford; Pergamon Press)
[16] Messiah A 1965 *Quantum mechanics* v 2 (Amsterdam: North Holland Publishing Company)
[17] Griffiths D J 1995 Introduction to quantum mechanics (New York: Prentice Hall Inc.)
[18] Berry M V 1985 *Quantum, classical and semiclassical adiabaticity Theoretical and Applied Mechanics*, ed Niordson F I and Olhoff N (Amsterdam: North-Holland, Elsevier Science Publishers) pp.83-96
[19] Makarov V A et al 2014 *Laser Phys.* **24** 085405
[20] Makarov V A et al 2015 *Quantum Electronics* **45** 35
[21] Makarov V A et al 2014 *Laser Phys. Lett.* **11** 115402
[22] Makarov V A et al 2015 *Physics of Wave Phenomena* **23** 96
[23] Makarov V A et al 2014 *Opt. Express* **22** 26607
[24] Kibler B et al 2012 *Scientific Reports* **2** 463
[25] Gradshteyn I S and Ryzhik I M 2007 *Table of integrals, series, and products* (New York: Academic Press)
[26] Hu X et al 2012 *Phys. Rev.* E **85** 056607
[27] Lou S Y et al 2014 *Chin. Phys. Lett.* **31** 070201

Journal of Physics: Conference Series **737** (2016) 012006 doi:10.1088/1742-6596/737/1/012006

Cnoidal waves, solitons and vortices in the flow of polaritons

I V Dzedolik and V Pereskokov

V. I. Vernadsky Crimean Federal University,
Physics and Technology Institute, Simferopol, Russian Federation

E-mail: dzedolik@crimea.edu

Abstract. We theoretically have investigated the properties of the phonon-polariton waves propagating in Kerr-type dielectric medium, and examined the conditions of appearance of the spatial solitons, cnoidal waves and polariton vortices. It is shown that the polariton flow can propagate in the form of spatial soliton, or can be split into several flows in the form of filaments depending on the field density and parameters of the nonlinear medium. In the polariton flow the vortices may appear at the presence of local inhomogeneities of medium refractive index causing the wavefront rotation.

1. Introduction

The properties of electromagnetic waves in a medium are mainly described in the literature by regarding the local and non-local linear and nonlinear response of the medium, or by the analysis of one-photon and multiphoton processes [1 - 6]. However, the electromagnetic wave while propagating in the medium always generates the transverse and longitudinal phonons. Thus, the hybridization of the electromagnetic wave and wave of optical phonons takes place, and the quasiparticles (named polaritons) are generated as a result of coupling of the photons and phonons according to the quantum approximation [7 - 20]. The hybridization of the electromagnetic wave with the wave of dipole excitations of the medium (phonons, magnons, plasmons, excitons) leads to the generation of the corresponding type of polaritons. According to the classical approximation, the hybrid phonon-polariton wave propagates in the medium in the form of a polariton flow. The polariton spectrum has a gap, and an electromagnetic wave with the frequency falling within this spectral gap, does not propagates in the dielectric medium. The nonlinear polariton wave having initially the flat wavefront due to transverse instability is transformed into a spatial soliton or cnoidal (non-linear periodic) wave [13 - 20].

The properties and dynamics of nonlinear waves in dielectric media attract attention of researchers in connection with the problems of the propagation and control of power laser beams in a medium, of the spectral analysis at crystalline and amorphous media, of the design and development of new optical devices and logic gates for optical communications, etc. [16, 19, 20]. The properties of vector solitons and cnoidal polariton waves, nonlinear vortex solitons, nonlinear wave splitting into separate flows named filaments are actively investigated now both theoretically and experimentally [6, 16, 17, 19 - 30].

In our paper we theoretically investigate the properties of the phonon-polariton waves propagating in the Kerr-type dielectric medium on the basis of the classical approximation. We have examined the conditions of occurrence of the spatial solitons, cnoidal waves and polariton vortices, and have shown that the polariton flow can be propagated in the form of spatial solitons, or it can split into several flows in the form of filaments depending on the field density and the nonlinear medium parameters. We have shown that a lattice of nonlinear vortices can appear in the presence of local inhomogeneities of medium refractive index that causes the rotation of the wave front of the polariton wave.

2. Vector equation for the phonon-polaritons

The polaritons are generated in the medium as the bound states of the transverse electromagnetic fields and optical phonons (high transverse oscillations of the ions and the electron shells), and they are propagated in the medium in the form of hybrid photon-phonon waves [10]. We can describe [16 - 20] this process using the system of equations consisting of the equation of motion of the ions in a unit cell of the crystal lattice, and the equation of motion of the outer electron shell of the ion,

$$\begin{Bmatrix} m_{eff} \\ m \end{Bmatrix} \frac{d^2}{dt^2} \begin{Bmatrix} \mathbf{R} \\ \mathbf{r} \end{Bmatrix} + \begin{Bmatrix} \Gamma_I \\ \Gamma_e \end{Bmatrix} \begin{Bmatrix} m_{eff} \\ m \end{Bmatrix} \frac{d}{dt} \begin{Bmatrix} \mathbf{R} \\ \mathbf{r} \end{Bmatrix} + \nabla_R \begin{Bmatrix} U_R \\ U_r \end{Bmatrix} = \begin{Bmatrix} e_{eff} \\ -e \end{Bmatrix} \left(\mathbf{E} + \frac{1}{c} \frac{d}{dt} \begin{Bmatrix} \mathbf{R} \\ \mathbf{r} \end{Bmatrix} \times \mathbf{B} \right), \tag{1}$$

and the electromagnetic field equations

$$\nabla \times \mathbf{B} = \frac{1}{c} \frac{\partial}{\partial t} (\mathbf{E} + 4\pi \mathbf{P}), \quad \nabla \times \mathbf{E} = -\frac{1}{c} \frac{\partial}{\partial t} \mathbf{B}, \tag{2}$$

where e_{eff}, m_{eff} are the effective charge and mass of the ions, $\mathbf{R} = \mathbf{r}_+ - \mathbf{r}_-$ is the displacement vector of positive and negative ions, $U_R = (Q_{1R}/2)R^2 + (Q_{2R}/3)R^3 + (Q_{3R}/4)R^4$ is the potential energy of the ion, $U_r = (Q_{1r}/2)r^2 + (Q_{2r}/3)r^3 + (Q_{3r}/4)r^4$ is the potential energy of the electron, $\Gamma_{I,e}$ are the attenuation coefficients, $\mathbf{P} = e_{eff} N_C \mathbf{R} - eN_e \mathbf{r}$ is the polarization vector of the medium, N_C is the number of cells per unit volume, N_e is the number of electrons per unit volume, Q_{jR} and Q_{jr} are the phenomenological elastic lattice parameters. We take into account the bonding of charges through the electromagnetic field in the system of equations (1) - (2).

Then we can neglect the charges response on the magnetic component of electromagnetic field in the medium in the equations of motion of the ions and electrons $|\mathbf{E}| >> |c^{-1}(d\mathbf{R}/dt) \times \mathbf{B}|$, $|\mathbf{E}| >> |c^{-1}(d\mathbf{r}/dt) \times \mathbf{B}|$ at an optical frequency. If the incident electromagnetic wave is a harmonic one $E \sim exp(-i\omega t)$, then the polarization vector of medium has the form

$$\begin{aligned} \mathbf{P} = \chi_1 \mathbf{E}_a\, exp(-i\omega t) + \chi_{20} E_a \mathbf{E}_a + \chi_{22} E_a \mathbf{E}_a\, exp(-i2\omega t) \\ + \chi_3 E_a^2 \mathbf{E}_a\, exp(-i\omega t) + \chi_{33} E_a^2 \mathbf{E}_a\, exp(-i3\omega t) + \ldots, \end{aligned} \tag{3}$$

where χ_ℓ are the linear and nonlinear susceptibilities of the medium. We neglect the generation of harmonics in the medium considering their nonsynchronous with the fundamental (first) harmonic, i. e. we neglect the energy exchange between the harmonics. Then the polarization vector of the medium (3) can be written as

$$\mathbf{P} = \chi_1 \mathbf{E}_a\, exp(-i\omega t) + \chi_3 E_a^2 \mathbf{E}_a\, exp(-i\omega t), \tag{4}$$

where the susceptibilities of the medium $\chi_1 = \frac{1}{4\pi}\left(\frac{\omega_e^2}{\tilde{\omega}_1^2} + \frac{\omega_I^2}{\tilde{\Omega}_1^2}\right)$ and $\chi_3 = -\frac{1}{4\pi}\left(\frac{e^2\alpha_{3r}\omega_e^2}{m^2\left(\tilde{\omega}_1^2\right)^3\left(\tilde{\omega}_1^2\right)^*} + \frac{e_{eff}^2\alpha_{3R}\omega_I^2}{m_{eff}^2\left(\tilde{\Omega}_1^2\right)^3\left(\tilde{\Omega}_1^2\right)^*}\right)$ are obtained by solving the equations of motion of electrons and ions. Here $\tilde{\omega}_1^2 = \omega_0^2 - \omega^2 - i\Gamma_{e1}\omega$, $\tilde{\Omega}_1^2 = \Omega_\perp^2 - \omega^2 - i\Gamma_{I1}\omega$, $\omega_e^2 = 4\pi e^2 N_e m^{-1}$, $\omega_I^2 = 4\pi e_{eff}^2 N_C m_{eff}^{-1}$, $\omega_0^2 = Q_{1r} m^{-1}$, $\Omega_\perp^2 = Q_{1R} m_{eff}^{-1}$, ω_e and ω_I are the electron and ion plasma frequencies, ω_0 is the electron resonance frequency, $\Omega_\perp$ is the resonance frequency of the lattice, $\alpha_{3r} = Q_{3r} m^{-1}$, $\alpha_{3R} = Q_{3R} m_{eff}^{-1}$ are the phenomenological parameters that depend on the nonlinear properties of the medium, Γ_e and Γ_I are the attenuation coefficients of electron and ion subsystems. The third-order susceptibility of the medium $\chi_{31} > 0$ and $\chi_{33} > 0$ are greater than zero, because the phenomenological parameters $\alpha_{3R} = -Q_{3R} m_{eff}^{-1} < 0$ и $\alpha_{3r} = -Q_{3r} m^{-1} < 0$ are negative. The phenomenological parameters are expressed in the terms of measurable quantities such as resonance frequency, etc.

The dynamics of polaritons are described in reviewed case by the vector equation obtained from equations of motion of ions, electrons and electromagnetic fields. We represent the equation for the electric vector of the polariton waves in the form of

$$-\nabla^2\mathbf{E} + \nabla(\nabla\mathbf{E}) + \frac{1}{c^2}\frac{\partial^2}{\partial t^2}\mathbf{E} = -\frac{4\pi}{c^2}\frac{\partial^2}{\partial t^2}\mathbf{P}, \tag{5}$$

where $\mathbf{P}$ is the polarization vector (4) of the medium. As it follows from the Maxwell equation $\nabla(\varepsilon\mathbf{E}) = 0$, the gradient of divergence of the electric field is non-zero in the Eq. (5) $\nabla(\nabla\mathbf{E}) \neq 0$, because $\nabla\mathbf{E} = -\varepsilon^{-1}(\mathbf{E}\nabla\varepsilon) \neq 0$. Generally, we cannot neglect the mixed derivatives in Eq. (5), because the permittivity of the medium depends on the coordinates in the presence of inhomogeneities.

3. Equations for the envelopes of polariton waves

Suppose that the electric field is polarized in the transverse plane $\mathbf{E} = \mathbf{1}_x E_x(x,y,z) + \mathbf{1}_y E_y(x,y,z)$, then from Eq. (5) we obtain the system of equations

$$\begin{aligned} &\frac{\partial^2 E_x}{\partial y^2} + \frac{\partial^2 E_x}{\partial z^2} - \frac{\partial^2 E_y}{\partial x \partial y} + \frac{\omega^2}{c^2}(1 + 4\pi\chi_1)E_x + \frac{4\pi\omega^2\chi_3}{c^2}\left(|E_x|^2 + |E_y|^2\right)E_x = 0, \\ &\frac{\partial^2 E_y}{\partial x^2} + \frac{\partial^2 E_y}{\partial z^2} - \frac{\partial^2 E_x}{\partial x \partial y} + \frac{\omega^2}{c^2}(1 + 4\pi\chi_1)E_y + \frac{4\pi\omega^2\chi_3}{c^2}\left(|E_x|^2 + |E_y|^2\right)E_y = 0. \end{aligned} \tag{6}$$

In this case, the fundamental harmonic of polariton wave can be represented as $E_{x,y} = \tilde{E}_{x,y}(x,y,z)exp(ik\,z)$, where $\tilde{E}_{x,y}(x,y,z)$ are the slowly varying amplitudes of the transverse components of the electric field, k is the wave vector. In case when the amplitude of the electric vector of the polariton wave varies slowly along the axis z, it is possible to determine its dependence on the longitudinal coordinate as $\tilde{E}_j = e_j(x,y)exp(iqz)$ by entering a constant phase shift q, $j = x, y$. Then from Eq. (6) we obtain the system of equations for the envelopes of the electric field $e_j(x,y)$ of the polariton wave, depending only on the transverse coordinates

$$\frac{\partial^2 e_x}{\partial y^2} - \frac{\partial^2 e_y}{\partial x \partial y} + \alpha_1 e_x + \alpha_3 \left(e_x^2 + e_y^2\right) e_x = 0,$$
$$\frac{\partial^2 e_y}{\partial x^2} - \frac{\partial^2 e_x}{\partial x \partial y} + \alpha_1 e_y + \alpha_3 \left(e_x^2 + e_y^2\right) e_y = 0, \tag{7}$$

where $\alpha_1 = c^{-2}\omega^2(1 + 4\pi\chi_1) - k^2 - 2kq$, $\alpha_3 = 4\pi c^{-2}\omega^2\chi_3$; $\alpha_3 > 0$ at $\chi_3 > 0$ in self-focusing medium, and $\alpha_3 < 0$ at $\chi_3 < 0$ in self-defocusing medium.

4. Circularly polarized polariton waves

The system of equations (7) represented in the Cartesian coordinates can be transformed into a system of equations for the polariton waves of circular polarization in the rotating coordinate system [17, 19, 20]. We introduce the vector $\mathbf{e}_+ = \mathbf{e}_x + i\mathbf{e}_y = e(\mathbf{1}_x + i\mathbf{1}_y)/\sqrt{2} = e\mathbf{1}_+$ for the polaritons with the right-handed helicity at the left-handed circular polarization wave, and the vector $\mathbf{e}_- = \mathbf{e}_x - i\mathbf{e}_y = e(\mathbf{1}_x - i\mathbf{1}_y)/\sqrt{2} = e\mathbf{1}_-$ for polaritons with the left-handed helicity at the right-handed circular polarization wave, and rotating two-dimensional coordinates $\xi = x + iy$ and $\eta = x - iy$. Then from Eqs. (7) we obtain the system of complex equations for the envelopes of polariton waves with the right $\mathbf{e}_+$ and left $\mathbf{e}_-$ helicity in the rotating coordinates (ξ, η),

$$\frac{\partial^2 \mathbf{e}_+}{\partial \xi^2} + \frac{\partial^2 \mathbf{e}_+}{\partial \eta^2} + i\frac{\partial^2 \mathbf{e}_-}{\partial \xi^2} - i\frac{\partial^2 \mathbf{e}_-}{\partial \eta^2} - 2\frac{\partial^2 \mathbf{e}_-}{\partial \xi \partial \eta} - \left(\alpha_1 + \alpha_3 e^2\right)\mathbf{e}_- = 0,$$
$$\frac{\partial^2 \mathbf{e}_-}{\partial \xi^2} + \frac{\partial^2 \mathbf{e}_-}{\partial \eta^2} + i\frac{\partial^2 \mathbf{e}_+}{\partial \xi^2} - i\frac{\partial^2 \mathbf{e}_+}{\partial \eta^2} - 2\frac{\partial^2 \mathbf{e}_+}{\partial \xi \partial \eta} - \left(\alpha_1 + \alpha_3 e^2\right)\mathbf{e}_+ = 0. \tag{8}$$

Equating to zero the determinant $Det(\mathbf{e}_+, \mathbf{e}_-)$ of the system of vector equations (8), we obtain the equation for the envelope of the polariton wave $e(\xi, \eta)$,

$$(1 - i)\frac{\partial^2 e}{\partial \xi^2} + (1 + i)\frac{\partial^2 e}{\partial \eta^2} + 2\frac{\partial^2 e}{\partial \xi \partial \eta} + \left(\alpha_1 + \alpha_3 e^2\right)e = 0. \tag{9}$$

The Eq. (9) in the rotating coordinates (ξ, η) describes the polariton flow both with the right and left polariton helicity. Obtaining the solution of Eq. (9) we can return to the Cartesian coordinates for describing the polariton wave with linear polarization.

We can obtain from Eq. (9) an equation for the envelope of "scalar" polariton flow with the right $e_+ = e(\xi)$ helicity, or with the left $e_- = e(\eta)$ helicity

$$\frac{d^2 e_\pm}{d\varsigma^2} + \alpha_1^{(\pm)} e_\pm + \alpha_3^{(\pm)} e_\pm^3 = 0, \tag{10}$$

where $\alpha_1^{(\pm)} = (1 \pm i)\alpha_1/2$, $\alpha_3^{(\pm)} = (1 \pm i)\alpha_3/2$, $\varsigma = \begin{Bmatrix} \xi \\ \eta \end{Bmatrix}$.

5. Solitons and cnoidal waves in the flow of polaritons

It is well known that a harmonic plane wave is unstable in a nonlinear medium [6]. Instability of the plane wave depends on the parameters of field and medium, and it leads to the transverse and / or longitudinal modulation and further transformation of the plane wave to the spatial soliton or cnoidal wave [16, 17, 19 - 25, 27 - 30].

Consider the formation of spatial soliton or transverse cnoidal wave in nonlinear infinite dielectric medium based on the polariton concept. The polariton flow takes the form of the spatial soliton due to the plane wave modulation in the transverse plane when the boundary conditions have the form $e \to 0$ and $de/d\varsigma = 0$ at $|\xi| \to \infty$, or $|\eta| \to \infty$ [6]. The boundary conditions for the soliton center $\varsigma = 0$ at the input of medium $e(0) = const$ and $de(0)/d\varsigma = 0$ enable us to determine the phase shift q. We can get the soliton solutions of Eq. (10) for polaritons with the right and with the left helicity [17, 19, 20] as

$$e_\pm = \left| \sqrt{\frac{2\overline{\alpha}_1^{(\pm)}}{\alpha_3^{(\pm)}}} \right| sch \left[\sqrt{\left|\overline{\alpha}_1^{(\pm)}\right|}\varsigma - sch^{-1}\left| e_\pm(0)\sqrt{\frac{\alpha_3^{(\pm)}}{2\overline{\alpha}_1^{(\pm)}}} \right| \right] exp(iq_\pm z) \tag{11}$$

in the self-focusing medium at $\alpha_1 = -\overline{\alpha}_1$, $\alpha_3 > 0$, or in the self-defocusing medium at $\alpha_1 = -\overline{\alpha}_1$, $\alpha_3 < 0$, where $q_\pm = \left[\alpha_3 e_\pm^2(0)/2 + c^{-2}\omega^2(1 + 4\pi\chi_1) - k^2\right]/2k$ is the phase shift. The distribution of polaritons $A_\pm = \left(Re^2 e_\pm + Im^2 e_\pm\right)^{1/2}$ at the cross-section of the flow in the form of spatial soliton is shown in figure 1; here the flow propagates along the axis z.

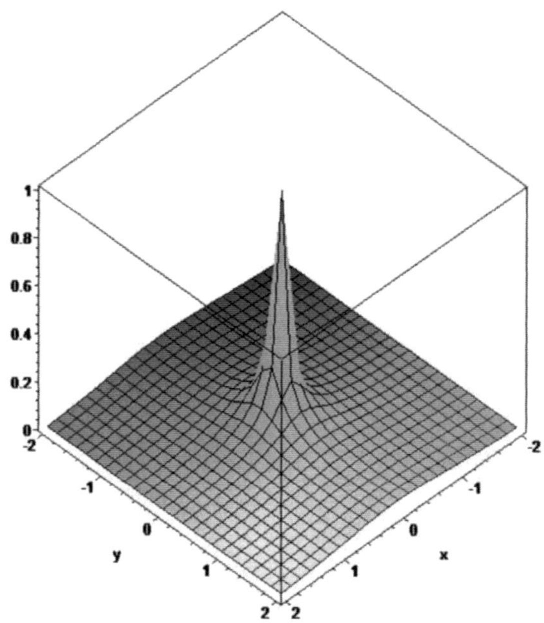

Figure 1. Polariton flow in the form of spatial soliton.

In addition to the spatial soliton formation, the cnoidal (non-linear periodic) wave can be formed in nonlinear medium at wavelength modulation in the transverse plane. Cnoidal polariton wave occurs when the boundary conditions are $e = const$ and $de/d\varsigma = 0$ at $|\xi| \to \infty$, or $|\eta| \to \infty$, i. e. the cnoidal wave does not vanish at infinity. Thus, for generation of the cnoidal waves the medium should be more transparent in comparison with the medium for the soliton generation.

The envelopes of cnoidal polariton waves have the form of elliptic cosine

$$e_{\pm} = \tilde{e}_{0\pm}\, cn\left[\left(\alpha_{\pm}'^{2}/4 + C_{\pm}'\right)^{1/4}\sqrt{\alpha_3}\varsigma - K\left(\tilde{k}_{\pm}\right),\ \tilde{k}_{\pm}\right]\ exp\left(iq_{\pm}z\right), \tag{12}$$

where $\tilde{e}_{0\pm} = \left[\alpha_{\pm}'/2 + \left(\alpha_{\pm}'^{2}/4 + C_{\pm}'\right)^{1/2}\right]^{1/2}$, $\alpha_{\pm}' = 2\overline{\alpha}_1^{(\pm)}/\left|\alpha_3^{(\pm)}\right|$, $C_{\pm}' = 2C_{\pm}/\alpha_3^{(\pm)}$, $C_{\pm} = \alpha_3 e_{\pm}^4(\infty)/2 - \overline{\alpha}_1^{(\pm)} e_{\pm}^2(\infty)$, $q_{\pm} = \left[\alpha_3^{(\pm)} e_{\pm}^2(0)/2 + c^{-2}\omega^2\left(1 + 4\pi\chi_1\right) - k^2 - C_{+}e^{-2}(0)\right]/2k$; $K\left(\tilde{k}_{\pm}\right)$ is the complete elliptic integral; $\tilde{k}_{\pm} = \left[2 + \alpha_{\pm}'\left(\alpha_{\pm}'^{2}/4 + C_{\pm}'\right)^{-1/2}\right]^{1/2}/2$ is the elliptic integral module. The polariton flow is split into several flows in the transverse plane, and they are propagated along the axis z (figure 2), i. e. the filamentation of the polariton flow takes place [17, 19, 20]. The cnoidal wave which is described by the Eq. (12) can be transformed into the spatial soliton $cn(\varsigma,1) \to sch(\varsigma)$ at $\tilde{k}_j \to 1$, when the wave attenuation takes place $e_{\pm}(\infty) \to 0$, i. e. $C_{\pm} \to 0$.

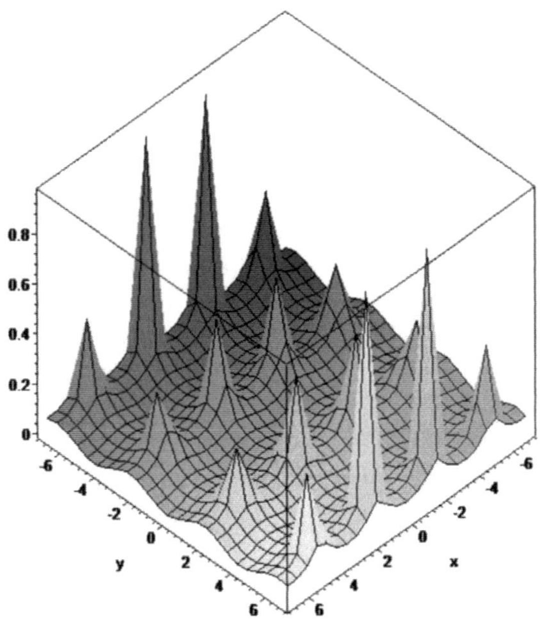

Figure 2. The filamentation of polariton flow at $\tilde{k}_{\pm} = 0.5$.

6. Polariton vortices

The optical vortices arise in the areas of the medium with inhomogeneities where the wavefront acquires the rotation around the singular points [6, 16, 17, 19 - 21, 23 - 26, 28]. Such areas are, for example, the dislocations of crystal lattice, the local inhomogeneities of refractive index of the medium, etc. In this case, the singular points with screw phase dislocations appear on the surface of the wavefront, and we call them the optical vortices.

The wave amplitude vanishes in the singular point $e = 0$, and the phase ϕ is not defined. The optical vortex appears around the singular point. In the linear medium the vortex is described as

$\left[(Ree)^2 + i(Ime)^2\right]^{|\ell|/2} exp(i\ell\phi)$, where $\phi = arctan(Ime/Ree)$, $\ell = \pm 1, \pm 2, \ldots$ is the topological charge of the vortex. The phase of the wave is changed to $\oint d\mathbf{r}\nabla\phi = 2\pi\ell$ while the path tracing around the singular point in which $Ree = 0$ and $Ime = 0$. The phase distribution of the cnoidal polariton wave (12) is described by the function $\phi(x, y) = arctan[Ime(x, y)/Ree(x, y)]$, that allows us to visualize the location of the vortices on the wavefront in cross-section of the polariton flow $z = const$ in the nonlinear medium (figure 3).

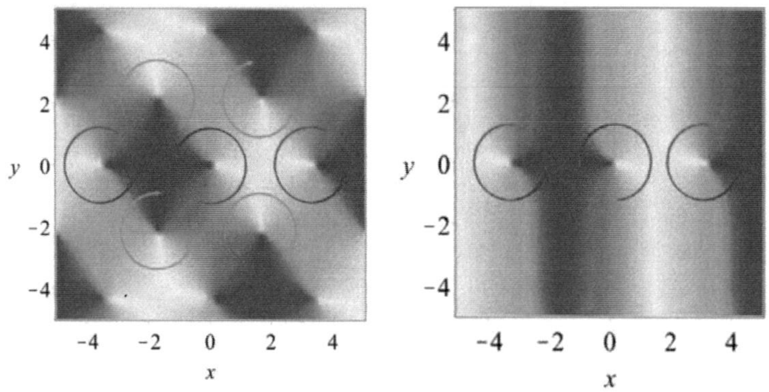

(a) (b)

Figure 3. Vortices in the cross-section of polariton flow at its filamentation in the nonlinear medium: (a) $\tilde{k}_{\pm} = 0.5$, (b) $\tilde{k}_{\pm} = 0.01$; the arrows show the topological charge of vortex, clockwise $\ell = -1$, counterclockwise $\ell = +1$.

The phase of polariton wave is described by the ratio of the elliptic functions, whose values depend on the modulus $\tilde{k}_{\pm}$ of the elliptic integral. In its turn the modulus $\tilde{k}_{\pm}$ depends on the boundary conditions that determine the possibility of generation of the vortices in nonlinear polariton wave. The boundary conditions in inhomogeneous medium can also be formulated as $e = 0$ and $de/d\varsigma = const$ at $|\xi| \to \infty$, or $|\eta| \to \infty$, for describing of the polariton vortex appearance.

7. Conclusion

Electromagnetic wave hybridizes with an optical phonon wave in the dielectric medium that leads to generation of the phonon-polaritons. The plane polariton wave is a nonstable wave in the nonlinear medium, and its transverse envelope can be transformed into the spatial soliton or cnoidal wave depending on the cross-modulation and the medium transparency. In the first case, the polaritons are propagated as a single flow. In the second case, the polariton wave splits into several flows and the filaments appear in the cross-section of the flow. The boundary conditions determine the value of modulus of the elliptic function describing the soliton or cnoidal polariton waves. The cnoidal wave can be transformed into the spatial soliton as a result of dissipation in the medium.

In the presence of local inhomogeneities of the refractive index in dielectric medium causing the wavefront rotation, the singular points arise on the wavefront of polariton wave, where the vortices are formed. The vortex location in the cross-section of polariton flow depends on the configuration of inhomogeneities in the medium.

References

[1] Bloembergen N 1965 *Nonlinear Optics* (New-York: Benjamin)
[2] Loudon R 1973 *The Quantum Theory of Light* (Oxford: Clarendon Press)
[3] Zernike F and Midwinter J E 1973 *Applied Nonlinear Optics* (New-York: John Wiley & Sons)
[4] Hasegawa A and Kodama Yu 1995 *Solitons in Optical Communications* (Oxford Series in Optical and Imaging Sciences) (Oxford: Clarendon Press)
[5] Boyd R W 2003 *Nonlinear Optics* (Amsterdam: Academic Press)
[6] Kivshar Yu S and Agrawal G P 2003 *Optical Solitons: from Fibers to Photonic Crystals* (San Diego: Academic Press)
[7] Tolpygo K B 1950 *J. Exp. Theor. Phys. (USSR)* **20** 497
[8] Huang K 1951 *Proc. Roy. Soc. A* **208** 352
[9] Hopfield J J 1958 *Phys. Rev.* **112** 1555
[10] Kittel C 1963 *Quantum Theory of Solids* (New-York: John Wiley & Sons)
[11] Madelung O 1972 *Festkörpertheorie I II* (Berlin: Springer-Verlag)
[12] Agranovich V M and Ginzburg V L 1979 *Crystal Optics with Spatial Dispersion and the Theory of Excitons* (Moscow: Nauka)
[13] Klyshko D N 1980 *Quantum and Nonlinear Optics* (Moscow: Nauka)
[14] Agranovich V M and Mills D L 1982 *Surface Polaritons: Electromagnetic Waves at Surfaces and Interfaces* (Amsterdam: North-Holland)
[15] Sukhorukov A P 1988 *Nonlinear Wave Interactions in Optics and Radiophysics* (Moscow: Nauka)
[16] Dzedolik I V 2007 *Polaritons in Optical Fibers and Dielectric Resonators* (Simferopol: DIP, 2007)
[17] Dzedolik I V and Karakchieva O 2013 *J. Opt. Soc. Am. B* **30** 843
[18] Dzedolik I V and Pereskokov V 2014 *Appl. Opt.* **53** 3318
[19] Dzedolik I V 2016 *Solitons and Nonlinear Waves of Phonon-Polaritons and Plasmon-Polaritons* (New York: Nova Science Publishers)
[20] Shulika O and Sukhoivanov I 2016 *Contemporary Optoelectronics: Materials, Metamaterials and Device Applications* (Dordrecht: Springer Series in Optical Sciences vol 199)
[21] Rosas D, Law C T and Swartzlander G A 1997 *J. Opt. Soc. Am. B* **14** 3054
[22] Xu J, Shandarov V, Wesner M and Kip D 2002 *Phys. Stat. Sol. A* **189** R4
[23] Desyatnikov A S, Kivshar Yu S and Torner L 2005 *Optical Vortices and Vortex Solitons* (Progress in Optics vol 47) ed Wolf E (Amsterdam: Elsevier) chapter 5 pp 291-391
[24] Ouyang S and Guo Q 2009 *Opt. Exp.* **17** 5170
[25] Kazantseva E V, Maimistov A I and Ozhenko S S 2009 *Phys. Rev. A* **80** 043833
[26] Dennis M R, O'Holleran K and Padgett M J 2009 *Singular Optics: Optical Vortices and Polarization Singularities* (Progress in Optics vol 53) ed Wolf E (Amsterdam: Elsevier) chapter 5 pp 293-363
[27] Chekalin S V and Kandidov V P 2013 *Phys. Usp.* **56** 123
[28] Kartashov Y V, Malomed B A, Shnir Y and Torner L 2014 *Phys. Rev Lett.* **113** 264101
[29] Kamchatnov A M and Pavloff N 2015 *Eur. Phys. J. D* 69: 32
[30] Shandarov V M 2016 *Russian Phys. J.* **58** 1378

Journal of Physics: Conference Series **737** (2016) 012007 doi:10.1088/1742-6596/737/1/012007

Degenerate four-wave mixing in transparent two-component medium considering spatial structure of the pump waves

V V Ivakhnik and M V Savel'ev

Samara National Research University, 1 Akademika Pavlova Street, Samara 443011, Russia

E-mail: ivakhnik@samsu.ru, belchonokenot@mail.ru

Abstract. In this paper we investigate spatial selectivity of the degenerate four-wave radiation converter in transparent liquid containing nanoparticles considering spatial structure of the pump waves. The bandwidth of the most efficiently converted spatial frequencies is associated with the rotation and divergence of the pump waves.

1. Introduction

Accounting for spatial structure of the pump waves in the study of radiation (image) conversion quality by four-wave mixing is the necessary condition. The spatial structure of the pump waves fully determines the four-wave converters resolution [1, 2] or it is one of the main factors leading to the deterioration of the image conversion quality [3].

In recent years the ability to use media containing micro- and nanosized particles (colloidal solutions, suspensions, etc.) for realization of the four-wave mixing is actively discussed [4-10]. If liquids containing nanoparticles are used as media in four-wave mixing then in such media electrostriction and Dufour effect influence significantly on the characteristics of four-wave radiation converter filtering high spatial frequencies of the object wave [11, 12], which may be preferable in correction systems of small-scale phase inhomogeneities [13].

The aim of this work is to study influence of the divergence and rotation of the pump waves on spatial selectivity of the degenerate four-wave radiation converter in transparent liquid containing nanoparticles.

2. Relation of the interacting waves spatial spectra

Let us consider a plane layer thickness ℓ of transparent two-component medium (liquid in which nanoparticles are located). In liquid in the Z-axis direction two pump waves with complex amplitudes A_1 and A_2 towards each other and signal wave with complex amplitude A_3 propagate. Diffraction of the second pump wave (A_2) on the induced dynamic phase grating in the medium leads to emergence of the object wave with complex amplitude A_4.

Degenerate four-wave mixing $\omega+\omega-\omega=\omega$ in transparent two-component medium is described by the Helmholtz equation [1]

$$\left(\nabla^2 + k^2 + \frac{2k^2}{n_0}\frac{\mathrm{d}n}{\mathrm{d}T}\delta T\right)\left(A + A^*\right) = 0 \tag{1}$$

where $A=\sum_{j=1}^{4}A_j$, δT is a temperature variation, $k=\frac{\omega n_0}{c}$, n_0 is the average refractive index.

Equation (1) is complemented by the material equation for the temperature variation [11]

$$\nabla^2\delta T=\frac{\gamma D_{12}}{D_{11}D_{22}}\nabla^2 I. \tag{2}$$

Here, $I=AA^*$, D_{11} is the thermal conductivity coefficient, D_{22} is the diffusion coefficient, D_{12} is a coefficient which describes Dufour effect, γ is the electrostriction coefficient.

Four-wave mixing is considered under the following conditions:

- the approximation of the specified field by pump waves is valid $\left(\left|A_{1,2}\right|>>\left|A_{3,4}\right|\right)$;
- the reflection (conversion) coefficient is small $\left(\left|A_4\right|<<\left|A_3\right|\right)$.

The intensity of radiation which propagates in nonlinear medium can be written as follows

$$I=I_0+A_1A_3^*+A_1^*A_3.$$

Here, $I_0=A_1A_1^*+A_2A_2^*$.

Then the temperature variation can be expressed as the sum of the quickly (δT_{31}) and slowly (δT_0) varying components which depend on the coordinates

$$\delta T=\delta T_0+\delta T_{31}+\delta T_{31}^*.$$

In view of the foregoing, the Helmholtz equation (1) is divided into four equations

$$\begin{gathered}\left(\nabla^2+k^2+\frac{2k^2}{n_0}\frac{\mathrm{d}n}{\mathrm{d}T}\delta T_0\right)A_{1,2,3}=0,\\ \left(\nabla^2+k^2+\frac{2k^2}{n_0}\frac{\mathrm{d}n}{\mathrm{d}T}\delta T_0\right)A_4+\frac{2k^2}{n_0}\frac{\mathrm{d}n}{\mathrm{d}T}\delta T_{31}A_2=0\end{gathered} \tag{3}$$

and material equation (2) is divided into two equations

$$\nabla^2\delta T_0=\frac{\gamma D_{12}}{D_{11}D_{22}}\nabla^2 I_0,\quad \nabla^2\delta T_{31}=\frac{\gamma D_{12}}{D_{11}D_{22}}\nabla^2 A_1A_3^*. \tag{4}$$

We expand the interacting waves in plane waves

$$A_j(\vec{r})=\int_{-\infty}^{\infty}\tilde{A}_j(\vec{\kappa}_j,z)\exp\left(-\mathrm{i}\vec{\kappa}_j\vec{\rho}-\mathrm{i}k_{jz}z\right)\mathrm{d}\vec{\kappa}_j,\quad j=1-4 \tag{5}$$

and expand the quickly varying component of temperature in harmonic gratings

$$\delta T_{31}(\vec{r})=\int_{-\infty}^{\infty}\delta\tilde{T}_{31}(\vec{\kappa}_T,z)\exp\left(-\mathrm{i}\vec{\kappa}_T\vec{\rho}\right)\mathrm{d}\vec{\kappa}_T. \tag{6}$$

Here, $\tilde{A}_j$ is spatial spectrum of *j*th wave, $\delta\tilde{T}_{31}$ is spatial spectrum of the temperature grating, $\vec{\kappa}_j$ and k_{jz} are the transverse and longitudinal components of the wave vector $\vec{k}_j$, $\left|\vec{k}_j\right|=k$, $\vec{\kappa}_T$ is the grating wave vector, $\vec{\rho}(x,y)$ and z are the transverse and longitudinal components of the radius vector $\vec{r}$.

Substituting (5) – (6) in the equations (3) – (4) we obtain the equations describing the changes in spatial spectra of the interacting waves and temperature grating along the z coordinate

$$\left(\frac{\mathrm{d}}{\mathrm{d}z}+\mathrm{i}\frac{k^2}{k_{1,2,3z}n_0}\frac{\mathrm{d}n}{\mathrm{d}T}\delta T_0\right)\tilde{A}_{1,2,3}\left(\vec{\kappa}_{1,2,3},z\right)=0,$$

$$\left(\frac{\mathrm{d}}{\mathrm{d}z}+\mathrm{i}\frac{k^2}{k_{4z}n_0}\frac{\mathrm{d}n}{\mathrm{d}T}\delta T_0\right)\tilde{A}_4\left(\vec{\kappa}_4,z\right)= \tag{7}$$

$$=-\mathrm{i}\frac{k^2}{k_{4z}n_0}\frac{\mathrm{d}n}{\mathrm{d}T}\int_{-\infty}^{\infty}\delta\tilde{T}_{31}\left(\vec{\kappa}_\mathrm{T}=\vec{\kappa}_4-\vec{\kappa}_2,z\right)\tilde{A}_2\left(\vec{\kappa}_2,z\right)\exp\left[-\mathrm{i}\left(k_{2z}-k_{4z}\right)z\right]\mathrm{d}\vec{\kappa}_2,$$

$$\left(\frac{\mathrm{d}^2}{\mathrm{d}z^2}-\kappa_\mathrm{T}^2\right)\delta\tilde{T}_{31}\left(\vec{\kappa}_\mathrm{T},z\right)=$$

$$=\frac{\gamma D_{12}}{D_{11}D_{22}}\int_{-\infty}^{\infty}\tilde{A}_1\left(\vec{\kappa}_1,z\right)\tilde{A}_3^*\left(\vec{\kappa}_3=\vec{\kappa}_1-\vec{\kappa}_\mathrm{T},z\right)\exp\left[-\mathrm{i}\left(k_{1z}-k_{3z}\right)z\right]\left[\kappa_\mathrm{T}^2+\left(k_{1z}-k_{3z}\right)^2\right]\mathrm{d}\vec{\kappa}_1. \tag{8}$$

Equations (7) and (8) are written under the condition that $\vec{\kappa}_\mathrm{T}=\vec{\kappa}_1-\vec{\kappa}_3=\vec{\kappa}_4-\vec{\kappa}_2$.

Using the boundary conditions on spatial spectra of the interacting waves $\tilde{A}_{1,3}\left(\vec{\kappa}_{1,3},z=0\right)=\tilde{A}_{1,30}\left(\vec{\kappa}_{1,3}\right)$, $\tilde{A}_2\left(\vec{\kappa}_2,z=\ell\right)=\tilde{A}_{20}\left(\vec{\kappa}_2\right)$, $\tilde{A}_4\left(\vec{\kappa}_4,z=\ell\right)=0$ and the condition that the temperature is invariable at the nonlinear layer edges $\left(\delta T_{31}\left(z=0\right)=\delta T_{31}\left(z=\ell\right)=0\right)$ we find changes in spatial spectra of the pump waves and the signal wave by nonlinear layer thickness

$$\tilde{A}_{1,3}\left(\vec{\kappa}_{1,3},z\right)=\tilde{A}_{1,30}\left(\vec{\kappa}_{1,3}\right)\exp\left[-P(z)\right],\ \tilde{A}_2\left(\vec{\kappa}_{1,3},z\right)=\tilde{A}_{20}\left(\vec{\kappa}_2\right)\exp\left[-P(\ell)+P(z)\right]$$

then the spatial spectrum of the temperature grating and the spatial spectrum of the object wave on the front edge of the nonlinear layer

$$\delta\tilde{T}_{31}\left(\vec{\kappa}_\mathrm{T},z\right)=\frac{\gamma D_{12}}{D_{11}D_{22}}\int_{-\infty}^{\infty}\mathrm{d}\vec{\kappa}_1\tilde{A}_{10}\left(\vec{\kappa}_1\right)\tilde{A}_{30}^*\left(\vec{\kappa}_3=\vec{\kappa}_1-\vec{\kappa}_\mathrm{T}\right)\left(\frac{1}{\sinh\kappa_\mathrm{T}\ell}\left\{\exp\left[-\mathrm{i}\left(k_{1z}-k_{3z}\right)\ell\right]\sinh\kappa_\mathrm{T}z-\right.\right.$$
$$\left.\left.-\sinh\kappa_\mathrm{T}\left(z-\ell\right)\right\}-\exp\left[-\mathrm{i}\left(k_{1z}-k_{3z}\right)z\right]\right), \tag{9}$$

$$\tilde{A}_4\left(\vec{\kappa}_4,z=0\right)=-\mathrm{i}\frac{k\gamma D_{12}}{n_0D_{11}D_{22}}\frac{\mathrm{d}n}{\mathrm{d}T}\exp\left[-P(\ell)\right]\int_{-\infty}^{\infty}\int_{-\infty}^{\infty}\mathrm{d}\vec{\kappa}_1\mathrm{d}\vec{\kappa}_2\tilde{A}_{10}\left(\vec{\kappa}_1\right)\tilde{A}_{20}\left(\vec{\kappa}_2\right)\tilde{A}_{30}^*\left(\vec{\kappa}_3=\vec{\kappa}_1-\vec{\kappa}_\mathrm{T}\right)\times$$
$$\times\left\{\left[\frac{1}{2\sinh\left|\vec{\kappa}_4-\vec{\kappa}_2\right|\ell}\left(\left\{\exp\left[-\mathrm{i}\left(k_{1z}-k_{3z}\right)\ell\right]-\exp\left(-\left|\vec{\kappa}_4-\vec{\kappa}_2\right|\ell\right)\right\}\times\right.\right.\right.$$
$$\times\frac{\exp\left\{\left[\left|\vec{\kappa}_4-\vec{\kappa}_2\right|-\mathrm{i}\left(k_{2z}-k_{4z}\right)\right]\ell\right\}-1}{\left|\vec{\kappa}_4-\vec{\kappa}_2\right|-\mathrm{i}\left(k_{2z}-k_{4z}\right)}+\left\{\exp\left[-\mathrm{i}\left(k_{1z}-k_{3z}\right)\ell\right]-\exp\left(\left|\vec{\kappa}_4-\vec{\kappa}_2\right|\ell\right)\right\}\times \tag{10}$$
$$\left.\left.\times\frac{\exp\left\{-\left[\left|\vec{\kappa}_4-\vec{\kappa}_2\right|+\mathrm{i}\left(k_{2z}-k_{4z}\right)\right]\ell\right\}-1}{\left|\vec{\kappa}_4-\vec{\kappa}_2\right|+\mathrm{i}\left(k_{2z}-k_{4z}\right)}\right)-\mathrm{i}\frac{\exp\left(-\mathrm{i}\Delta\ell\right)-1}{\Delta}\right].$$

Here, $P(z)=\mathrm{i}\frac{k}{n_0}\frac{\mathrm{d}n}{\mathrm{d}T}\int_0^z\delta T_0\left(z_1\right)\mathrm{d}z_1$, $\Delta=\left(\vec{k}_1+\vec{k}_2-\vec{k}_3-\vec{k}_4\right)_z$ is the projection of the wave mismatch on the *Z*-axis. Expression (10) is written at quasicollinear propagation of waves $\left(kk_{1,3z}^{-1}\approx-kk_{2,4z}^{-1}\approx1\right)$ and fully describes the relation of the interacting waves spatial spectra in transparent two-component medium.

3. Discussion of the results

We consider the signal wave emanating from a point source located on the front edge of the nonlinear layer $\left(\tilde{A}_{30}\left(\vec{\kappa}_3\right)=1\right)$. Let the first pump wave spatial spectrum changes according to the Gaussian law and the second pump wave is plane

$$\tilde{A}_{10}\left(\vec{\kappa}_1\right)=\exp\left(-\frac{\kappa_1^2}{\kappa_0^2}\right),\quad \tilde{A}_{20}\left(\vec{\kappa}_2\right)=\delta\left(\vec{\kappa}_2-\vec{\kappa}_{20}\right) \tag{11}$$

where κ_0 is a parameter which characterizes the first pump wave divergence, $\vec{\kappa}_{20}$ is a vector determining the propagation direction of the second pump wave.

In view of (11) in the paraxial approximation the expression for the spatial spectrum of the object wave (10) can be written as follows

$$\begin{aligned}
\tilde{A}_4\left(\vec{\kappa}_4\right)=&-\mathrm{i}\frac{\pi\kappa_0^2 k\gamma D_{12}}{n_0 D_{11}D_{22}}\frac{\mathrm{d}n}{\mathrm{d}T}\left\{\frac{1}{2\sinh\left|\vec{\kappa}_4-\vec{\kappa}_{20}\right|\ell}\left(\left\{\exp\left[-\mathrm{i}\frac{\left(\vec{\kappa}_4-\vec{\kappa}_{20}\right)^2}{2k}\ell-\frac{\left(\vec{\kappa}_4-\vec{\kappa}_{20}\right)^2\kappa_0^2\ell^2}{4k^2}\right]-\right.\right.\right.\\
&-\exp\left(-\left|\vec{\kappa}_4-\vec{\kappa}_{20}\right|\ell\right)\Bigg\}\frac{\exp\left(\left|\vec{\kappa}_4-\vec{\kappa}_{20}\right|\ell-\mathrm{i}\frac{\kappa_{20}^2-\kappa_4^2}{2k}\ell\right)-1}{\left|\vec{\kappa}_4-\vec{\kappa}_{20}\right|-\mathrm{i}\frac{\kappa_{20}^2-\kappa_4^2}{2k}}+\left\{\exp\left[-\mathrm{i}\frac{\left(\vec{\kappa}_4-\vec{\kappa}_{20}\right)^2}{2k}\ell-\frac{\left(\vec{\kappa}_4-\vec{\kappa}_{20}\right)^2\kappa_0^2\ell^2}{4k^2}\right]-\right.\\
&-\exp\left(\left|\vec{\kappa}_4-\vec{\kappa}_{20}\right|\ell\right)\Bigg\}\frac{\exp\left(-\left|\vec{\kappa}_4-\vec{\kappa}_{20}\right|\ell-\mathrm{i}\frac{\kappa_{20}^2-\kappa_4^2}{2k}\ell\right)-1}{\left|\vec{\kappa}_4-\vec{\kappa}_{20}\right|+\mathrm{i}\frac{\kappa_{20}^2-\kappa_4^2}{2k}}\Bigg)-\frac{\sqrt{\pi}k}{\left|\vec{\kappa}_4-\vec{\kappa}_{20}\right|\kappa_0}\exp\left[-\left(\frac{\vec{\kappa}_{20}\left(\vec{\kappa}_4-\vec{\kappa}_{20}\right)}{\left|\vec{\kappa}_4-\vec{\kappa}_{20}\right|\kappa_0}\right)^2\right]\times\\
&\times\left(\mathrm{erf}\left[\frac{\left|\vec{\kappa}_4-\vec{\kappa}_{20}\right|\kappa_0\ell}{2k}-\mathrm{i}\frac{\vec{\kappa}_{20}\left(\vec{\kappa}_4-\vec{\kappa}_{20}\right)}{\left|\vec{\kappa}_4-\vec{\kappa}_{20}\right|\kappa_0}\right]+\mathrm{erf}\left[\mathrm{i}\frac{\vec{\kappa}_{20}\left(\vec{\kappa}_4-\vec{\kappa}_{20}\right)}{\left|\vec{\kappa}_4-\vec{\kappa}_{20}\right|\kappa_0}\right]\right)\Bigg\}
\end{aligned} \tag{12}$$

where $\mathrm{erf}(x)=\frac{2}{\sqrt{\pi}}\int_0^x\exp\left(-t^2\right)\mathrm{d}t$ is the error function.

In figure 1 we provide characteristics plots of the normalized module of the spatial spectrum of the object wave $\left(A_4^n\left(\vec{\kappa}_4\right)=\left|\tilde{A}_4\left(\vec{\kappa}_4\right)\right|\cdot\left|\tilde{A}_4\left(\kappa_4=0.1k,\kappa_0=0,\vec{\kappa}_{20}=0\right)\right|^{-1}\right)$ obtained using the expression (12). If the pump waves propagating accurately toward each other are plane then the spatial spectrum module of the object wave with the growth of spatial frequency $\left(\vec{\kappa}_4\right)$ increases and it reaches a constant value. However, accounting pump waves divergence reduces the wave conversion efficiency at high spatial frequencies without changing the form of the spatial spectrum on the low spatial frequencies. In the pump wave flat $\left(\left(\vec{\kappa}_4\vec{\kappa}_{20}\right)=\kappa_4\kappa_{20}\right)$ the second pump wave rotation shifts the band of spatial frequencies cut out by four-wave radiation converter on rotation value and narrows the spatial spectrum of the object wave. In the flat that is perpendicular to the pump waves flat $\left(\left(\vec{\kappa}_4\vec{\kappa}_{20}\right)=0\right)$ the pump wave rotation does not influence the spatial selectivity of four-wave radiation converter.

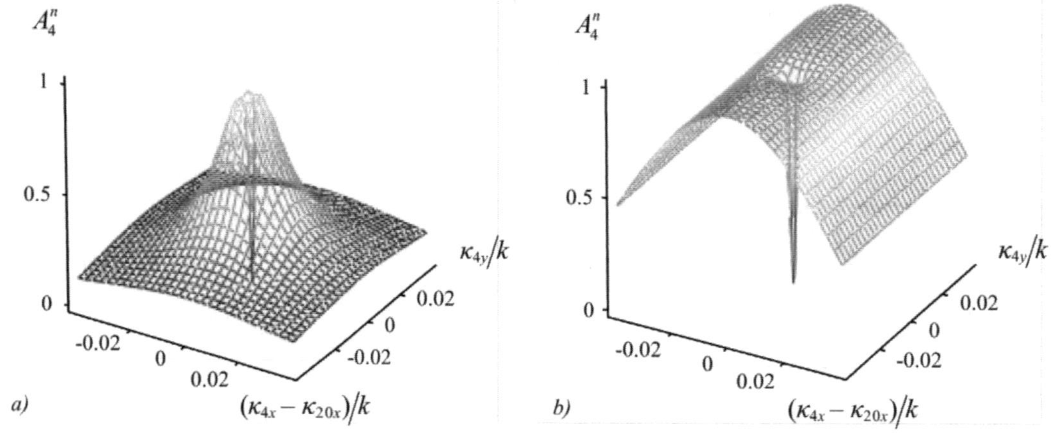

Figure 1. Spatial spectra modules of the object wave at $k\ell = 5\times10^3$, $\kappa_{20y} = 0$, $\kappa_{20x}k^{-1} = 2\times10^{-2}$, $\kappa_0 k^{-1} = 5\times10^{-2}$ (a), 1×10^{-8} (b).

Four-wave radiation converter with the rotated second pump wave filters low spatial frequencies along with the high spatial frequencies due to the presence of the non-zero wave mismatch projection included explicitly in the last term of the expression (10). A similar behaviour of the spatial spectrum module form with changing of the wave mismatch is observed for nondegenerate four-wave radiation converter in transparent two-component medium [12].

When the second pump wave is plane and propagates accurately along the Z-axis the spatial spectrum of the object wave is symmetrical about an axis passing through the zero spatial frequency and depends on $\kappa_4 = |\vec{\kappa}_4|$ (figure 2). To compare in the same figure we present the spatial spectrum module of the object wave of four-wave radiation converter on the Kerr nonlinearity calculated using the expression [1]

$$\tilde{A}_4(\vec{\kappa}_4) = \mathrm{i}G\,\mathrm{erf}\left(\frac{\kappa_4\kappa_0}{2k}\ell\right). \tag{13}$$

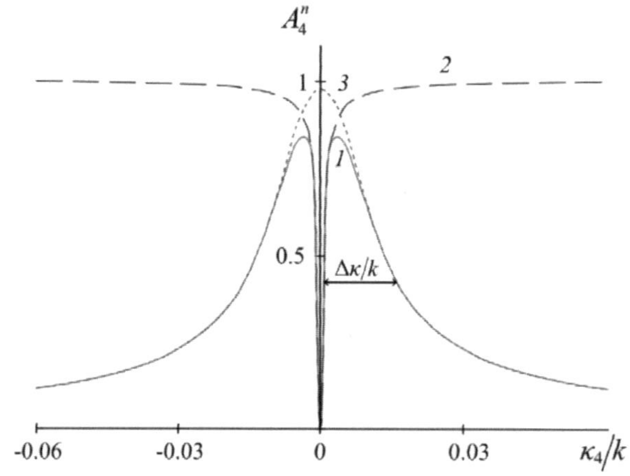

Figure 2. Dependence of the spatial spectra modules from the spatial frequency value at $k\ell = 5\times10^3$, $\vec{\kappa}_{20} = 0$, $\kappa_0 k^{-1} = 5\times10^{-2}$ (1, 3), 1×10^{-8} (2), transparent two-component medium (1, 2), medium with Kerr nonlinearity (3).

Here, G is the parameter that determines the efficiency of four-wave radiation converter.

In constructing graphs we equated the spatial spectra modules of the object waves of two four-wave radiation converters at spatial frequency $\kappa_4 = 5\kappa_0$. As seen, the spatial selectivity of four-wave radiation converters in transparent liquid containing nanoparticles and medium with Kerr nonlinearity is the same at high spatial frequencies.

In the area of spatial frequencies cut out by four-wave radiation converter the object wave amplitude is independent on the divergence and pump wave rotation.

To characterize the spatial selectivity of four-wave radiation converter in the pump waves flat we use the bandwidth of the most efficiently converted spatial frequencies defined as difference between spatial frequencies at which the maximum value of spatial spectrum module of the object wave is reduced by half (figure 2) [12].

Without the second pump wave rotation the reduction of the bandwidth of the most efficiently converted spatial frequencies with increasing of the first pump wave divergence (figure 3) is well described by the expression of the form

$$\frac{\Delta\kappa}{k} = \alpha\left(\frac{\kappa_0}{k}\right)^{-1} + \beta. \qquad (14)$$

Here, α and β are the parameters which are inversely proportional to the multiplication of the wave number on nonlinear layer thickness. As for the values $k\ell = 1.5\times10^3$, 5×10^3, 1.5×10^4 parameters in the expression (14) take the values $\alpha = 2.4\times10^{-3}$, 7.2×10^{-4}, 2.4×10^{-4} and $\beta = 5\times10^{-3}$, 2×10^{-3}, 5×10^{-4}.

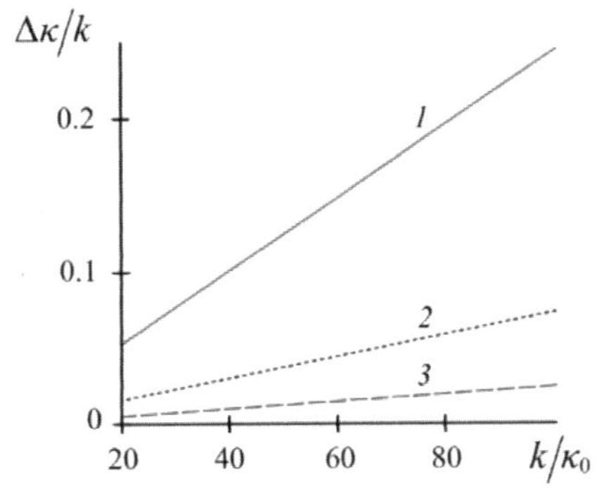

Figure 3. Dependence of the bandwidth of spatial frequencies from the pump wave divergence at $\vec{\kappa}_{20} = 0$, $k\ell = 1.5\times10^3$ (1), 5×10^3 (2), 1.5×10^4 (3).

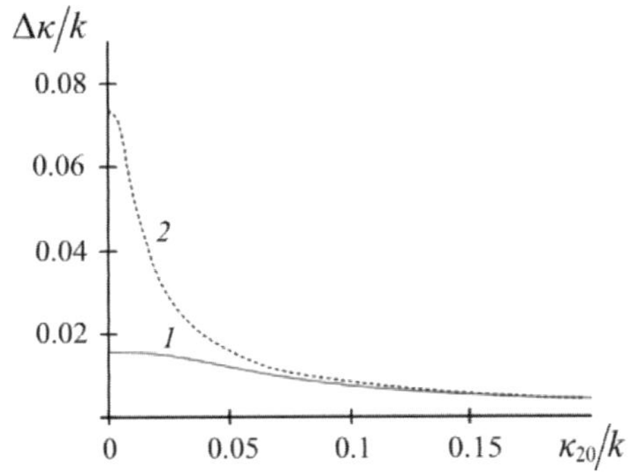

Figure 4. Dependence of the bandwidth of spatial frequencies from the pump wave rotation at $k\ell = 5\times10^3$, $\kappa_0 k^{-1} = 5\times10^{-2}$ (1), 1×10^{-2} (2).

Within the bandwidth of spatial frequencies $\Delta\kappa$ accounting the pump waves divergence $\left(\kappa_0/k < 5\times10^{-2}\right)$ insignificantly changes the spatial spectrum phase of the object wave.

When $\kappa_0 \neq 0$ the second pump wave rotation leads to additional reduction of the wave conversion efficiency at high spatial frequencies in the pump waves flat. With the growth of the second pump wave rotation the bandwidth of spatial frequencies reduces monotonically (figure 4). When $\kappa_{20} \leq \kappa_0$ the dependence of bandwidth of the most efficiently converted spatial frequencies from the second pump wave rotation is well described by a Gaussian law

$$\Delta\kappa = \Delta\kappa_0 \exp\left(-b\frac{\kappa_{20}^2}{k^2}\right). \qquad (15)$$

Here, $\Delta\kappa_0$ is the bandwidth of the most efficiently converted spatial frequencies without the second pump wave rotation, b is the coefficient that depends on the value of the first pump wave divergence. When the first pump wave divergence $\kappa_0 k^{-1} = 1\times10^{-2}$ and 5×10^{-2} coefficient b takes the values 50 and 10 respectively.

When $\kappa_{20} > 2\kappa_0$ the bandwidth of spatial frequencies most efficiently converted at four-wave mixing is independent on the first pump wave divergence and is fully determined by the second pump wave rotation.

4. Conclusion

As a result of the paper, shown that in the area of spatial frequencies cut out by the degenerate four-wave radiation converter the object wave amplitude is indepentent on the spatial structure of the pump waves. The divergence and pump waves rotation reduce the wave conversion efficiency at high spatial frequencies without changing the spatial spectrum form of the object wave at low spatial frequencies. We obtained the expressions relating the bandwidth of spatial frequencies most efficiently converted at degenerate four-wave mixing with the divergence and pump waves rotation.

References

[1] Ivakhnik V V 2010 *Wavefront reversal at four-wave interaction* (Samara: Samara University Publisher)
[2] Dmitriev V G 2003 *Nonlinear optics and wavefront reversal* (Moscow: Fizmatlit Publisher)
[3] Ivakhnik V V and Kharskaya T G 2009 *Computer Opt.* **33** 17-26
[4] Livashvili A I, Kostina G V and Yakunina M I 2013 *J. Opt. Technol.* **80** 124-6
[5] Rusconi R, Isa L and Piazza R 2004 *JOSA* B **21** 605-15
[6] Mahilny U V, Marmysh D N, Stankevich A I, Tolstik A L, Matusevich V and Kowarschik R 2006 *Appl. Phys.* B **82** 299-302
[7] Afanas'ev A A, Rubinov A N, Mikhnevich S Yu and Ermolaev I E 2005 *JETP* **101** 389-400
[8] Lopez-Mariscal C, Gutierrez-Vega J C, McGloin D and Dholakia K 2007 *Opt. Exp.* **15** 6330-5
[9] El-Ganainy R, Christodoulides D N, Rotschild C and Segev M 2007 *Opt. Exp.* **15** 10207-18
[10] Ivanov V I and Okishev K N 2006 *Tech. Phys. Lett.* **32** 967-8
[11] Ivakhnik V V and Savelyev M V 2015 *Phys. Procedia* **73** 26-32
[12] Ivakhnik V V and Savelyev M V 2015 *Computer Opt.* **39** 486-91
[13] Ivakhnik V V, Petnikova V M and Shuvalov V V 1981 *Sov. J. Quantum Electron.* **11** 467-9

Electromagnetic field distribution in a quasi-1D rhombic waveguide array

A I Maimistov[1,2], **V A Patrikeev**[1,3]

[1]National Research Nuclear University MEPhI (Moscow Engineering Physics Institute), 31 Kashirskoe shosse, 115409 Moscow, Russia
[2]Moscow Institute of Physics and Technology, 9 Institutskiy per., Dolgoprudny, Moscow Region, 141700, Russia
[3]Med-SK, Moscow, 123290, Russia

E-mail: aimaimistov@gmail.com, sugrobs@yandex.ru

Abstract. The exact solution of the coupled mode equations is presented in the case of the continuous electromagnetic wave propagation in the quasi-one-dimensional rhombic array of the waveguides. In general case of the boundary conditions the discrete diffraction is described by these solutions. The flat band mode propagation is found as the special case.

1. Introduction

Recently the optical simulations of the different phenomena of solid state have been developed. There is one interested example. The investigation of two dimensional electron systems demonstrated that presence of the third atom in the elementary cell as well as long range interaction in the lattice leads to emerging of a flat sheet (flat band) between conventional zones. Similar optical lattices can be realized by means of waveguides as nodes of the lattice. Some kinds of the optical lattices that demonstrate the photonic spectrum with flat band have been discussed in [1,2,3]. In [1] the waveguide array consisting from three parallel linear chain of waveguides was considered. The central waveguide chain is shifted according to either chains at half lattice period. Resulting configuration seams as linear chain of the rhombus. This array of waveguides was named as the quasi-one-dimensional rhombic array [1,4,5].

In this paper, we consider the electromagnetic field distribution in the quasi-one-dimensional rhombic array of the waveguides. All waveguides are supposed as linear waveguides. The exact solution of the coupled mode equations is found. These solutions demonstrate both the flat band mode and the discrete diffraction phenomenon.

2. Base equations

System of equations describing coupled waves in this quasi-one-dimensional rhombic array of waveguides has the following form [1,4,8]

$$i\frac{\partial A_n}{\partial \zeta} = (B_n + B_{n-1}) + (C_n + C_{n-1}),$$

$$i\frac{\partial B_n}{\partial \zeta} = (A_{n+1} + A_n), \qquad i\frac{\partial C_n}{\partial \zeta} = (A_{n+1} + A_n). \tag{1}$$

Here $\zeta = Kz$ is normalized coordinate, K is coupling constant. The fields A_n, B_n and C_n are dimensionless slowly varying amplitudes of the electric fields in n-th waveguide. The sub-indices n are markers of the elementary cells. Phase matching condition is assumed to be satisfied, and coupling constants between waveguides are equal to unit. Since system of equation (1) is linear, the dispersion relation can be found in a standard way:

$$q_{\pm}(s) = \pm 2\sqrt{2}\,|\cos(\pi s / M)|, \quad q_0(s) = 0. \tag{2}$$

Here $q(s)$ is transversal wave number of the mode with index s, where $s = -M,\ldots,-1,0,1,\ldots,M$, $N = 2M + 1$ is number of the elementary cells of waveguide array.

3. Solution of the base equations

We can use the generation function method. Let us introduce the following functions

$$P_A(\zeta, y) = \sum_{n=-\infty}^{\infty} A_n(\zeta)e^{iny}, \quad P_B(\zeta, y) = \sum_{n=-\infty}^{\infty} B_n(\zeta)e^{iny}, \quad P_C(\zeta, y) = \sum_{n=-\infty}^{\infty} C_n(\zeta)e^{iny}.$$

From (1) the following system of equations can be obtained

$$i\frac{\partial P_A}{\partial \zeta} = \kappa^*(P_B + P_C), \quad i\frac{\partial P_B}{\partial \zeta} = \kappa P_A, \quad i\frac{\partial P_C}{\partial \zeta} = \kappa P_A, \tag{3}$$

where

$$\kappa = \kappa(y) = (1 + e^{-iy}) = 2\cos(y/2)\exp(-iy/2).$$

With taking into account the boundary conditions at $\zeta = 0$

$$P_{A0} = \sum_{n=-\infty}^{\infty} A_n(0)e^{iny}, \quad P_{B0} = \sum_{n=-\infty}^{\infty} B_n(0)e^{iny}, \quad P_{C0} = \sum_{n=-\infty}^{\infty} C_n(0)e^{iny},$$

the solution of (3) can be written as

$$P_A(\zeta, y) = P_{A0}\cos\Omega\zeta - i\beta(P_{B0} + P_{C0})\sin\Omega\zeta, \tag{4}$$

$$P_B(\zeta, y) = \frac{1}{2}\left[P_{B0} - P_{C0} + (P_{B0} + P_{C0})\cos\Omega\zeta - \frac{i}{\beta}P_{A0}\sin\Omega\zeta\right], \tag{5}$$

$$P_C(\zeta, y) = \frac{1}{2}\left[P_{C0} - P_{B0} + (P_{B0} + P_{C0})\cos\Omega\zeta - \frac{i}{\beta}P_{A0}\sin\Omega\zeta\right], \tag{6}$$

where

$$\Omega^2 = 8\cos^2(y/2), \quad \beta = \frac{\kappa^*}{\Omega} = \frac{1}{\sqrt{2}}\exp(-iy/2).$$

By using the orthogonality condition

$$\int_{-\pi}^{\pi} \exp[iy(n-m)]dy = 2\pi\delta_{nm}$$

the amplitudes $A_n(\zeta)$, $B_n(\zeta)$ and $C_n(\zeta)$ can be determined from 4)-(6):

$$2\pi A_n(\zeta) = \int_{-\pi}^{\pi} P_A(\zeta, y)\exp(-iyn)dy, \quad 2\pi B_n(\zeta) = \int_{-\pi}^{\pi} P_B(\zeta, y)\exp(-iyn)dy,$$

$$2\pi A_n(\zeta) = \int_{-\pi}^{\pi} P_A(\zeta, y)\exp(-iyn)dy.$$

4. Particular solution of the base equations

Let us consider some particular solutions of the base equations (1). The particular solutions are defined by the different boundary conditions. Supposing the conditions, which are correlated to situation where the radiation is initially input only to waveguides of one elementary cell in array.

$$A_n(0) = A_0\delta_{n0}, \quad B_n(0) = B_0\delta_{n0}, \quad C_n(0) = C_0\delta_{n0}.$$

Thus, $P_{A0} = A_0$, $P_{B0} = B_0$, $P_{C0} = C_0$. With taking into account (4)-(6) the express the amplitudes $A_n(\zeta)$, $B_n(\zeta)$ and $C_n(\zeta)$ can be written as

$$A_n(\zeta) = A_0(-1)^n J_{2n}(\eta) - \frac{iR_0}{\sqrt{2}}(-1)^n J_{2n+1}(\eta), \tag{7}$$

$$B_n(\zeta) = \frac{1}{2}S_0\delta_{n0} + \frac{1}{2}R_0(-1)^n J_{2n}(\eta) - \frac{iA_0}{\sqrt{2}}(-1)^n J_{2n+1}(\eta), \tag{8}$$

$$C_n(\zeta) = -\frac{1}{2}S_0\delta_{n0} + \frac{1}{2}R_0(-1)^n J_{2n}(\eta) - \frac{iA_0}{\sqrt{2}}(-1)^n J_{2n+1}(\eta), \tag{9}$$

where the constants $S_0 = B_0 - C_0$ and $R_0 = B_0 + C_0$ are introduced, and $\eta = 2\sqrt{2}\zeta$. The details of the calculations are presented in [8]. Theses expressions describe the electromagnetic radiation spreading along array.

In the case of boundary conditions

$$A_n(0) = 0, \quad B_n(0) = -C_n(0) = B_0\delta_{n0},$$

we have $R_0 = 0$, but $S_0 = 2B_0$. The distribution of the electromagnetic fields in waveguide array is

$$A_n(\zeta) = 0, \quad B_n(\zeta) = B_0\delta_{n0}, \quad C_n(\zeta) = -B_0\delta_{n0} \tag{10}$$

In this case the discrete diffraction is absent. It corresponds for the excitation of the flat-band modes [4,5]. However, if the radiation will input into one of the waveguide of the central part of array, i.e., to use the boundary condition

$$A_n(0) = A_0\delta_{n0}, \quad B_n(0) = -C_n(0) = B_0\delta_{n0},$$

then the distribution of the amplitudes $A_n(\zeta)$, $B_n(\zeta)$ and $C_n(\zeta)$ takes the following form

$$A_n(\zeta) = A_0(-1)^n J_{2n}(\eta), \tag{11}$$

$$B_n(\zeta) = B_0\delta_{n0} - \frac{iA_0}{\sqrt{2}}(-1)^n J_{2n+1}(\eta), \tag{12}$$

$$C_n(\zeta) = -B_0\delta_{n0} - \frac{iA_0}{\sqrt{2}}(-1)^n J_{2n+1}(\eta), \tag{13}$$

The discrete diffraction takes place in this case of the boundary conditions.

5. Conclusion

The propagation of the electromagnetic continues wave in the quasi-one-dimensional rhombic array of the waveguides is investigated. The exact solution of the coupled mode equations (1) is found. In general case of the boundary conditions the discrete diffraction is described by these solutions. However, the without diffraction regime of the wave propagation at the particular boundary condition exists.

The obtained exact solution could be correct in the case of slowly varying envelopes of the electromagnetic solitary waves, if the group velocities for each waveguide in array are equal.

If the rhombic waveguide array is composed of the unit cells containing three waveguides: two positive refraction index waveguides and one negative refraction index-waveguide, then the electromagnetic wave spectrum contains two gap modes and one gapless mode. However, the gapless mode approximately corresponds to flat one, thus we can say about quasi-flat band.

Acknowledgements

We are grateful to Prof. I. Gabitov and Dr. C. Bayun for enlightening discussions. This investigation is funded by Russian Science Foundation (project 14-22-00098).

References

[1] Longhi S 2014 *Opt. Lett.* **39** 5892
[2] Maimistov A 2015 *J.Phys. Conference Series* **613** 012011
[3] Maimistov A I and Gabitov I R 2016 *J.Phys. Conference Series* **714** 012013
[4] Mukherjee S and Thomson R R 2015 *Opt. Lett.* **40** 5443
[5] Mukherjee S, Spracklen A, Choudhury D. Goldman N, Ohberg P, Andersson E and Thomson R.R. 2015 Phys.Rev.Lett. **114** 245504
[6] Vicencio R A, Cantillano C, Morales-Inostroza L, Real B, Mejia-Cortes C, Weimann S, Szameit A and Molina M I 2015 *Phys.Rev.Lett* **114** 245503
[7] Zong Y, Xia Sh, Tang L, Song D, Hu Yi, Pei V, Su J, Li Y and Chen Zh .2016 *Opt.Express* **24** 8877
[8] Maimistov A I, Patrikeev V A 2016 *Preprint* arXiv:1606.03697 [physics.optics]

Negative curvature hollow core fibers for Raman lasing in the mid IR spectral range

A N Kolyadin[1], G K Alagashev[1], A D Pryamikov[1], A V Gladyshev[1], A F Kosolapov[1], A S Biriukov[1,2], IA Bufetov[1]

[1] Fiber Optics Research Center of Russian Academy of Sciences, 38 Vavilov street, Moscow, 119333, Russia
[2] Moscow Institute of Physics and Technology (State University), Institutskii per. 9, 141700 Dolgoprudnyi, Moscow region, Russia

E-mail: kolyadin@fo.gpi.ru

Abstract. In this paper we consider a problem of using negative curvature hollow-core fibers for creation of Raman lasers in the mid IR spectral range. New designs of this type of fibers with cladding formed by one layer of double nested capillaries are discussed and their optical properties are investigated numerically and experimentally. It will be shown that it is possible to reduce the Raman generation threshold by decreasing an effective mode area in such fibers using nested capillaries in the cladding.

1. Introduction

Gas lasers based on negative curvature hollow-core fibers (NC-HCFs) have many advantages over traditional gas cells, for example, the long interaction length, high efficiency, low threshold. Moreover, the size of such laser systems can be reduced to a great extent compared with other types of the gas cells and damage threshold and nonlinear limits of hollow-core fibers (HCFs) are much higher compared to usual solid fibers.

As it is known the first fabricated microstructured HCF was hollow core photonic crystal fiber (HC-PCF) [1]. Several years ago a new type of HCF was proposed and fabricated in several laboratories [2], [3], [4], [5]. The key feature of these HCFs was so called 'negative curvature' (NC) of the core – cladding boundary (when the normal vector to the inner surface of the core – cladding boundary is oppositely directed to a radial unite vector). Such design of the core – cladding boundary allowed to localize 99.993% of the air core mode radiation in the air [6]. Due to very weak interaction of the air core modes with the core – cladding boundary NC-HCFs have dispersion curves with very low slopes and values [7]. At the present time, there are several well known types of NC-HCFs, namely, revolver NC-HCF - fibers with cladding consisting of one row of capillaries [4], NC-HCF with elements in cross – section in the form of 'ice cream cone' [5] and hypocycloid-core Kagome lattice hollow core fiber [3]. The strong localization of the radiation in the NC-HCFs made of silica glass compared to HC-PCFs allowed to transmit light in the mid IR spectral range where material loss of silica glass is very high[8]. Moreover, the strong light localization in the NC-HCF air core allows to transmit high power radiation [9].

Mid-infrared spectral region (mid-IR) is very interesting for development of a new laser technologies. Different laser sources can operate in this spectral range such as quantum cascade lasers or systems based on nonlinear conversion. But unlike these laser sources, gas hollow core fiber lasers much more robust and have tremendous potential for scalability in power and wavelength in the mid-IR.

Stimulated Raman scattering in HC-PCF was firstly reported in [10]. A Raman laser based on NC-HCF was firstly reported in [11]. In this work stimulated Raman scattering (SRS) at 1.9 μm was generated in a 6.5 m long hydrogen-filled at 23 bars NC-HCF , pumped by a 1064 nm microchip laser. 1500 W peak power of pure vibrational Stokes SRS at 1907 nm was measured with a corresponding quantum conversion efficiency above 48 %. Another Raman laser based on NC-HCF filled with hydrogen with higher pump-to-Stokes conversion quantum efficiency 60% was demonstrated in [12]. In this work the steady-state (pump pulse τ = 125 ns) 1.06->1.9 μm SRS conversion in a hydrogen-filled revolver-type hollow-core fiber was demonstrated. The average output Stokes power was 330 mW, corresponding to 3 kW of peak power at a Stokes pulse duration of 93 ns.

It worth noting, that mid-IR radiation can be generated also using stimulated transitions in gas-filled HCFs. In [13] 3.1-3.2 μm mid-infrared emission with in the single pass of silica revolver NC-HCF filled with acetylene gas was reported. The maximum power conversion efficiency of the single-pass amplified spontaneous emission (ASE) was approximate 30%, with respect to the absorbed pump power in a 10.5 m length of fiber at pressure of 0.7 mbar. The minimum pump pulse energy required was less than 50 nJ (pump pulse duration τ =20 ns). Also 3.1-3.2 μm mid-infrared gas laser based on two types of NC-HCFs filled with acetylene with a ring cavity was demonstrated in [14].

Traditional electrical gas discharge for gas lasers are also being explored in gas laser with NC-HCF filled with Helium-Xenon[15].

2. Comparison of different structures

The aim of this work is the development and creation of the hollow-core fiber for efficient Raman conversion of laser radiation in gas-filled hollow core. Problems related to non-linear effects require a high-intensity radiation. One way to increase the radiation intensity in case of using hollow-core fibers is to decrease diameter of the hollow core and, respectively, decrease the effective mode area. It is known that waveguide losses in the hollow-core fibers (like inset b1 in Fig. 1b) depend on core diameter approximately at least as D^{-4} [16], i.e. reduction of the hollow-core diameter strongly increase the waveguide losses. Accordingly, the authors had to compare different variants of the hollow fiber cross-section structures to select one that would ensure a minimum level of the waveguide loss with reducing the effective mode area. Since the experimental search of the optimal design is very complex, initially numerical simulation of the most promising designs was conducted. We used the finite-element method (Comsol Multiphysics) to perform our numerical analysis.

As a starting point in our analysis, we used structure of a revolver NC-HCF (revolver fiber 1 – RF1) in which we had already achieved an efficient Raman conversion [12]. Cladding of RF1 consists of eight capillaries, the core diameter is 57 μm (Fig. 1b1), the loss at 1.06 and 1.9 μm was found to be 0.25 and 0.1 dB/m respectively (Figure 1a, black line). Simple reduction of the fiber core diameter to 29 μm with the same number of capillaries (Fig. 1b2) leads to a significant increase in the waveguide loss (Fig. 1a, red line). Reducing the number of capillaries to six with the same diameter of the hollow core leads to a slight reduction of the level of losses (Fig. 1a, green line), but their level is still significantly higher than in the original fiber (Figure 1a, black line). Further reduction of losses in such a structure with the same diameter of the hollow core is possible if its cladding will consist of double nested capillaries as was proposed and made in [17], thoroughly discussed in [18] and made in [19]. Maximal distance between inner and outer capillaries (distance between capillaries' walls in radial direction) should be approximately a half of the core diameter[18]. We used approximately the same values of this parameter in structures of computational and real experimental fibers. For convenience of manufacturing, it is desirable that the thickness of the internal capillary wall would be twice thinner than the external capillary wall (Fig. 1b4). The loss spectrum of such optical fiber was calculated.

As it can be seen from Fig. 1a (dark blue line), loss spectrum of this optical fiber is strongly distorted and has areas with both lower and higher losses compared with the loss spectrum shown in Figure 1a, green line. Such a distortion, obviously, occurs due to appearance of additional resonant electromagnetic field states excited in the walls of internal capillaries and shift of the boundaries of their transmission bands with respect to the transmission bands of the outer capillaries. Thus, to match their transmission bands it is necessary to match their thickness. Loss spectrum of a fiber with six double nested capillaries with the same wall thickness is shown in Fig. 1a, blue line. It was observed that a decrease in the number of capillaries in this case leads to a reduction of loss level. The loss spectrum of hollow-core fiber with a core diameter of 29 μm and five double nested capillaries in the cladding (Fig. 1b6) is represented in Fig. 1a, purple line. Optical loss in this fiber around the considered spectral region is lower than in the case of the fibers mentioned above. Moreover, its loss on the average is much lower than the losses in the original fiber with a core diameter of 57 μm. Thus, such fiber structure allows to reduce by half the core diameter and keep the total loss at an acceptable level.

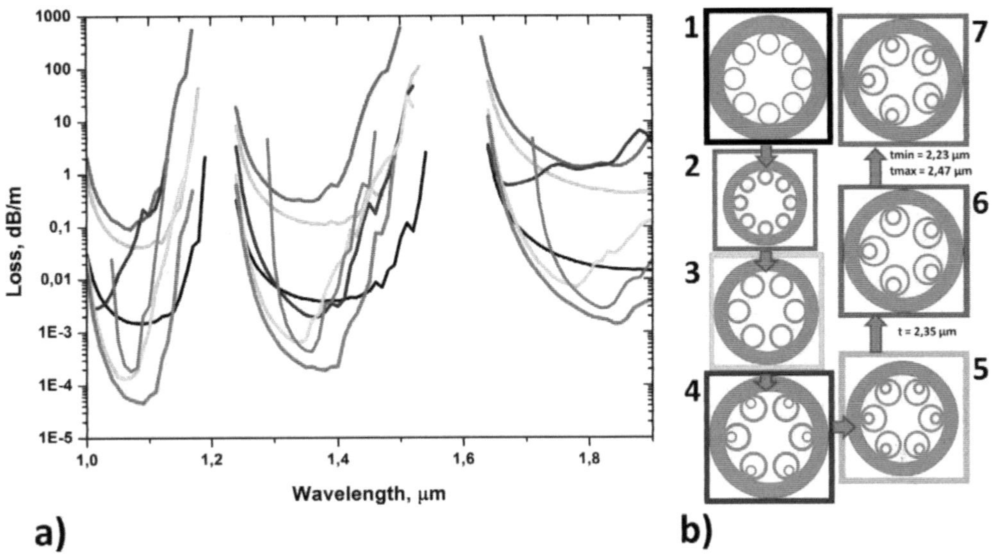

Figure 1. a) optical losses calculated for different fiber cladding structures shown in b); b) different fiber cross-sections and the order of calculation. The color of the frame corresponds to the color of the line in the graph. In cross-section b1 – D_{out}=155 μm (outer diameter of the fiber), D_{core}=57 μm. In cross-sections b2-b7 – D_{core}=29 μm.

However, the optical loss in real fibers is always higher than the calculated one. This is due to the fact that the cladding structure of the real fiber is changed during the drawing process and become different from the ideal used in the calculations (wall thickness of the capillary and distance between them varies, cross-section of the capillary differs from a circle). Because it is important to analyze in what extent the losses of the selected structure are sensitive to deviations of cladding geometrical parameters mentioned above, we tried to take account of these factors in the calculations.

We identified the variations in the fiber cladding geometrical parameters by careful analysis of the cross-section of the real fibers. However, our model only takes into account variations over the cross section and does not account variations in the length, which also occur in the drawing process and also increase the loss level. The resulting loss spectrum of fiber with non-ideal cladding and its cross section are shown in Fig. 1a, olive line and Fig. 1b7 respectively. As can be seen from the graph, the introduction of inhomogeneities in the structure increases the level of losses and makes the transmission bands narrower, but wavelengths we are interested in (1.06 and 1.9 μm) are still fall into the transmission bands and its losses are approximately the same level as that in the original fiber (Fig. 1a, black line).

On the basis of the calculation given above, corresponding preform was assembled and drawn. The loss spectrum and cross-section of the resulting revolver fiber (RF2) are shown in Fig.2 (Fig. 2a, dark blue line and 2b in dark blue frame). The cladding of RF2 is comprised of five double nested non-touching capillaries, the core diameter is Dc = 25 μm, the outer diameter of the fiber is 110 μm, outer diameter of the large and small capillaries 29 and 15 μm correspondently, the average wall thickness of the capillaries is 2.3 μm. SEM image of a cross section of RF2 is shown in Fig. 2b in dark blue frame. Fiber losses were measured by «cut-back» method using a fiber piece of 57 m length. The measurements were carried out using monochromator, a source of supercontinuum radiation (Fianium) was used as a light source and InGaAs detector was taken as a detector.

Loss measured in real fiber was higher than loss calculated for model of this fiber in which the non-ideal structure was taken into account. We attribute this to the fact that fiber cladding can also has non-uniformity in length, that increase the level of losses but which are difficult to take into account in the calculations. In addition, uncontrolled excitation of higher modes is possible in method used for loss measurements, which can also lead to higher values of losses.

It is interesting to compare the spectra of optical loss of two hollow-core fibers: with double nested capillaries in the cladding (RF2 in Fig. 2b, dark blue frame) and fiber with a single layer of capillaries in the cladding (RF1 in Fig. 2b, green frame). The capillaries wall thicknesses in both fibers are close (about 2.3 μm) and positions of their bandwidths are approximately the same (Fig. 2a). Their minima optical losses are also close: ~100 dB/km at a wavelength of about 2 μm in RF1 and 74 dB/km at a wavelength of 1.8 microns in RF2. But core diameter of RF2 is 25 μm what is more than two times smaller than the core diameter of RF1 (57 μm). According to estimations given in [16], with this reduction of the core diameter, loss in the hollow-core fiber with one layer of single capillaries in the cladding (revolver NC-HCF) would become ~27 times higher. Therefore, the introduction of double nested capillaries allows to significantly reduce the effective mode area without significantly increasing of the optical loss.

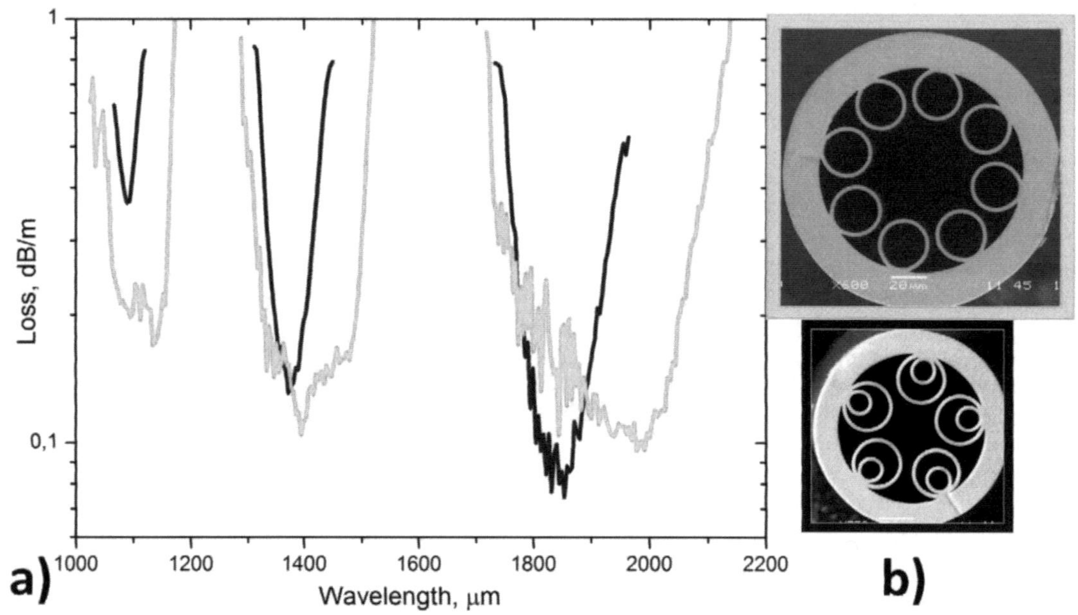

Figure 2. a) measured loss spectra for revolver NC-HCFs with single (RF1) and double nested capillaries (RF2). b) SEM images of the corresponding fibers(in one scale). Green frame - revolver NC-HCF (RF1) with single capillaries in the cladding, D_{core}=57 μm; dark blue frame - revolver NC-HCF(RF2) with double nested capillaries in the cladding, D_{core}=26 μm[19]. The color of the frame corresponds to the color of the line in the graph.

3. Bend loss

To measure and compare bend loss we used two different revolver fibers with double nested capillaries in the cladding (not the same as above). Transmission under different bend radii was measured for both fibers (Fig. 3). For the experiment 5 meter long fiber was taken and 4 meter of the fiber was always under bend. Halogen lamp was taken as a light source and a spectrum analyzer as a light detector. Measurements were carried out with fibers with five capillaries in the cladding. First one (RF3) has core diameter 50 μm and the capillary wall thickness 3.7 μm, second one (RF4) has core diameter 30 μm and the capillary wall thickness 2.6 μm. Figure 3 shows a strong difference in the bend losses between two fibers. Such a strong difference arises due to difference in the core diameter, capillary wall thickness and probably because of different gap between double nested capillaries. Bend losses increase with increase of order of the resonance in the capillary wall. In the fiber on Fig. 3a order of this resonance higher than in the fiber on Fig. 3b.

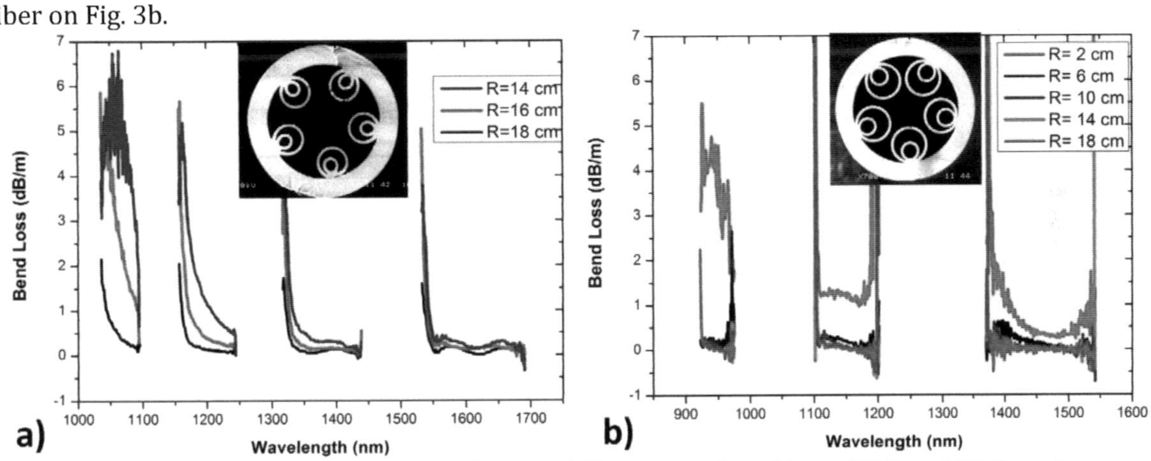

Figure 3. Measured bend loss spectra for two different revolver fibers (RF3 and RF4) under different bend radii. a) measured bend loss spectra for RF3; b) measured bend loss spectra for RF4 (curves for R=18,14,10 cm are coincide). Insets – SEM images of the according fibers.

4. Conclusions

In conclusion, we investigated optical properties of several negative curvature hollow core fibers to obtain an optimal geometry of the cross–section for Raman lasing at the Stokes wavelength $\lambda = 1.9$ μm. It was shown that minimal total loss level can be obtained for the revolver negative curvature hollow core fiber with double nested capillaries and with uniform thickness of the cladding capillaries walls. Such fiber allows to achieve a low value of the effective mode area at the pump and 1-st Stokes wavelengths compared with the conventional design of the revolver fiber with single capillaries and strong resistivity to the bending. Our investigation have demonstrated that the negative curvature hollow core fibers have great potential for laser applications in the mid IR spectral range.

Acknowledgments

This work was supported in part by the Russian Foundation for Basic Research, grant # 15-02-99688 A and #15-02-08840 A.

References

[1] Russell P S J 2006 *Photonic-Crystal Fibers* **24** 4729–4749
[2] Wang Y Y, Couny F, Roberts P J and Benabid F 2010 *Lasers Electro-Optics Quantum Electron. Laser Sci. Conf. (QELS), 2010 Conf.* CPDB4
[3] Wang Y Y, Wheeler N V, Couny F, Roberts P J and Benabid F 2011 *Optics Letters* **36** 669–671

[4] Pryamikov A D, Biriukov A S, Kosolapov A F, Plotnichenko V G, Semjonov S L and Dianov E M 2011 *Optics Express* **19** 1441–1448
[5] Yu F, Wadsworth W J and Knight J C 2012 *Opt. Express* **20** 11153–11158
[6] Alagashev G K, Pryamikov A D, Kosolapov A F, Kolyadin A N, Lukovkin A Y and Biriukov A S 2015 *Laser Physics* **25 (5)** 055101
[7] Kolyadin A N, Alagashev G K, Pryamikov A D, Mouradian L, Zeytunyan A, Toneyan H, Kosolapov A F and Bufetov I A 2015 *Physics Procedia* **73** 59–66
[8] Kolyadin A N, Kosolapov A F, Pryamikov A D, Biriukov A S, Plotnichenko V G and Dianov E M 2013 *Optics Express* **21** 9514–9519
[9] Urich A, Maier R R J, Yu F, Knight J C, Hand D P and Shephard J D 2013 *Biomedical Optics Express* **4 (2)** 193-205
[10] Benabid F, Knight J C, Antonopoulos G and Russell P S J 2002 *Science* **298** 399–402
[11] Wang Z, Yu F, Wadsworth W J and Knight J C 2014 *Laser Physics Letters* **11** 105807
[12] Gladyshev A V, Kolyadin A N, Kosolapov A F, Yatsenko Y P, Pryamikov A D, Biryukov A S, Bufetov I A and Dianov E M 2015 *Quantum Electronics* **45** 807–812
[13] Wang Z, Belardi W, Yu F, Wadsworth W J and Knight J C 2014*Optics Express* **22** 21872–21878
[14] Hassan Muhammad Rosdi A, Yu F, J. Wadsworth W and Knight J C 2016 *Optica* **3** 218–221
[15] Bateman S A, Belardi W, Yu F, Webb C E and Wadsworth W J 2014 *CLEO 2014 Postdeadline Pap. Dig.* 50–51
[16] Zheltikov A M 2008 *PHYS-USP* **51 (6)** 591–600
[17] Belardi W and Knight J C 2014 *Opt. Fiber Commun. Conf. OSA Tech. Dig. (Optical Soc. Am. 2014),* Pap. Th2A.45
[18] Poletti F 2014 *Optics Express* **22** 23807–23828
[19] Kosolapov A F, Alagashev G K, Kolyadin A N, Pryamikov A D, Biryukov A S, Bufetov I A and Dianov E M 2016 *Quantum Electronics* **46** 10–14

Journal of Physics: Conference Series **737** (2016) 012010 doi:10.1088/1742-6596/737/1/012010

Compensation of asymmetry of diffraction characteristics of holographic chirped photonic structures in PDLC by the impact of spatially non-uniform electric field

A O Semkin, S N Sharangovich
Tomsk state university of control systems and radioelectronics, 40 Lenina av., Tomsk, Russia

E-mail: shr@tusur.ru

Abstract. In this work we have proposed the method of asymmetry compensation of diffraction characteristics of chirped photonic structures in PDLC by the impact of electric field, which has different forms of spatial inhomogeneity. It's shown that for each area of external electric field values the function of inhomogeneity has its optimal form.

1. Introduction

Recently many of practical challenges of photonics, non-linear optics and optical telecommunication systems may be solved by chirped photonic structures (ChPS) using. Earlier in [1] the possibility to increase the efficiency of light input to the polymer optical fiber by using ChPS, holographically formed in photopolymer materials, was shown. Besides, in [2] the possibility to compensate the inhomogeneity of holographic chirped diffraction grating by amplitude profiles of recording beams was shown. The liquid crystal (LC) molecules introduction to photopolymer material allows us to make a composite material – polymer-dispersed liquid crystal (PDLC). Optical properties of this material are manageable due to possibility of LC molecules to reorient in external electric field. In [3] we have shown, that the impact of external electric field on ChPS in PDLC causes the distortion of its diffraction characteristics.

The aim of this work is to theoretically investigate the compensation of ChPS diffraction characteristics asymmetry by the impact of external electric field which has the special form of spatial inhomogeneity.

2. Theoretical model

For description of light beam diffraction on holographic ChPS in PDLC we will use results of [4-7]. We will consider a two-dimensional Bragg diffraction of quasi-monochromatic beam $E_p(\mathbf{r})$, corresponding to an extra-ordinary wave, on inhomogeneous photonic structure (figure 1a). The direction of a periodic perturbation of the dielectric constant is given by the vector $\mathbf{K}$, which is due to inhomogeneity of ChPS represents a function of the coordinates $\mathbf{K} = \mathbf{K}(\mathbf{r})$. The spatial inhomogeneity of the external electric field is set by a special topology of the electrode structure (figure 1a). Different forms of external field inhomogeneity we will describe by the following expression:

$$E(y) = \mathrm{ch}[c \cdot (s \cdot y - t)]^{-1} - g , \quad (1)$$

where y – coordinate oriented along the grating vector; c, s, t, g – approximation coefficients.

Form of external field inhomogeneity $E(y)$ for incident beam aperture $w=1$ cm and for different approximation coefficients is shown on figure 1b. Curve 1 corresponds to uniform field, curve 2 – linearly non-uniform, in analogue to [3]. Function (1) takes the form of increasing, downward convex function (curve 3 on figure 1b) at $c=2.7$, $s=-2$, $t=-2.1$, $g=1$. Curve 4 on figure 1b corresponds to increasing, upward convex function, $c=0.46$, $s=-1$, $t=-0.5$, $g=0.1$. The bell-shaped form (curve 5) function (1) takes at $c=1.8$, $s=-0.5$, $t=-0.01$, $g=0.1$.

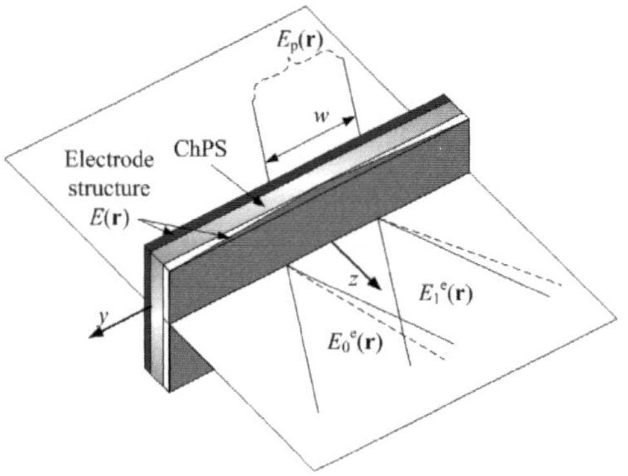

a)

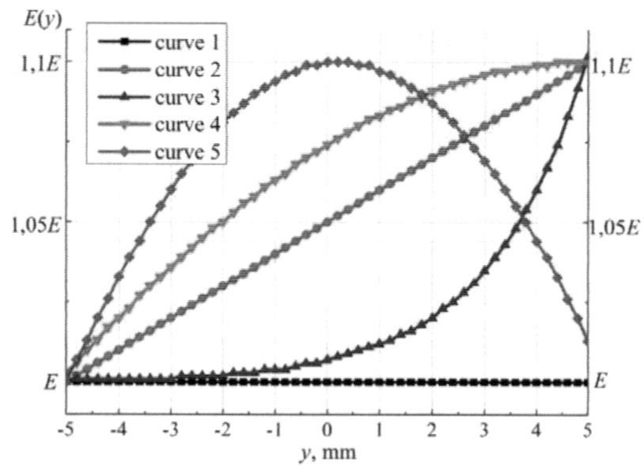

b)

Figure 1. Non-uniform electric field: a) – diffraction geometry and electrode structure; b) – forms of electric field inhomogeneity functions

The perturbation of the sample dielectric tensor because of its smallness in relation to $\hat{\varepsilon}_0$ we will present as the following sum:

$$\hat{\varepsilon}(\mathbf{r})=\hat{\varepsilon}_0+\Delta\hat{\varepsilon}_s(\mathbf{r})+\Delta\hat{\varepsilon}_e(\mathbf{r}), \tag{2}$$

where $\hat{\varepsilon}_0$ – unperturbed dielectric tensor; $\Delta\hat{\varepsilon}_s(\mathbf{r})$ – periodical perturbation, caused by the inhomogeneous ChPS recording process; $\Delta\hat{\varepsilon}_e(\mathbf{r})$ – smooth perturbation, caused by the impact of spatially non-uniform electric field. $E(y)$ (figure 1b).

$$\Delta\hat{\varepsilon}(\mathbf{r})=0.5\Delta\hat{\varepsilon}[U_0U_m(\mathbf{r})\exp[i\cdot\varphi(\mathbf{r})]], \tag{3}$$

where U_0, $U_m(\mathbf{r})$ – amplitude of perturbation and normalized amplitude profile; $\varphi(\mathbf{r})$ – phase profile, which in the case of ChPS can be represented as $\varphi(\mathbf{r})=\mathbf{K}(\mathbf{r})\cdot\mathbf{r}$; $\Delta\hat{\varepsilon}$ – perturbation of $\hat{\varepsilon}_0$, corresponding to $U_0=1$; $\mathbf{r}$ – radius-vector.

Changing of the dielectric constant caused by the impact of external field is due to the reorientation of LC director in the electric field under the action of Fredericks effect [8-10]. As in the considered medium the statistical distribution of the LC molecules preferred orientation in the "droplets" is observed [9] we will introduce the corresponding changing of the dielectric constant in the form of statistically averaged tensor:

$$\langle\Delta\hat{\varepsilon}_e(\mathbf{r},E)\rangle=-\Delta\varepsilon\int_0^{\pi}\int_{-1}^{1}\mathbf{C}(\mathbf{r},E)\mathbf{C}(\mathbf{r},E)p(\alpha)q(\phi)d\alpha d\phi, \tag{4}$$

where $\mathbf{C}(\mathbf{r},E)$ – spatial distribution of LC director orientation; $\langle...\rangle$ means statistical averaging.

In (4) we introduced the Gaussian functions of director distribution in the quasi-ellipsoidal LC "droplet" [9]:

$$p(\alpha)=A\exp\left[-(\alpha-\bar{\alpha})^2/2\sigma_\alpha^2\right],\quad q(\phi)=B\exp\left[-(\phi-\bar{\phi})^2/2\sigma_\phi^2\right], \tag{5}$$

where α, ϕ – orientation angles of LC molecules in the "droplet"; $\overline{\alpha}$, $\overline{\phi}$ – averages; σ_α, σ_ϕ – standard deviations.

LC director orientation changing in each spatial point under the influence of an electric field will be given by the angle of declination from its initial direction. As the considered medium is a set of LC "droplets" suspended in the polymer, to determine the distribution of the "droplets" rotation angle in the sample we will use the theoretical model, developed in [8]:

$$\gamma(\mathbf{r},E) = \psi(\mathbf{r}) + \frac{1}{2}\cdot arctg\left[\frac{\sin[2\vartheta_0(\mathbf{r})]}{e^2+\cos[2\vartheta_0(\mathbf{r})]}\right], \quad (6)$$

where $\psi(\mathbf{r})$ – orientation angle of external electric field vector $\mathbf{E}$; $\vartheta_0(\mathbf{r}) = \gamma_0 - \psi(\mathbf{r})$ – initial angle between $\mathbf{E}$ and $\mathbf{C}$ vectors; $e = |\mathbf{E}|R\sqrt{\Delta\overline{\varepsilon}/K(5.7\delta^2+2.1\lambda)}$ – parameter, which characterizes the impact of electric field on bipolar LC droplet; R – radius of the droplet; δ – eccentricity of the droplet; $\lambda = RW_a/K$ – parameter of surface adhesion; W_a – azimuth surface adhesion coefficient; K – Frank's elasticity coefficient; $\Delta\overline{\varepsilon}$ – effective dielectric anisotropy of bipolar droplet.

Critical electric field at which the Fredericks effect will occur is given by [8]:

$$Ec = \frac{1}{R}\cdot\left(\frac{5.7K\delta^2+2.1W_aR}{\Delta\overline{\varepsilon}}\right)^{1/2}. \quad (7)$$

In the considered case of Bragg diffraction of the extraordinary waves on grating (2), the total optical field in the sample can be represented as [4]:

$$E_s(\mathbf{r},E) = \sum_{i=0,1}\mathbf{e}_i E_i^e(\mathbf{r},E)\exp[i(\omega_i t - \int \mathbf{k}_i\cdot d\mathbf{r})], \quad (8)$$

where $\mathbf{e}_i$, $\mathbf{k}_i$, ω_i – polarization vectors, wave vectors and frequencies of transmitted $E_0^e(\mathbf{r},E)$ and diffracted $E_1^e(\mathbf{r},E)$ beams.

Amplitude distributions $E_{0,1}^e(\mathbf{r},E)$ can be determined by solving the system of coupled waves equations (CWE) [4]:

$$\mathbf{N}_{r0}^e\cdot\nabla E_0^e(\mathbf{r},E) = -iC_1^e(\mathbf{r},E)U_m^*(\mathbf{r})E_1^e(\mathbf{r},E)\exp(+i\int_0^r \Delta K(\mathbf{r},E)dr), \quad (9)$$

$$\mathbf{N}_{r1}^e\cdot\nabla E_1^e(\mathbf{r},E) = -iC_0^e(\mathbf{r},E)U_m(\mathbf{r})E_0^e(\mathbf{r},E)\exp(-i\int_0^r \Delta K(\mathbf{r},E)dr), \quad (10)$$

where $C_i^e(\mathbf{r},E) = \frac{1}{4}\frac{\omega\cdot U_0}{c_c n_i^e(\mathbf{r},E)}\mathbf{e}_1^e(\mathbf{r})\cdot\Delta\hat{\varepsilon}(\mathbf{r},E)\cdot\mathbf{e}_0^e(\mathbf{r})$ – amplitude coupling coefficients; $\mathbf{N}_{r0,1}^e$ – group normals; c_c – speed of light; $\Delta K(\mathbf{r})$ – length of spatially non-uniform phase mismatch vector $\Delta\mathbf{K}(\mathbf{r},E) = \mathbf{k}_0(\mathbf{r},E) - \mathbf{k}_1(\mathbf{r},E) + \mathbf{K}(\mathbf{r})$, characterizing the diffraction geometry changing by the impact of external electric field and ChPS period varying. Parameter $\Delta K(\mathbf{r})$ can be determined as in [4].

Refraction indexes $n_{0,1}^e(\mathbf{r},E)$ for the diffracted light beams are given by:

$$n_i^e(\mathbf{r},E) = n_o n_e\left[n_e^2\cdot\sin^2(\gamma(\mathbf{r},E)\pm\theta_i^e(\mathbf{r})) + n_o^2\cdot\cos^2(\gamma(\mathbf{r},E)\pm\theta_i^e(\mathbf{r}))\right]^{-1/2}, \quad (11)$$

where $\theta_i^e(\mathbf{r})$ – angles of incidence and diffraction of extra-ordinary waves.

Solution of the diffraction problem we will seek for a uniform amplitude profile $U_m(\mathbf{r}) = 1$ and quasi-quadratic phase front $\varphi(\mathbf{r})$. We will consider the case when the $\nabla\varphi(\mathbf{r})$ is directed along the grating vector $\mathbf{K}$ (figure 1a), which corresponds to the ChPS. In this case, the phase profile of the ChPS included in (3) we can represent as a Taylor series:

$$\varphi(\mathbf{r}) = \varphi_0 + \nabla\varphi \cdot \mathbf{r} + 0.5\nabla^2\varphi \cdot \mathbf{r}^2, \qquad (12)$$

where $\nabla\varphi = K_0$ – average length of vector $\mathbf{K}$; $0.5\nabla^2\varphi$ – deviation of $\mathbf{K}$ length, caused by the inhomogeneity of ChPS phase profile.

Substituting (12) to (3), considering a uniform amplitude profile, we will find the solution of CWE system (9)-(10) in analogy with [4] in aperture coordinates ξ_0, ξ_1, without taking into account the attenuation of light. Then the distribution of the diffracted light field amplitude at the output of the PDLC sample can be written as follows:

$$E_1(\eta) = -i\frac{C_1 U_0 L}{2\upsilon_1}\int_{-1}^{+1}\left\{ \begin{array}{l} \exp\left[\delta m(1-y) + \delta^2 n(1-y^2)\right]\times \\ \times\left[\Phi\left(\frac{d}{a}, 1; a\delta^2\frac{\upsilon_1}{\upsilon_0}(1-y)^z\right)E_0\left(\frac{\delta(1-y)}{\upsilon_0} - \frac{\eta}{\upsilon_1}\right)\right] \end{array} \right\} dy, \qquad (13)$$

where $\Phi(a,c,z)$ – confluent hypergeometric function of the first kind; L – sample thickness; the remaining symbols are shown in [4].

For analysis of observed solution and dependences of diffracted field amplitude on the form of external impact inhomogeneity we will consider the incidence of a plane monochromatic wave of unit amplitude on ChPS. For this, we will introduce the relative phase mismatch $\Delta = f(\delta\theta)$, where $\delta\theta = \theta - \theta_B$ – the deviation of the plane wave components incidence angle from the Bragg angle. For a quantitative estimation of a ChPS characteristic asymmetry compensation компенсации by the different forms of external impact (figure 1b), in dependence on impact value, we will estimate the symmetry of $|E_1(\Delta, E)|^2$ relatively to the line $\Delta = 0$ by the following expression:

$$\delta(E) = \left(\int_{-20}^{0}|E_1(\Delta,E)|^2 d\Delta - \int_{0}^{20}|E_1(\Delta,E)|^2 d\Delta\right)\cdot\left(\int_{-20}^{0}|E_1(\Delta,E)|^2 d\Delta\right)^{-1}\cdot 100\%. \qquad (14)$$

3. Numerical simulations and discussions

Figure 2a shows the results of numerical simulations of $|E_1(\Delta, E)|^2$ dependencies on the relative phase mismatch for the amplitude-uniform photonic structures with quadratic phase profile relative to $0.5\nabla^2\varphi = 2.5\cdot 10^5$ for external functions $E(y)$ of different forms at $E = 1.1E_c$. The incident beam lightwave $\lambda = 1.55$ um. Figure 2b shows the results of numerical simulations of $\delta(E)$ for such ChPS at different functions' form $E(y)$ and different external impact values. Curves numeration on figure 2 corresponds to function types shown on figure 1b.

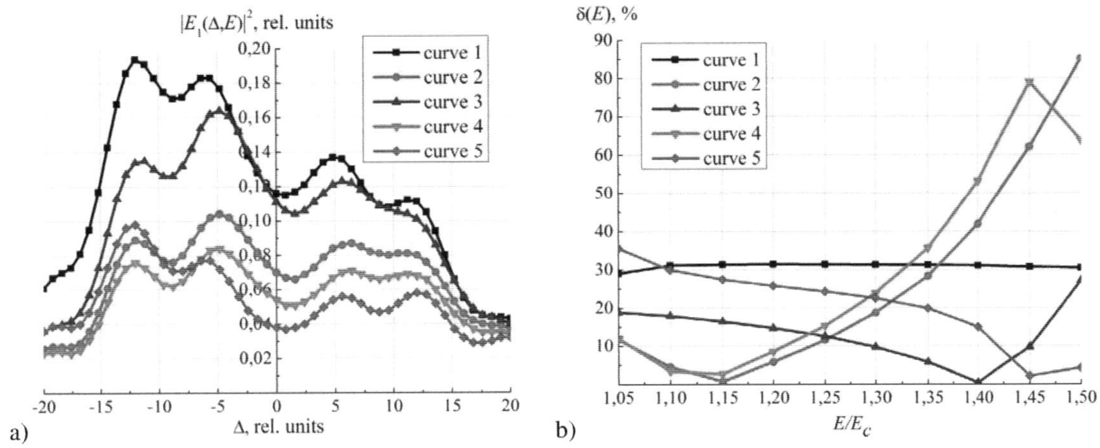

Figure 2. Numerical simulations results: a) – diffraction characteristic of ChPS under the impact of each form of external electric field ($E = 1.1E_c$); b) – the characteristic asymmetry value under the impact of each form of external electric field

Figure 2 analysis shows that uniform impact of different values (curves 1) causes the characteristic asymmetry at about 31% (figure 2b). This asymmetry compensation to the level about 10% provide linearly non-uniform impact (curves 2) and increasing, upward convex function (curves 4). However, this level of compensation is observed only at external field values $E = 1 \div 1.25E_c$. At $E > 1.35E_c$ the degree of inhomogeneity of the ChPS characteristics for these forms of impact increases rapidly. On the other hand, the same level of asymmetry compensation at values $E = 1.3 \div 1.5E_c$ and more provides increasing, downward convex (curves 3) and bell-shaped (curves 5) functions.

4. Conclusion

Thus, in this work we've shown that function of external electric field inhomogeneity $E(y)$ has its optimal form for each area of field values. This effect can be explained by inhomogeneity of ChPS phase profile.

Acknowledgments

The work is performed as a base part of Government Task of Russian Ministry of Education (project № 3642).

References

[1] Dovolnov E A, Sharangovich S N, Sheridan J T 2005 *Photorefractive Effects, Materials, and Devices 2005 (PR05), OSA Trends in Optics and Photonics Series (TOPS)* **99** 337.
[2] Dovolnov E A, Sharangovich S N 2006 *Proc. Of SPIE* **6187** 6187L-1.
[3] Semkin A O, Sharangovich S N 2015 *Proceedings of the IX International Conference of young scientists and professionals "Optics-2015"* St. Petersburg. Russia 245.
[4] Kushnarev I N, Sharangovich S N 1993 *Technical physics* **63** №2 24
[5] Semkin A O, Sharangovich S N 2013 *Bulletin of the Russian Academy of Sciences. Physics* **77** №12 1722.
[6] Semkin A O, Sharangovich S N 2013 *Izvestiya vuzov. Fizika* **56** № 9/2 21.
[7] Semkin A O, Sharangovich S N 2014 *Doklady TUSURa* **1(31)** 136.
[8] Afonin O A, Nazvanov V F 1998 *Technical physics letters* **24** 451
[9] Dovolnov E A, Sharangovich S N 2008 *Izvestiya vuzov. Fizika* **51** № 9/3 171
[10] Sonin A S *Introduction to the physics of liquid crystals* (Science Press, Moscow) 1983

Journal of Physics: Conference Series **737** (2016) 012011 doi:10.1088/1742-6596/737/1/012011

Laser beam-induced bistability of concentration in nanofluids

A I Livashvili,[1] V V Krishtop,[1] Y M Karpets,[1] T N Bryuhanova[2] and N M Kireeva[1]

[1] Far Eastern State Transport University, 47 Seryshev Street, Khabarovsk, Russia.
[2] Pacific National University, 136 Tihookeanskaya Street, Khabarovsk, Russia.

E-mail: Krishtop@list.ru

Abstract. We conduct a theoretical study ofthe dynamics of the concentration of nanoparticles in liquid-phase media under the influence of a laser light field. An exact solution of the nonlinear diffusion equation in the form of switching wavesis found. It is shown that, in the conditions of a fixed temperature and a nonlinear medium thermal conductivity, nanofluid becomes bistable.

1. Introduction

Colloidal suspensions, or nanofluids as they are called today, are widely used in various fields of modern technology. For example, the magnetic fluid is used for polishing optical components [1], the slurry of silica particles in the liquid crystal substantially improves the characteristics of the optical drive [2]. Note also their use in chemical processes (catalysis), the creation of new drugs, lubricantsand so forth. With the increasing performance of electronic devices and the development of high technologies, it is necessary to create effective cooling systems and manage large heat flows.

One of the ways to intensify heat transfer is to increase fluid thermal conductivity by adding solid particles with high thermal conductivity. Of particular interest in the creation of such suspensions are nanoparticles [3–6]. Further, as shown by recent studies [7–10], the liquid phases of the medium, in which the componentssuch as the dispersed nanoparticles are taken from wide bandgap semiconductors and dielectrics, are very effective for a number of nonlinear optical effects. In these media, unlike in homogeneous media, the nonlinear optical response occurs due to light wave-induced changes in the refractive index and absorption coefficient, thermal diffusion phenomena due to electrostriction and particles.

At the same time, the physical mechanisms involved, in particular the optical nonlinearity and the processes of heat and mass transfer in these environments, require further study in our view.

2. Theoretical model

The aim of our work is to conduct a theoretical study of the dynamics of nanoparticle concentration in the liquid-phase medium to be subjected to laser radiation of constant intensity, taking into account the thermal conductivity of the medium depending on its concentration.

We believe that the particle size satisfies the condition: $a_0 << \hat{\lambda}$, where a_0 is their linear dimension and $\hat{\lambda}$ is the wavelength of light. Thus, we do not consider diffraction and scattering processes. We also assume that the density of the liquid and the particles are nearly the same, giving the option to exclude sedimentation processes from consideration.

Consider the liquid-phase medium with microparticles irradiated with a light beam evenly distributed over the cell intensity. As a result of the light field in the medium, temperature and concentration gradients occur, causing heat and mass transfer processes. These phenomena are described by the system of balance equations for temperature and particles [11]:

$$C_p\rho\frac{\partial T}{\partial t} = \lambda\nabla^2 T + \alpha I_0, \tag{1}$$

$$\frac{\partial C}{\partial t} = D\nabla^2 C + D_T\nabla\left[C(1-C)\nabla T\right]. \tag{2}$$

Note that in the heat equation (1), we have omitted the term responsible for the Dufour effect because of its smallness, and in the diffusion equation (2), a term corresponding to the action of gradient forces from the light field was not considered at this stage of the research. The following designations are accepted: ∇ is the Laplace operator, T is the ambient temperature, $C = C(r,t) = \frac{m_0}{m}$ is the mass concentration of particulate matter (m_0 is the mass of particles, m is the mass of the entire environment) C_p, ρ, λ is the constant thermal fluid, I_0 is the light intensity, α is the optical absorption coefficient of the medium and D and D_T are the diffusion and thermal coefficients respectively.

In such a general statement of the system, (1)–(2) are unlikely to be solved. Therefore, we make some simplifying assumptions: we consider the one-dimensional case and exclude the contribution of the convective term, which occurs in the equation (2). Then we take into account the fact that the processes establishing the temperature goes much faster diffusion. This provides the opportunity to learn the latest on the background of fixed temperature: $\frac{\partial T}{\partial t} = 0$.

In equation (1), we assume that the thermal conductivity depends on the concentration and that this dependence has the form:

$$\lambda(C) = \lambda_0 + \beta C = \lambda_0(1 + pC), \tag{3}$$

where $p = \frac{\lambda_0}{\beta}$ and β is the proportionality factor [5, 6]. Note that this type of dependence, $\lambda(C)$, was observed in a number of experiments [9, 10]. Given the above, the diffusion equation (2) can be written as:

$$\frac{\partial C}{\partial t} = D\frac{\partial^2 C}{\partial x^2} - D_T\frac{\alpha I_0}{\lambda(C)} C\,(1-C). \tag{4}$$

Proceeding in this equation to the dimensionless variables: $\tau = \frac{S_T}{\lambda_0} I_0 \alpha D t$, $y = x\sqrt{\frac{S_T}{\lambda_0} I_0 \alpha}$, ($S_T = \frac{D_T}{D}$ – Soret coefficient), and using the approximation $\frac{1}{\lambda(C)} \approx \lambda_0(1 + pC)$, we obtain the problem:

$$\frac{\partial C}{\partial \tau} = \frac{\partial^2 C}{\partial y^2} - pC^3 + (1+p)\,C^2 - C, \tag{5}$$

$$0 \le C \le 1, \quad \infty \le y \le \infty, \quad 0 \le \tau < \infty. \tag{6}$$

Such parabolic equations with cubic nonlinearity were considered in [12, 13] for modeling dissipative media. Obviously, the zeros of the source function in equation (5): $C_1 = o$, $C_2 = \frac{1}{p}(p>1)$, $C_3 = 1$ correspond to spatially homogeneous stationary states.

As is known, the kinetics strongly dissipative system depends on the stability of the stationary states. In this case, the statesof $C = C_{1,3}$ are stable (they derive from the source $F'(C) < 0$) and the state $C = C_2$ is unstable. Thus, we found that the medium is bistable. Following [12], the solution of equation (3) will be found by substituting the Cole-Hopf:

$$C(y,\tau) = \frac{u_y \mu}{u(y,\tau) + u_0}, \tag{7}$$

where $u(y,\tau)$ isa new feature, and μ and u_0 arepermanent.

Substituting (7) into (5) and equating the coefficients of the same powers $(u + u_0)$ to zero, we obtain a system of linear equations to determine $u(y,\tau)$:

$$u_\tau = 3u_{yy} - (1+p)\, u_y, \tag{8}$$

$$2u_{yyy} - (\operatorname{sgn}\mu)(1+p)\, u_{yy} + u_y = 0, \tag{9}$$

a $\mu = \pm\sqrt{2p}$. The characteristic equation corresponding to (9) can be written as:

$$k\,[2k^2 - (\operatorname{sgn}\mu)(1+p)k + 1] = 0, \tag{10}$$

whose rootsare: $k_0 = 0$, $k_1 = \frac{1}{\sqrt{2p}}$ and $k_2 = \sqrt{\frac{p}{2}}$. Further, given the symmetry of equation (5) with respect to replacing $y \to -y$, we restricted ourselves to a positive value μ. Therefore, the function $u(y,\tau)$ we have:

$$u(y,\tau) = a_0 + a_1(\tau)\exp(k_1 y) + a_2(\tau)\exp(k_2 y). \tag{11}$$

Substituting (11) into (9) we find:

$$a_0 = A_0,\ a_i(\tau) = A_i \exp(\eta_i \tau),\ \eta_i = 3k_i^2 - k_i(1+p)\sqrt{\frac{p}{2}},\quad i = 1, 2, \tag{12}$$

where A_i are the constants determined from the initial conditions.

Thus, the exact solution of equation (5) will have the form:

$$C(y,\tau) = \sqrt{2p}\,\frac{N_1 k_1 \exp(k_1 y + \eta_1\tau) + N_2 k_2 \exp(k_2 y + \eta_2 \tau)}{1 + N_1 \exp(k_1 y + \eta_1 \tau) + N_2 \exp(k_2 y + \eta_2 \tau)}, \tag{13}$$

where $N_i = \frac{A_i}{a_0 + z_0}$.

Note that such a solution was obtained by another method in [12].

3. Analysis of results and discussion

Obviously, the solution (13) is continuous for all values of y and τ, if $N_i > 0$. We note that the solution (13) describes the dynamics of a bistable system, in which a wave of switching state of the system is formed by a two-wave mechanism.

Passing from the exponent in (13) to the dimensional variables, we can obtain the expression for the wave velocities:

$$v_{1,2} = \sqrt{\frac{S_T I\alpha}{\lambda_0}} \left[(1+p)\sqrt{\frac{p}{2}} - k_{1,2} \right] D. \tag{14}$$

Given that $p > 1$, as well as the expressions for $k_{1,2}$, it is evident that this "two-phase" solution is a concentration of two plane waves moving in the same direction and $v_1 > v_2$. Thus, the expression (13) describes the interaction of the two waves of switching from the intermediate unstable state, with $C = C_2$, intothe stable C_1 and C_3. Thus, the velocity of these waves $(v_1$ and $v_2)$ depends solely on the thermal conductivity (λ_0, β).

From (13) we can obtain an initial condition in which this mode is implemented:

$$C(y,0) = \sqrt{2p}\,\frac{N_1 k_1 \exp k_1 y + N_2 k_2 \exp k_2 y}{1 + N_1 \exp k_1 y + N_2 \exp k_2 y}. \tag{15}$$

Note that, as shown in [12], the two-wave solution (13) is true for other initial conditions. Further, since $\eta_1 > \eta_2$, then we have at long times:

$$C(y,\tau) = \sqrt{2p}\,\frac{N_1 k_1 \exp(k_1 y + \eta_1 \tau)}{1 + N_1 \exp(k_1 y + \eta_1 \tau)}.$$

Note that if we consider the diffusion processes against a background of stationary temperature under constant thermal conductivity, equation (3) is transformed into a well-known Kolmogorov-Petrovskii-Piskunov equation [14], which, as we know, has only a single-wave solution.

Of course, our approach is to some extent a model; however, as the authors hoped, we managed to discover under which conditions the irradiated nanofluids acquire the properties of a dissipative bistable medium that can be distributed switching wave. We believe that not all of the issues have been studied exhaustively; for example, an inverse relationship between temperature and the concentration of nanoparticles has not been considered. A detailed examination of this topic will be the subject of our further research.

References

[1] Bardakhanov S P, Novopashin S A and Serebrjakova M A 2012 *Nanosystems: Phys. Chem. Math.* **3** 27–33
[2] Kreuzer M, Tschudi T, de Jeu W H and Eidenschink R 1993 *Appl. Phys. Lett.* **62** 17121–17130
[3] Rosensweig R E 1985 *Ferrohydrodynamics* (Cambridge: Cambridge University Press)
[4] Hong T, Yang H and Choi C J 2005 *J. Appl. Phys.* **97** 064311-1–064311-4
[5] Putnam S A,Cahill D G and Braun P V 2006 *J. Appl. Phys.* **99** 084308
[6] Rudyak V J, Belkin A A and Tomilina E A 2010 *Technical Physics Letters* **36 (7)** 660-662
[7] Kul'chin Y N, Shcherbakov A V, Dzyuba V P, Voznesenskiy S S and Mikaelyan G T 2008 *Quantum Electronics* **38** 154–158
[8] Livashvili AI, Krishtop V V and Yakunina M I 2013 *Adv. Condens. Matter Phys.* **2013** 591087
[9] Livashvili A I, Krishtop V V, Bryukhanova T N and Kostina G V 2015 *Physics Procedia* **73** 156–158
[10] El-Ganainy E, Christodoulides D N, Rotschild C and Segev M 2007 *Optics Express* **15** 12207-10218
[11] De Groot S R and Mazur P 1964 *Non-equilibrium thermodynamics* (Moscow: Mir)
[12] Ognev A V, Petrovsky S V and Prostokishin V M 1995 *ZHTF* **65** 1–8
[13] Danilov V G and Subochev P Yu 1991 *Theoretical and Mathematical Physics* **89** 25–47
[14] Kolmogorov A N, Petrovskii I G and Piskunov N S 1937 *Bull. Moscow Univ. Math.* ser. A 1 1–25

The spectrum simulation of the soft X-ray for vacuum spark with laser initiation

A P Melekhov, E D Vovchenko
National Research Nuclear University MEPhI (Moscow Engineering Physics Institute), Kashirskoe shosse 31, 115409 Moscow, Russia

E-mail: apmelekhov@mephi.ru

Abstract. The spectra recovery technique based on measurement of soft X-ray radiation in a low-energy vacuum spark (discharge energy $Ec \leq 20$ J) with using of pair of pin-diodes and a set of removable *K*-edge filters is presented. The numerical model of soft x-ray spectrum and algorithm of selection of spectral distributions for coordination with experimental data are offered. Results of processing are presented in the form of the simulation spectra.

1. Introduction

Studying of temporary and spectral characteristics of the soft x-ray radiation of a vacuum spark plasma with laser initiation gives important and useful information on the processes proceeding in such hot plasma [1-5]. This information allows to create sources of the soft x-ray radiation with necessary spectral characteristics.

A simple technique of recovery of soft x-ray spectrum region for the low-inductive vacuum spark with small energy of discharge (≤ 20 J) is reported in this article. As input data for simulation of spectrum, we use measurements of soft x-ray in the pairs of matched pin-diodes with the removable *K*-edge filters corresponding to different regions of spectrum [6]. The offered algorithm we developed for rapid analysis of the experimental data obtained using X-ray pin-diodes and representation of result in the form of the simulated spectra.

2. Experimental setup

The scheme of the experimental setup is described in our articles [5,6]. The experimental setup consists of a vacuum interaction chamber (the residual pressure $P \approx 10^{-5}$ Torr), a high-voltage discharge system, initiating *Nd:YAG* laser (λ=1,06 μm, the pulse energy E_{laser}=30÷60 mJ and pulse duration$\tau \approx 15$ ns) and the diagnostic system. In turn, the high-voltage discharge system consists of conical electrodes, low inductance capacitor bank (capacitance C=0.22 uF), high voltage *DC* (U=10÷15 kV). The cathode is made from brass and anode is made from iron. The gap between the electrodes is $d_{AC} \approx 5.3$ mm. In our experiments the laser radiation was focused on the anode.

Two matched *pin*-diodes in combination with three pairs of *K*-edge filters (Al(9 μm) and Mg(17 μm), Ti(5 μm) and Al(36 μm), Fe(10 μm) and V(20 μm)) for the registration of soft x-ray radiation were applied. The *pin*-diodes were placed inside the vacuum chamber at the distance of 20 cm from the discharge area. With each pair of filters was done the individual series of measurements.

3. Experimental results

Figure 1 shows typical recorded waveforms of signals from x-ray *pin*-diodes for the two pairs of *K*-edge filters produced under the same conditions. It also shows synchronous signals of the discharge current from Rogowski coil.

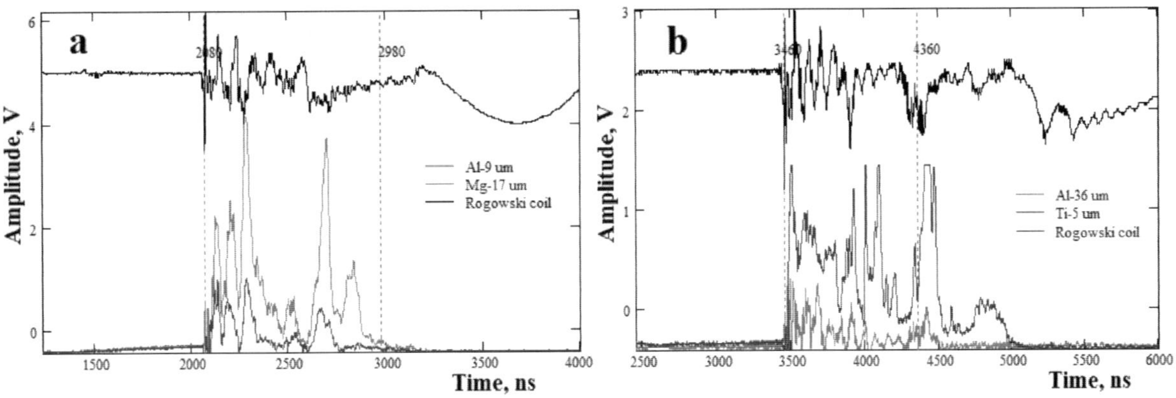

Figure 1. Waveforms of soft x-ray signals and signal from Rogowski coil: ***a***- Al (9 um) and Mg (17 um) film *K*-edge filters with the value of the stored energy *Ec*= 17.21 J; ***b***- Al (36 um) and Ti (5 um) filters with the value of the stored energy *Ec*= 16.83 J.

Figure 2 shows recorded waveforms of soft x-ray signals for Fe (10 um) and V (20 um) *K*-edge filters and signals from the Rogowski coil for different discharges in the same series of measurements.

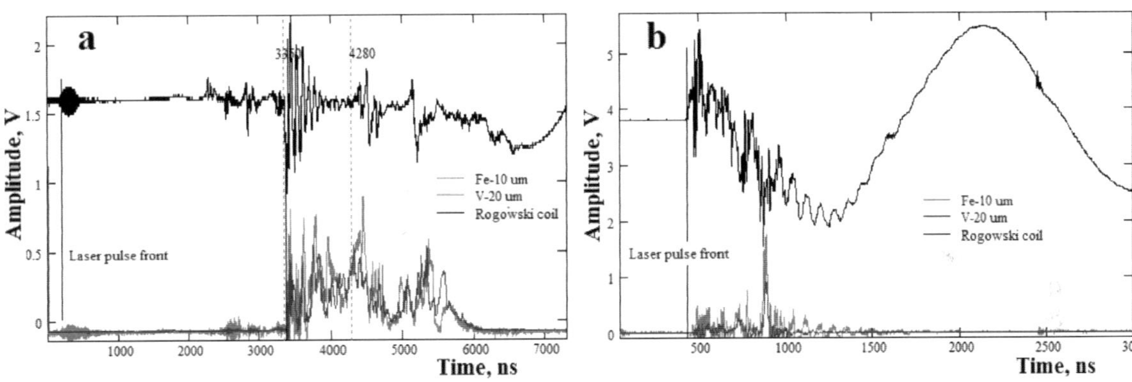

Figure 2. Waveforms of soft x-ray signals for Fe (10 um) and V (20 um) *K*-edge filters and signals from the Rogowski coil: as named a *long* pulse of emission of soft x-ray(***a***) and as named a *short* pulse of soft x-ray emission(***b***). Shown pulses of soft x-ray were obtained under identical conditions in a series of measurements. The energy stored in the discharge capacitor *Ec*=16.33 J.

As it can be seen from Figures 1a,b and 2a there is a significant delay duration of the start of the discharge and much longer duration of time of forming of the discharge than in the case of focusing of laser beam on the cathode [5]. In some cases the discharge with a small delay duration and with as named a *short* pulse of emission of soft x-rays as on figure 2b is possible. The proportion of such pulses can reach 20% in the series of measurements and they are not considered.

4. Processing of the experimental data

The algorithm of the processing of experimental results is constructed as follows. We find the amount of soft x-ray emission *Nm* during the pulse of soft x-ray for each of three pairs of *K*-edge filters. It is the time integral within the vertical dotted red lines shown in Figures 1-2a. Table 1 shows the values of Nm_n and Nm_m for five combinations of filters. Thus two pairs of filters Al (9 um) + Al (36 um) and Fe (10 um) + Al (36 um) were add to the pairs of filters, which had been listed above. In the fourth column of Table 1 the experimental values $\eta_{n,m} = Nm_n/Nm_m$ are shown and physically similar calculated value of $\delta_{n,m} = Nm^S_n/Nm^S_m$ is shown in the fifth column (explanation is below).

Table 1. Quantitative soft x-ray evaluations for different filters.

Pairs of Filters	Amount of soft x-ray emission Nm_n ,n=1÷5	Amount of soft x-ray emission Nm_m, m=6÷10	$\eta_{n,m} \pm \Delta\eta_{n,m}$ (experiments)	$\delta_{n,m} = Nm^S{}_n / Nm^S{}_m$ (evaluations)
Al(9 um), Mg(17) (Figure 2a)	1201	291	4.1±1.0	3.3
Ti(5 um), Al(36 um) (Figure 2b)	809	159	5.1±1.3	3.3
Fe(10 um),V(20 um) (Figure 3a)	297	252	1.2±0.3	1.5
Al(9 um), Al(36 um)	1201	159	7.6±2.0	6.8
Fe(10 um), Al(36 um)	297	159	1.9±0.5	0.5

Full spectrum of soft x-rays emission *Sfull* can be represented as follows:

$$Sfull = P(T_e)\cdot p + S\_L(Fe) + S\_L(brass) + S\_K(Fe) + S\_K(brass), \tag{1}$$

where *P(Te)* is *bremsstrahlung* spectrum, *p* is the varied parameter, *Te* is the electron temperature of plasma[6]. Other components in equation (1) determine the contribution of *K*, *L*- recombination radiation series for anode (Fe) and cathode (brass) materials. For example we write the contribution to the full spectrum of *K*-series of recombination radiation for cathode (brass) in the following form:

$$S\_K(brass)_i = A_{K(brass)} \cdot \exp\left(\frac{-(\varepsilon_i - \varepsilon_{K(brass)})^2}{\Delta\varepsilon^2_{K(brass)}}\right), \tag{2}$$

where $S\_K(brass)$ is the form of the function which can be used to describe of *K*-series of recombination radiation for Cu and Zn ions, $\varepsilon_i = i \cdot 20$ eV (i = 0 ÷ 500) is value of quantum energy of soft x-rays, $\varepsilon_{K(brass)}$ is most probable energy and $\Delta\varepsilon_{K(brass)}$ is deviation from this energy (their selection is based on physical considerations), $A_{K(brass)}$ is varied parameter. For example $\varepsilon_{K(brass)} = 8400 \div 8600$ eV and $\Delta\varepsilon_{K(brass)} = 500 \div 700$ eV, that corresponds to the area of location and width of *K*-series of ions Cu and Zn (brass). Other components of *Sfull* are formed in the same manner.

We introduce the quantity Nm^S_n as the sum over the range of energies of the simulated spectrum of soft x-rays *Sfull* for the *k*-th filter (*k=n,m*) and taking into account the sensitivity of the *pin*- diode:

$$Nm^S_k = \sum_i Sfull_i \cdot F_{i,k} \cdot Sd_i, \tag{3}$$

where $F_{i,k}$ is the spectrum transmission of the *k*-th filter, Sd_i is the spectral sensitivity of *pin*- diodes. We see that the calculated value Nm^S_k is physically similar to the experimental value Nm_k . If our the simulated spectrum *Sfull* is close to the real spectrum of soft x-rays in the investigated energy range $\varepsilon_{x\text{-}ray}$=0.5÷10 keV then the value $\delta_{n,m} = Nm^S_n / Nm^S_m$ can be correctly compared with the experimental value $\eta_{n,m}$ for the corresponding filters from the Table 1.

We can write the following inequalities using data from the fourth and fifth columns of Table 1:

$$\left|\delta_{n,m} - \eta_{n,m}\right| \le \Delta\eta_{n,m}, \text{ (n=1÷5, m=6÷10)}, \tag{4}$$

where $\Delta\eta_{n,m}$ is the experimental variation of the $\eta_{n,m}$ (see Table 1). For three pairs of filters there are five independent inequalities (4) which can be written in the form of system. in the system (4). We assume that this the system of inequalities (4) is compatible.

After that we achieve the *fulfillment* of all the inequalities of system (4) by varying the parameters $p, A_{K(Fe)}, A_{L(Fe)}, A_{L(brass)}, A_{K(brass)}$ and adjusting the power parameters in (2) and electron temperature *Te*. In the future this procedure can be automated.

As the system (4) can have more than one solution we introduce the criterion of optimization for solution found in system of inequalities (4). The idea of this criterion is to determine the proportion of soft x-rays with the energy of a photon $\varepsilon_{x\text{-ray}} \leq 2$ keV and to compare it with a similar calculated value. We introduce the parameter *Cor* which is calculated as follows:

$$Cor = \frac{|\mu^S - \mu|}{\mu}, \tag{5}$$

where values $\mu^S = \dfrac{|Nm_1^S - Nm_6^S|}{Nm_1^S}$ and $\mu = \dfrac{(Nm_1 - Nm_6)}{Nm_1}$ are calculated for Al and Mg *K*-edge filters (indexes 1 and 6 respectively). The value of μ=0.758.

Let us explain. For example, we found two different sets of coefficients $p, A_{K(Fe)}, A_{L(Fe)}, A_{K(Cu)}, A_{L(Cu)}, A_{K(Zn)}, A_{L(Zn)}$ (et al.), which are solutions of the system (4). Then the best set of parameters is that one for which the value *Cor* is less. In our case the value *Cor* = 0.085.

5. Simulation of spectrum

As a result of this simulation we obtain the soft x-ray spectrum of vacuum spark for the range of photon energies $\varepsilon_{x\text{-ray}}$=0.5÷10 keV. Parameters $\delta_{n,m}$ which were calculated to this spectrum are listed in the right column of Table 1. We optimized the values $\delta_{n,m}$ only for 4 pairs of filters. The value $\delta_{5,10}$ (pair of filters Fe (10 um) and Al (36 um)) is significantly less than the range of allowable values for $\eta_{5,10}$ that indicates the energy imbalance for this pair of filters. It can be explained by different conditions of the laser initiation of vacuum spark in experiments with different filters. Figure 3 shows the synthesized soft x-ray spectrum of vacuum spark which was simulated for the electron temperature of the plasma *Te* <1000 eV [4].

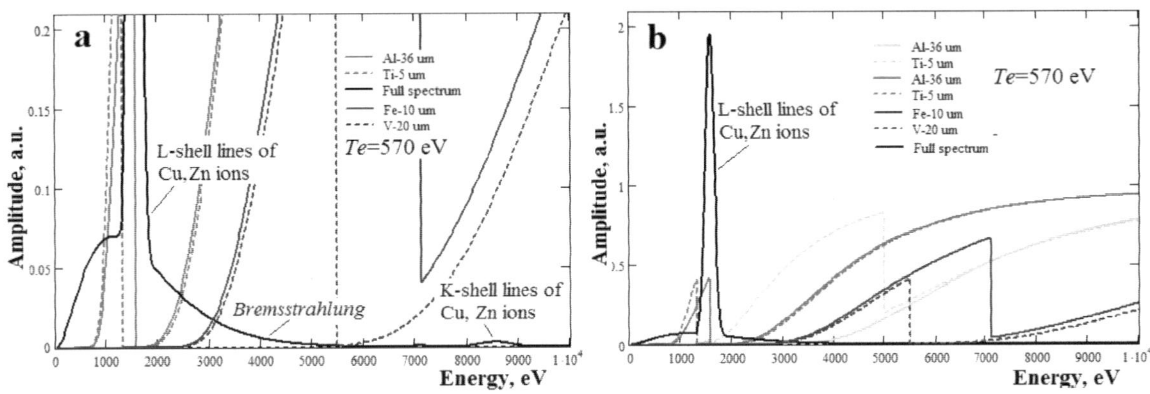

Figure 3. Detailed synthesized soft x-ray spectrum of vacuum spark and transmission spectra of used *K*-edge filters [7] (***a***). Full synthesized soft x-ray spectrum and transmission spectra of used *K*-edge filters (***b***). The value of discharge energy *Ec*=16.8 ± 0.5 J.

In Figure 3 we can see the emergence of strong peaks, presumably recombination *L*-series of ions Cu and Zn (brass) in the range $\varepsilon_{x\text{-ray}}$=1200÷1800 eV. Soft x-rays *L*-series of ions Fe on the low-energy edge of energy range is absent. This is because the sensitivity of used *pin*-diodes and the transmittance of Al and Mg filters in the range of quanta's energy $\varepsilon_{x\text{-ray}}\leq$1000 eV are small.

Conclusions

This work presents the algorithm and the results of recovery of soft x-rays spectrum of vacuum spark with focusing of laser initiated beam on the anode of the discharge system. Perhaps due to the instability of the laser radiation the vacuum spark has not stable dynamics of plasma formation. Similarly unstable is a spectrum of soft x-ray.

When calculations are automated, the proposed method will allow to produce express-analyze of the spectral structure of x-ray pulse at the time of experiment. It is convenient and essential for understanding the physics of the processes occurring in the vacuum spark discharge.

References

[1] Serbanescu C G, Sandolache G 2001 *Nukleonika* **46 (1)** pp 17-20

[2] Korobkin Y V et. al. 2008 *Plasma Physics and Controlled Fusion* **50** 065002

[3] Dolgov A N, Klyachin N A, Prokhorovich D E 2014 *Pis'ma v Zh. Èksper. Teoret. Fiz.* **99 (9)** pp 595-596

[4] https://umexpert.um.edu.my/file/publication/00004250_99577.pdf

[5] Alkhimova M A et. al. 2015 *Nukleonika* **60 (2)** pp 221-227

[6] http://www.sciencedirect.com/science/article/pii/S187538921501113X

[7] http://henke.lbl.gov/optical_constants/filter2.html

The influence of the conditions of laser initiation on soft X-rays emission of vacuum spark

E D Vovchenko, A P Melekhov
National Research Nuclear University MEPhI (Moscow Engineering Physics Institute), Kashirskoe shosse 31, 115409 Moscow, Russia

E-mail: apmelekhov@mephi.ru

Abstract. For exploring the influence of the conditions of laser initiation on soft x-ray emission of the laser-triggered vacuum spark we performed a series of experiments under different conditions of the laser beam focus on the cathode(anode) of the discharge system.

1. Introduction

The emission of soft x-ray radiation from vacuum spark discharge under the focusing of a laser pulse on the cathode is the subject of many publications [1-4] and recently the theoretical model of such a discharge was developed [5]. What caused our interest to the laser initiation of vacuum spark discharge under the focusing of the laser beam on the anode? Experiments have shown that the principal role for the dynamics of the development of the vacuum discharge plays the position of the focal point of the laser beam at the vertex of the cone-shaped electrode. A series of comparative experiments with different polarity electrode irradiated by laser and different positions of a focal point of laser radiation was conducted.

Depending on the focus of the laser on the cathode or the anode of the discharge system, we say, respectively, about the *cathode* or the *anode* laser-initiated discharge.

2. Experimental setup

The scheme of the experimental setup is shown in figure 1[6].

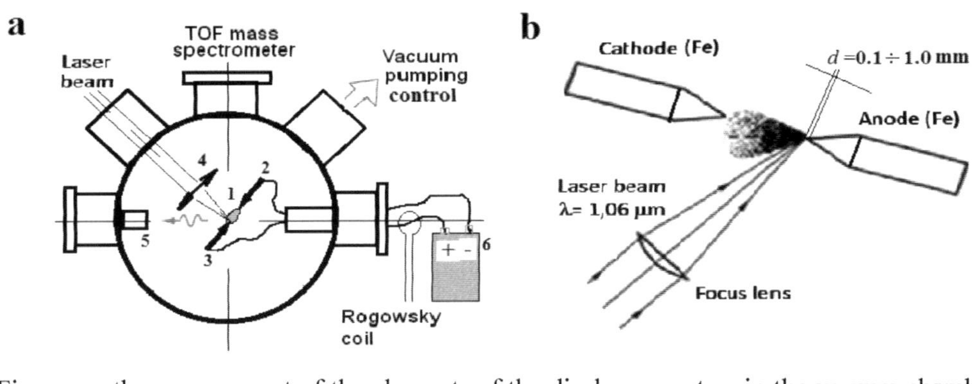

Figure 1. Figure ***a*** - the arrangement of the elements of the discharge system in the vacuum chamber: 1) the area of the discharge; 2) anode (cathode); 3) cathode (anode); 4) optical lens; 5) two identical *pin*-diodes with *K*-edge filters; 6) capacity bank C=0.22 μF. Figure ***b***-detailed area of spark discharge.

The experimental setup consists of a vacuum interaction chamber (pressure P ≈ 10^{-5} Torr), a high-voltage discharge system, initiating *Nd: YAG* laser (λ=1,06 μm, the pulse energy E_{laser}=30÷60 mJ and pulse durationτ≈15 ns) and the diagnostic system. In turn, the high-voltage discharge system consists of conical electrodes, low inductance capacitor bank (capacitance *C*=0.22 uF), high voltage *DC* (*U*=10÷15 kV). The electrodes are made from iron. The gap between the electrodes d_{AC}≈5.3 mm. The distance from the focus point of the laser beam to the top of the cone-shaped electrode *d*=0.1÷1.1 mm (see figure 1b). The size of the focusing spot of the laser radiation *Df*≈100 μm. Two identical pin-diodes with a matched pair of *K*-edge filters (Al – 9 μm and Mg – 17 μm) were applied for the registration of soft X-ray radiation. They were placed inside the vacuum chamber at the distance of 20 cm from the discharge area.

3. Experimental results

The following are the results of a series of comparative experiments in order to determine the influence of laser focus settings for initiating of a vacuum spark x-ray emission.
Figure 2 shows the oscillograms of signals from the soft x-ray pin-diodes for *d*≈0.1 mm. In both cases the magnitude of the stored energy *Ec*=18.1 J at voltage *Uc*=12.8 kV.

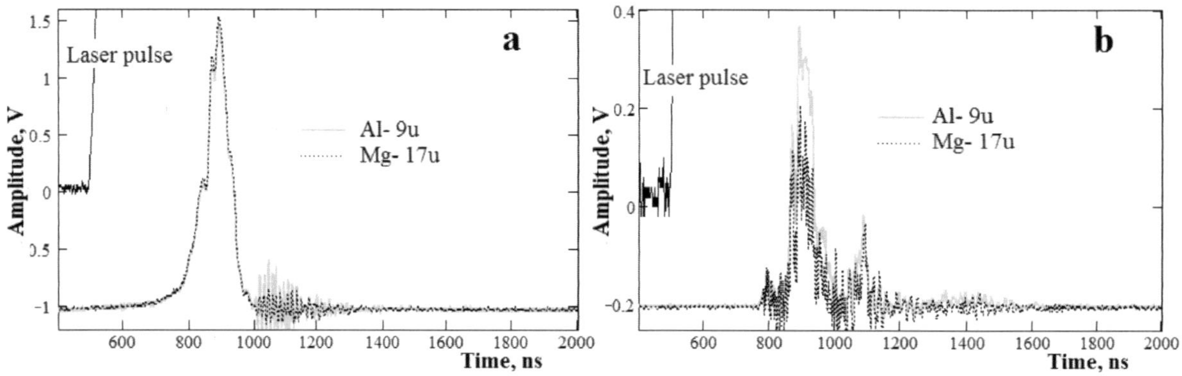

Figure 2. Oscillograms of signals of x-ray pin-diodes at d≈0.1 mm: ***a*** - *cathode* discharge; ***b*** - *anode* discharge. The magnitude of the stored energy *Ec*=18.1 J at voltage *Uc*=12.8 kV.

Figure 3 shows the results of processing of a series of signals from *pin*-diodes for the condition *d*≈0.1 mm.

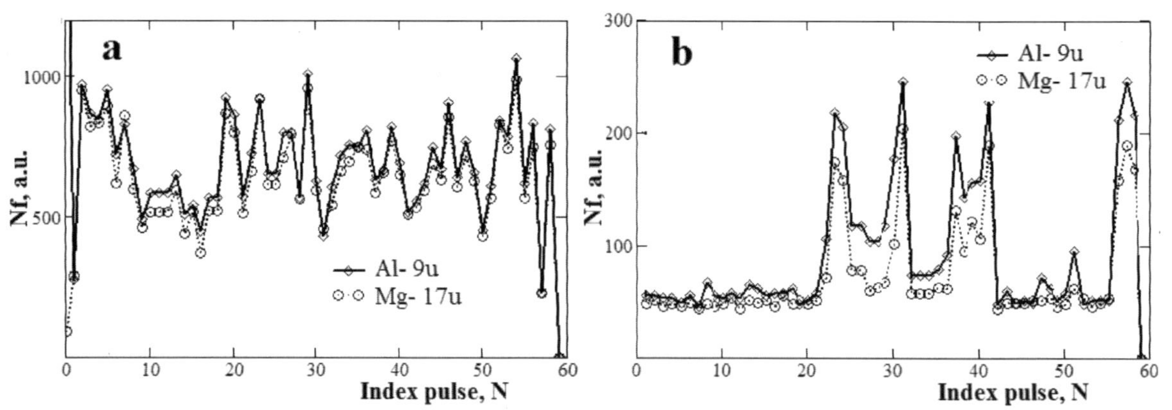

Figure 3. Integral emission of the soft *X*-ray of the vacuum spark plasma when d≈0.1 mm: ***a*** - *cathode* discharge; ***b*** - *anode* discharge. The magnitude of the stored energy *Ec*=18.1 J at voltage *Uc*=12.8 kV.

Here Nf_m are values of integral over the duration of soft x-ray pulses are registered by the *pin*- diodes with Al(m=1) and Mg(m=2) *K*-edge filters. The magnitudes Nf_m (m=1,2) are proportional to the number of quanta of soft x-ray. There are about 60 pulses in each experiment for a given polarity.

Figure 4 shows the oscillograms of signals from the x-ray pin-diodes for $d \approx 1.0$ mm.

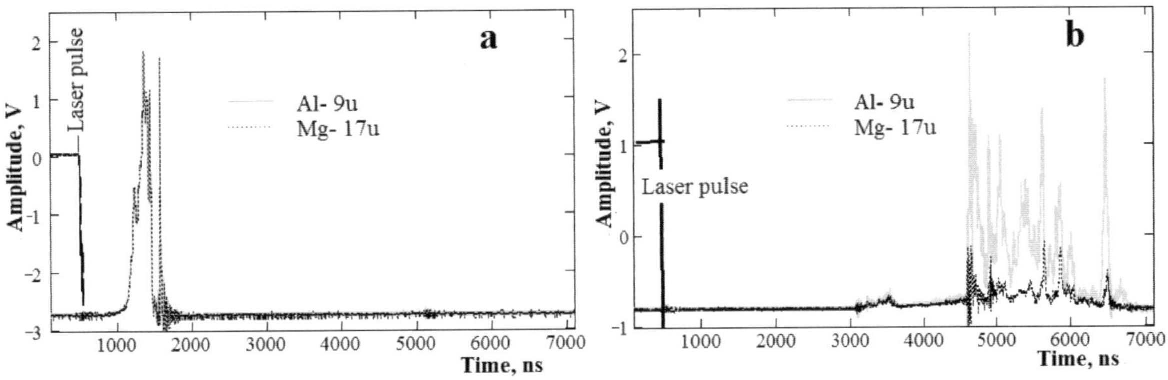

Figure 4. Oscillograms of signals of x-ray pin-diodes at $d \approx 1.0$ mm: ***a*** - cathode discharge; ***b*** - anode discharge. The magnitude of the stored energy Ec=18.2 J at voltage Uc=12.9 kV.

Figure 5 shows the results of processing of a series of signals from the x-ray *pin*-diodes for $d \approx 1.0$ mm.

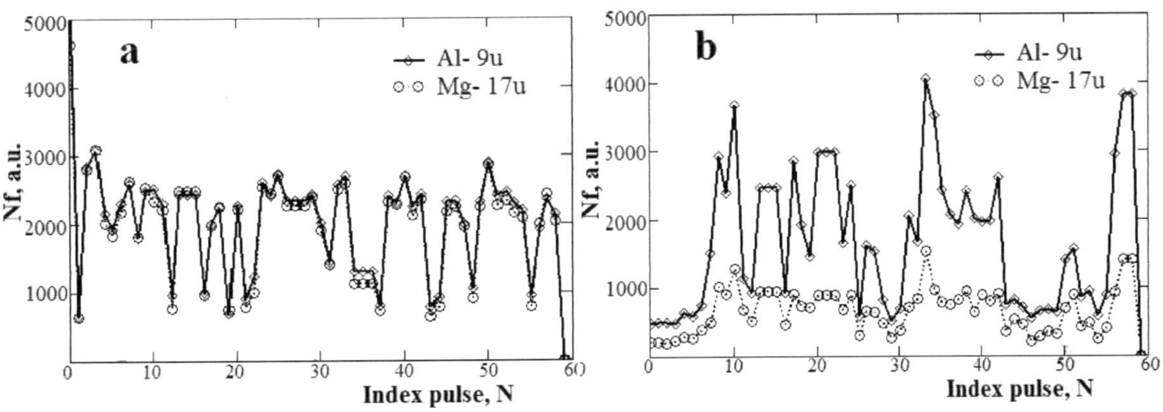

Figure 5. Integral emission of soft *X*-ray of the vacuum spark plasma when $d \approx 1.0$ mm: ***a*** - cathode discharge; ***b*** - anode discharge. The magnitude of the stored energy Ec=18.2 J at voltage Uc=12.9 kV.

Experiments showed that for the *cathode* discharge for any values d temporal and spectral characteristics of the soft x-ray emission are approximately preserved. These characteristics differ only in the amplitude of the signals, most likely due to the instability of the laser radiation (see figure 3a and figure 5a). With increasing the distant parameter d the delay duration of the anode discharge initiation increases about 3-10 times. From figure 3 it can be seen that when $d \approx 0.1$ mm the plasma *anode* discharge emits of soft x-ray significantly weaker than the plasma *cathode* discharge. When $d \approx 1.0$ mm the soft x-ray emission is comparable for both cases (figures 5a,b for 9 μm Al filter).

In Table 1 provides the estimates of the average values $\langle Nf_m \rangle$ (m=1,2) for the various experimental conditions of the vacuum spark initiation. Using the data from the right column Table 1 and the algorithm proposed in [Vovchenko E D, Melekhov A P 2016 The spectrum simulation of the soft x-ray for vacuum spark with laser initiation *Journal of Physics: Conference Series* (to be published)] we make simulation of the soft x-ray spectra for the *cathode* discharge and the *anode* discharge. Figure 6 shows the simulated spectra of the soft *X*-ray of the vacuum spark plasma for $d \approx 1.0$ mm. Spectra of vacuum spark was simulated for the electron temperature of the plasmas $Te <$ 1000 eV. In figure 6b we see the emergence of a strong, presumably, *L*-shell series of characteristic radiation of Cu and Zn ions in range $\varepsilon_{x\text{-ray}}=1200 \div 1800$ eV. It can be seen that *bremsstrahlung* and *K*-series radiation are more intensive in the case of the *cathode* discharge in figure 6a. Given simulations spectra of soft x-ray are correct for the spectral range $500 \leq \varepsilon_{x\text{-ray}} \leq 2000$ eV.

Table 1. Estimates of values $<Nf_m>$ (m=1,2).

The conditions (laser focus settings) of the experiment	$<Nf_1>$ (the average value Nf_1 for Al filter)	$<Nf_2>$ (the average value Nf_2 for Mg filter)	$<Nf_1>/<Nf_2>$ (*the ratio of* the average values Nf_1 *and* Nf_2)
d≈0.1 mm, *cathode* discharge (Figure 3a)	770	700	1.1
d≈0.1 mm, *anode* discharge (Figure 3b, for range of index pulses 20-42)	150	90	1.67
d≈1.0 mm, *cathode* discharge (Figure 5a)	2540	2390	1.07
d≈1.0 mm, *anode* discharge (Figure 5b)	2520	1000	2.52

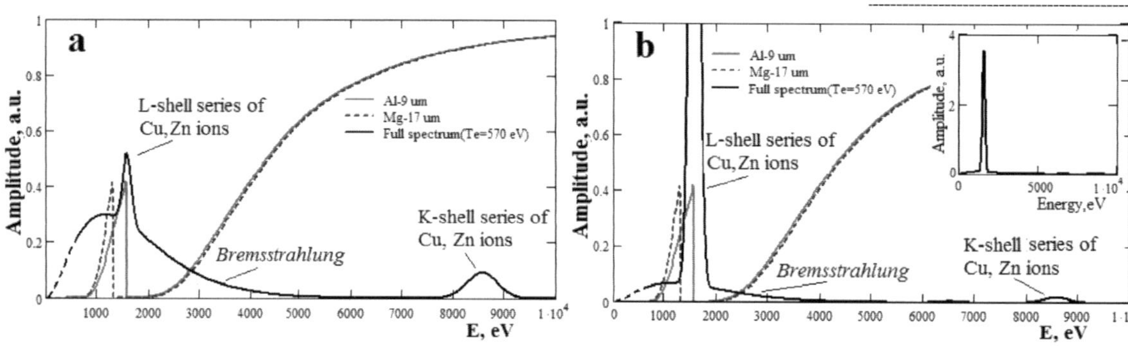

Figure 6. The simulated soft *x*-ray spectra of the vacuum spark plasma for *d*≈1.0 mm: ***a*** - *cathode* discharge; ***b*** - *anode* discharge. The magnitude of the stored energy *Ec*=18.2 J at voltage *Uc*=12.9 kV. Also on the figures are shown the transmission spectra of Al and Mg film filters. The inset on figure 6b shows the full spectrum of *anode* discharge.

It is obvious that the significant role in the dynamics of the development of the discharge plays the presence of strong inhomogeneous external electric field. In the case of the *anode* discharge the electric field will slow the discharge initiating electrons. Almost all of electrons remain near the anode because the electron temperature of laser plasma is $Te \leq 30$ eV.

The authors suggest that the defining role for the start of the development of the *anode* discharge play the *stimulated* field-emission of electrons from the cathode [8]. This electron emission is initiated by the radiation, ions, neutrals and neutral particles (droplets) from the laser plasma (anode).

4. Conclusions

In the case of the *anode* discharge the change of the position of the laser beam focus in relation to the edge of the electrode (*d*≈0.1 mm up to *d*≈1.0 mm) leads to the significant change of temporal and spectral characteristics of a vacuum spark. Currently some important details of the mechanism of formation of the *anodic* discharge are not well understood. For explaining the experimental results it is necessary now to develop an adequate model of discharge incorporating the distant parameter *d*.

References

[1] Romanov I V et. al. 2013 *Letters Technical Physics* **39 (4)** pp 388–392
[2] Serbanescu C G, Sandolache G 2001 *Nukleonika* **46 (1)** pp 17-20
[3] Korobkin Y V et. al. 2008 *Plasma Physics and Controlled Fusion* **50** 065002
[4] Dolgov A N, Klyachin N A, Prokhorovich D E 2014 *Pis'ma v Zh. Èksper. Teoret. Fiz.* **99 (9)** pp 595-596
[5] http://keldysh.ru/papers/2013/prep2013_78.pdf
[6] Alkhimova M A et. al. 2015 *Nukleonika* **60 (2)** pp 221-227
[7] http://www.sciencedirect.com/science/article/pii/S187538921501113X
[8] Litvinov E A, Mesuats G A, Proskurovsky D I 1983 *Soviet Physics-Uspekhi* **139 (2)** pp 265-302

Journal of Physics: Conference Series **737** (2016) 012014 doi:10.1088/1742-6596/737/1/012014

Surface nanogratings of abnormal orientation in universal polariton model of laser-induced damage of condensed media

V S Makin[1], Yu I Pestov[1], R S Makin[2]

[1]Public Limited Company Scientific Researxh Institute for Optoelectronic Device Instrumentation, Sosnovy Bor, 188540 Leningrad oblast, Russia
[2]National Research Nuclear University MEPhI (Moscow Engineering Physics Institute), Kashirskoe shosse 31, 115409 Moscow, Russia

E-mail: makin@sbor.net

Abstract. An experimental results about formation of small-scale gratings of anomalous orientation under the interaction of polarized laser radiation with condensed media were analyzed. The model based on surface polaritons participation in interference process was put forward to explain observed results.

Sufficient number of experimental publications during last decade have been devoted to the formation of ordered micro- and nanostructures on boundaries of media and inside transparent materials under one beam laser radiation interaction with condensed media. Most frequently just the resonant structures the period of which is proportional to the wavelength of incident radiation (λ) were the subject. For long pulse durations radiation the model has been developed and verified based on surface polaritons and incident radiation interference [1]. Recently the universal polariton model [2] has been extended together with nonlinear mathematical model of the formation of spatial structures period's multiple to the laser radiation wavelength (λ) [3] to explain experimental results in wide range of durations (down to ultrashort one [4]), energy density and physical properties of considered media. The universal polariton model accounts for the grating (*g*) formation with orientation $g \parallel E$ and periods d~λ, where E is the electric field strength of incident laser radiation. The explanation was based on the accounting the interference phenomenon with surface plasmon polariton participation excited by polarized laser radiation. Bellow such structures will be named as normal one.

At the same time, in experiments at later stages of normal resonant structures evolution the formation of additional new structures were frequently observed of orthogonal orientation $g \perp E$, the period of which usually was $d \ll \lambda$ [4-14]. Below this structures will be called as anomalous one. Important to note that the anomalous structures are not the peculiarity of ultrashort regime of laser-matter interaction and were observable also for regimes with long pulse durations.

To explain the origin of anomalous structures formation which were observed at the boundaries with materials of different physical properties the set of model was proposed among which we select following. The first one included the formation of a near-surface waveguide in silicon in the process of its electron depletion due to emission under the action of ultrashort pulse (USP) radiation [4]. The second model is based on the phenomenon of metal spallation [5]. The third model proposed for titanium [6] was based on generating the third harmonic of laser radiation in a TiO_2 layer with an optimum thickness. The first model is applicable only for some semiconductors. In addition, according to estimates, there exist difficulties in reaching the threshold thickness of the depleted semiconductor layer for the existence of the lower waveguide mode. As for the second model, the creators of the metal spallation model specifically stipulated that it does not include phenomena caused by excitation and interference of surface polaritons [15]. The third model was proposed only for titanium. However, the formation of saturated titanium oxide (TiO_2) with an optimal thickness is very problematic for USP irradiation regimes (in the air). We have proposed a model [16, 17] based on excitation and mutual interference of wedge (channel) surface plasmon polaritons [18, 19]. The optical scheme in which the model is usually implemented in experiments is presented in Fig. 1. Under the effect of linearly polarized laser radiation on the metal surface, resonance linear relief (or dielectric permittivity) gratings shown in Fig. 1 can be formed in the frameworkof the universal polariton model [2]. The ridges and valleys of this relief are, respectively, waveguiding structures for wedge (channel) surface plasmon polaritons (WSPPs): counterpropagating WSPPs interfere with the formation of a periodic grating with the direction $g \perp E$. The WSPPs are most efficiently excited by a wave the electric field of which is parallel to the direction of the ridge (valley) of the main resonance relief. There exists an opinion [20] that the phenomenon of the formation of small-scale ($d << \lambda$) gratings with anomalous orientation is typical only for the interaction between the USP radiation and condensed media. In experiments on the interaction between linearly polarized radiation (λ = 1.064 μm, τ = 10 ns, f = 17 Hz) and surface of VT1-0 titanium, we observed structurization of ridges of the formed main resonance relief (d_0 = 1.02 μm) with a period d = 0.21 μm [21]. Gratings, in particular, anomalous ones, are observed in condensed media with significantly different physical properties: metals, semiconductors, and dielectrics. This is related to metallization of semiconductors and dielectrics during the generation of nonequilibrium electrons in them due to multiphoton (or tunnel) ionization. Let us consider the formation of anomalous structures on metal surfaces. Figure 2a presents the image of a niobium surface destructed with the formation of anomalous gratings [7]. The structures with normal orientation of the grating and periods $d = \lambda/\xi \approx 600$ nm and $d = \lambda/2\xi \approx 300$ nm forming under the action of a linearly polarized radiation (λ = 800 nm, τ = 100 fs, f = 1 kHz) at $Q \approx 0.5$ J/cm^2 and $N \approx$ (1–2) $\times$ 10^3 are caused by the interference of the incident radiation and WSPP and mutual interference of the WSPPs with opposite propagation directions, respectively. Here, ξ is the real part of the refractive index of the medium interface for surface plasmon polaritons. At low energy densities ($Q \approx$ 0.29 J/cm^2), $N \approx 1100$, and $N \approx 1900$, anomalous nanostructures appear on the niobium surface. Their period can be described by the formula:

$$d = \frac{\lambda}{2k\eta}, \qquad (1)$$

where k = 1, 2, 4, 8, ..., η is the real part of the refractive index of the medium interface for the WSPPs.

First, there appear structures with period $d = \lambda/(2k\eta) \approx 65$ nm (k = 4, $\eta \approx 1.53$); then, with a larger period $d = \lambda/(2k\eta) \approx 145$ nm (k = 2, $\eta \approx 1.38$). Larger values of the period correspond to a larger value of N.

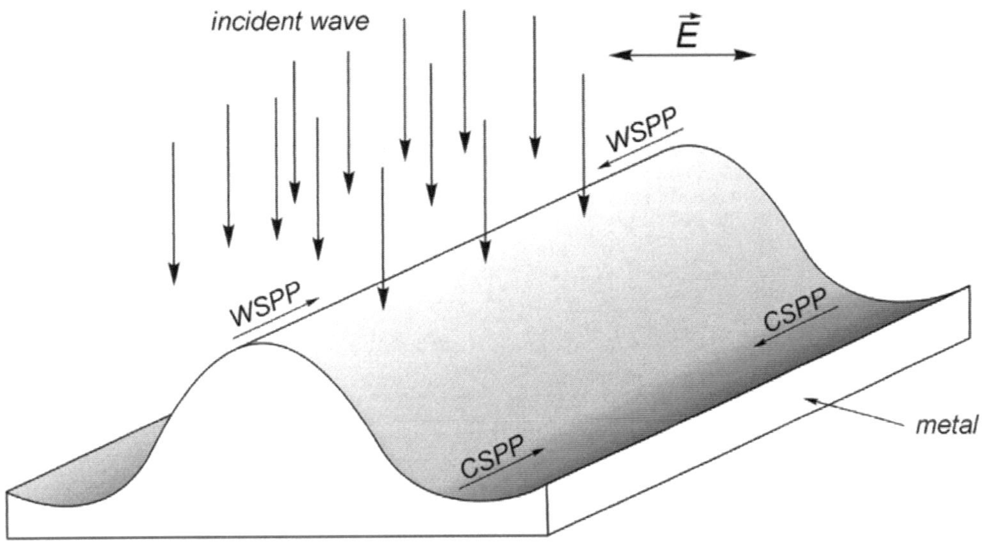

Figure 1. Optical scheme illustrating the mechanism of formation and orientation of anomalous gratings. The incident laser radiation excites counterpropagating WSPPs along the extended ridge (valley) of the main resonance relief formed in the process of laser radiation interference with WSPPs. The direction of laser radiation polarization is shown by the double arrow. The direction of the forming nanograting g is perpendicular to the tangential projection of the strength vector of the laser radiation electric field.

Anomalous structures are localized mostly on ridges of the main resonance relief, in particular, at initial stages of its formation (see peripheral zones of the scanning region), which corresponds to the interference with the participation of wedge SPPs. Values of quantity η turned out to be rather large, which may be related to the influence of the gas solved in the nearsurface metal layer, as well as of the oxide layer, both formed under the action of the radiation. With an increase in N, nanorelief ridges were transformed into rounded nanodroplets. At larger energy densities, about 47 J/cm^2, and $N \approx 300$, anomalous structures were localized only in valleys of the main relief (see the insert in Fig. 2a) and their period was $d = \lambda/(2k\eta) \approx 300$ nm ($k = 1$, $\eta \approx 1.3$), which corresponds to the interference with the participation of WSPPs. The next example is tungsten subjected to a series of femtosecond pulses of a linearly polarized radiation ($\lambda = 800$ nm, $\tau = 180$ fs, $f = 1$ kHz) with energy density $Q = 0.4$ J/cm^2 at $N = 1000$ pulses. In Fig. 2b, one can clearly see the formation of the main resonance relief with a period of about $d_0 = \lambda / \xi \approx 600$ nm (and normal orientation of the grating g || E) and orthogonal gratings superposed on it with a characteristic period $d = \lambda/(2k\eta) \approx 70$ nm ($k = 4$, $\eta \approx 1.4$). Here, the experimental results were interpreted using conclusions of the nonlinear mathematical model for the formation of spatial periods of structures [3] in the framework of the universal polariton model.

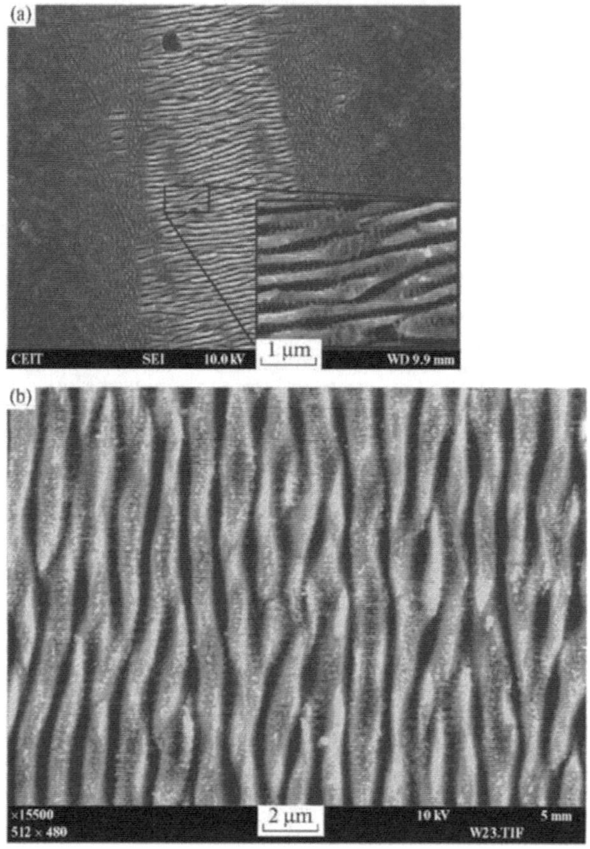

Figure 2. Illustration of anomalous nanostructure formation by a femtosecond linearly polarized radiation under the following conditions: (a) scanning with velocity $v = 0.1$ mm/s in the direction of the laser radiation electric field vector ($Q = 32$ mJ/cm2, $\lambda = 800$ nm, $\tau = 100$ fs, and $f = 1$ kHz) with periods of 145 and 65 nm, respectively, in the peripheral and central (see insert) irradiated regions on the surface of niobium [7] and (b) with a period of ≈60 nm on the surface of mechanically polished tungsten by a series of pulses ($Q = 0.2$ J/cm^2, $\lambda = 800$ nm, $\tau = 160$ fs, $f = 1$ kHz, $N = 700$).

Note that detailed experimental data on the formation of anomalous nanostructures on the titanium surface were published in [5, 6, 10, 22, 23]. Figure 3a illustrates the formation of anomalous gratings on a SiC semiconductor under the action of a linearly polarized femtosecond radiation and beam scanning in the direction of electric field strength vector E [12]. One can see the formation of main resonance structures with a period $d_0 = 500$ nm, as well as the formation of nanochannels (nanocracks). Each nanocrack turns out to be spatially modulated by a grating in the form of nanoholes with period $d = 60$ nm, which corresponds to $k = 4$ and $\eta \approx 1.25n$ of formula (1), where $n = 1.33$ is the refractive index of water. The formation of a grating in the form of nanoholes is related to the specificity of the SiC surface destruction: in regions of intensity maximums, destruction of the material occurs with the removal of decomposition products from the affected area. Anomalous gratings of nanostructures were experimentally found for the first time on the surface of silicon in the region of its transparency ($\lambda = 1.25$ μm, $\tau = 10$ fs) [4]. However, they were studied experimentally in most detail in [11] using irradiation of silicon immersed in water ($\lambda = 800$ nm, $\tau = 150$ fs, and $f = 1$ kHz). The irradiation process consisted of two stages: after the formation of main resonance structures ($d0 = \lambda/(n\xi)$), the direction of linear polarization was changed by 90° and the irradiation continued with a somewhat decreased power density. As a result, ridges of the main resonance relief turned out to be modulated with $d = \lambda/(kn\eta) = 800/(1.33 \times 4 \times \eta) \approx 90$ nm for $\eta \approx 1.3$ due to WSPP interference. The polarization change provided effective WSPP excitation and comparatively deep modulation of the relief ridge. The gratings at adjacent ridges were not phased, which is related to strong transverse (along the surface) localization of WSPP fields. Note that the small component of vector E in the direction of WSPP propagation in the scheme of Fig. 1 appeared due to radiation focusing. The contact with water ensured effective cooling of the irradiation zone. With a further increase in the number of radiation pulses, regions of local nanoridges were transformed into nanocolumns with a high (1 : 8) aspect ratio.

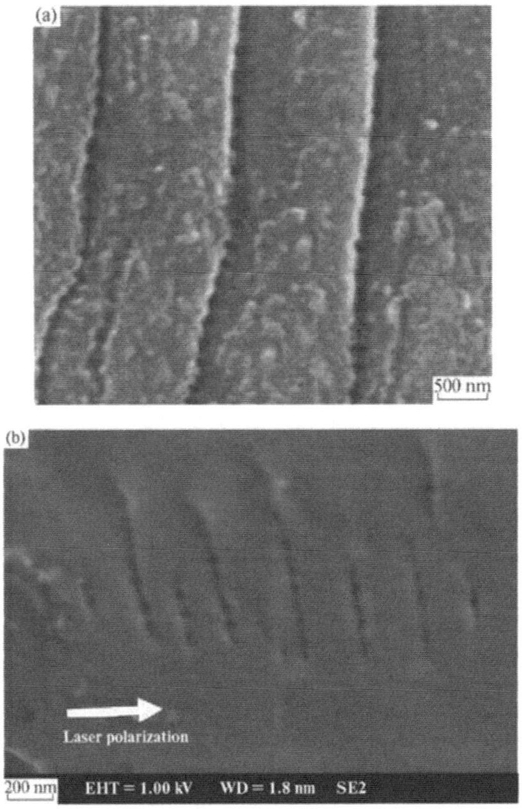

Figure 3. Illustration of the formation of anomalous nanostructure gratings in valleys (nanocracks) of the main resonance microrelief after the action of a series of femtosecond linearly polarized radiation pulses on surfaces of condensed media: (a) SiC (P =15 mW, λ = 800 nm, τ = 150 fs, and f = 1 kHz) and scanning in the direction of strength vector E of the laser radiation electric field with velocity v = 1500 μm/s [12] and (b) silicate glass (E = 0.5 mJ, λ = 1030 nm, τ = 300 fs, f = 200 kHz, v = 200 μm/s), laser radiation polarization is perpendicular to the direction of scanning [14]. Vectors g of the forming nanostructure gratings in the form of nanoholes are perpendicular to the direction of laser radiation polarization.

The regime of anomalous grating formation in mode of nano-holes on mono-silicon surface has been realized in standard scheme of experiment (see Fig.1) [24]. The irradiation was made by liner polarized laser radiation (λ=800 nm, τ =130 fs, f =1 kHz) and scanning in direction perpendicular to the vector **E** direction (in air). The laser energy density was near material ablation threshold. The main resonant structures were the prerequisite for anomalous grating formation as stressed by authors [24]. The period of anomalous grating can be estimated using their images as ~200 nm (authors of [24] did not indicate the period value).

Journal of Physics: Conference Series **737** (2016) 012014 doi:10.1088/1742-6596/737/1/012014

The different scheme of experiment have been realized in [25] also on silicon surface. The anomalous gratings were observed in conditions of radiation scanning (v~ 1mm/s) in direction of laser radiation polarization. In such conditions the gratings of normal orientation, if produced, must have orientation perpendicular to the scanning direction. But produced anomalous gratings were oriented along electric field strength vector direction. Their formation were due to channel SPP and incident laser radiation interference (the grating was localized in valleys of relief formed by scanned radiation), and their period $d=\lambda/\xi\approx720$ nm. The productions of two or more anomalous gratings were observed with increasing laser energy density separated by distance of the order of 2 μm. This residual grating causes the selective absorption of radiation from white light and structured surface coloration.

In one of first works on the formation of gratings with normal orientation in a volume of glass (dielectric) and periods …, $\lambda/(m\eta)$, $\lambda/(2m\eta)$, $\lambda/(4m\eta)$, …, traces of the formation of a grating in the form of nanoholes were observed along regions of nanocracks; on the ridges, nanostructure gratings were observed [26]. Figure 3b [14] more explicitly illustrates the formation of an anomalous grating on valleys and ridges of the relief with normal orientation ($d_1 = \lambda/(2n\xi) \approx 230$ nm and $d_2= \lambda/(4n\xi)$) on the silicate glass with periods $d = \lambda/(2k\eta) \approx 50$ nm ($k = 4$, $\eta \approx 1.34$) (wedge SPPs) and in valleys (nanocracks) with $d \approx 58$ nm ($k = 4$, $\eta \approx 1.16$). According to the model that we propose, the formation of the last structures is caused by interference of WSPPs localized and propagating along linear nanocracks and explains the physical cause of the formation of nanopores (see the discussion in [14]).

Fluorite (CaF_2) [13] and TiO_2 [27] serve as additional examples of the formation of anomalous gratings on dielectrics. Structures with anomalous orientation on silica glass were found also in the case of the action of axisymmetrically polarized laser radiation [28].

Conclusion

Thus, experimental data on the formation of gratings of structures with anomalous orientation under the action of linearly polarized laser radiation pulses in condensed media have been analyzed. To explain the causes of the formation of anomalous gratings, a model predicting spatial periods and orientation of gratings has been proposed. The model is based on (i) mutual interference of WSPPs propagating in opposite directions and (ii) a nonlinear mathematical model predicting the formation of spatial nanograting periods multiple to the wavelength. The proposed model well describes the available experimental data on anomalous structures on media with different physical properties: metals, semiconductors, and dielectrics. The model is applicable not only to USP radiation, but also to nanosecond radiation. Like periods of gratings with normal orientation, periods of anomalous gratings obey the inverse Feigenbaum universality of doubling the spatial period. The spatial period of anomalous gratings can be considerably less that the diffraction optical limit (e.g., for stainless steel, $d \approx 30$ nm $\lambda = 515$ nm) and physically restricted by processes of thermal conductivity "smoothing." Note that the process of the decrease in the spatial scale of structures can continue due to interference of WSPPs excited on ridges (valleys) of already formed anomalous gratings. Traces of this process are found in a series of irradiated regions. The excitation of WSPPs can be considered as a new channel of transforming the laser radiation into heat under the interaction of powerful USP radiation with condensed media. Anomalous gratings make an additive contribution to the absorptive capacity of the so-called «black metals».

References

[1] Bonch-Bruevich A M, Libenson M N, Makin V S, Trubaev V V 1992 *Optical Engineering* **31** **(4)** 718-730
[2] Makin V S and Vorob'ev A Ya 2007 *Vestn. SPbO Akad. Inzh. Nauk* **3**, 456
[3] Makin V S, Makin R S, Vorob'ev A Ya and Guo Ch 2009 *Nonlinearity in Modern Natural Sciences,* (LKI, Moscow) 309
[4] Ganeev R A 2014 *Optics and Spectroscopy* **117** 320-340
[5] Nathala C S R, Ajami A, Ionin A A, Kudryashov S I, Makarov S V, Ganz T, Assion A and Husinsky W *Opt. Express* 2015 **23** 5915

[6] Li X-F, Zhang C-Y, Li H, Dai Q-F, Lan F and Tie S-L 2012 *Opt. Express* **22** 28086
[7] Pan A, Dias A, Gomez-Aranzadi M, Olaizola S M and Rodriguez A 2014 *J. Appl. Phys.* **114** 173101
[8] Gecys P, Vinciunas A, Gedvilas M, Kasparaitis A, Lazdinas R and Raciukaitis G 2015 *J. Laser Micro/Nanoeng.* **10** 129
[9] Hashida M, Gemini L, Nishii T, Miyasaka Y, Sakagami H, Shimizu M, Inoue S, Limpouch J, Mocek T and Sakabe S 2014 *J. Laser Micro/Nanoeng.* **9** 234
[10] Wu N, Wang Z, Wang X, Shimotsuma Y, Nishi M, Miura K and Hirao K 2011 *J. Ceram. Soc. Jpn.* **119** 898
[11] Shen M, Carey J E., Crouch C H., Kandyla M, Stone H A and Mazur E 2008 *Nano Lett.* **8** 2087
[12] Khuat V, Si J, Chen T and Hou X 2015 *Opt. Lett.* **40** 209
[13] Reif J, Varlamova O and Costache F 2008 *Appl. Phys.Lett.* **92** 1019
[14] Lancry M, Poumellec B, Canning J, Cook K, Poulin J-C and Brisset F 2013 *Laser Photon. Rev.* **7**, 953
[15] Anisimov S I, Inogamov N A, Petrov Y V, Khokhlov V A, Zakhlovsky V V, Nishihara K, Agranat M B, Ashitkov S I and Komarov P S 2008 *Appl. Phys. A* **92** 797
[16] Makin V S, Pestov Yu I, Makin R S and Vorob'ev A Ya 2009 *J. Opt. Technol.* **76** 555
[17] Makin V S 2014 *Proceedings of the International Conferece on Photooptics (*Lisboa, Portugal) 131
[18] Novikov I V and Maradudin A A 2002 *Phys. Rev. B* **66** 035403
[19] Moreno E, Rodrigo S G, Bozhevolnyi S I, Martin-Moreno L and Garcia-Vidal F 2008 *J. Phys. Rev. Lett.* **100** 023901
[20] Huang M, Xu Z 2014 *Laser Photon. Rev.* **8** 633
[21] Makin V S, Pestov Yu I and Privalov V E 2011 *Proceedings of the 19th International Conference on Laser Informational Technology in Medicine, Biology and Geoecology* (Tipogr. Variant, Novorossiisk) 41
[22] Oktem B, Pavlov I, Ilday S, Kalaycıoğlu H, Rybak A, Yavas S, Erdoğan M and Ilday F O 2013 *Nature Photon.* **7**, 897
[23] Bonse J, Holm S, Rosenfeld A and Kroger 2013 *J. Appl. Phys. A: Mater. Sci. Process* **100** 547
[24] Shimotsuma Y, Kazansky P G, Qin J R and Hirao K 2003 *Phys. Rev. Lett.* **91**, 247405
[25] Yuan D Q, Zhou M, Wu Q R, Xu J T, Yang H F 2015 *J. Opt. Technol.* **82 (6)** 31
[26] Zhang C-Y, Yan J W, Lin H-Y, Dai Q-F, Wu L-J, Lan S, Trofimov V A, Lysak T M 2012 *Optics Lett.* **37 (6)** 1106
[27] Shimotsuma T, Tsukamoto M, Nishii R, Io Y, Nagai A, Yamashita K, Hanawa T, Matsushita N, Guoqiang X and Abe N 2012 *Trans. JWRI* **41** 25
[28] Makin V S and Makin R S 2013 *Opt. Spectrosc.* **115** 591

Journal of Physics: Conference Series **737** (2016) 012015 doi:10.1088/1742-6596/737/1/012015

Techniques of surface optical breakdown prevention for low-depths femtosecond waveguides writing

M A Bukharin[1,2], N N Skryabin[1,2], D V Ganin[3,4], D V Khudyakov[2,3] and S.K. Vartapetov[2,3]

[1]Moscow Institute of Physics and Technology, 9 Institutskiy per., Dolgoprudny, Moscow Region, 141700, Russia
[2]Optosystems Ltd., Troitsk, Moscow, 142190, Russia
[3]Physics Instrumentation Center of Prokhorov General Physics Institute RAS, Russian Academy of Sciences, 142190, Troitsk, Moscow, Russia
[4]National Research Nuclear University MEPhI (Moscow Engineering Physics Institute), 31 Kashirskoe shosse, 115409 Moscow, Russia

E-mail: mikhail.bukharin@phystech.edu

Abstract. We demonstrated technique of direct femtosecond waveguide writing at record low depth (2-15 μm) under surface of lithium niobate, that play a key role in design of electro-optical modulators with low operating voltage. To prevent optical breakdown of crystal surface we used high numerical aperture objectives for focusing of light and non-thermal regime of inscription in contrast to widespread femtosecond writing technique at depths of tens micrometers or higher. Surface optical breakdown threshold was measured for both x- and z-cut crystals. Inscribed waveguides were examined for intrinsic microstructure. It also reported sharp narrowing of operating pulses energy range with writing depth under the surface of crystal, that should be taken in account when near-surface waveguides design. Novelty of the results consists in reduction of inscription depth under the surface of crystals that broadens applications of direct femtosecond writing technique to full formation of near-surface waveguides and postproduction precise geometry correction of near-surfaces optical integrated circuits produced with proton-exchanged technique.

1. Introduction

Under the influence of tightly focused ultrafast laser pulses refractive index of optical crystals could be permanently changed. Translating the crystal relatively to focus of the lens one can directly inscribe optical waveguide in the bulk crystal [1-3]. This technique is maskless in contrast to standard fabrication techniques of subsurface waveguides formation [4-6]. One of the most important challenges of the technology nowadays is inscription of waveguides at low depth (< 15 um) under the surface of electro-optical crystals, due to the opportunity to produce fast electro-optical (e.-o.) modulators with low operating voltage.

Such well known and widespread technologies as Ti-diffusion, ion implantation and proton exchange can provide production of subsurface waveguides, but their properties do not meet all requirements of practical applications: mode field diameter of designed waveguides is significantly asymmetric due to diffusion character of the techniques, production process is multistage and complex. In addition, Ti-diffused waveguides has high photorefraction [7, 8], and proton exchanged ones maintain only of one linear polarization state [9].

On the other hand, femtosecond waveguide writing could meet the challenge: it provides low optical losses, low influence on waveguide`s core material (there is no observable changes in photorefraction as compared to undoped $LiNbO_3$), and maintenance of all polarization states of propagating light, that is crucial for development of fast electro-optical modulators for laser emission with broad spectrum, based on electro-optical polarization rotation [10, 11]. According to present knowledge, there is no experimentally written waveguides at depths lower than 20 μm [1, 12, 13], which could provide only high operating voltage for e.-o. modulators (>15V compared with <1.5V for modulators, based on Ti-diffused or proton-exchanged waveguides [1]). Thus, actuality of the investigation consists in development of femtosecond writing technique for near surface waveguides production.

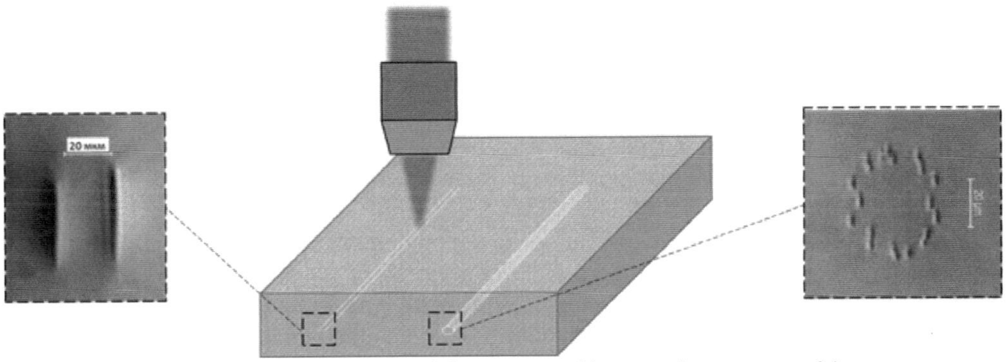

Figure 1. Schematic of femtosecond laser written waveguides.

Fundamental problem resolved in the investigation consists in suppression of optical thermos-mechanical breakdown of surface, that impeding femtosecond inscription of waveguides at low depths. The technique of femtosecond writing proposed in the paper doesn`t eliminate negative factors, described above, but sufficiently decreases its effect.

In this work we experimentally shown femtosecond inscription of waveguides at record low depths (2-15 μm) under the surface of lithium niobate (both z- and x-cut), that could be used with surface electrodes to form electro-optical modulator.

2. Experimental setup

For direct femtosecond waveguide writing we used experiential setup, represented in Fig.2. Ultrashort laser pulses (340 fs, 1040 nm) from oscillator without additional power amplifier (HighQ femtoTRAIN, 0.1-10 MHz) were focused into lithium niobate sample (both x- and z-cut), which was mounted on precise motorized 3D-translational stage and moved with velocity 10 μm/s to form extended tracks through the full length of the sample (25 mm).

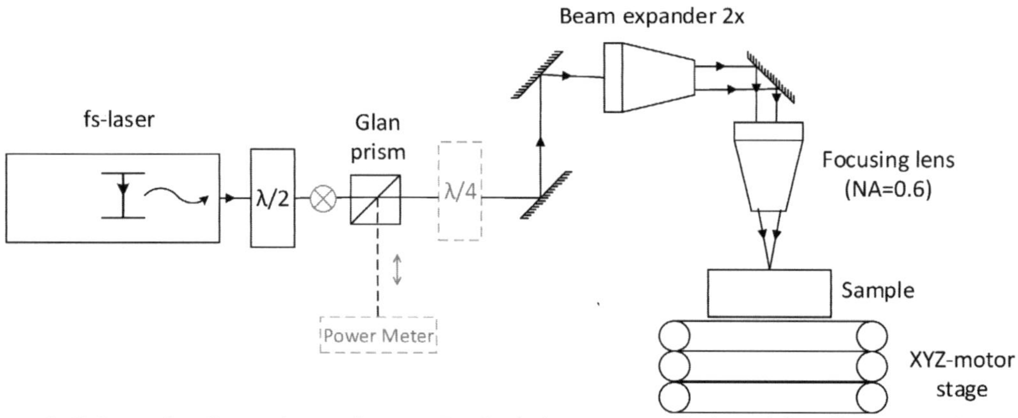

Figure 2. Schematic of experimental setup. Dashed elements are designed for alignment of sample regarding to lens focus and placed in the optical path only for the time of adjustment and investigation of surface optical breakdown threshold.

2.1. Technique of femtosecond writing

To write tracks with induced refractive index at record low depth under surface of crystal we proposed technique of negative factors reduction that eliminates optical and thermomechanical breakdown of the surface. To prevent optical breakdown we used high numerical aperture aspheric lenses for focusing of light (NA=0.6). As a development of our previous investigation [14], we have shown that to write near-surface waveguides there is no need in expensive high numerical aperture objectives (NA>0.8). Moreover, for the purpose under consideration utilization of oil-immersed objectives is prohibited due to low breakdown threshold of immersion oil.

Varying pulse repetition rate from 0.1 MHz up to 10 MHz we experimentally shown, that widely used heat accumulation regime of femtosecond writing is unsuitable for inscription of waveguides at low depth due to formation of steep temperature gradients [15], which could break up crystal surface even at pulse energies lower than required for refractive index induction. According to estimation described in details in [16], critical repetition rate for transition to thermal cumulative regime equals $f_{PRR}^{critical} = D/w_0^2 = 1.4MHz$, where $D = 1.9\cdot 10^{-6}\, m^2/s$ - thermal diffusivity and $w_0 = 1.3\mu m$ - focal spot radius. Experiments confirm the estimation: at repetition rate higher than 2 MHz written tracks with induced refractive index were not observed under surface breakdown threshold at depths lower than 15 μm (thermomechanical breakdown of the surface occurred at lower pulse energies). To prevent the negative thermal accumulation effect we proposed to write the structures in non-thermal regime in spite of lower achievable induced refractive index value compared to the thermal cumulative one. In the investigation laser pulse energy in focal region was 40-60 nJ and pulse repetition rate was f_{PRR}=0.1 MHz.

Under the influence of tightly focused femtosecond laser emission refractive index of lithium niobate permanently changes due to several photochemical processes and formation of macroscopic defects that depend on number of laser pulses, accumulated in the interaction volume. Thus, to compensate reduced number of accumulated pulses per focal spot due to low repetition rate and to increase achievable value of induced refractive index, writing exposition was proportionally increased by reducing of translational velocity down to 10 μm/s. As a result, induced refractive index was of the order of $3\cdot 10^{-3}$, which is sufficient for formation of waveguide`s cladding. The value of induced refractive index was measured on the base of widely used quantitative phase microscopy technique, described in details in [16-19].

2.2. *Surface optical breakdown threshold*

Before writing of waveguides the crystal should be precisely aligned in relation to focus of the lens. One of the most practic and precise techniques consists in analyzing of back-reflected operating emission from the surface of treated crystal, as shown in Fig.2 (dashed elements).

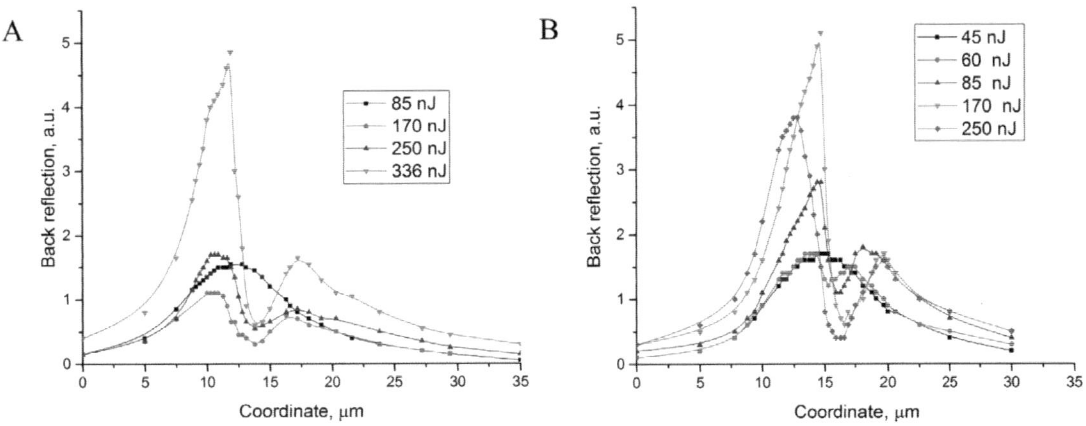

Figure 3. Reduction of back-reflection signal due to local ablation of x-cut (A) and z-cut (B) lithium niobate crystal surface at high pulse energies while z-scanning around focusing point at surface (near 15 µm coordinate).

The alignment setup also provides an opportunity to estimate surface optical breakdown threshold by reduction of back-reflection signal due to local ablation in the focal spot even at low pulse energies, as shown in Fig.3. Each experimental graph was fitted with theoretical z-scan curve, representing part of back-reflected light in the presence of central ablated (non-reflected) region for Gaussian beam profile. It was found that critical peak energy density for x-cut $LiNbO_3$ was nearly twice higher than the one for z-cut $LiNbO_3$ ($I_{x-cut}^{critical} \approx 2 \cdot I_{z-cut}^{critical}$) and under experimental focusing conditions achieved at 56 nJ pulse energy. The relation strongly depends on surface polishing quality. In the experiments we used $LiNbO_3$ wafer (Crystal Technology Inc.) with surface quality scratch #10, dig #5.

3. Results

Using the proposed technique of femtosecond inscription, we have produced a number of parallel tracks (from 34 to 50, see Fig.4) with induced refractive index ($\Delta n=3\cdot 10^{-3}$), that form a waveguides with mode field diameter from 8 µm up to 13 µm at record low depths (2-15 µm) under the surface of $LiNbO_3$ crystal (both x- and z-cut). Each track has cross sectional dimensions of 1.1µm×6.0µm. In comparison to previous published results [14], that were written with higher NA=0.8, aspect ratio of the tracks significantly increased from the value of 2.5 up to 5.5. It connected with 1.8 times higher Rayleigh length in current investigation and higher writing pulse energy, that increases plasma defocusing in the focal region and also increases aspect ratio of written structures. Typical microscopic images of single near-surface tracks are shown in Fig.4a and 4b. Cross section of designed waveguide and corresponding modelling trajectory are shown in Fig.4c and 4d. It should be noted, that there is no negative macroscopic cracks that are typically appear in widespread high energy and low numerical aperture technique of femtosecond writing.

In series of experiments it was shown that operating pulse energy range strongly depend on writing depth under the surface of crystal (see Fig.5). While the lower boundary of the range do not changed with writing depth (with amount nearly 50 nJ), the upper boundary rapidly decreases from more than 110 nJ at 8 µm down to 60 nJ at 2 µm depth. Thus, operating range rapidly narrows and complicates process of near-surface waveguide formation due to necessity of dynamic adjustment of operating energy.

Journal of Physics: Conference Series **737** (2016) 012015 doi:10.1088/1742-6596/737/1/012015

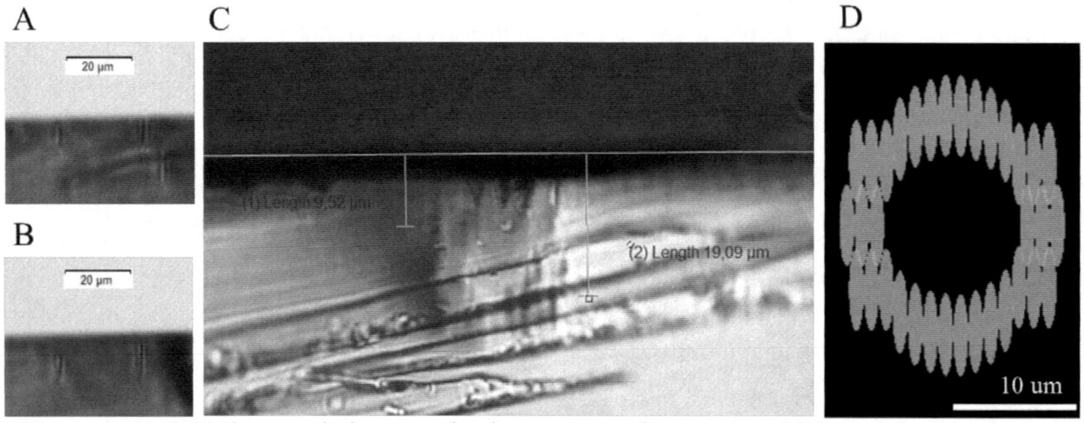

Figure 4. (A, B) Microscopic images of written near-surface tracks with induced refractive index; (C) facet view of inscribed waveguide with mode field diameter 10 µm at low depth under the surface of crystal, white pointers indicates top and bottom borders of waveguide`s core; (D) schematic of written cladding of waveguide from Fig. 4c.

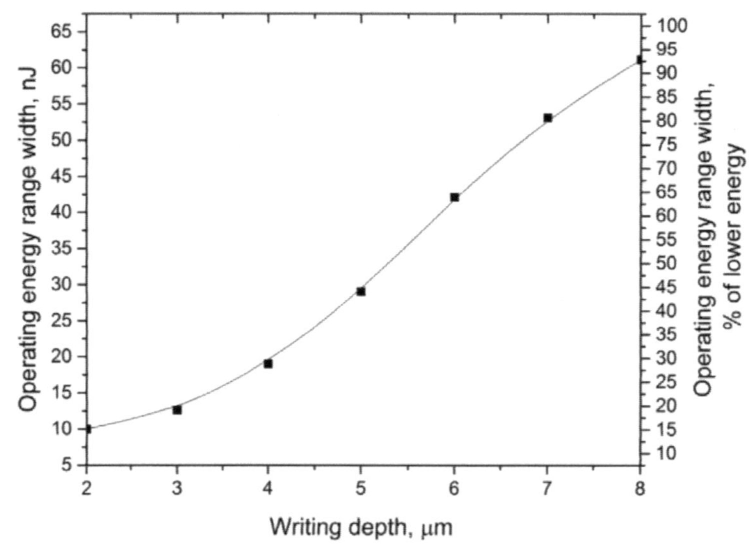

Figure 5. Narrowing of operating pulses energy range with writing depth under the surface of crystal.

4. Discussion

Waveguides produced by the proposed technique possess both advantages of low optical losses and maintaining of all polarization states of propagation light [14]. Taking into account the opportunity to use commercially available aspheric lens to directly write three-dimensional near-surface waveguides instead of costly multilens objectives with high NA or multistep lithographic techniques, it provides a new avenue in production of fast electro-optical modulators with non-interferometric architectures, based on electro-optical polarization rotation, that is crucial for fast modulation of femtosecond laser emission in fully integral schemes.

Journal of Physics: Conference Series **737** (2016) 012015 doi:10.1088/1742-6596/737/1/012015

On the one hand, increased aspect ratio of written tracks (5.5 instead of 2.5 for high NA femtosecond writing) negatively effect on written waveguide`s properties, increasing its asymmetry and impels to compensate it with more complex multilayer cladding, as shown in Fig.4d. But on the other hand, narrow and tall (1µm×6µm in cross-section) tracks with induced refractive index provides a new application of proposed femtosecond writing technology. Using single tracks with such geometry one can carry out postproduction precise geometry correction of near-surfaces optical integrated circuits (Y-branching power dividers, for instance), produced with proton-exchanged technique [20]. Proton exchanged waveguides have longitudinal dimension (in depth) of the same 6 µm. Thus, femtosecond written tracks could cover all of its cross-section in single path. This result provides an opportunity to reduce thermal sensitivity of proton-exchanged Y-branching power divider (Fig.6a) due to reduction of mode coupling in the transitional region of branching without increasing of optical losses (as in the case of correction with optical breakdown, shown in Fig.6b), but only with increasing of refractive index contrast between waveguide`s core and cladding, as shown in Fig.6c.

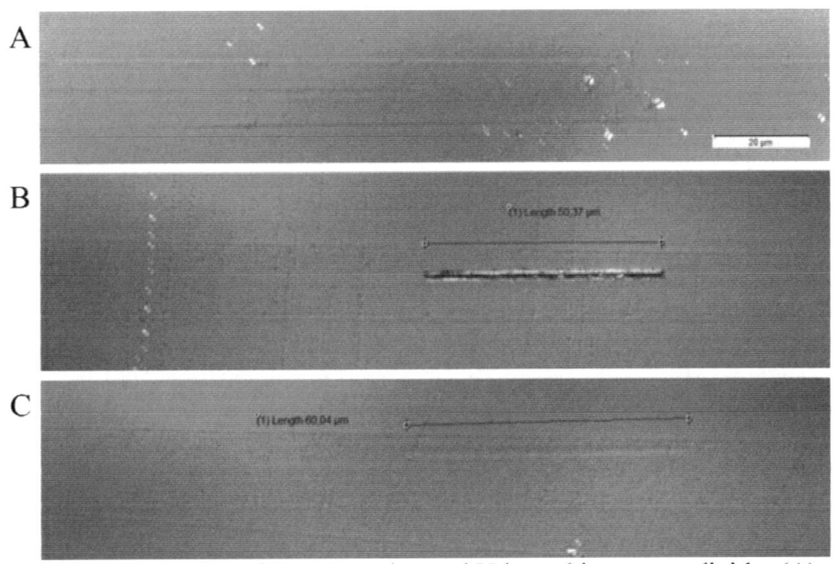

Figure 6. Microphotographs of proton-exchanged Y-branching power divider (A) with mode coupling via optical breakdown (B) and femtosecond induced track (C).

5. Conclusion

In the paper we have reported an advanced technique of direct femtosecond inscription of near-surface waveguides at record low depth in lithium niobate (both x- and z-cut). Using high-NA focusing lens and non-thermal regime of femtosecond writing with increased exposure for optical and thermomechanical surface breakdown prevention it was experimentally shown the opportunity to write full cladding of single-mode waveguide in depth range 2-15 µm. The result opens a new opportunities in fabrication of fast electro-optical modulators with low operating voltage and postproduction precise geometry correction of near-surfaces optical integrated circuits produced with proton-exchanged technique. As compared to previous investigations in the field of waveguide writing in $LiNbO_3$ for the purposes of modulators development (structures at depths more than 50 µm with buried electrodes [1]), the obtained waveguides could be used with easy-to-produce on-surface electrodes.

Acknowledgements

The work was partially supported by Foundation for Assistance to Small Innovative Enterprises in Science and Technology (4806ГУ1/2014 №0008340 and 8611ГУ/2015 №0018824).

References

[1] Osellame R, Cerullo G, Ramponi R 2012 *Femtosecond Laser Micromachining Photonic and Microfluidic Devices in Transparent Materials* (London: Springer)
[2] Okhrimchuk A G, Shestakov A V, Khrushchev I and Mitchell J 2005 *Opt. Lett.* **30 (17)** 2248–2250
[3] Streltsov A M and Borrelli N F 2001 *Opt. Lett.* **26 (1)** 42–43
[4] Williams R J, Kramer R G, Nolte S, Withford M J 2013 *Opt. Lett.* **38 (11)** 1918–1920
[5] McMillen B, et al. 2014 *Opt. Lett.* **39 (12)** 3579–3582
[6] Dubov M, Boscolo S and Webb D J 2014 *Opt. Mater. Express* **4 (8)** 1706–1716
[7] Chung H P, Huang K H, Yang S L, Chang W K, Wu C W, Setzpfandt F, Pertsch T, Neshev D N and Chen Y H 2015 *Opt. Exp.* **23 (24)** 30641-30650
[8] Zhang M, Chen W, Chen L, Quan B and Liu Y 2014 *Opt. Quant. Electron.* **46** 1529-1538
[9] Kostritskii S M, Korkishko Yu N and Fedorov V A 2013 *Proc. SPIE* **9065** 90650E
[10] Palfalvi L, Hebling J, Almasi G, Peter A and Polgar K 2003 *J.Opt.A: Pure Appl. Opt.* **5** S280-S283
[11] Kroesen S, Horn W, Imbrock J and Denz C 2014 *Opt. Exp.* **22 (19)** 23339-23348
[12] Kowalevicz A M, Sharma V, Ippen E P and Fujimoto J G 2005 *Opt. Lett.* **30 (9)** 1060-1062
[13] Rodenas A and Kar A K 2011 *Opt. Exp.* **19 (18)** 17820-17833
[14] Bukharin M A, Skryabin N N, Khudyakov D V and Vartapetov S K 2016 *Proc. of SPIE* ***9891*** 989102
[15] Eaton S M, Ng Mi Li, Osellame R and Herman P R 2011 *Journal of Non-Crystalline Solids* **357** 2387
[16] Bukharin M A, Khudyakov D V and Vartapetov S K 2015 *Appl. Phys. A* **119 (1)** 397–403
[17] Bukharin M, Khudakov D and Vartapetov S 2015 *Phys. Proc.* **71** 272–276
[18] Paganin D and Nugent K A 1998 *Phys. Rev. Lett.* **80 (12)** 2586–2589
[19] Barty A, Nugent K A, Paganin D and Roberts A 1998 *Opt. Lett.* **23 (11)** 817–819
[20] Kostritskii S M, Korkishko Yu N, Fedorov V A 2013 *Proc. of SPIE* **9065** 90650E-1

Investigation of spatially nonuniform nonlinear response of a lithium niobate crystal sample at low light intensity

E Dmitriev, E Beresina, V Krad`ko, V Ryabchenok, A Perin, and V Shandarov

Tomsk State University of Control Systems and Radioelectronics, Lenin Prospect 40, Tomsk, 634050, Russia

E-mail: shandarovvm@svch.rk.tusur.ru

Abstract. The spatial distribution of nonlinear optical response over a bulk of lithium niobate sample is experimentally studied through the distortions of the two-dimensional light beam intensity patterns at the sample output surface caused by the beam spatial self-action. The compensation of these distortions and the linear light beam divergence by means of the pyroelectric effect contribution into the nonlinear optical response of the crystal are also studied. The results obtained for the light wavelength of 532 nm and beam waist diameter of 13 μm demonstrate the partial or total compensation of the beam divergence depending on light power and a temperature increase at the sample heating.

1. Introduction

The homogeneity of physical properties of dielectric and semiconductor materials is a critical requirement for production of large aperture optical and acousto-optical components, electro-optical modulators and switches, wavelength converters and many other optical elements [1 - 4]. Light beam propagation in nonlinear optical medium may result in the transformation of its spatial and spectral structure due to the self-action effects. In turn, the efficiency and magnitude of distortions appearing during light propagation in the medium depend on the local concentrations of impurity atoms, on the presence of defects and on small variations of the crystal structure itself. In some known electrooptic crystals like lithium niobate ($LiNbO_3$), lithium tantalate ($LiTaO_3$) and strontium – barium niobate ($Sr_xBa_{1-x}Nb_2O_6$) the spatial self-action of light is caused by the photorefractive effect and may be observed at the light powers of a microwatt range that is very useful for the experimental investigations [5 - 7]. The high optical nonlinearity results in the slow speed of the medium nonlinear response that also brings the advantages to the distortion accumulation study because it requires using simple and not too expensive equipment.

The most interesting among the spatial conversion effects of optical beams is the light localization in the form of spatial solitons or light field parts which demonstrate diffraction-free propagation in nonlinear optical media. The spatial soliton effects make possible design of all-optical photonic elements and circuits with light controlled by light. Another result of similar effects relates to generation inside the nonlinear medium optical waveguides and regular diffractive elements which also can be used for the light control needs. It should be noted that the boundaries of the nonlinear medium or interfaces between various materials may sufficiently vary the self-action characteristics of light fields. For example, generation of regular light patterns due to the self-action effect becomes possible in nonlinear optical cavities where the optical feedback comes into play. Similar effects have been observed in photorefractive crystals and other nonlinear media with both, self-focusing and self-defocusing nonlinear response [8 - 10].

In this work, we study the spatial distributions of nonlinear optical properties within $LiNbO_3$ bulk sample testing the distortions of light beam profiles in conditions of normal room temperature, the photorefractive effect influence and at the contribution of thermo-optic effect at different elevated temperatures and light wavelength of 532 nm.

2. Experimental setup and conditions

The schematic of the experimental setup used to study the uniformity of linear and nonlinear properties of the $LiNbO_3$ sample over its cross-section is shown in Figure 1. A continuous-wave YAG:Nd^{3+} laser (1) with radiation wavelength of 532 nm and output power up to 50 mW is used as the light source in our experiments. The light polarization corresponds to the extraordinary wave of a crystal. The light beam is focused onto the sample input surface by a lens (2) with focal length of 40 mm. It provides the diameter of light spot at the input surface about 13 μm for the half-maximum intensity level. The dimensions of $LiNbO_3$ sample (3) are 4×10×4 mm^3 along X, Y and Z axes. The sample is allocated on the Peltier element (4) which provides the crystal heating up to 100°C. Its temperature is measured with a thermocouple or a non-contact infrared thermometer. Light beam propagates within the crystal along its Y-axis running the distance ~10 mm. The spherical lens (5) images light pictures at the input or output surfaces of the crystal sample onto the sensor plane of a laser beam analyzer (6). The corresponding light fields at these surfaces are studied visually and their images are saved and analyzed with a computer.

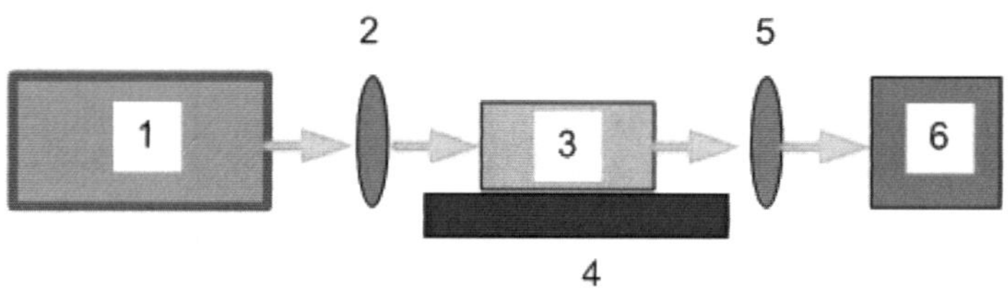

Figure 1. Schematic of the experimental setup: 1 – laser; 2, 5 – lenses; 3 – $LiNbO_3$ crystal; 4 – Peltier heater; 6 – laser beam analyzer.

We test in experiments distortions of focused light beam probing different regions of a crystal bulk. The schematic of the incident beam positions (white circles) over the sample entrance surface is shown in Figure 2. To precisely move the sample with respect to the light beam, the three-coordinate positioning stage is used. Light beam freely diffracts while propagation through the sample in the linear regime, e.g. at very low light intensity and short time periods after light switching on. At the first set of experiments we study the light beam patterns at the exit surface of the sample that informs on the uniformity of crystal sample optical properties together with a quality of optical polishing of its input and output surfaces. The spatial inhomogeneity of nonlinear optical properties of the sample is studied due to comparison of light beam images at its output surface in the initial moment after light switching on and after development of beam distortions because of the spatial self-action effects. The different mechanisms of nonlinear optical response may contribute to the beam self-action in $LiNbO_3$. These are the photorefractive effect, the light absorption and the thermo-optic effect which includes the pyroelectric effect and the sample thermal expansion. The thermo-optic contribution may vary by the sample heating and for $LiNbO_3$ such a heating can compensate the photorefractive beam distortions [11, 12]. We use the sample heating procedure to check the possible compensation of light beam divergence as at low as at higher light intensity.

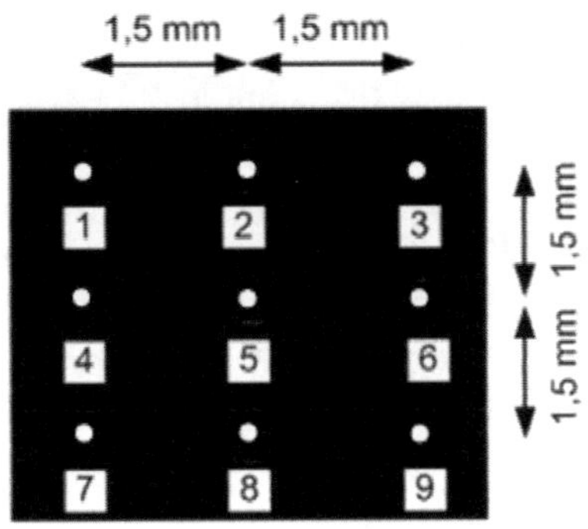

Figure 2. Schematic of probing beam locations over $LiNbO_3$ sample input surface.

3. Experimental results and discussions

The results of the linear diffraction of light beam at its propagation within the crystal sample, the distortions of the beam cross-section at the sample output surface due to the nonlinear diffraction caused by the photorefractive effect, and the total compensation of linear and nonlinear diffraction of the probing light beam are illustrated by Figure 3. The light power is 0.2 mW in this experiment. The first column shows the light pattern at the input surface of the sample and the intensity profile of this field along direction parallel to the crystal optical axis (Z axis). The linear diffraction broadens this beam about four times on the sample length and corresponding light pattern and intensity profile are shown in the second column. In some time the diffraction divergence of this beam becomes larger due to the uncontrolled photorefraction of the crystal. The crystal sample is nominally un-doped with special photorefractive impurities like iron or copper [5 - 7]. However, the photorefraction within it is notable due to the existence of intrinsic defect states and some dose of uncontrolled impurities. Thus, the space-charge electric field, directed primarily along the crystal polar axis, develops within the sample because of the light influence. The photorefractive nonlinearity of $LiNbO_3$ is of self-defocusing kind and this electric field induces a negative nonlinear lens within the illuminated area due to the linear electrooptic effect. It results in the anisotropic distortions of intensity distribution at the output surface of the sample with strong broadening of light pattern along the optical axis (Figure 3, column 3). This is a nonlinear diffraction broadening of a light beam.

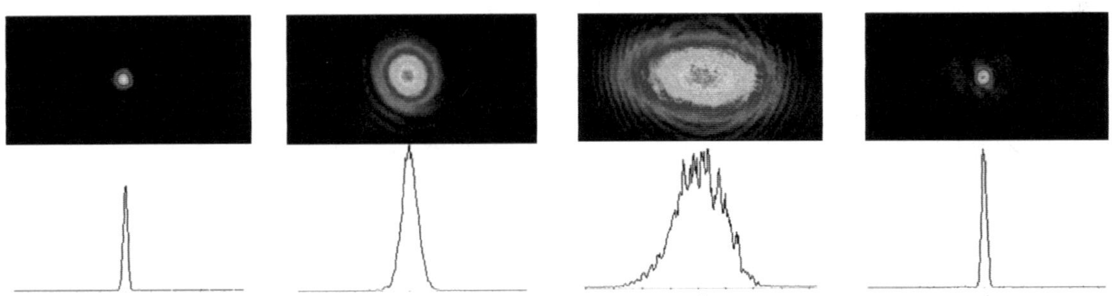

Figure 3. Light patterns (top row) and intensity profiles (bottom row) at the sample input surface (1), output surface in the linear regime (2), output surface in 10 minutes after light switching on (3) and at output surface and sample heating to 60°C (4).

Both, the linear and nonlinear diffraction may be suppressed in experiments by the crystal heating using Peltier element. Light pattern and its intensity profile (Figure 3, column 4) confirm the total compensation of the beam diffraction which is obtained in this experiment at the crystal temperature ~60°C. The effect of the beam diffraction compensation is generally caused by the pyroelectric field E_{pyro} arising within a ferroelectric crystal bulk due to its uniform heating accompanied with the spontaneous polarization $\boldsymbol{P}_S$ change [12, 13]:

$$E_{pyro} = -\frac{1}{\varepsilon_0 \varepsilon_r} \frac{dP_S}{dT} \Delta T,$$

where ε_0 and ε_r are the dielectric constants of vacuum and the crystal, and ΔT is a temperature change. This field decreases the refractive indices of the whole crystal due to the linear electro-optic effect with the decreasing value $\Delta n = 0.5 n^3 r_{eff} E_{pyro}$. Here n is average refractive index, and r_{eff} is effective electro-optic constant. However, the pyroelectric field E_{pyro} becomes screened in the illuminated region of the crystal due to the redistribution of charge carriers existing within illuminated bulk because of the photoexcitation. Thus, the interplay of the photorefractive and the pyroelectric effects may result at some conditions in the formation of two-dimensional bright spatial soliton in the crystal bulk. The inhomogeneity of light beam distortions is experimentally studied in further our experiments in conditions of low and higher light intensity that can change the results of interplay of photorefractive and pyroelectric effects.

3.1. Light beam distortions at low optical intensity

As the result of $LiNbO_3$ sample testing using its probing with a low intensity laser beam, we observe a map of light images at the output surface of the crystal sample which points to inhomogeneous distribution of internal optical properties over the crystal bulk (Figure 4). The optical power is 0.1 mW and the exposure time is ten minutes in this case. We probe the input surface in nine points with distances of 1.5 mm between them in directions parallel to X- and Z- axis (see Figure 2). These images correspond to photorefractive-distorted initially radial symmetric light patterns with dimensions observed at the light switching on moment (see Figure 3, column 2). Their comparison demonstrates slightly different nonlinear diffraction broadening of initial images corresponding to the linear diffraction of probe beam in different points of the input surface. It points to inhomogeneity of photorefractive properties over the sample bulk. Two main subjects may reason that. First, it is the inhomogeneous distribution of impurities and defects responsible for the crystal photorefractive nonlinearity. And the second subject is possible drift of some ions within the sample bulk along direction of crystal polar axis caused by some cycles of sample heating, cooling and illumination in different previous experiments [15].

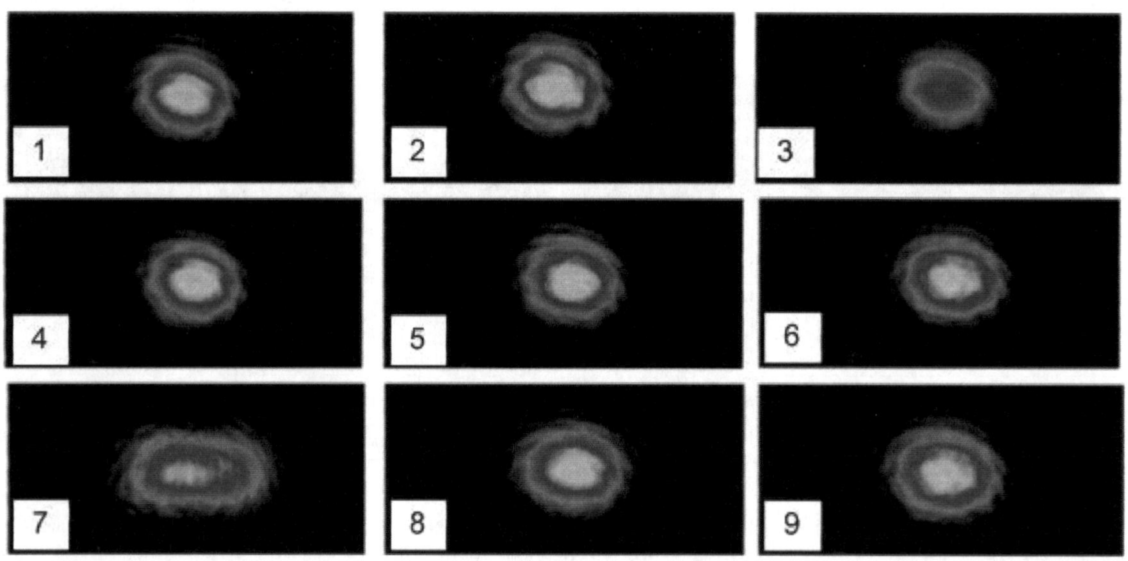

Figure 4. Light beam images at the output surface of the crystal sample in 10 minutes after light «switching on» at light power 0.1 mW.

The sample heating to the temperature 35°C results in the partial compensation of the light beam diffraction. It is illustrated by intensity profiles of light patterns at the output surface (Figure 5) after 10 minutes exposure at room temperature (PR) and after sample heating to 35°C (H). It should be noted that we observe only compensation of nonlinear diffraction of light beam in these conditions because the pyroelectric field value is not enough to compensate both, linear and nonlinear diffraction [13].

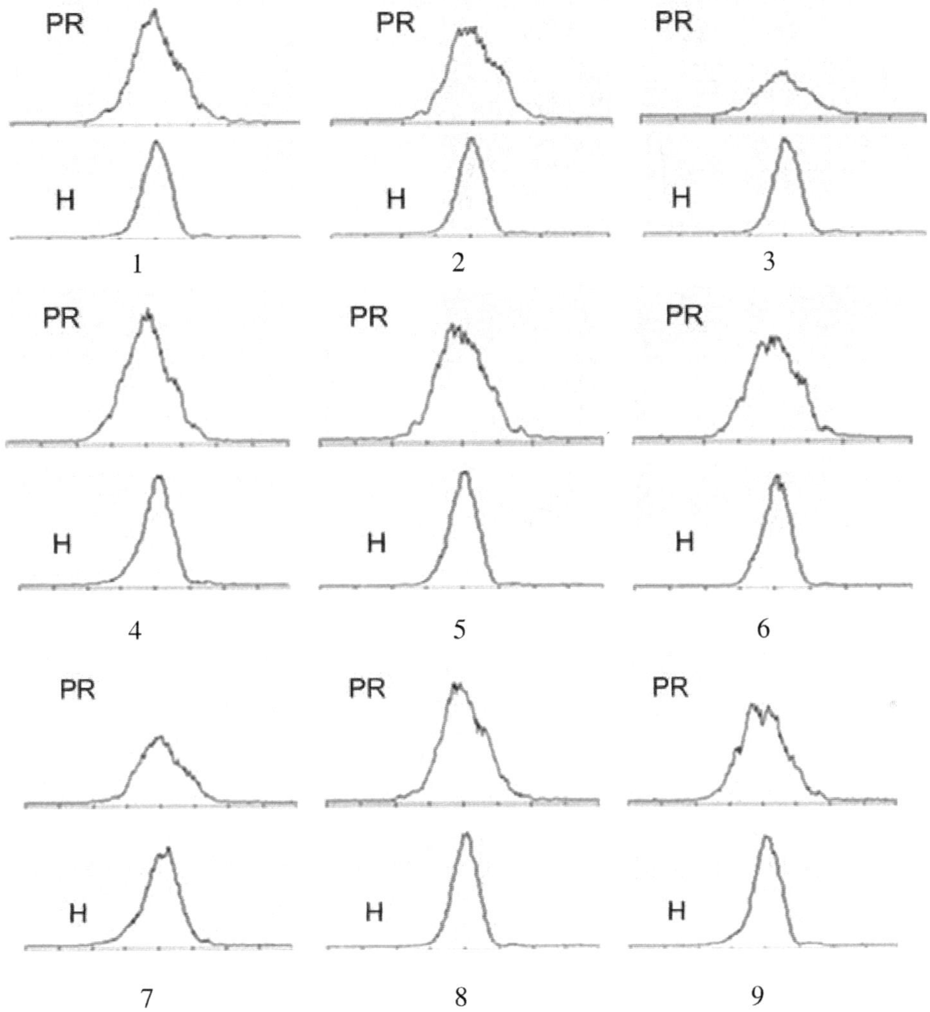

Figure 5. Light intensity profiles at the crystal output surface for optical power of 0.1 mW: profiles PR correspond to beam broadening due to the photorefractive effect; profiles H show partial compensation of light broadening in conditions of crystal heating to 35°C.

3.2. Light beam distortions at high optical intensity

We also use the same approach to test the influence of the thermo-optic effect on the light beam diffraction in conditions of the high optical intensity. Light power is 5.5 mW in this case and the crystal sample is heated to 60°C temperature. As in the case of low light intensity and sample temperature of 35°C, the total compensation of light beam divergence is not obtained at these conditions. The experimental results for the point (7) of laser beam location at the sample input surface (see Figure 2) are shown in Figure 6. The only nonlinear contribution to the beam divergence is compensated here at the crystal sample heating. Obviously, the pyroelectric field value is not enough in this case as well to compensate as linear as nonlinear diffraction of light beam [13, 14]. Similar results are observed for other locations of probing optical beam at the input surface.

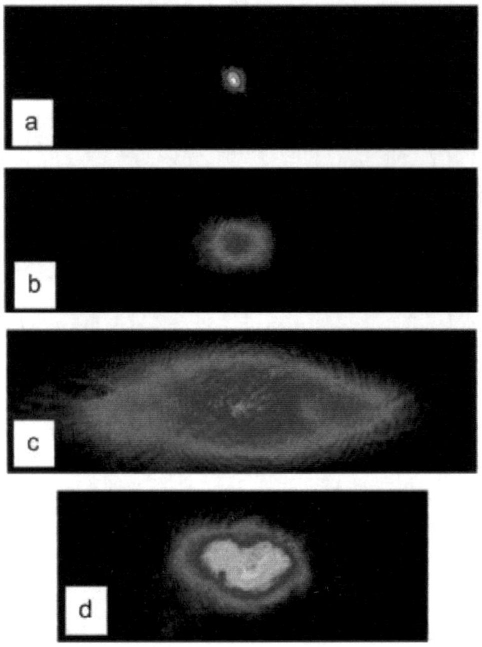

Figure 6. Light fields: at the input sample surface (a); at its output surface (b); at the output surface with photorefractive broadening (c) and after partial compensation of nonlinear diffraction broadening at crystal sample heating to 60°C (d).

3.3. Discussions

The comparison of light beam intensity profiles at the output sample surface in conditions of uncontrolled photorefractive effect and in regime of its compensation by the pyroelectric effect points to the possible improvement of light field distributions due to the crystal heating. Indeed, beam intensity profiles with photorefractive induced distortions demonstrate noisy envelopes even at near to perfect optical polishing of the sample surfaces. Ten degrees temperature increase transforms these noisy envelopes to smooth Gaussian-like shapes. Experiments demonstrate that the total compensation of narrow optical beam divergence including their linear and nonlinear parts requires taking into account the crystal quality, composition and content of different impurities.

Conclusion

The inhomogeneity of photorefractive optical nonlinearity can sufficiently distort the light beam profiles even in linear regime within lithium niobate samples used for operation in the green range of visible, especially at requirements of large apertures of crystal elements. These distortions can be suppressed at the controlled elevation of crystal sample temperature for some degrees. It is important that contribution of pyroelectric effect may result in the diffraction free propagation of light beams with diameters of micrometer range value in this crystal that may be used in creation of all-optical elements with light controlled by light.

Acknowledgments

This study was carried out with the financial support of Ministry of Education and Science of Russia (the task N 3.878.2014/K of the project part).

References

[1] Yeh Chai 1994 *Applied photonics* (Elsevier - Academic Press)
[2] Whitman R L and Korpel A 1969 *Applied Optics* **8** 1567
[3] Korpel A 1981 *Proc. IEEE* **69** 48
[4] Henderson D M and Abrams R L 1970 *Optics Communications* **5** 223
[5] Kip D 1998 *Appl. Phys. B* **67** 131
[6] Petrov M P, Stepanov S I and Khomenko A V 1991 *Photorefractive Crystals in Coherent Optical Systems* (Berlin: Springer-Verlag)
[7] Sturman B I and Fridkin V M 1992 *The photovoltaic and photorefractive effects in noncentrosymmetric materials* (Gordon and Breach Science Publishers)
[8] Carmon T, Buljan H and Segev M 2004 *Opt. Express* **12** 3481
[9] Smirnov E, Stepic M, Shandarov V and Kip D 2006 *Appl. Phys. B* **85** 135
[10] Zhu Nan, Guo Ru, Liu Simin, Liu Zhaohong and Song Tao 2006 *J. Opt. A: Pure Appl. Opt.* **8** 149
[11] Safioui J, Devaux F and Chauvet M 2009 *Optics express* **17** 205
[12] Popescu S T, Petris A and Vlad V I 2013 *J. Appl. Phys.* **113**, 213110
[13] Kostritskii S M, Chauvet M, Aillerie SM and Sevostyanov O G 2015 *Journal of Applied Spectroscopy* **82** 479
[14] Safioui J, Devaux F, Huy K P and Chauvet M 2013 *Optics Communications* **294** 294
[15] Kösters M, Sturman B, Werheit P, Haertle D, and Buse K 2009 *Nature Photonics* **3** 510

Journal of Physics: Conference Series **737** (2016) 012017 doi:10.1088/1742-6596/737/1/012017

Quantum efficiency of energy transfers in non-uniformly doped crystals of Er, Yb: $LiNbO_3$

N N Nalbantov, E V Stroganova and V V Galutskiy
Kuban State University, Krasnodar, Stavropolskaya str. 149, Russian Federation

Abstract. Spatial and temporal distribution of quantum efficiency of energy transfers in crystals of lithium niobate with non-uniform concentration profiles of Er^{3+} and Yb^{3+} ions has been studied for the cases of Yb^{3+}-Er^{3+} Foerster transfer and radiative decay of upper energy levels of erbium ions (green and red luminescence).

1. Introduction

Previous studies [1-4] of non-uniformly doped laser crystals, which are being grown with technique developed by authors of this work [5], shows that efficiency of energy transfers can be tuned and increased by varying concentrations of optical centers along the active medium. In this paper we report the impact of non-uniform doping on dynamics of some major energy transfers in crystals of Er, Yb: $LiNbO_3$ grown with concentration profiles of active ions which are depicted on Figure 1. Curves can be approximated with following equations:

$$C_{Yb}(z) = C_{1\%} \cdot (az^3 + bz^2 + cz + d) \tag{1}$$

$$C_{Er}(z) = C_{1\%} \cdot (a'z^3 + b'z^2 + c'z + d') \tag{2}$$

where: $C_{1\%}$ – concentration of dopant in $LiNbO_3$ lattice equal to 1 % at. ($C_{1\%}$=$1.88\cdot10^{20}$ cm^{-3});
a, a', b, b' c, c', d, d' – numerical coefficients (Table 1).

Table 1: Coefficients for equations of concentration profiles

a	*b*	*c*	*d*	*a'*	*b'*	*c'*	*d'*
0.0205	-0.1118	-0.0372	1.2089	-0.0098	0.0511	0.0221	0.0205

2. Foerster resonance energy transfer in system of Yb^{3+}-Er^{3+} ions

In energetic system of Yb^{3+} - Er^{3+} ions erbium generally acts as an acceptor of electron excitation energy, while ions of ytterbium provide the most of it as donors. Main channel of energy transition between these ions is a non-radiative FRET (Foerster resonance energy transfer) process between Yb: $^2F_{7/2}$ and Er: $^4I_{11/2}$ states, which is more effective for excitation of erbium ions than direct pumping at 980 nm wavelength. General scheme of energy levels of Yb^{3+} - Er^{3+} system is shown on Figure 2 where direct FRET process is marked with coefficient C_{25}. Optical pumping of crystal was provided with GaAs laser diode with 980 nm wavelength.

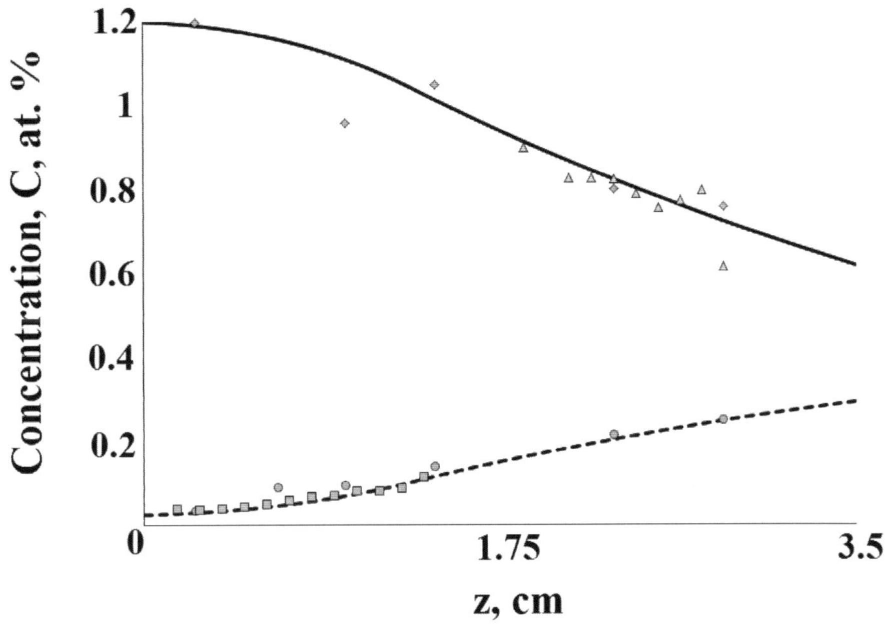

Figure 1 Concentration profiles of active ions (Yb^{3+} (solid line) and Er^{3+} (dashed line)) and their relations with absorption and emission processes: Er^{3+} absorption at 980 nm (circles), Yb^{3+} absorption at 980 nm (rhombs), Er^{3+} emission at 1.5 μm (triangles), Er^{3+} emission at 550 nm (squares).

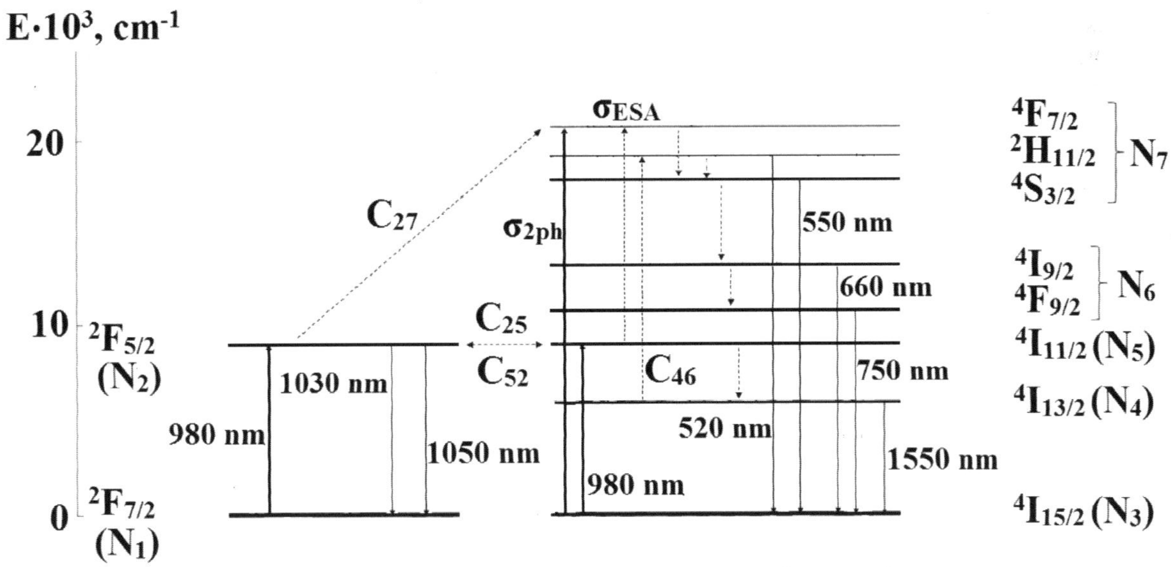

Figure 2 Scheme of energy levels in system of Er^{3+} and Yb^{3+} ions

Efficiency of FRET can be evaluated with application of mathematical formalism of rate equations. For system of Yb^{3+} - Er^{3+} ions it comes down to system of five equations [6], one for each of excited states (Figure 2 – N_2, N_4, N_5, N_6, N_7). After solving rate equations with Runge-Kutta method of fourth-order we expanded this mathematical model with spatial inhomogeneity of concentrations of dopants [7, 8], and resulting equation for efficiency of FRET process obtained in following form:

$$\eta_{Yb-Er}(z) = \frac{\frac{A_5}{A_5 + C_{52}N_1(z)}}{\sigma_{Yb_{abs}}N_1(z) + \sigma_{Er_{abs}}N_3(z)} \cdot \left(\frac{C_{25}N_3(z)\sigma_{Yb_{abs}}N_1(z)}{\frac{C_{25}N_3(z)A_5}{A_5 + C_{52}N_1(z)} + A_2} + \sigma_{Er_{abs}}N_3(z) \right), \tag{3}$$

where: A_i – decay probability constant of i-th excited state (Figure 2), s^{-1};

C_{ij} – coefficient of energy transfer rate between i-th and j-th states, $cm^3 \cdot s^{-1}$;

σ_{Ybabs}, σ_{Erabs} – absorption cross-sections of ytterbium and erbium ions at 980 nm wavelength, cm^2;

$N_i(z)$ – concentration of ions at i-th excited state, cm^{-3}.

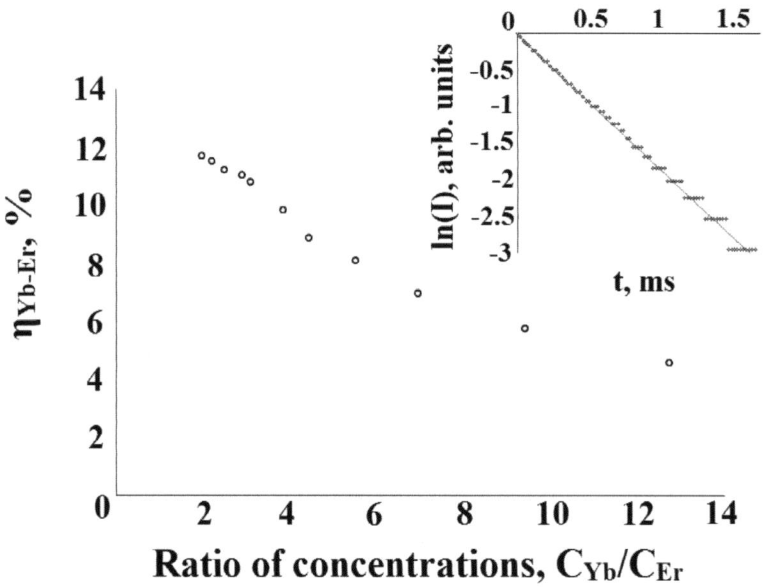

Figure 3 Efficiency of direct FRET in non-uniformly doped $LiNbO_3$:Yb, Er crystal and kinetics of luminescence of ytterbium ions at 970 nm (inset)

Figure 3 shows results experimental data of luminescence quenching of ytterbium ions at 970 nm wavelength and evaluated values of efficiency of FRET process in dependence of ratio of concentrations of active ions. Ytterbium ions were excited by YAG: Nd laser (1064 nm), measurements of quenching rate were conducted in several points along the crystal with different ratios of ytterbium and erbium concentrations, $C_{Yb}(z)/C_{Er}(z)$. It was obtained that the efficiency of FRET reaches higher values at the regions with low ratios of concentrations. Studies of luminescence of erbium ions at 1.5 μm wavelength showed that intensity of radiation measured along the non-uniformly doped crystal of $LiNbO_3$:Yb,Er correlates with the form of concentration profile of ytterbium ions (triangles on Figure 1). That proves leading role of FRET process in populating of excited states of erbium ions and shows correlation between efficiency of this energy transfer and intensity of luminescence at 1.5 μm wavelength.

3. Quantum efficiency of visible light luminescence

Solved rate equations also can be applied for evaluation of quantum efficiency of radiative transitions. For example, mathematical expressions for efficiency of green (η_g; 520 and 550 nm) and red (η_r; 660 and 750 nm) luminescence can be obtained as:

$$\eta_g(t,z) = \frac{A_{73}N_7(t,z)}{C_{27}N_2(t,z)N_5(t,z) + (\sigma_{ESA} + \sigma_{2ph})\varphi(t,z)N_3(t,z)} \quad (4)$$

$$\eta_r(t,z) = \frac{A_{63}N_6(t,z)}{C_{46}N_2(t,z)N_4(t,z) + (A_{76} + W_{76})N_7(t,z)} \quad (5)$$

where: A_{ij} – probability constant of radiative transition from i-th to j-th state (Figure 2), s^{-1};

$\varphi(t,z)$ – flux density of optical pumping, $cm^{-2}s^{-1}$;

σ_{2ph}, σ_{ESA} – cross-sections of two-photon absorption (2ph) and excited-state absorption (ESA), cm^2.

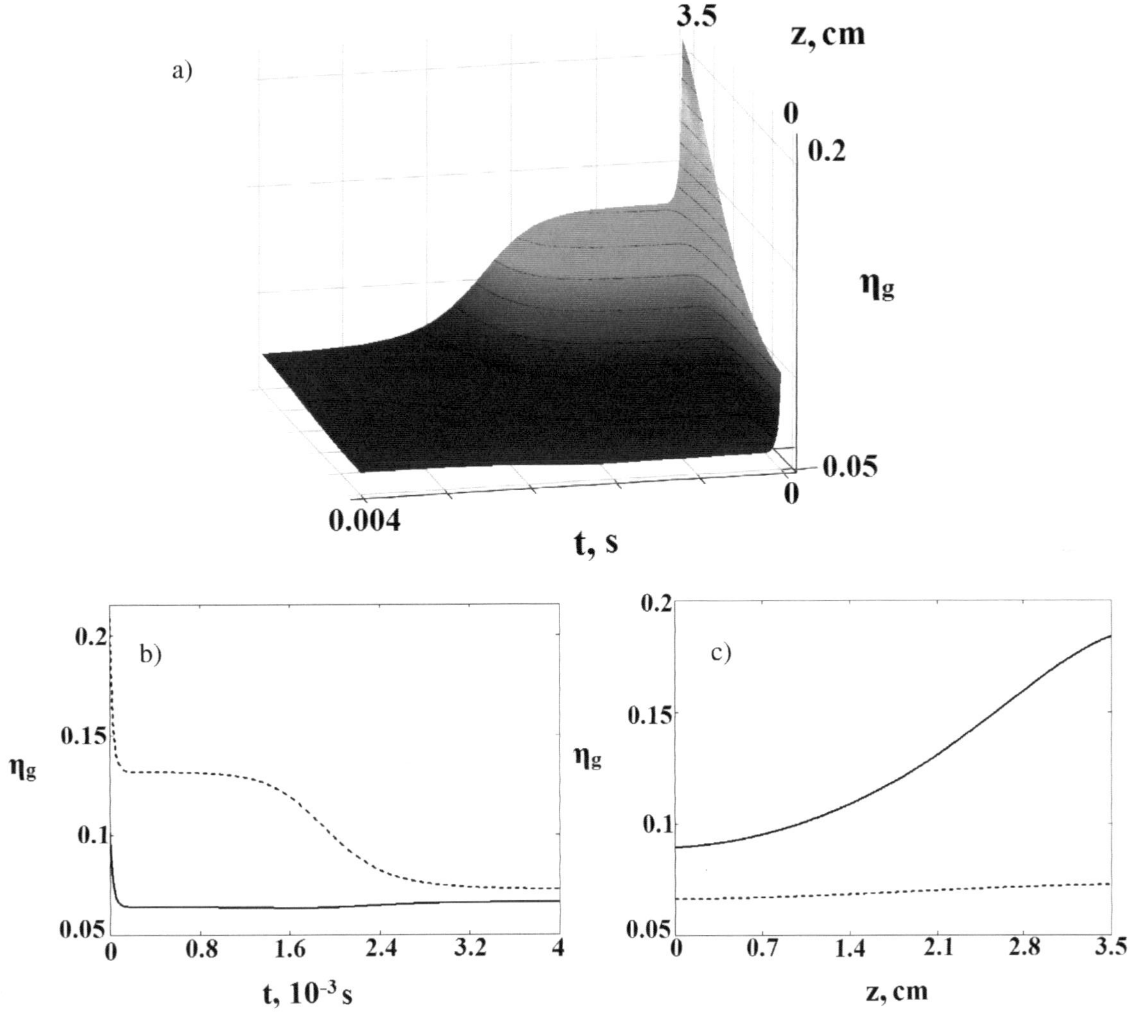

Figure 4 Quantum efficiency of green luminescence of erbium ions:
a) spatio-temporal dynamics, $\eta_g(t,z)$;
b) temporal dynamics, $\eta_g(t)$ (solid line – at z=0 cm, dashed line – at z=3.5 cm);
c) spatial dynamics $\eta_g(z)$ (solid line – $t=8\cdot10^{-6}$ c, dashed line – $t=4\cdot10^{-3}$ c)

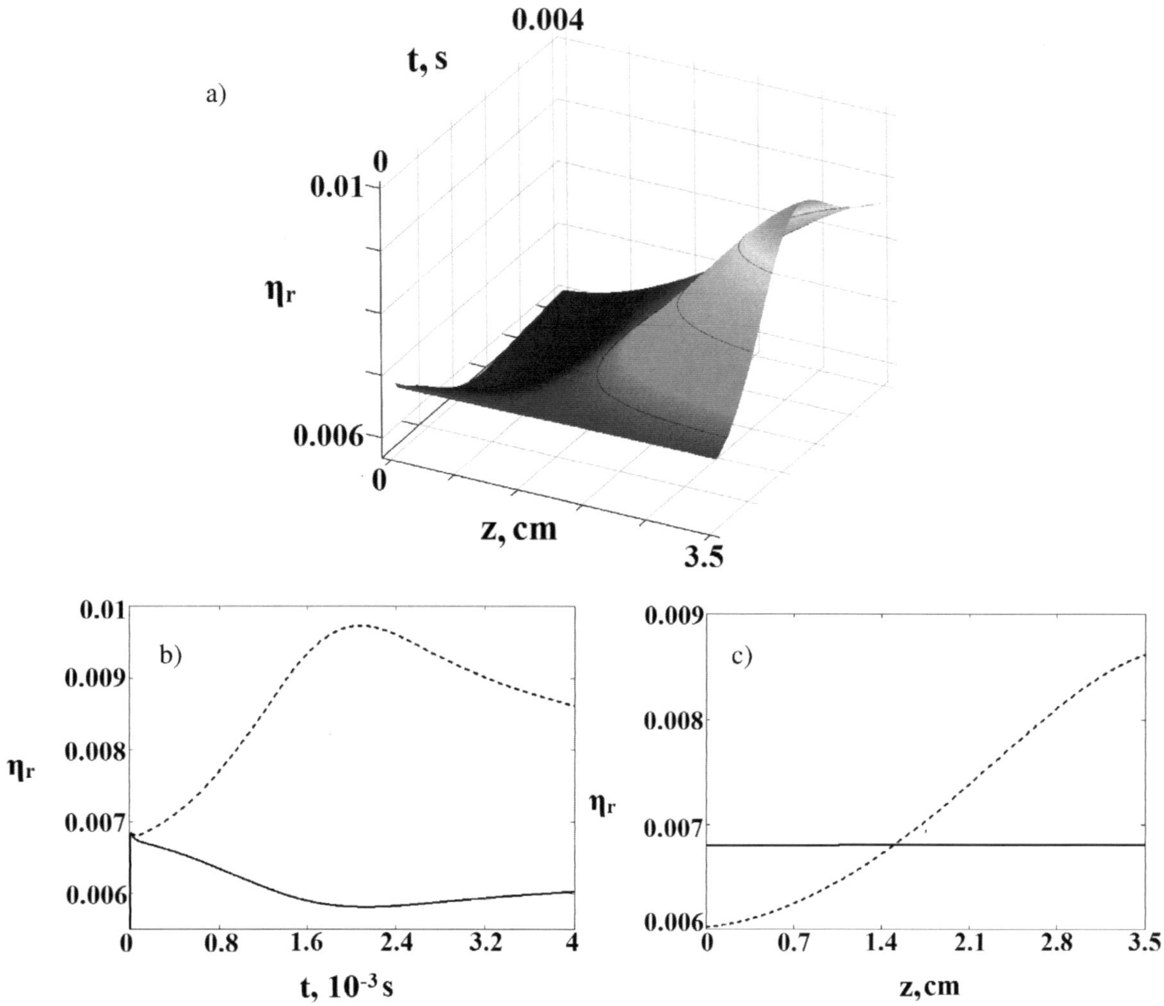

Figure 5 Quantum efficiency of red luminescence of erbium ions: a) spatio-temporal dynamics, $\eta_r(t, z)$; b) temporal dynamics, $\eta_r(t)$ (solid line – at z=0 cm, dashed line – at z=3.5 cm); c) spatial dynamics $\eta_r(z)$ (solid line – $t=8\cdot10^{-6}$ c, dashed line – $t=4\cdot10^{-3}$ c)

Equations (4-5) allow to evaluate ratio of rate of radiative decay of corresponding excited states in the volume of the crystal (numerator, $cm^{-3}s^{-1}$) to rate of population of these energy levels in the volume of the crystal (denominator, $cm^{-3}s^{-1}$). Results of evaluation for green and red luminescence are shown of Figures 4 and 5. Figure 4 (a) represents combined spatio-temporal dynamics of quantum efficiency of green luminescence in non-uniformly doped crystal. Temporal profiles $\eta_g(t)$ at both edges of the crystal (z=0 and z=3.5 cm) are shown on Figure 4 (b): from the start of optical pumping (t=0 s) until stationary mode of generating luminescence ($t=4\cdot10^{-3}$ s). Inflection points correlate with lifetimes of Yb: $^2F_{5/2}$ (Figure 2: N_2) and Er: $^4I_{11/2}$ (N_4) excited states, which are participating in transition of energy to emitting manifolds $^2H_{11/2}$ and $^4S_{3/2}$ (N_7). Figure 4 (c) depicts spatial profiles of quantum efficiency $\eta_g(z)$ in two moments of time: during transient state ($t=8\cdot10^{-6}$ s) and in stationary mode ($t=4\cdot10^{-3}$ s). In stationary mode distribution of efficiency is almost uniform along the crystal (changes from 6.6 to 7.3 %), while in transient state populations of short-lived upper levels increase noticeably with changes of erbium and ytterbium concentrations.

Manifolds $^4F_{9/2}$ and $^4I_{9/2}$ (N_6) demonstrate a very different dynamics of quantum efficiency of luminescence (Figure 5 (a)). The main reason behind that is a mechanism of population of these energy levels: under pumping with wavelength 980 nm it can't be populated directly, only by slow non-radiative decays of upper levels, $^2H_{11/2}$ and $^4S_{3/2}$ (N_7). Secondly, both $^4F_{9/2}$ and $^4I_{9/2}$ states has lifetimes in order of 10^{-6} s which is shorter than lifetimes of all other energy levels in the system by 2-3 orders of magnitude. In result, the main factor of shaping quantum efficiency of red luminescence is not a rate of population but a concentration of erbium ions and its rate to ytterbium ions concentration. Figures 5 (b) and (c) illustrate this thesis. Efficiency of red luminescence at the low-doped end of crystal (z=0 cm) reacts to lifetimes of feeding energy levels, including FRET-related excited states (Figure 5 (b), solid line), but at high-doped end of crystal (z=3.5 cm; dashed line) up-conversion from $^4I_{13/2}$ state becomes a dominant factor in population of "red" levels. It leads to a formation of a significant peak close to lifetime $^4I_{13/2}$ ($t=2.7\cdot10^{-3}$ s). Spatial dynamics on Figure 5 (c) shows that distribution of efficiency is less uniform in stationary mode than in transient state. This peculiarity is also can be explained with strong dependence of efficiency to concentration of erbium ions: highest rate of decay leads to a difference of population of 7-8 orders of magnitude with adjacent excited states by the time of the start of stationary mode, so quantum efficiency of luminescence directly correlates with concentration profile of erbium ions.

4. Conclusion

We have studied relation of efficiency of FRET process between Yb^{3+} and Er^{3+} ions to ratio of their concentrations in non-uniformly doped crystal of lithium niobate. It was obtained that efficiency of luminescence at 1.5 μm wavelength also depends on this ratio and correlates with concentration profile of ytterbium ions.

Spatio-temporal dynamics of quantum efficiency of green and red luminescence of erbium ions were obtained. It was shown that these transitions demonstrate a highly different dynamics because of major distinctions in lifetimes and populating mechanisms, and it leads to different correlations with concentration profiles of erbium and ytterbium ions.

Acknowledgments

This work was supported by the RFBR-South (16-42-230214 (16/110)) and the State Assignment 2014/75 R&D №1291 (14/200-т).

References

[1] Stroganova E V, Galutskiy V V, Tkachev D S, Tsema A A, Nalbantov N N and Yakovenko N A 2014 *Opt. Spectrosc* **117** 1012-7

[2] Nalbantov N N, Stroganova E V and Galutskiy V V 2013 *Proc. III All-Russian Conf. on Optics (Saint-Petersburg)* (Saint-Petersburg: ITMO Press) p 168-170

[3] Nalbantov N N, Stroganova E V and Galutskiy V V 2014 *Proc. III All-Russian Conf. on Photonics and Information Optics (Moscow)* (Moscow: MEPhI Press) p 97-98

[4] Nalbantov N N, Stroganova E V and Galutskiy V V 2015 *Proc. XXI All-Russian Conf. on Optics and Spectroscopy of Condensed Matter (Krasnodar)* (Krasnodar: KubSU Press) p 287-90

[5] Galutskiy V V, Vatlina M I and Stroganova E V 2009 *J. Cryst. Growth* **311** 1190–4

[6] Cantelar E and Cusso F 2003 *J. Lumin.* **102-103** 525–31

[7] Nalbantov N N, Stroganova E V and Galutskiy V V 2016 *Proc. V Int. Conf. on Photonics and Information Optics (Moscow)* (Moscow: MEPhI Press) p 39-40

[8] Cantelar E, Munoz J A, Sanz-Garcia J A and Cusso F 1998 *J. Phys.: Condens. Matter.* **10** 8893-903

Changes in the optical absorption induced in the $Bi_{12}TiO_{20}$:Al crystal by exposition to short- and long-wavelength radiation

V G Dyu[1], M G Kisteneva[1], S M Shandarov [1], S V Smirnov[1], A S Akrestina[1] and Yu F Kargin[2]

[1]State University of Control System and Radioelectronics, 40, Lenin Avenue, Tomsk 634050, Russia
[2] A.A. Baikov Institute of Metallurgy and Materials Science RAS, Moscow 119991, Russia

E-mail: m-kisteneva@mail.ru

Abstract. The impurity absorption model is used to describe the experimentally observed reversible changes in the optical absorption spectrum induced in the $Bi_{12}TiO_{20}$:Al crystal by subsequent exposition to short-wavelength radiation with λ_i = 532 nm and longer-wavelength radiation with λ_n changing from 588 to 1064 nm. The model takes into account the photoinduced transitions between two metastable states of a deep defect center leading to the change of its position in the crystal lattice under conditions of strong lattice relaxation.

1. Introduction

$Bi_{12}TiO_{20}$ (BTO) crystals belonging to the sillenite family are promising for application in dynamic holography based on counter interaction of light waves in reflective geometry [1 – 4]. This is due to high values of the space charge field of reflective holograms formed in the BTO crystals due to considerable concentration of photoactive defect centers inherent in the crystals through which the photoinduced redistribution of the electric charge occurs [5, 6]. The special feature of the BTO crystals is the strong photochromic effect also observed at room temperatures; it consists in reversible changes of the optical absorption induced by exposition to visible radiation [7, 8]. The main reason for the increase in the optical absorption in the BTO crystals is considered to be the capture of photoexcited charge carriers (electrons) from deep donor centers to less deep traps characterized by the photoionization cross section larger than in the initial state [9 – 11]. Such process in which the nonequilibrium charge carriers participate can lead to an increase in the photorefractive sensitivity of the BTO crystals in the near-IR range of the spectrum [12, 13]. Another reason for the increased absorption in the BTO crystals that is not associated with electron photoexcitation to the conduction band can be generation of defects for which intracenter transitions occur as a result of exposition to light [11]. In [14] the intracenter character of transitions in the BTO:Al crystal in the wavelength range from 500 to 900 nm characterized by a resonant dependence of the photo- and thermoinduced changes in the light absorption was suggested.

Detailed experimental investigations performed in [15, 16] demonstrated that exposition to short-wavelength radiation with the wavelength λ_i = 532 nm caused an increase in the optical absorption in the BTO:Al crystal and resulted in the spectrum whose shape in the wavelength range 470–1000 nm was independent of the initial crystal state. The subsequent exposition to a longer-wavelength radiation with λ_n changing from 588 to 1064 nm led to the enhanced transmittance of the BTO:Al crystal in the entire examined spectral range. In this case it was established that the enhanced transmittance fixed in the saturation region depended nonmonotonically on the wavelength and reached a maximum at $\lambda_n \sim 670$ nm. It should be noted that the observed changes in the optical absorption spectra for the BTO:Al crystal during enhanced-diminished transmittance cycles were reversible and well reproduced experimentally.

In the present work, to describe the experimentally observed reversible changes in the optical absorption spectrum induced in the BTO:Al crystal by subsequent exposition to short- and long-wavelength radiation, the model is considered that takes into account strong relaxation of the lattice during photoinduced transitions between two metastable states of a deep defect center. The redistribution of center concentrations over the states that differ by the photoabsorption cross sections arising during such transitions leads to changes in the optical absorption additional to changes caused by the photoinduced electron redistribution over the deep donors via the conduction band and by generation and decay of defects for which intracenter transitions occured.

2. Theoretical model

As is well known [17 – 19], localized optically induced changes in the electronic configuration of the deep defect centers of some types in semiconductor and dielectric materials are accompanied by considerable reconstruction of such centers manifested via changes in their positions in the lattice. This effect called the photoinduced lattice relaxation [17] leads to the possibility of photocontrollable change of the properties of crystals comprising such centers and of their application as recording media [18]. The configuration coordinate diagram [17, 19] best described the change in the geometry of the defect caused by the change in its electronic state.

Let us consider the deep defect center that can be in two metastable states A and C with the potential energy minima shifted along the configuration coordinate Q (Fig. 1). The parabolas $E_{d1}(Q)$ and $E_{d2}(Q)$, described by the well-known relations [20]

$$E_{d1}(Q) = a_1(Q - Q_1)^2, \tag{1}$$

$$E_{d2}(Q) = E_{02} + b_2(Q - Q_2)^2, \tag{2}$$

characterize the energies of the ground (1) and excited (2) states of the centers with the parameters a_1 and b_2 of the parabolic functions that differ from each other taking into account the thermal excitation energy E_{02}.

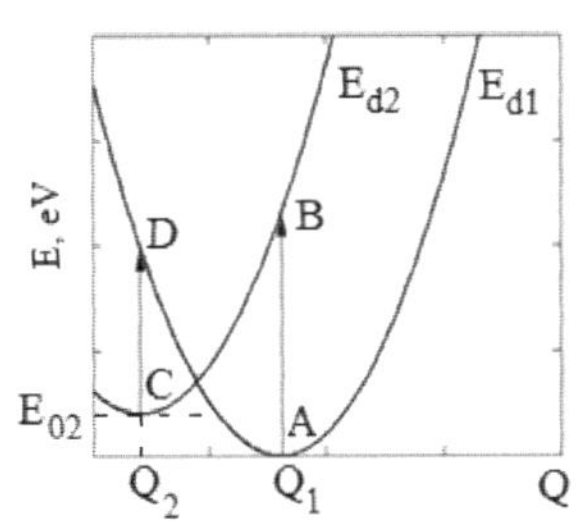

Figure 1. Configuration coordinate diagram of a defect center.

For vertical transitions that can be induced only by light radiation, the configuration coordinates Q_{v1} and Q_{v2} corresponding to the quantum energy $\hbar\omega_{12}$ causing optical transitions from the ground to the excited state

$$\hbar\omega_{12} = E_{d2}(Q_{v1}) - E_{d1}(Q_{v1}), \tag{3}$$

and to the energy $\hbar\omega_{21}$ of optical transitions back to the ground state

$$\hbar\omega_{21} = E_{d1}(Q_{v2}) - E_{d2}(Q_{v2}), \tag{4}$$

can be obtained from Eqs. (1)–(4) in the following form:

$$Q_{v1}(\hbar\omega) = \frac{a_1Q_1 - b_2Q_2}{a_1 - b_2} - \sqrt{\left(\frac{a_1Q_1 - b_2Q_2}{a_1 - b_2}\right)^2 - \frac{E_{02} - \hbar\omega + a_1Q_1 - b_2Q_2}{a_1 - b_2}}, \quad (5)$$

$$Q_{v2}(\hbar\omega) = \frac{a_1Q_1 - b_2Q_2}{a_1 - b_2} - \sqrt{\left(\frac{a_1Q_1 - b_2Q_2}{a_1 - b_2}\right)^2 - \frac{E_{02} + \hbar\omega + a_1Q_1 - b_2Q_2}{a_1 - b_2}}. \quad (6)$$

Assuming that the energy distributions for the defect center concentrations in the ground state, $N_{d1}(E_{d1})$, and in the excited state, $N_{d2}(E_{d2})$, are described by the Boltzman distributions and considering relations (1) and (2), we find their dependences on the corresponding configuration coordinates

$$N_{d1}(Q) = N_{d10}\exp\left(-\frac{a_1(Q-Q_1)^2}{k_BT}\right), \quad (7)$$

$$N_{d2}(Q) = N_{d20}\exp\left(-\frac{b_2(Q-Q_2)^2}{k_BT}\right), \quad (8)$$

where N_{d10} and N_{d20} are particle concentrations in the ground state for $Q = Q_1$ and in the excited metastable state for $Q = Q_2$, respectively; k_B is the Boltzmann constant, and T is the absolute temperature.

Taking into account Eqs. (7) and (8), the expressions determining the spectral dependence of the absorption coefficient $k_{d1}(\omega)$ caused by the optical transitions from the ground state to the excited state, and of the absorption coefficient $k_{d2}(\omega)$ caused by the reverse transitions can be written in the form

$$k_{d1}(\omega) = \hbar\omega S_{d1}N_{d10}\exp\left(-\frac{a_1(Q_{v1}-Q_1)^2}{k_BT}\right), \quad (9)$$

$$k_{d2}(\omega) = \hbar\omega S_{d2}N_{d20}\exp\left(\frac{b_1(Q_{v2}-Q_2)^2}{k_BT}\right), \quad (10)$$

where S_{d1} and S_{d1} are the photoabsorption cross sections for the ground state and excited state, respectively.

Thus, the optical absorption in the crystal caused by the examined deep defect center is spectrally dependent, and its absorption coefficient $k_d(\omega)$ is determined by the sum

$$k_d(\omega) = k_{d1}(\omega) + k_{d2}(\omega), \quad (11)$$

depending, according to Eqs. (9), (10) and (5), (6), on the concentrations of the centers in the ground and excited states, the thermal excitation energy E_{02}, differences between their photoabsorption cross sections and the energy parameters a_1 and b_2, and the shift between the ground and excited states on the configuration coordinate $Q_1 - Q_2$. The optical transitions induced by exposition of the crystal to radiation with a certain frequency ω_n lead to redistribution of the center concentration between these states. An analysis demonstrates that the equilibrium condition defined by equality of the numbers of direct and reverse transitions is established when the condition

$$k_{d1}(\omega_n) = k_{d2}(\omega_n) \quad (12)$$

is satisfied.

3. Approximation of the spectral dependences of the optical absorption

The spectral dependences of the optical absorption in the BTO:Al crystal observed experimentally in [15] can be approximated by the impurity absorption model that takes into account the contribution of electrons photoexcited from deep donor centers to the conduction band and the intracenter transitions [11] supplemented by the model of the defect center interacting with the lattice that is based on the application of the configuration coordinate diagram considered above. Below we restrict ourselves by an analysis of the approximation of three experimental dependences from [15] shown by open circles in Fig. 2.

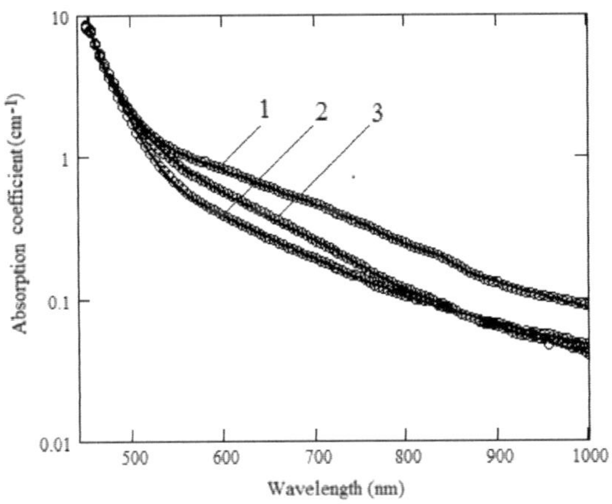

Figure 2. Experimental and calculated spectral dependences of the optical absorption coefficient for the $Bi_{12}TiO_{20}$:Al crystal exposed first to laser radiation with a wavelength of 532 nm (1) and then to radiation with wavelengths of 663 (2) and 780 nm (3), respectively, until saturation. The open circles are for the experimental data, and the solid curves are for the calculated dependences

Dependence 1 in this figure shows the absorption spectrum for the BTO:Al crystal exposed in the initial stage of each experiment to laser radiation with the wavelength λ_i =532 nm and the intensity I_i = 15.4 mW/cm^2 for 1930 s. This dependence was reproduced well irrespective of the initial state of the BTO:Al sample [15]. Spectral dependences 2 and 3 shown in Fig. 2 illustrate the subsequent exposition of the crystal to continuous laser radiation with the wavelength λ_n = 663 nm (I_n = 270 mW/cm^2 and the exposure time until saturation t_n = 960 s) and λ_n = 780 nm (I_n = 4 mW/cm^2 and t_n = 6420 s), respectively, in the second stage of the experiments leading to the enhanced transmittance of the crystal.

The approximation of the experimental spectral dependences of the optical absorption in the BTO:Al crystal yielded the following values of the model parameters for the absorption coefficient caused by deep defect centers and described by Eqs. (9) and (10): $a_1 = 5.426$ eV, $b_2 = 6.480$ eV, $E_{02} = 0.103$ eV; $Q_1 = 0.8$, $Q_2 = 0.2$; S_{d1} = 3.94·10^{-5} cm^2·eV^{-1} and S_{d2} = 1.0·10^{-4} cm^2·eV^{-1}. The parameters N_{d10} and N_{d20}, depending on the illuminating radiation wavelength ensuring the fulfillment of balance condition (12) for transitions between the deep center states, are presented in Table 1 for laser radiation with three wavelengths used in experiments by Dyu et al. [15] and for nonmonochromatic radiation of a semiconductor light-emitting diode with the average wavelength λ_n = 588 nm. For nonmonochromatic radiation, the experimental technique was analogous to that used in [15]. In the first stage, the BTO:Al sample was irradiated by laser radiation with the wavelength λ_i = 532 nm that caused the diminished transmittance of the crystal; the subsequent sample exposition to incoherent radiation with λ_n = 588 nm and intensity I_n = 0.3 mW/cm^2 in the second stage caused the enhanced transmittance of the crystal with saturation at t_n = 2760 s.

Table 1. Parameters of the deep defect center model depending on the irradiated wavelength

λ, nm	N_{d10}, m^{-3}	N_{d20}, m^{-3}
532	$4.68 \cdot 10^{23}$	$7.33 \cdot 10^{23}$
588	$8.55 \cdot 10^{23}$	$3.45 \cdot 10^{23}$
663	$10.7 \cdot 10^{23}$	$1.33 \cdot 10^{23}$
780	$11.5 \cdot 10^{24}$	$0.50 \cdot 10^{23}$

To approximate satisfactorily the spectral dependences, three intracenter transitions with the Gaussian spectral characteristics and maxima at quantum energies of 1.49, 1.62, and 1.77 eV must be taken into account together with the electron photoexcitation to the conduction band from four donor centers with average ionization energies of 1.08, 1.56, 1.96, and 2.75 eV. The energy parameters of the defect centers that characterized these processes in the BTO:Al crystal were close to those used in [11]. The approximated spectral dependences shown in Fig. 2 by solid curves demonstrate good agreement with the available xperimental data.

An analysis of the model parameters used to approximate the experimental spectral dependences of the optical absorption in the BTO:Al crystal demonstrated the following. The increased optical absorption induced in the BTO:Al crystal by exposition to short-wavelength radiation with the wavelength $\lambda_i = 532$ nm was due to the increased contribution of the intracenter transitions to the radiation absorption, photoinduced increase in the degree of occupation by electrons of the centers with ionization energies of 1.08 and 1.96 eV, and photoexcitation of deep defect centers and their transition from the ground state to the excited metastable state (see Fig. 1). The subsequent exposition to long-wavelength radiation led to the reduced contribution of intracenter transitions to the absorption during occupation by electrons of the centers with ionization energies of 1.08 and 1.96 eV. In this case, the degree of occupation by electrons of the center with ionization energy of 1.56 eV increased, and the concentration of the deep defect centers in the excited state decreased due to their photoinduced transitions to the ground state.

Thus, the observed nonmonotonic dependence of the enhanced transmittance for the $Bi_{12}TiO_{20}$:Al crystal, having the diminished transmittance, on the wavelength λ_n of radiation leading to the enhanced transmittance can be caused not only by quantitative changes of contributions to the impurity absorption described in [11], but also by photoinduced transitions of the deep defect center between the ground and excited states, accompanied by strong electron-phonon interaction with distortion of the crystal lattice. For energy of the radiation quantum with $\lambda_i = 532$ nm corresponding to the vertical transition from the state A (Fig. 1) with maximum concentration of the centers to the ground state, they are effectively translated to the excited state (AB transition), leading to the diminished transmittance of the crystal. The enhanced transmittance of the crystal will be maximum for radiation with energy of the quantum at which the CD transition occurs (in the examined case, at the wavelength $\lambda_n \approx 670$ nm).

4. Conclusions

Thus, the reversible changes in the optical absorption spectrum for the BTO:Al crystal observed experimentally in [15] and induced by the subsequent exposition to short- and long-wavelength radiation are well described by the impurity absorption model, considering the photoexcitation of electrons from the deep centers to the conduction band and the intracenter transitions [11], complemented by the model of deep defect centers with two metastable states interacting with the lattice. The photoinduced transitions between these states are accompanied by changes in the configuration coordinates. The observed photoinduced changes in the optical absorption in the BTO:Al crystal can be caused not only by the photoinduced electron redistribution over the deep donors via the conduction band and by generation and decay of defects for which the intracenter transitions occur, but also by redistribution of the deep center concentration between two metastable states with different photoabsorption cross sections.

Acknowledgments
This work was supported in part by the State Assignment of the Ministry of Education and Science of the Russian Federation for 2015–2016 and R&D under the Project of State Assignment No. 3.878.2014/IK.

References
[1] Kukhtarev N, Chen B S, Venkateswarlu P, Salamo G and Klein M 1993 *Optics Communications* **104** pp 23–28
[2] Kamshilin A A, Romashko R V, Kulchin Y N 2009 *Journal of Applied Physics* **105** p 031101
[3] Kolegov A A, Shandarov S M, Simonova G V, Kabanova L A, Burimov N I, Shmakov S S, Bykov V I and Kargin Y F 2011 *Quantum Electronics* **41** pp 847–852
[4] Shandarov S M, Shmakov S S, Zuev P V, Burimov N I, Kargin Y F, Shepelevich V V, Ropot P.I and Gudelev V.G 2013 *Journal of Optical Technology* **80** pp 409–414
[5] Shandarov S M, Burimov N I, Kul'chin Y N, Romashko R V, Tolstik A L and Shepelevich V V 2008 *Quantum Electronics* **38** pp 1059–1069
[6] Shandarov S M, Shmakov S S, Burimov N I, Syuvaeva O S, Kargin Y F and Petrov V M 2012 *JETP Letters* **95** pp 618–621
[7] Mersch F, Buse K, Sauf W, Hesse H and Krätzig E 1993 *Physica Status Solidi A* **140** pp 273–281
[8] Matusevich A, Tolstik A, Kisteneva M, Shandarov S, Matusevich V, Kiessling A and Kowarschik R 2008 *Applied Physics B* **92** pp 219–224
[9] Kobozev O V, Shandarov S M, Kamshilin A A and Prokofiev V V 1999 *Journal of Optics A* **1** pp 442–447
[10] Marinova V, Lin S H, Sainov V, Gospodinov M and Hsu K Y 2003 *Journal of Optics A* **5** pp S500–S506
[11] Kisteneva M G, Khudyakova E S, Shandarov S M, Akrestina A S, Dyu V G and Kargin Y F 2015 *Quantum Electronics* **45** pp 685–690
[12] Odoulov S G, Shcherbin K V and Shumeljuk A N 1994 *Journal of the Optical Society of America B* **11** pp 1780–1785
[13] Dos Santos P V, Frejlich J and Carvalho J F 2005 *Journal of Applied Physics B* **81** pp 651–655
[14] Kisteneva M G, Shandarov S M, Akrestina A S, Popugaeva V V and Smirnov S V 2010 *Izv. Vyssh. Uchebn. Zaved. Fiz.* **9/3** pp 145–146
[15] Dyu V G, Kisteneva M G, Shandarov S M, Khudyakova E S, Smirnov S V and Kargin Y F 2015 *Physics Procedia* **73** pp 131–135
[16] Dyu V G, Khudyakova E S and Kisteneva M G 2015 *XIII Vserossiiskii molodezhnyi Samarskii konkurs-konferentsiya nauchnykh rabot po optike i lasernoi fizike: sbornic konkursnykh dokladov* (Moscow: P.N. Lebedev Physical Institute of the RAS) pp 66–71
[17] Shinozuka Y 1993 *Japanese Journal of Applied Physics* **33** pp 4560–4570
[18] Ryskin A I and Fedorov P P 1997 *Physics of the Solid State* **39** pp 943–947
[19] Alkauskas A, McCluskey M D and Van de Walle C G 2016 *Journal of Applied Physics* **119** p 181101
[20] Ridley B K 1982 *Quantum Processes in Semiconductors* (Oxford University Press) p 304

Journal of Physics: Conference Series **737** (2016) 012019 doi:10.1088/1742-6596/737/1/012019

Speckle-noise suppression using electro-optical cell with helix-free ferroelectric LC

A L Andreev[1], T B Andreeva[1], I N Kompanets[1, 2], N V Zalyapin[2], R S Starikov[2]

[1]P.N. Lebedev Physical Institute of the Russian Academy of Sciences 53 Leninskiy Prospekt, 119991, Moscow, Russia
[2]National Research Nuclear University MEPhI (Moscow Engineering Physics Institute), 31 Kashirskoe shosse, 115409 Moscow, Russia

E-mail: kompan@sci.lebedev.ru

Abstract. The authors develop an original and promising method of suppressing the speckle-noise in images generated by a laser beam by means of a compact despeckler based on an electro-optical cell with the smectic ferroelectric liquid crystal (FLC), realizing spatially inhomogeneous phase modulation of light. The mechanisms of destruction of the phase relations in the laser beam passing through a cell with helix-free FLC are discussed. The electric field induces the light scattering and small-scale randomly distributed gradients of the refractive index in FLC layer. The features and benefits of a despeckler using helix-free FLC compared to helix FLC are indicated.

1. Introduction

In works [1, 2] a method of spatially inhomogeneous light phase modulation using a simple electro-optical cell filled with liquid crystal (LC), namely ferroelectric LC (FLC) of the smectic type with a helix structure was proposed. The alternating electric field of the order of 2 V/μm applied simultaneously at a low frequency (450 Hz) and high frequency (3500 Hz) caused the spatial deformations of a helix and resulted in the formation in LC layer of about 18 μm thickness the small-scale spatially inhomogeneous and quickly changeable structures with random distribution of the refractive index gradients. As a result, caused by the alternating field action the spatially inhomogeneous phase modulation of the light passing through a cell was realized in FLC layer due to chaotic changes of the position of the scattering indicatrix, and modulation depth was of the order and more than π, and moreover the high spatial resolution was achieved (order of the helix pitch, i.e. fractions of a micrometer). This allowed to destroy the phase relationships in the laser beam and its ability to interference, and, consequently, to suppress the speckle-noise in images generated by a laser.

In [3-5] the optical modulator - despekler based on an electro-optical cell with helix-free FLC was described. Helical twisting of the director in a layer of this FLC was compensated due to the interaction of chiral additives with opposite signs of optical activity. The effect of the spatially inhomogeneous (over the aperture) phase light modulation in FLC layer was achieved by simultaneous action of the high-frequency (10 kHz) and low frequency (1000 Hz) voltage pulses, i.e. the interval of light modulation frequencies at the electric field of the order of 2 V/μm was increased twice in comparison with [1], that increases the possible applications of the modulator. Its main advantages are the absence of distortions in the spectral composition of the modulated optical radiation and the lack of

Journal of Physics: Conference Series **737** (2016) 012019 doi:10.1088/1742-6596/737/1/012019

light scattering when you turn off the electric field. In addition, the same shape (square wave) of voltage pulses on the low and high frequencies allowed simplifying the electronic control circuit.

Below we discuss the mechanism and parameters of spatially inhomogeneous phase light modulation in the helix-free FLC- materials (in comparison with helix materials) and the prospects for the practical use of the method in laser display devices.

2. The mechanism of light scattering and spatially inhomogeneous phase modulation in helix FLC

Spatially inhomogeneous phase modulation is a consequence of chaotic change of the position of FLC scattering indicatrix, when the control voltage pulses of low and high frequency simultaneously are applied to an electro-optical cell.

In our previous works [6, 7] the light scattering on the dynamic domain structure arising in the nonlinear process of reorientation of the helix FLC director in an alternating electric field was described. Scattering was characterized by high speed (switching on and off times of the scattering process do not exceed 150 μs at the field tension of 5 V/μm) and good efficiency - contrast ratio was about 100: 1 or higher.

It should be noted that the deformed helix FLC structure changes the spectral composition of light. Moreover, after electric field switching off the residual light scattering occurs due to the helix. Besides, when the electric field strength is lower than 3 V/μm the frequency of light modulation in an electro-optical cell at the regime of spatially inhomogeneous phase modulation is limited (less than 500 Hz) that reduces the possible applications of a despeckler.

3. The mechanism of light scattering in helix-free FLC

Below we consider a new type of light-scattering FLC materials. These materials - specially designed helix-free FLC, capable to operate effectively in the regime of electrically controlled light scattering.

In an electro-optical cell with helix-free FLC in the absence of an electric field at a certain ratio between FLC rotational viscosity γ_φ, spontaneous polarization P_S and modulus of elasticity determining the deformation along smectic layers K_φ, when $Ps < 50$ nC / cm^2, $K = (1 \div 3)\cdot 10^{-12}$ N and $0.3 < \gamma_\varphi < 1.0$ Poise, the periodic deformation of smectic layers arise, resulting in periodic changes in the position of FLC director along each layer [9]. Smectic layers are treated as periodic ordering of the centers of molecule mass in the direction of the director with a period of the order of FLC molecule length.

In the work we used helix-free FLC with the following parameters: $P_S = 40$ nC / cm^2, $\gamma_\varphi = 0.7$ Poise, tilt angle of the molecules in smectic layers $\Theta_0 = 23^0$ (at 20 ^{0}C) and temperature range of the existence of the ferroelectric (chiral smectic C*) phase from 2 ^{0}C to 70 ^{0}C.

The alternating electric field applied along the smectic layers, interacting with the spontaneous polarization changes the distribution of the director in each smectic layer. The development of this process, due to the transition to the Maxwell mechanism of energy dissipation, causes the appearance of a soliton, which is the wave packet with a periodic wave localized in it (soliton train) [3, 9].

The appearance of waves with a stationary profile results in the formation of the structure of transient domains - spontaneously ordered regions, where light scattering occurs on the borders. Movement of solitons reorients the director in all volume of FLC layer. In this case, if the polarization plane of the incident light is along the direction of FLC director (along the main optical axis), the light transmittance of an electro-optical cell is maximum. Invert the sign of the electric field (the polarity of a control voltage pulse) induces the formation of new soliton wave that causes the appearance of refractive index gradients along smectic layers and accompanied with intensive light scattering.

Under certain conditions, the scattering process can have a bistable character (Figure 1).

There is an optimal ratio between the period of deformation of smectic layers and the thickness of an electro-optical cell, when at a certain electric field tension the velocity of soliton waves motion is maximum (the electro-optical response time is minimum correspondently), and the light modulation frequency is maximum also.

 doi:10.1088/1742-6596/737/1/012019

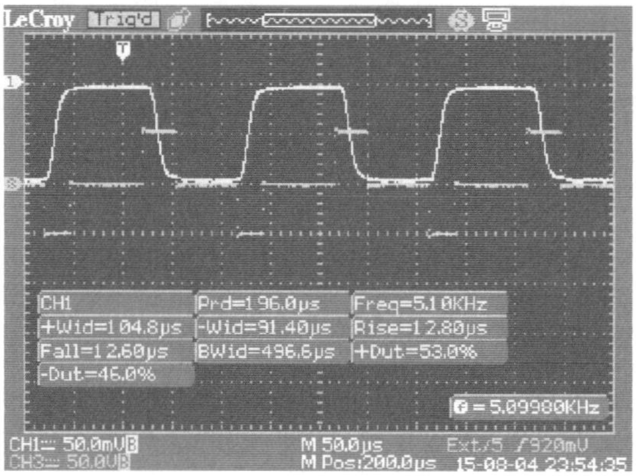

Figure 1. Oscillograms of the control voltage (channel CH3) and electro-optical response (channel CH1) at the amplitude of voltage pulses of ± 48 V and light modulation frequency of 5 kHz (a). The thickness of a cell is 18 μm. The upper level of the electro-optical response – a scattering state, low level – transmitting state; the zero level of light transmission - the arrow with the digit 1. The electro-optical response time on the front edge - Rise, trailing edge - Fall.

For used FLC with the deformation period of about 2 μm the optimal thickness of an electro-optical cell is of about 20 μm. For the concrete cell with FLC thickness of 18 μm the maximum light modulation frequency is about 5 kHz (time $\tau_{0.1\text{-}0.9}$ does not exceed 12 μs) at the electric field strength of 2.7 V / μm (Figure 1).

Switching by bipolar voltage pulses provides the operation of an electro-optical cell in bistable mode with light transmission of about 80% and a contrast ratio of over 200:1. However, unlike the helix FLC, both optical states in a cell with helix-free FLC can be memorized for several tens of seconds after switching off an electric field or until the arrival of a pulse of the opposite polarity [3].

4. The deformation of smectic layers

The presence of periodical deformations of smectic layers means that FLC molecules initially inclined by the angle Θ_0 relative to the normal to a layer at some point, are deflected additionally by the angle ψ with respect to z-axis (Figure 2). Because of this the projection of the director on the xy plane changes. The thickness of the curved smectic layer in the projection on the axis z: $l = l_0 \cos \Theta_0 / \cos \psi$.

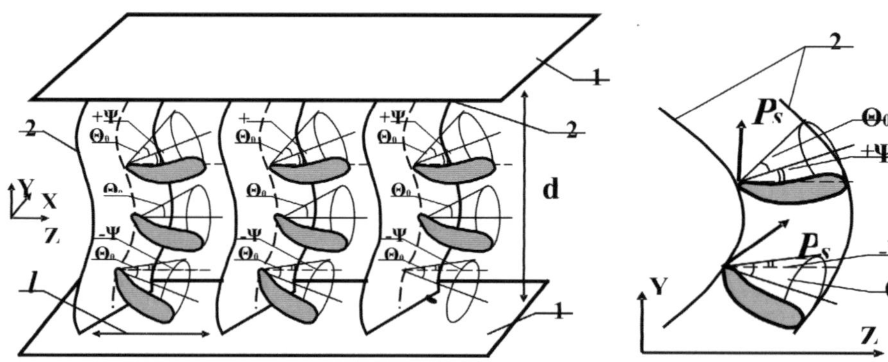

Figure 2. Helix-free FLC with a planar orientation of the director in an electro-optical cell. 1 - glass substrates with transparent conductive coating, 2 - smectic layers, d - cell thickness, Θ_0 - molecules tilt angle in smectic layers, ψ – incline angle of a smectic layer, P_S - vector of spontaneous polarization, l - the thickness of a smectic layer.

Reorientation of FLC director due to the interaction of the alternating electric field E (the field is applied along the coordinate y) with the spontaneous polarization P_S can occur both at changing the azimuth angle φ of the director orientation by 180^0, when the director is reoriented along a cone surface with the cone angle $2\Theta_0$, and at changing the distribution of the angle ψ, describing the deformation of smectic layers. In the first case the dissipative coefficient is a rotational viscosity γ_φ, while in the second case - the viscosity γ_ψ at the shear deformation [10].

If the frequency of changing the external electric field $f \sim 1/\tau_m$ (here τ_m - Maxwellian relaxation time, i.e. the time during which a shear tension after deformation caused by the electric field returns to the unperturbed state), the viscosity is defined by the shear modulus μ [11]: $\gamma \sim \tau_m \cdot \mu$.

Otherwise, if the time τ_m is frequency independent, the viscosity γ_φ also does not depend on the frequency. In this case the director reorientation time $\tau_R \sim \gamma_\varphi / P_S \cdot E$ [12]. Accordingly, if τ_m varies with the frequency, the viscosity of the molecular system will change with regarding to the variation of frequency as well. As a result, the dissipative coefficient is γ_ψ – the viscosity at the shear deformation.

The character of the process of FLC director reorientation under the influence of the alternating electric field depends on which of two dissipative coefficients (rotational or shear viscosity) predominates. The transition to Maxwellian mechanism of energy dissipation (when the shear viscosity γ_ψ becomes the dissipative coefficient) results in the appearance of a soliton, which is a localized wave packet with the periodic wave in it [3].

The maximal speed of the soliton motion can be written as:

$$V = 2\Theta_0 \sqrt{\alpha K}/(\gamma_\psi d) = (\Theta_0/\gamma_\psi)\sqrt{2K(M + P_S E \cos\varphi_0)}. \quad (1)$$

Characteristic time of the director reorientation due to the movement of the orientation bend is as follows [3]:

$$\tau_C = \frac{\gamma_\psi d^2}{K\xi} = \frac{2\gamma_\psi}{\Theta_0^2 (P_S E \cos\varphi_0 + M)} \quad (2)$$

At the initial azimuth angle of the director orientation φ_0 = 30°, the spontaneous polarization P_S = 50 nC/cm², the energy of smectic layers deformation $M = 4 \cdot 10^3$ erg/cm³, elasticity coefficient $K = 5 \cdot 10^{-12}$ N, the electric field E = 3 V/μm, the molecules tilt angle in smectic layers Θ_0 = 23° and γ_ψ = 0,2 Poise the velocity of the soliton center of V = 0.65 cm/s, and the time of the director reorientation is $\tau_C \approx$ 150 μs [3].

5. Spatially inhomogeneous light scattering structures in helix-free FLC

The transition to Maxwellian mechanism of energy dissipation is accompanied with a strong frequency dependence of the time $\tau_{0.1\text{-}0.9}$ (Figure 3, curve 1). Increasing this time on the first stage (not more than 25%) is due to the presence of both dissipative coefficients γ_φ and γ_ψ, and probably the value of shear viscosity γ_ψ in this case does not exceed 0.2 Poise (for used FLC with γ_φ = 0.7 Poise). When shear viscosity begins to predominate the time $\tau_{0.1\text{-}0.9}$ decreases a few times. At farther frequency increasing, when the transition to the soliton mode takes place, the optical response time is defined by the velocity (1) of soliton waves motion, and the frequency dependence of $\tau_{0.1\text{-}0.9}$ practically disappears.

The transition to Maxwellian mechanism of energy dissipation is accompanied also with diminishing the birefringence index Δn (Figure 3, curve 2). The maximum change Δn (about one and half times) takes place at the transition to the soliton mode.

Light scattering in helix-free FLC is due to the spatial inhomogeneity of the optical anisotropy: scattering occurs on the boundaries of spontaneously ordered regions, which are formed in FLC, when waves of a stationary profile – solitons appear. This results in forming a structure of transient domains (scattering centers), when inhomogeneous distribution of electric polarization appears along the direction of smectic layers. The action of the high-frequency electric field increases the length of domain boundaries and forms a sufficiently regular structure of circular domains that in turn results in an increasing the density of scattering centers. Changing the direction of the electric field induces the formation of a new domain structure that causes the appearance of refractive index gradients along smectic layers, and this process is accompanies again with intense light scattering.

When operating in light scattering mode the polarizers are not required, that increases the light transmission of an electro-optical cells up to 80%. This transmission is limited mainly by the light transmission of transparent conductive coatings deposited on glass substrates. To estimate the efficiency of scattering the contrast ratio is commonly used, which is defined as the ratio of intensities of light radiation propagating in a straight direction without scattering and scattering in a sufficiently small solid angle.

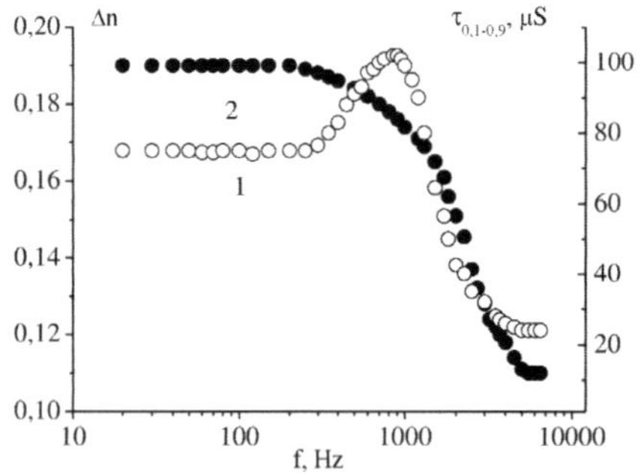

Figure 3. Frequency dependence of the electro-optical response (1) and the index of birefringence Δn (2). The thickness of an electro-optical cell $d = 13$ μm. The electric field $E = 6$ V / μm. Bipolar control voltage is of the rectangular shape (meander), amplitude is ± 35 V.

The efficiency of the light scattering and light transmission of an electro-optical cell is defined by the frequency and amplitude of the control voltage. At a fixed electric field tension (amplitude of bipolar pulses of the control voltage), the maximum efficiency of light scattering (i.e. contrast ratio) and the maximum light transmission of an electro-optical cell are achieved at different pulse duration (Figure 4). This means that the time of forming a regular structure of transient domains and the time of complete disappearance of this structure differ. So, for switching to a state with maximum light transmission, which corresponds to the complete absence of transient domains, the pulses of 1.5 ÷ 2 times larger duration are required (Figure 4, curve 2). If the pulse duration is less than the minimum time required for complete disappearance of transient domains, then the light transmittance of a cell decreases.

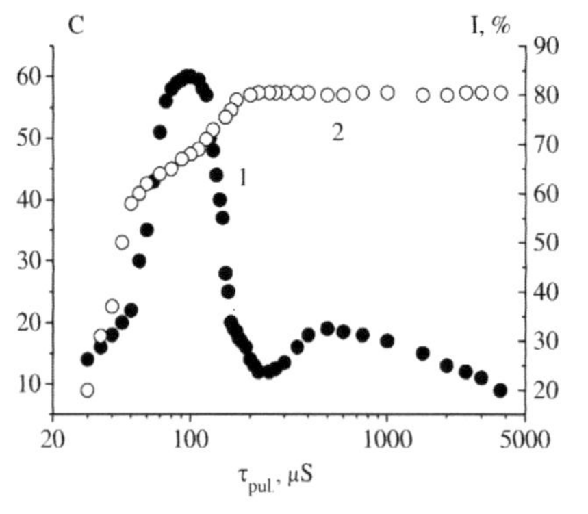

Figure 4. The dependence of the light scattering efficiency (curve 1) and light transmission (curve 2) on the duration of bipolar voltage pulses at the fixed amplitude. The amplitude of pulses is ± 35 V. The thickness of an electro-optical cell is 13 μm.

At the inversion of electric field sign (i.e. pulse polarity) not enough time of field action (i.e. small pulse duration) does not allow forming a regular structure of transient domains that results in decreasing the density of scattering centers, and as a consequence the efficiency of light scattering decreases. Namely, a regular structure of circular domains as scattering centers fairly uniformly distributed throughout the volume of FLC layer corresponds to the maximum efficiency of light scattering. The dependence of the scattering efficiency – contrast ratio on voltage pulse duration and thickness of an electro-optical cell can show a few maximums of the scattering efficiency (Figure 5). The appearance of the second and third maximums of the scattering efficiency occurs at increasing the thickness of an electro-optical cell (compare Figure 4 and Figure 5).

Increasing the pulse duration results in increasing the length of domain boundaries and the transition to irregular scattering structures. Because of this, the density of scattering centers reduces that in turn results in decreasing the scattering efficiency.

The transitions between the light-scattering modes (maximums of the light scattering efficiency correspond to them), while feeding voltage pulses with the duration corresponding to different maximums, result in the chaotic change in the position of the scattering indicatrix. As a result of short-term (less than 50 μs) switch on, light scattering in FLC the structures are formed with a random distribution of the refractive index gradients over the volume of FLC layer that in turn causes the spatially inhomogeneous (in the cross- section of the light beam) phase modulation of light in an electro-optical cell.

Thus, chaotic light scattering in helix-free FLC layer is the result of the appearance of refractive index gradients at the inversion of electric field sign (polarity of control voltage pulses), and the reason of this is nonlinear mechanism of FLC director reorientation due to soliton waves motion.

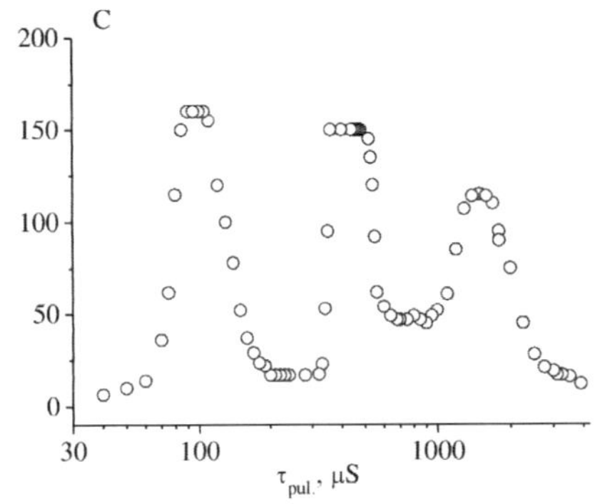

Figure 5. The dependence of the light scattering efficiency on the duration of bipolar voltage pulses at a fixed amplitude. The amplitude of pulses is of ± 50 V. The thickness of an electro-optical cell is 18 μm.

6. Spatially inhomogeneous light modulation and speckle-noise suppression

Spatially inhomogeneous modulation of a phase delay, the depth of which is of the order and more than π allows to destroy phase relationships in a laser beam passing through an electro-optical cell, and, as a consequence, to suppress the speckle-noise in images [4].

Phase light modulation with a high level of inhomogeneity is reached at the amplitude modulation of bipolar voltage pulses of the rectangular shape (meander) and frequency of 1 kHz by high-frequency voltage (10 kHz) of the same shape at the electric field tension of the order of 2 V/μm (Figure 6). The amplitude, duration and pulse repetition frequency are selected so that the light scattering efficiency changes for the duration of each pulse of pulse sequence, but it does not reach its maximum value. The consequence of the selected voltage regime is the spatially inhomogeneous light phase modulation with the depth of up to 4 π that destroys phase relations in a laser beam passing through FLC cell and suppresses the speckle-noise in images.

Figure 7 shows the photo of intensity distribution - speckle patterns (a) in the cross section of the laser beam behind FLC electro-optical cell and its measured profile along the indicated line (b) in the absence of the control voltage. Figure 8 shows the photo of intensity distribution (a) and its measured profile (b) when the control voltage is applied to FLC electro-optical cell. One can see that speckle noise in Figure 11 is suppressed significantly. The level of suppression of speckle noise in Figure 8 can be estimated from the measured value of the contrast function C [13], which is defined as the ratio of the standard deviation σ of intensity fluctuations to the average value of the intensity $<I>$:

$$C = \frac{\sigma}{<I_k>} = \frac{\sqrt{<I_k>^2 - <I_k^2>}}{<I_k>} = \frac{\sqrt{\left[\frac{1}{M*N}\sum_{m=1}^{M}\sum_{n=1}^{N} I_k\right]^2 - \frac{1}{M*N}\sum_{m=1}^{M}\sum_{n=1}^{N} I_k^2}}{\frac{1}{M*N}\sum_{m=1}^{M}\sum_{n=1}^{N} I_k} \quad (3)$$

where M - the width of the speckle pattern in pixels, N - the height of speckle pattern in pixels, I_k - intensity value in frame point k with coordinates $\{m, n\}$.

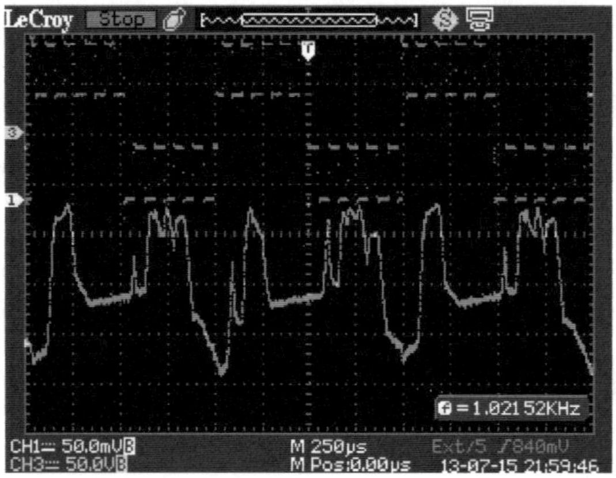

Figure 6. Waveforms of the control voltage (channel CH3) applied to FLC electro-optical cell of 18 µm thickness, and the optical response - modulation of the phase delay (channel CH1). The low-frequency signal (meander) – 1 kHz, amplitude ± 35 V. The high-frequency signal (meander) – 10 kHz, amplitude ± 35 V.

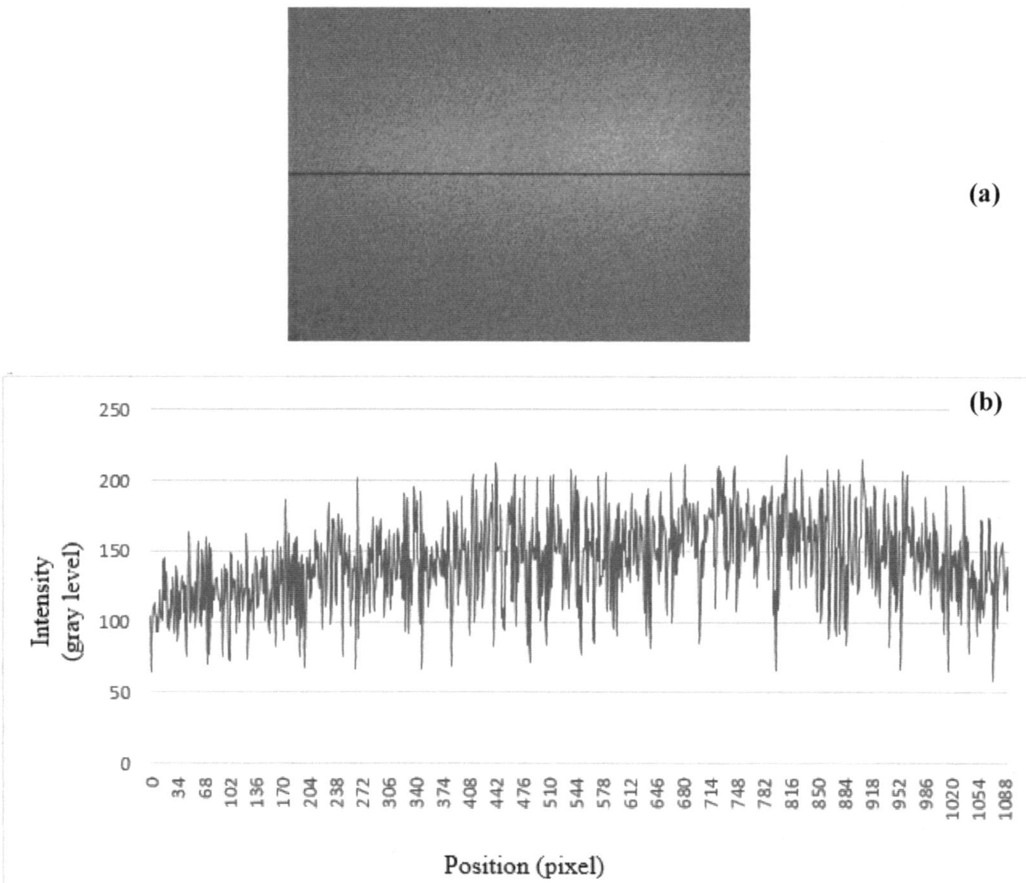

Figure 7. Photo of the intensity distribution in the cross-section of a laser beam passing through FLC cell in the absence of a control voltage (a) and the corresponding intensity profile along the indicated line in the dependence on a number of analyzed pixels (b), like in Figure 7 (a). The laser wavelength - 0.65 µm; FLC layer thickness - 18 µm.

The theoretical value of the speckle-pattern contrast varies in the range from 0 to 1. The maximal possible contrast value ($C = 1$) can be achieved only in a fully developed speckle-fields observed at the diffraction of broad laser beams on strongly rough surface or strongly scattering random transparency [14].

The effectiveness of reducing the speckle contrast was calculated as the ratio R [dB] between the contrast C_1 observed at switched off voltage and the contrast C_2 observed at switched on control voltage, namely [14, 15]:

$$R = 10\ln\left(\frac{C_1}{C_2}\right). \tag{4}$$

The contrast of the speckle pattern, shown in Figure 7 (a) was 0.72. The contrast of the speckle pattern, shown in Figure 8 (a) - 0.21. According to measurement data, the contrast decreased more than 3 times, and the effectiveness of reducing the contrast of presented speckle patterns is 5.4 dB. This result corresponds to theoretical concept and it is quite good.

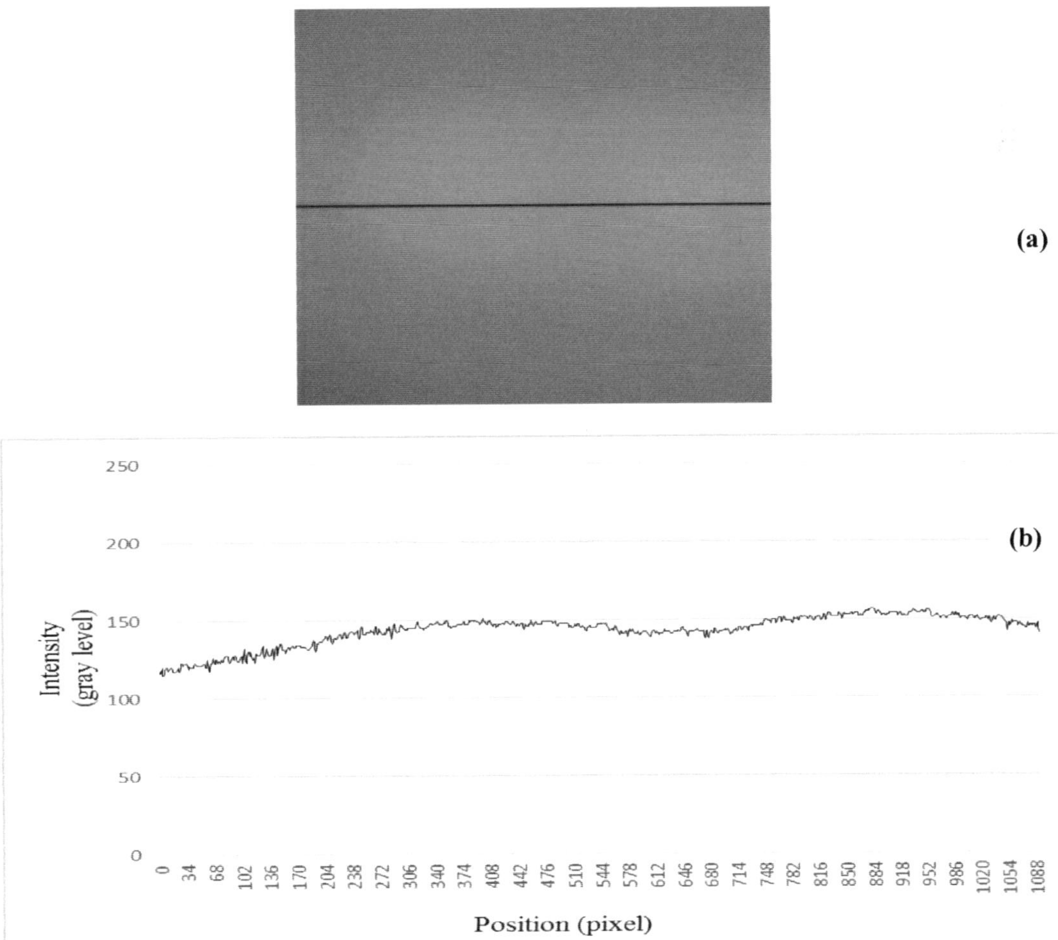

Figure 8. Photo of the intensity distribution in the cross-section of a laser beam passing through FLC cell upon application of a control voltage (a) and the corresponding intensity profile along the indicated line in the dependence on a number of analyzed pixels (b), like in Figure 8 (a). The laser wavelength - 0.65 μm; FLC layer thickness - 18 μm.

7. Conclusion

Electrically controlled spatially inhomogeneous phase modulation of light in the helix-free FLC allowed to realize the effective suppression of the speckle-noise at the electric field tension of about 2 V/μm and frequency of about 2 kHz. Experiments showed that a despeckler based on an electro-optical cell with helix-free FLC has advantages over a despeckler using helix FLC.

The modulation frequency in helix-free FLC is four times higher than in helix FLC, and the shape of low-frequency and high-frequency signals is the same that simplifies a control scheme. Since a helix is absent, there are no distortions in the spectral composition of modulated radiation and no light scattering in the absence of the electric field. Spontaneous polarization of used helix-free FLC is less than 50 nC/ cm^2, therefore there are no diffraction and light scattering due to ferroelectric domains (at electric field switching off), which appear at the spontaneous polarization more than 70 nC/ cm^2.

This work was supported by the Competitiveness Programm of the National Research Nuclear University MEPHI (Moscow Engineering Physics Institute). The authors are grateful also for the support from the Russian Foundation for Basic Research (grant 14-07-00185_a) and the Program of the Presidium of Russian Academy of Sciences.

References

[1] Andreev A L, Andreeva T B, Kompanets I N, Minchenko M V, Pozhidaev E P 2008 *Quantum Electronics* **38 (12)** pp 1166-1170

[2] Andreev A L, Andreeva T B, Kompanets I N, Minchenko M V, Pozhidaev E P 2009 *Journal of the SID* **17 (10)** pp 801–807

[3] Andreev A L, Andreeva T B, Kompanets I N, Zalyapin N V, Xu H, Pivnenko M, Chu D 2016. *Applied Optics* **55 (13)** pp 3483-3492

[4] Andreev A L, Andreeva T B, Kompanets I N, Zalyapin N V 2014 *Quantum Electronics* **44(12)** pp 1136-1140

[5] Andreev A L, Andreeva T B, Kompanets I N 2015 *RF Patent* №2561307

[6] Andreev A L, Bobylev Yu P, Gubasaryan N A, Kompanets I N, Pozhidaev E P, Fedosenkova T B, Shoshin V M, Shumkina Ju P 2005 *Journal of Optical Technology* **72 (9)** pp 702-707

[7] Andreev A L, Andreeva T B, Kompanets I N, Bobylev Yu P, Shoshin V M, Minchenko M V 2010 *Journal of Optical Technology* **77 (12)** pp 776-783

[8] Andreev A L, Kompanets I N, Andreeva T B, Shumkina Yu P 2009 *Physics of the Solid State* **51 (11)** pp 2415–2420

[9] Andreev A L, Andreeva T B, Kompanets I N, Zalyapin N V 2014 *Journal of the SID* **22 (2)** pp 115-121

[10] Fedosenkova T B, Andreev A L, Pozhidaev E P, Kompanets I N 2002 *Bulletin of the Lebedev Physics Institute* **3** pp 36-42

[11] Landau L D, Lifshits E M 1987 *Theory of Elasticity* (Moscow, Publishing House "Nauka") pp 188-189

[12] Handschy M A, Clark N A, Lagerwall S T 1983 *Physics Review Letters* **51** pp 471-474

[13] Baranov S A 2014 *Diagnostics of laser-induced structural changes in the cartilage by speckle interferometry (Chem diss) (*Moscow) 182p

[14] Application Note: LSR-3000 & LSR-OEM Series Laser speckle reducer Update: 07.07.2011

[15] Taganov A O, Gusho Y P, Dergunov N I 2014 *INTERMATIC-2014* **4** p 84-87.

Journal of Physics: Conference Series **737** (2016) 012020 doi:10.1088/1742-6596/737/1/012020

The study of photocurrent and power of THz radiation photoconductive antennas based on GaAs dependence on geometry of focusing and radiation parametres of femtosecond laser

A A Chistyakov, K I Kozlovskii, G E Kotkovskii, Yu A Kuzishchin, V A Krivenkov, Yu. A Mityagin[1], I N Piryazev

[1]National Research Nuclear University MEPhI (Moscow Engineering Physics Institute), 31 Kashirskoe shosse, 115409 Moscow, Russia
[2]P.N. Lebedev Physical Institute of the Russian Academy of Sciences 53 Leninskiy Prospekt, 119991, Moscow, Russia

E-mail: cozlowskij2013@yandex.ru

Abstract The results of experimental study of the effect of the focal spot location and the power density of radiation of femtosecond (FS) laser within the electrode gap of LT-GaAs photoconductive antenna on the photocurrent and average THz emission power are presented.

1. Introduction

The recent advent of high-power pulsed lasers, in particular, those with femtosecond pulse duration, opened the way to develop compact generators and detectors of broadband terahertz (THz) radiation, based on the interaction of laser radiation with matter [1]. Photoconductive antennas[2] or nonlinear optical crystals[3] are commonly used as active conversion elements. From the viewpoint of the efficiency of the optical–terahertz conversion, the former case appears more preferable [4, 5]. Currently, gallium arsenide grown by molecular-beam epitaxy at lower temperatures T < 300◦C is the most actively studied material for photoconductive antennas. The best structures based on "low-temperature" gallium arsenide (LT-GaAs) exhibit subpicosecond lifetimes of nonequilibrium carriers, relatively good mobilities (~103 cm2/V·s), high dark resistivities and breakdown fields (~105 V/cm) [6, 7].

To achieve the best performance characteristics of photoconductive antennas it is important to provide the optimal focusing of the pump FS laser radiation into the area between antenna electrodes. Previously we reported the temporal and spectral characteristics of various types of photoconductive antennas [8,9]. Here we report results of the effect of laser focusing conditions on the THz emission power of several photoconductive antennas.

2. Experimental setup

The scheme of the experimental setup is shown in figure 1. The experimental setup consists of a femtosecond laser (pulse duration 100 fs, pulse repetition rate 80 MHz). The average laser power focused to the antenna did not exceed 100 mW. The setup also included visible and THz focusing optics, pyroelectric detector and oscilloscope. The antennas studed were commercial ones of Zomega company as well as those manufactured by LETI and in microelectronics technology center of MEPhI. Al antennas have the same electrode geometry shown in Figure 2. The gap between electrodes was 200 μm. The focusing optics allowed to var the focus spot diameter as well as the spot position within the electrode gap.

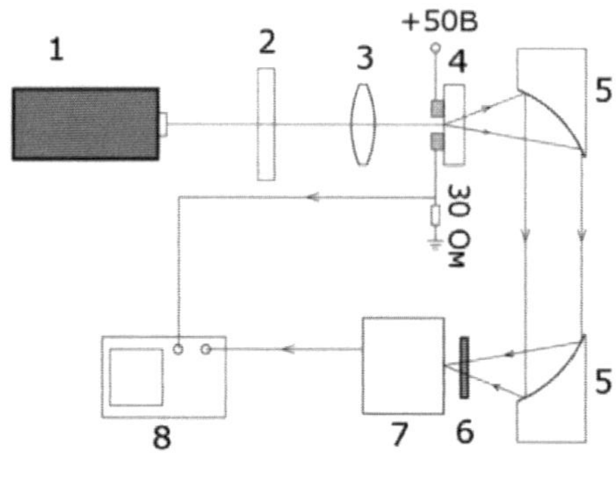

Figure 1. Experimental setup: 1- femtosecond laser, 2 – attenuator, 3 – spherical / cylindrical lens, 4 – photoconductive antenna, 5 – off-axis parabolic mirrors, 6 – THz filter, 7 – THz detector (pyroelectric), 8 – oscilloscope Tektronix TDS 2012B.

3. Experimental results

Firstly, it was studied the effect of the laser focus spot position within the electrode gap. The spot diameter was set to 25 μm and the dependence of the photocurrent and THz output was measured as a function of the spot shift from the center of a gap. The results are shown in Fig. 3 and Fig4. The maximum photocurrent was observed when the focus spot of FS laser radiation was placed exacty in the middle of the inter-electrode gap. On the contrary the maximum average power of THz radiation occurs when the spot is shifted by about 30 μm in the direction of the positive electrode of the photoconductive antenna.

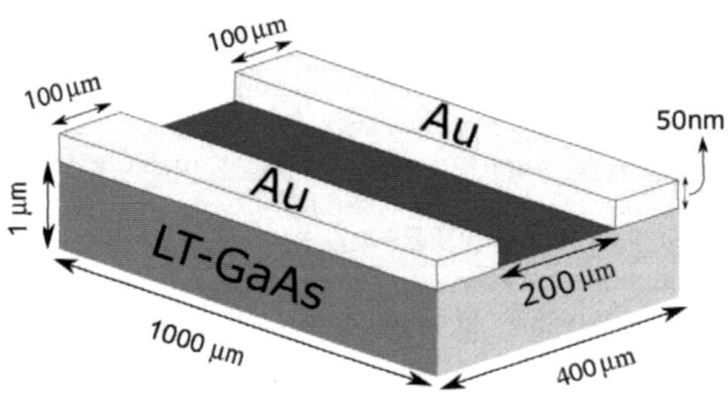

Figure 2. Geometry of the photoconductive antennas.

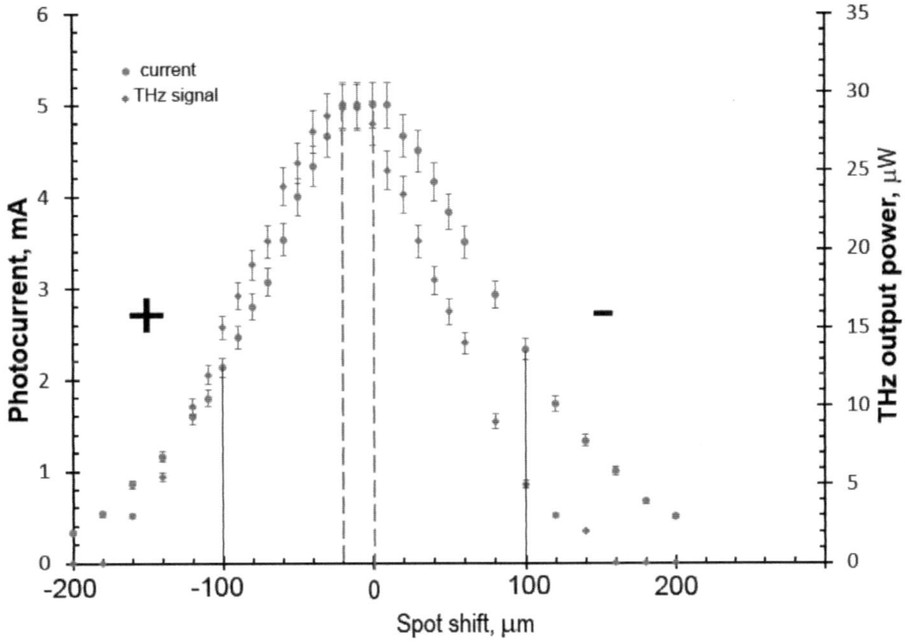

Figure 3. The dependencies of photocurrent and output THz power on the spot shift from the center of inter-electrode gap. Antenna Zomega, spot diameter 25 μm, average FS laser power 10 mW

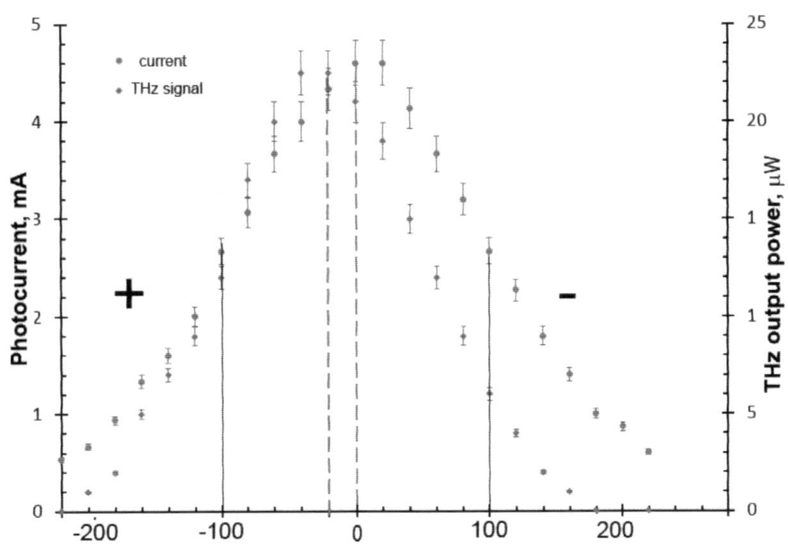

Figure 4. The dependencies of photocurrent and output THz power on the spot shift from the center of inter-electrode gap. F=Antenna Zomega, spot diameter 25 μm, average FS laser power 7 mW

The output THz average power for different focus spot diameters was measured while total excitation FS laser power was kept constant. The results are shown in Fig.5. The tendency is clearly seen of increasing THz power with more sharp focusing.

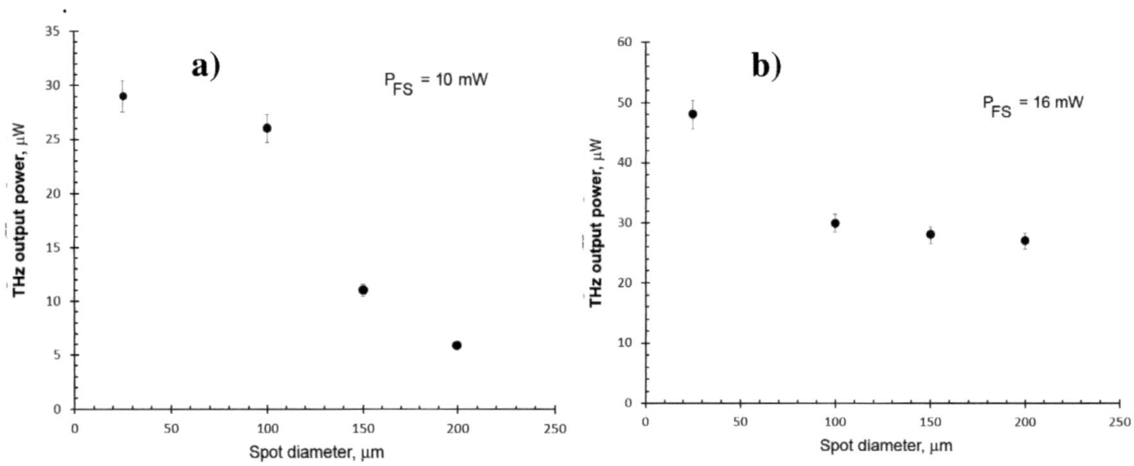

Figure 5. Output THz average power as a function of the focus spot diameter at excitation FS power of 10 mW (a) and 16 mW (b)..

So the increase of the excitation power density resulting in increasing density of excited photocarriers (at the same total FS laser power) allows to enhance the effectiveness of the antenna(Fig.6). For efficient THz generation is necessary to create a sufficient density of photocarrier, which provides power density of FS radiation on the surface of the photoconductive antenna ~ 10^9 W/cm^2

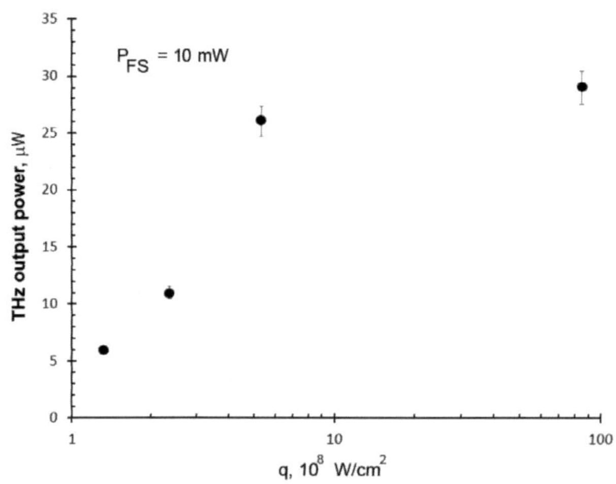

Figure 6. Output THz average power as a function of the excitation FS power density (excitatiom FS power = 10 mW).

It was found also that the average power of THz radiation are about three times higher in the case when the plane of polarization of the FS radiation is parallel to the electrodes, relative to the case when FS radiation is polarized normally to the electrodes.

4. Conclusions

It is shown that the maximum THz power of the photoconductive antenna is achieved when the focus spot of the exciting FS laser is shifted toward the anode electrode. The sharp focusing of the FS laser beam allows to improve the antenna output without increasing total FS power. Polarization of the exciting FS laser radiation with respect to electrodes is also of importance, and correct its selection can increase the THz output of the antenna.

References

[1] Dragoman D. and Dragoman M. 2004 *Quantum Electron.* **28** 1
[2] Auston D H, Cheung K P and Smith P R 1984 *Appl. Phys. Lett.* **45** 284
[3] Wu Q and Zhang X-C 1996 *IEEE J. Sel. Top. Quantum Electron.* **2** 693
[4] Cai Y, Brener I, Lopata J, et al 1998 *Appl. Phys. Lett.* **73** 444
[5]. Shen Y C, Upadhya P C, Beere H E, et al 2004 *Appl. Phys. Lett.* **85** 164
[6]. Gupta S, Frankel M Y, Valdmanis J A, et al 1991 *Appl. Phys. Lett.* **59** 3276
[7]. Hou Land Shi W 2013 *IEEE Trans. Electron Devices* **60** 1619
[8]. Savinov S A, Mityagin Yu A, Chistyakov A A, Kozlovsky K I, Kuzishchin Yu A, Krivenkov V A, Egorkin V I, Kazakov I P 2015 *Physics Procedia* **73** pp 54-58
[9] Gorbatsevich A A, Egorkin V I, Kazakov I P, Klimenko O A, Klokov A Yu, Mityagin Yu A, Murzin V N, Savinov S A, Tsvetkov V A 2015 *Bulletin of the Lebedev Physics Institute* **42(5)** pp 121-126

Journal of Physics: Conference Series **737** (2016) 012021 doi:10.1088/1742-6596/737/1/012021

Infrared and terahertz transmission properties of germanium single crystals

I A Kaplunov[1], P A Nikitin[2], V B Voloshinov[2], A I Kolesnikov[1], B A Knyazev[3]
[1]Tver State University, Zhelyabova str., 33, Tver, 170100, Russia
[2] Moscow State University, Vorob'evy gory, Moscow, 119991, Russia
[3] Novosibirsk National Research State University, Pirogova str., 2, Novosibirsk, 630090, Russia

e-mail: Kaplunov.IA@tversu.ru

Abstract. Experimental transmission spectra of samples fabricated of germanium single crystals doped with stibium were registered in the infrared 2.5-25 μm and terahertz 130 μm regions of spectrum. It is shown that doping concentration and treatment of the crystals surface have a noticeable influence on the samples absorption.

1. Introduction

Terahertz (THz) radiation is widely used in modern and traditional fields of science and technology such as optics, material science, medicine, biology, biochemistry, and etc. [1, 2]. As for optical devices operating in the THz spectral domain, it is very important to ensure development of their basic elements including radiation sources, detectors and transparent media. Special requirements on the THz optical elements are related to the peculiar character of the spectral range.

Usually crystalline silicon, crystals of quartz and sapphire as well as the polymers (polymethylpentene (TPX), polyethylene (PE), teflon (PTFE)) are applied in THz technology [3]. High-resistivity single crystal silicon grown by zone melting without crucible technology is the example of a transparent and well investigated medium. Germanium as well as silicon are also typical semiconductor materials applied in the THz spectral region. As known, crystalline germanium is widely used at the wavelengths 2.5-14 μm in mid-infrared (IR) optics, in photodetectors, detectors of ionized radiation and photovoltaic transducers [4-6]. P-type germanium served as a matrix of a monopolar THz laser [7]. The only acousto-optic crystalline material that has been used at the THz frequencies is the single crystal germanium. A wide-angle THz beam deflector has been developed and successfully tested on the base of this crystal [8].

Investigations of optical properties of germanium in the THz region is of a considerable interest in optics and acousto-optics. In this paper, a comparative analysis of IR and THz transmission spectra measured in the germanium crystals is carried out. The germanium samples were characterized by various concentrations of stibium and different treatment of their surfaces.

2. Objects of investigations and experimental techniques

In this research, a detailed investigation of germanium single crystals grown along the direction [111] by the Czochralski method was performed. We also examined properties of the germanium single crystals doped with stibium Sb or with both stibium and gallium Ga (compensated crystals). The treatment of the samples included polishing (N < 1; ΔN < 0.5) and grinding (by abrasive M10). Thickness of all samples was equal to l = 1.0 cm. Transmission spectra of the samples in the THz range were measured using the Novosibirsk free-electron laser operating at the wavelength $\lambda = 130$ μm. We also employed the Golay cell as a detector of the THz radiation. In the IR spectral domain 2.5-25 μm, the study was carried out by means of the Fourier transform spectrometer Tensor-27.

3. Results and discussions

The detailed characteristics of the samples as well as their determined absorption coefficients α and the refractive indexes n at the wavelengths 10.6 μm and 130 μm are summarized in Table 1. The recorded transmission spectra T in the frequency range from $\nu = 4500$ cm^{-1} to $\nu = 375$ cm^{-1} are shown in Fig. 1. The curves in Fig. 1 are numbered from 1 to 5 according to the sample numbers in Table 1. It was found that the transmission of the samples in the THz range depended on the concentration of doping. As for the IR transmission properties, they are mainly related to the free-carrier absorption.

As follows from data in Table 1, absorption of the samples in the THz range is substantially stronger than in the IR. For example, transmission is equal to $T = 0.47$ at the wavelength $\lambda = 10.6$ μm and it is about 0.15 at the wavelength $\lambda = 130$ μm. The minimal value of the absorption coefficient α = 1.04 см$^{-1}$ was registered in the sample number 3 having the smallest dopant concentration. In the reference [8], the reported absorption coefficient of a pure high-resistivity germanium crystal was equal to α = 0.75 см$^{-1}$. However, we found that the IR transmission of the doped germanium was similar to that in the THz region. We propose that this feature originates from high concentration of the free-carriers n_c even in the pure germanium. For example, the pure germanium samples have the intrinsic carrier concentration $n_c = 2.5 \cdot 10^{13}$ cm^{-3} and the resistivity of about $\rho = 47$ Ω·cm, whereas for the pure silicon, these parameters are the following $n_c = 5 \cdot 10^{10}$ cm^{-3} and $\rho = 2$ kΩ·cm.

Table 1. Determined absorption coefficients and detailed sample characters.

Property	λ, μm	Number of the sample (l = 1.0 cm)				
		1	2	3	4	5
Transmission T (n = 4.0000)	130	$1.3841 \cdot 10^{-5}$	$5.5882 \cdot 10^{-5}$	0.1469	0.050	0.1375
Absorption coefficient α, cm^{-1}		10.30	8.90	1.04	2.11	1.11
Transmission T (n = 4.0048)	10.6	0.135	0.364	0.456	0.451	0.447
Absorption coefficient α, cm^{-1}		1.122	0.207	0.024	0.032	0.041
Dopant concentration, cm^{-3} / impurity		$8.0 \cdot 10^{15}$ /Sb	$8.0 \cdot 10^{15}$ /Sb	$1.3 \cdot 10^{14}$ /Sb	$3.0 \cdot 10^{14}$ /Sb	$5.5 \cdot 10^{13}$* /Sb-Ga
Resistivity ρ, Ω·cm		0.26	0.26	13.5	5.0	33.0
Type of conductivity		n	n	n	n	p
Surface treatment		grinding	polishing	polishing	polishing	polishing
Dislocation density , cm^{-2}		< 10	< 10	10^4	10^4	10^4

* - concentration difference (Sb concentration $7.0 \cdot 10^{13}$ см$^{-3}$; Ga concentration $12.5 \cdot 10^{13}$ см$^{-3}$).

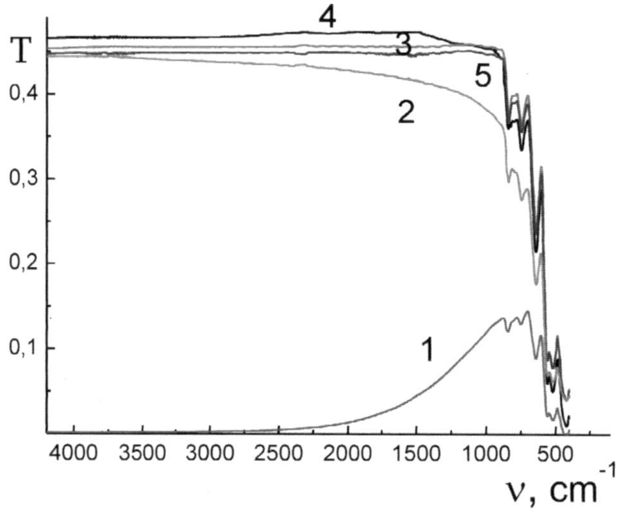

Fig. 1. IR transmission spectra of germanium alloys

As one can see from data presented in Table 1, the transmission of the polished sample (number 2) is substantially higher than the that in the grinding sample (number 1). The ratio of the transmission coefficients at the wavelength λ = 130 μm is equal to T_1/T_2 = 4. This feature may be explained in the following way. Optical path *d* of a THz beam is increased by an increment Δd due to oblique incidence of the radiation on each microfacet forming microtopography of a polished sample surface. In case of a high absorption this leads to a strong intensity reduction. The relative value $\Delta d/d$ = 15% was found using the Bouguer-Lambert law. Therefore, mean angle of incidence of the beam on the fracture-squares equaled to 29°. The averaged angle of microfacet orientation with respect to the vertical forming the surface microrelief of the polished sample was estimated making use of the law of refraction for n = 4.0 and λ = 130 μm. This angle was about 55° and it was close to the magnitude of the angle between the cleavage plane (111) in germanium and the optical input surface of the samples under investigation.

As follows from the Fig. 1, transmission of polished samples decreases at the absorption edge whereas that one of the grinding sample (number 1) increases. Under certain relations between the average angle of the microfacet and the refractive index of the sample, beams multiply reflected by the fracture-squares do not go backwards and give a contribution to the transmitted radiation mimicking the blooming effect of grinding surfaces the greater the less is refractive index. In the spectral range 5-10 μm, the decrease of refractive index revealed in our measurements. As for the blooming effect, it manifested itself right up to the phonon absorption edge.

4. Conclusion

Transmission of doped single crystal germanium in the THz range is sufficiently lower as compared to the IR range. This trend originates from high concentration of free-carriers even in the pure germanium. Roughness of the grinding surface in the samples leads to a stronger THz absorption at the wavelength 130 μm. It is concluded that this absorption is related to the increase of the optical path in the doped germanium. On the contrary, in the IR spectral range, the blooming effect takes place on grinding surfaces of germanium. The effect is due to multiple reflections accompanied by small absorption.

Journal of Physics: Conference Series **737** (2016) 012021 doi:10.1088/1742-6596/737/1/012021

Acknowledgments
The work was performed within the framework of the realization of the Federal Target Program «Investigations and elaboration in the priority directions of scientific–technological development of the complex of Russia for 2014 – 2020 years», agreement 14.577.21.0004 (RFMEFI57714X0004).

References

[1] Ziegel P H 2002 *IEEE Trans. Microwave theory and Tachniques* **50** 910
[2] Mickan S P and Zhang X.-C 2003 *Internat. Journ. of High Speed Electronics and Systems* **13** 601
[3] http://www.tydexoptics.com
[4] Clauws P 1996 *Materials science and Engineering* B **36** 213
[5] Podkopaev O I and Shimansky A F 2013 *Germanium single crystal growth with low density of dislocations and impuities* (Krasnoyarsk: Sib. Feder. Univ.) p 104
[6] Kaplunov I A, Smirnov Yu M, Kolesnikov A I 2005 *Journal of Optical Technology* **72** 214
[7] Patent US 5784397 21.07.1998 *Bulk semiconductor lasers at submillimeter/far infrared wavelengths using a regular permanent magnet*
[8] Voloshinov V B, Nikitin P A, Gerasimov V V, Knyazev B A and Choporova Yu Yu 2013 *Quantum Electronics* **43:12** 1139

Optical properties of lead germanate $Pb_5Ge_3O_{11}$ in terahertz range

M Yu Vlasov[1, 2], V F Losev[3], N A Nikolaev[1, 3, *], A A Mamrashev[1, 3]

[1] Institute of Automation and Electrometry SB RAS, Novosibirsk, 1 Ac. Koptyuga av., 630090, Russia
[2] Novosibirsk State University, Novosibirsk, 2 Pirogova str., 630090, Russia
[3] Institute of High Current Electronics, Tomsk, 2/3 Academichesky av., 634055, Russia

E-mail: *nazar@iae.nsk.su

Abstract. We measured ordinary and extraordinary refractive index and absorption of ferroelectric lead germanate in the range of 0.1–1.3 THz by the method of terahertz time-domain spectroscopy. We studied soft mode behaviour with the rise of temperature by fitting imaginary part of the complex dielectric permittivity with three Lorentz oscillators. We found that high birefringence and dichroism of lead germanate or its analogues can be used for manufacturing of effective terahertz polarizing filters.

1. Introduction

Lead germanate $Pb_5Ge_3O_{11}$ (PGO) is a uniaxial ferroelectric which crystalline structure belongs to P3 space group at room temperature. This crystal is one of the few ferroelectrics that exhibit birefringence and optical activity at the same time [1]. It demonstrates structural phase transition to non-polar state which crystalline structure belongs to $P\overline{6}$ space group at the critical temperature of $T_c \approx 446$ K. Linear and nonlinear optical properties of lead germanate and their dependence on the temperature are well studied at the visible wavelengths [2–4].

Behaviour of PGO optical modes above and below the phase transition point was also studied in the terahertz (THz) frequency range by the methods of Raman scattering spectroscopy [5,6] and spectroscopy based on backward wave oscillators [7]. It was determined that lead germanate is a displacement type ferroelectric at the temperatures far from the phase transition point, i.e. its polar axis is characterized by the presence of phonon soft mode. Its frequency f_0 is defined by the formula $f_0^2 \sim \frac{|T - T_c|}{T_c}$, which equals ~38 cm^{-1} at the room temperature. Complex dielectric permittivity $\tilde{\varepsilon}(\omega)$ can be well approximated by the model of one Lorentz oscillator $\tilde{\varepsilon}(f) = \frac{S}{f_0^2 - f^2 + if\gamma}$, where S – oscillator strength, f – frequency, γ – damping constant [7].

It is obvious that application of nonlinear optical medium for generation, detection and control of terahertz radiation requires knowledge of its optical properties at laser radiation wavelength as well as at terahertz frequencies. This work studies optical properties of lead germanate single crystals in the terahertz range.

2. Experimental setup

The samples under study were cut as plane-parallel plates with the thickness of ~250 μm from single crystals of lead germanate grown from solution-melt by the Czochralski process. The plates were cut to contain optical axis in the plane of the sample. The measurements were made by terahertz time-domain spectrometer. It measured signal proportional to the amplitude of terahertz electric field. Time-domain scan allowed measurement of terahertz pulses; Fourier transform allowed obtaining its spectrum. Lead germanate crystal has significant anisotropy – dichroism and birefringence – in the terahertz range. Due to this peculiarity it was important to control if two orthogonal components of electric field of terahertz pulses were detected after passing through the studied crystal. Owing to the fact that the terahertz pulses were observed in the time-domain it was possible to discern and reduce parasitic polarization component by carefully adjusting polarization optics of the spectroscopy system.

Typical measurement included scanning of two terahertz pulses – signal with the sample under study and reference without the sample. To allow for continuous series of measurements of signal and reference pulses at different temperatures we manufactured a symmetrical sample holder with two identical holes. One of the holes stayed free as the other one held the sample. The holes interchangeably were subjected to the terahertz radiation by a motorized translation stage. We then calculated the sample properties: refractive index, absorption, complex dielectric permittivity and thickness.

3. Results and discussion

The results of calculations of optical properties – refractive index and absorption – of lead germanate are shown on the **Figure 1** and **Figure 2**. Subscripts indicate material properties when polarization of terahertz field is perpendicular (*o*, ordinary) and parallel (*e*, extraordinary) to the *z* optical axis of the crystal. Several phonon modes for absorption α_o can be seen on the **Figure 1**. Positions of their maximum are in accordance with the frequencies of E modes previously measured by Raman scattering spectroscopy (see Table 4 in [6]): 0.933 THz (31.1 cm^{-1}), 1.11 THz (37 cm^{-1}). Phonon modes of higher frequencies (1.332 THz, 1.419 THz, 1.554 THz etc.) were not observed in our experiment directly. However their left wing increased average absorption in the frequency range under study. As we approached the phase transition temperature two phonon lines merged together, made up a shoulder and completely disappeared in paraelectric state. However, several phonon modes with higher frequencies didn't move and defined the level of absorption in the studied range. Refractive index n_o didn't change a lot in the range of 4.4–4.7 at the frequencies up to 0.8 THz as the temperature increased.

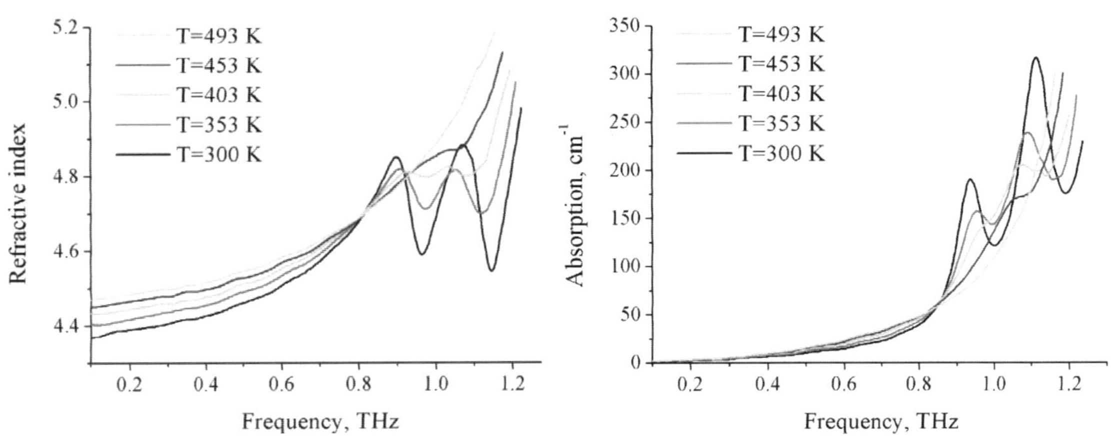

Figure 1. Refractive index n_o (left) and absorption α_o (right) of lead germanate at different temperatures

Journal of Physics: Conference Series **737** (2016) 012022 doi:10.1088/1742-6596/737/1/012022

Figure 2 shows spectral dependence of extraordinary refractive index n_e and absorption α_e. It can be seen that the properties greatly depend on the temperature which can be explained by the softening of a particular phonon mode. PGO as a displacement type ferroelectric is characterized by a phonon mode which frequency decreases when approaching critical temperature T_c.

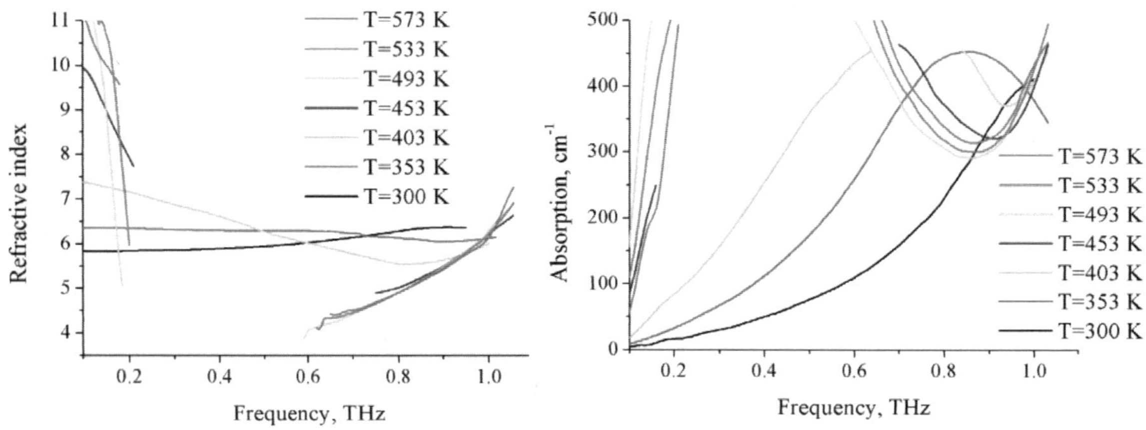

Figure 2. Refractive index n_e (left) and absorption α_e (right) of lead germanate at different temperatures.

In order to study characteristics of the soft mode with the change of the temperature we calculated complex dielectric permittivity $\tilde{\varepsilon}_e = \varepsilon'_e + i\varepsilon''_e$ from the complex refractive index $\tilde{n}_e = n_e + i\dfrac{\alpha_e c}{4\pi f}$, where c is the speed of light. Its imaginary component ε''_e was fit by a sum of three Lorentz oscillators. The frequency of the first one was zero, the second oscillator represented the soft mode, and the frequency of the third oscillator was fixed at 1.14 THz. The frequency of the third peak was taken from the results of Raman scattering spectroscopy (see Table 3 in [6]). **Figure 3** demonstrates good agreement between the experimental and modelling results at the temperature of 423 K.

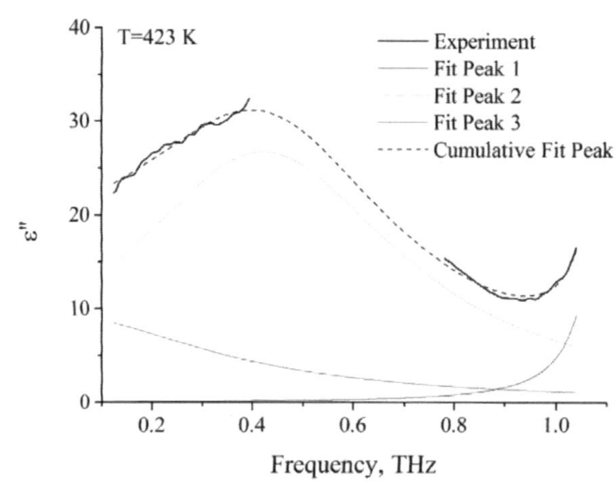

Figure 3. Imaginary part of complex dielectric permittivity ε''_e at the temperature of 423 K. Experimental results – solid line, approximation by three Lorentz oscillators – dashed line.

Experimental and modelled spectra of imaginary part of complex dielectric permittivity ε_e'' of lead germanate at different temperatures are depicted on **Figure 4**. Behaviour of the squared frequency of the softening phonon mode with the rise of the temperature is shown on the right **Figure 4**. As it was noted in the introduction squared frequency should be proportional to the difference T–T_c for displacement type ferroelectrics. Dashed line on the right **Figure 4** demonstrates this dependence. It can be seen that in the vicinity of the phase transition the frequency of the soft phonon mode does not turn to zero, therefore the model of displacement type semiconductor doesn't describe its behaviour well enough.

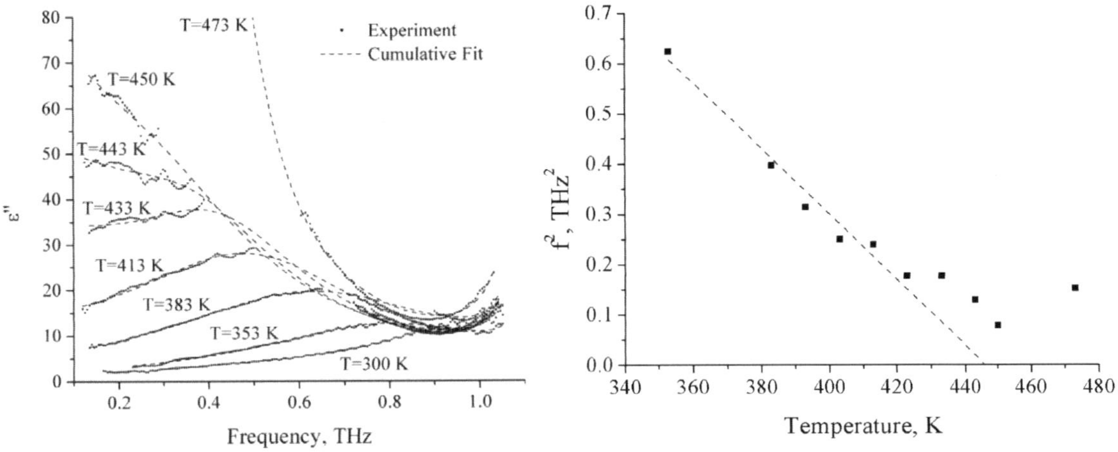

Figure 4. Experimental (solid line) and modelled (dashed line) spectra of imaginary part of complex dielectric permittivity ε_e'' of lead germanate at different temperatures (left). Behaviour of the squared frequency of the soft mode with the rise of the temperature (right). Experimental results – square marks, soft phonon mode model – dashed line.

The results depicted on the **Figure 1** and **Figure 2** show that the lead germanate crystal is highly birefringent and dichroic in the lower terahertz frequency range. For example, at the room temperature at the frequency of 0.5 THz the difference between refractive indices $n_e - n_o$ equals ~1.48 and absorption equals α_e=75 cm^{-1} and α_o=10 cm^{-1}. If the temperature rises to ~400 K absorption α_e gets higher than 400 cm^{-1} in the spectral range of 0.5–0.9 THz, absorption α_o changes by no more than 10 cm^{-1} at the same time. Lead germanate demonstrates that absorption for polarization where the polar mode is active is much higher than the absorption for polarization where the polar mode is not active. Thus lead germanate or its analogues can be used as effective polarizing filters of terahertz radiation. To make sure that the filter doesn't have sharp spectral peculiarities one should choose ferroelectric with a wide central polar mode, or order-disorder type ferroelectric, or ferroelectric with overdamped soft mode.

4. Conclusion

We investigated optical properties of lead germanate crystal in the terahertz range of 0.1–1.2 THz. Both ordinary and extraordinary refractive indices and absorption were measured. It was determined that PGO has significant birefringence and dichroism. High dichroism and its temperature behaviour may become the basis of creation of terahertz polarizes with record specifications.

Acknowledgments

The work was supported by Russian Science Fund, project 15-19-10021

Journal of Physics: Conference Series **737** (2016) 012022 doi:10.1088/1742-6596/737/1/012022

References

[1] Koňák Č, Fousek J and Kürsten H D 1978 *Ferroelectrics* **21** 347–8
[2] Iwasaki H 1972 *J. Appl. Phys.* **43** 4907
[3] Uchida N 1972 *J. Appl. Phys.* **43** 4933
[4] Miga S, Dec J, Molak A and Koralewski M 2006 *J. Appl. Phys.* **99** 124107
[5] Hosea T J, Lockwood D J and Taylor W 1979 *J. Phys. C Solid State Phys.* **12** 387–404
[6] Lockwood D J, Hosea T J and Taylor W 1980 *J. Phys. C Solid State Phys.* **13** 1539–53
[7] Kozlov G and Volkov A 1998 Coherent source submillimeter wave spectroscopy *Millimeter and submillimeter wave spectroscopy of solids* Topics in Applied Physics **74** pp 51–109

Femtosecond laser fabrication of linear graphitized microstructures in a bulk of polycarbonate samples

D V Ganin[1, 2], K E Lapshin[1], A Z Obidin[1] and S K Vartapetov[1]

[1]Physics Instrumentation Centre of Prokhorov General Physics Institute, Russian Academy of Sciences, Troitsk, 142190 Moscow, Russia
[2]National Research Nuclear University MEPhI (Moscow Engineering Physics Institute), 31 Kashirskoe shosse, 115409 Moscow, Russia

E-mail: ganin@optosystems.ru

Abstract. We have fabricated high aspect ratio straight and curved graphitized lines inside of polycarbonate samples by using a femtosecond laser. Use of a spherical lens with high NA to focusing femtosecond pulse in the bulk of material leads to self-diffraction of laser beam and formation a filamentary structure. We fabricated two kinds of graphitized lines. The first type is a straight line extended in the direction of the laser beam. This type of lines was created by femtosecond laser scanning without pulse overlapping. The second type of graphitized lines is curved lines, which was created by scanning with a significant overlapping of focal spot. We determined conditions of the formation of straight graphitized lines by one femtosecond pulse with diameter about 2 μm and length greater than 1 mm in polycarbonate samples. Mechanism of formation and potential applications of these structures are also discussed.

1. Introduction

Laser fabrication with high-energy femtosecond (fs) lasers has great advantages such as highest precision of micromachining, absence of heat affected zone, ability of 3D intra volume direct writing without surface damage, so femtosecond laser has become a powerful tool for fabrication of embedded optical, fluid and electrical components inside transparent materials. When a femtosecond pulse is tightly focused in the bulk of material, it will lead to nonlinear absorption and produce micro scale modification within the focal volume [1]. The structural modification inside of material can be different. The articles [2, 3, 4, 5] reported about voids, cracks, refractive index changes, graphitization and birefringent structures [6].

The shape of femtosecond laser induced structures can be strongly extended in direction of pulse propagation due to spherical aberration, nonlinear effects or their combination. The length of modification depends on the focusing depth of laser pulse, pulse peak power and focusing system. In our time, in most cases, the microchannels in the bulk of material are formed by transverse or longitudinal femtosecond laser scanning with a significant overlap of focus spots. But we assume that using nonlinear effects (Kerr self-focusing) or a special focusing system such as axicon or a lens with strong spherical aberration you can create linear microchannels with great lengths.

In this paper, we demonstrate high-aspect ratio microstructure fabrication by focusing of high energy femtosecond laser pulse with highly aberrated spherical lens. In experiments, we achieved straight and curved graphitizated lines inside polycarbonate samples with total length more than 1 mm.

Straight graphitized line with diameter about 2 μm and length 1 mm in the bulk of polycarbonate can be fabricated by one femtosecond pulse.

2. Scheme of experimental setup

Experimental setup is shown on figure 1. The polycarbonate (PC) sample was cut in a plate form of 50mmx20mmx3mm and four-surface polished. We used two femtosecond laser systems. First is Yb regenerative laser operating at the central wavelength of 1025 nm with pulse repetition rate of 2 kHz, pulse energy 150 μJ and pulse duration of 350 fs. Second laser system is Yb laser with central wavelength of 1030 nm, pulse duration about 300 fs and 26 μJ pulse energy at 80 kHz pulse repetition rate. To control the power and the pulse energy of the linear- polarized laser beam, we used a rotatable half-wave plate and thin film polarizer. Beam expander with 4x magnification was used to expand laser beam. We focused fs laser pulses by spherical lens (f=15 мм, clear aperture 22 mm). This lens has a strong spherical aberration. In experiments, we focused on bottom surface of sample. Sample was mounted on three-coordinate computer-controlled stage. Pictures of exposure of fs laser pulses were obtained using an optical microscope in transmitted light (Nikon LV100D).

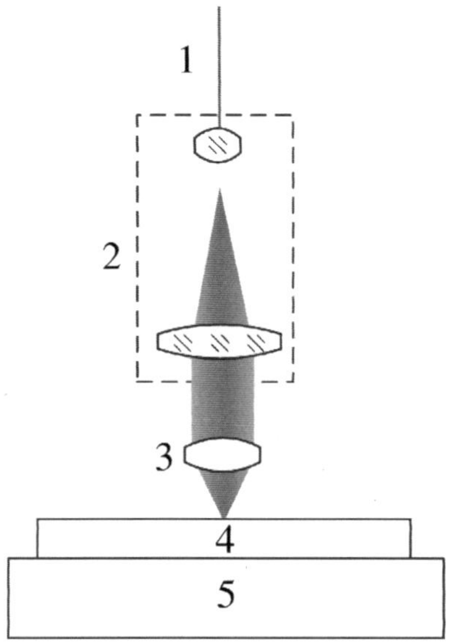

Figure 1. Scheme of experimental setup: (1) femtosecond laser beam; (2) beam expander; (3) focusing lens; (4) sample; (5) XYZ translation stage.

3. Experiment

3.1 *Beam intensity distribution conversion.*

Tightly focused fs laser pulse can lead to various nonlinear phenomena due to very high intensity of radiation. The most typical phenomenon is the laser pulse filamentation. Filamentation of laser pulse can lead to long track formation inside transparent material, but not only by filamentation can be created these tracks. In our experiments, initial Gaussian laser beam intensity distribution was transformed into a set of Fresnel type diffraction rings after passing through the focusing lens. Figure 2 shows the intensity distribution of focused beam in four positions. Each intensity distribution is a set of Fresnel type diffractive rings. This picture was achieved at metalized film surface after ablation by 1000 fs pulses with pulse energy 60 μJ. It is seen that diffraction ring with highest diameter and set of merging diffraction rings in center of intensity distribution are most intense.

It is obvious that each diffraction ring has its own focus, so diffraction rings will focused at the different depths in the sample according figure 3.

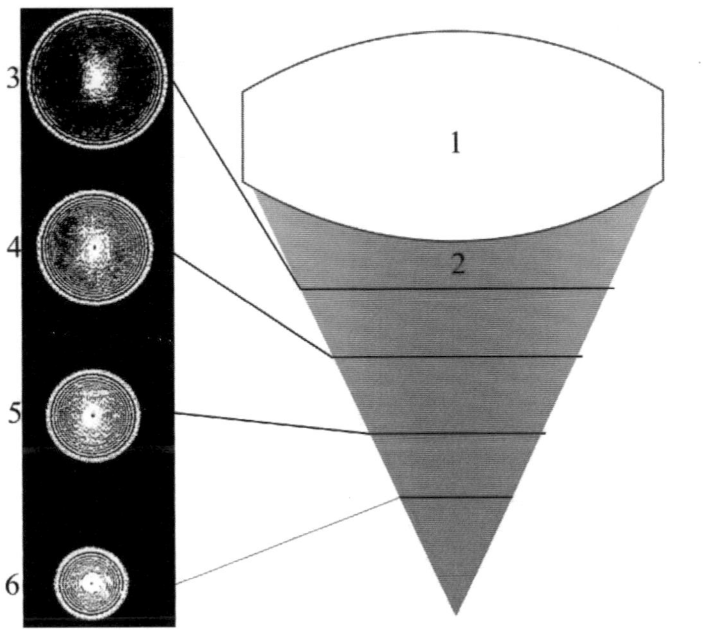

Figure 2. Picture of intensity distribution of focused beam in four positions. (1) Spherical lens; (2) focused beam; (3), (4), (5), (6) intensity distributions.

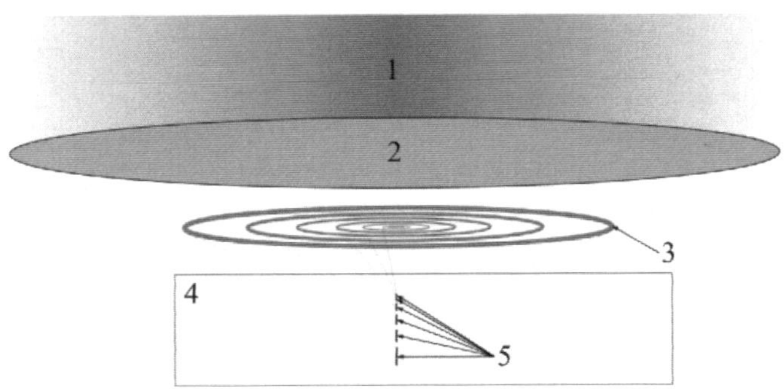

Figure 3. Sketch of focusing of diffraction rings inside transparent material. (1) Gaussian beam; (2) spherical lens; (3) set of diffraction rings; (4) sample; (5) set of focuses.

3.2 *Micromachining by fs laser pulses without focusing spots overlapping.*

Experiments on micromachining polycarbonate samples confirmed our assumptions (figure 4). In this experiment have been used the first laser source. Femtosecond laser pulse generates a set of linear micromodifications in the sample volume. Modifications are located on the fs laser pulse propagation axis. In this case, we scanned a sample by fs laser pulses with no overlap of focus spots. The scanning speed was constant and equal to 800 μm/s. Figure 4 a) shows microstructures obtained at 15μJ pulse energy. It is seen that one fs laser pulse generates a set of four linear microstructures that are located on pulse propagation axis. Number of linear microstructures in set increases with increasing pulse energy up to 60 μJ (figure 4 b). It is also worth noting two trends. The first is increasing the length of each microstructure. The second is merging nearby linear modifications into one.

When the energy of the fs laser pulse reached 150 μJ linear modifications in the set are merged into one (figure 5). The total length of the modification in this case exceeds 2 mm.

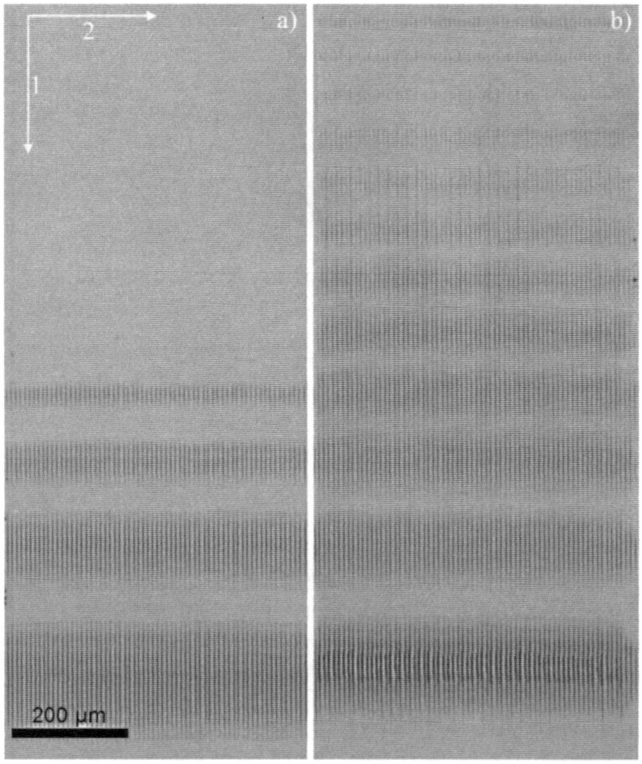

Figure 4. Pictures of micromodifications inside PC sample. a) pulse energy - 15 μJ; b) pulse energy - 60 μJ. (1) direction of pulse propagation; (2) scanning direction. Scanning speed is 800 μm/s. Pulse repetition rate 100 Hz.

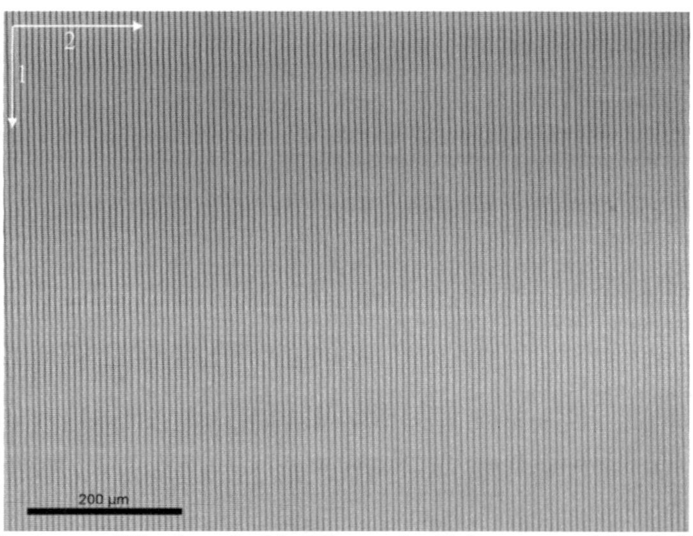

Figure 5. Picture of micromodifications inside PC sample. Pulse energy 150 μJ. (1) direction of pulse propagation; (2) scanning direction. Scanning speed is 800 μm/s. Pulse repetition rate 100 Hz.

3.3 *Micromachining by fs laser pulses with focusing spots overlapping.*

Scanning of material by powerful femtosecond laser in regime of generation one long cavity (figure 5) with focusing spot overlapping can be powerful tool for cutting. But in several cases it is impossible to obtain a uniform cut with good quality due to self-induced shift of the focusing spot towards the objective [7].

Figure 6 shows microstructures created in PC sample by fs laser pulse scanning with 50 percent overlap. In this case you can see curved lines produced inside the bulk of material in contrast to the expected continuous cutting. There are lines with different inclination relative to the direction of propagation of the pulses. We think that this phenomenon is caused by a gradual shift of focus towards the lens from pulse to pulse. In this experiment have been used the second laser source.

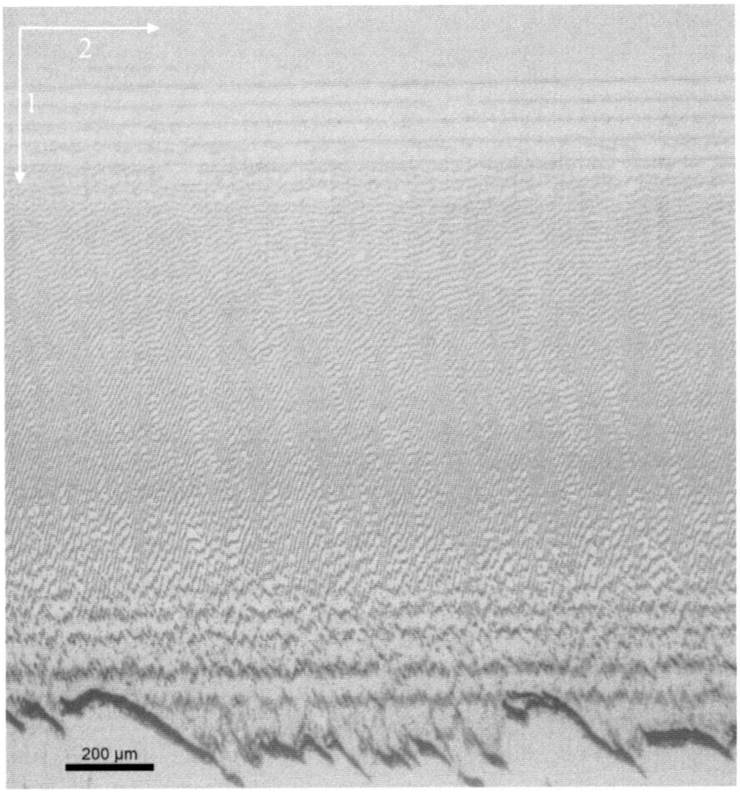

Figure 6. Picture of micromodifications inside PC sample. Pulse energy 26 μJ. (1) direction of pulse propagation; (2) scanning direction. Scanning speed is 40 μm/s. Pulse repetition rate 80 kHz. Pulse overlap 50%.

3.4 *Measurement of the electrical conductivity of the obtained microstructures.*

The blackening effect in the zone of the fs laser pulse impact in our view is evidence of material conversion to graphite-like form. A similar effect was observed in article [8]. The presence of graphite at the microstructure was confirmed by the results of electric conduction measurements in the layer of microscopic particles at the surface of the laser cut of the sample. The electric conduction of the layer is $\sigma \approx 1.0$ S/m^{-1}. It was estimated using the technique, analogous to [8].

4. Discussion of the results

Reasons for converting the Gaussian intensity distribution in the diffraction pattern at fs pulse passage through spherical lens may be different. In the first place it may be associated with the geometry of propagation of the beam through the lens. The second reason could be the non-linear interaction of powerful fs laser pulse with a lens material, analogous to [9]. This phenomenon requires additional research.

The formation of highly extended linear microstructures by focusing of fs laser pulses with a lens with strong spherical aberration may become a good tool in the production of high-aspect ratio microchannels and diffractive optics for various applications. Cutting of transparent materials by fs pulses in the long line generation mode is very promising method in terms of potential performance. However, as shown above, in some cases, it may not be applicable due to the effect of the self-induced displacement of the focus area.

The presence of the electrical conductivity of the obtained structures opens up opportunities for the creation of complex electro-optical devices in transparent materials.

5. Conclusions

Methods of producing of long length linear microstructures are presented. To create an extended microstructures were used a lens with strong spherical aberrations. In this case, the one femtosecond pulse generates a set of linear micromodifications in the bulk of material due to the conversion of a Gaussian beam into a set of diffraction rings. At energy of fs laser pulse about 150 μJ a set of linear micromodifications is combined into one line with a length of more than 2 mm.

Scanning of polycarbonate by fs laser pulses with 50% overlap of focusing spots leads to the formation of curved lines inside the material.

References

[1] Gattass R G , Mazur E 2008 *Nat. Photonics* **2** 219
[2] Schaffer C B, Brodeur A, Garcia J F, and Mazur E 2001 *Opt. Lett.* 26 93-95
[3] Vartapetov S, Ganin D, Lapshin K, Obidin A Z 2015 *Quantum Electronics* **45 (11)** 1029-1036
[4] Juodkazis S, Nishimura K, Tanaka S, Misawa H, Gamaly E G, Luther-Davies B, Hallo L, Nicolay P and Tikhonchuk V T 2006 *Phys. Rev. Lett.* **96 (16)** 166101.
[5] Kononenko T V, Zavedeev E V 2016 *Quantum Electronics* **46 (3)** 229.
[6] Ohfuchi T, Yamada Y, Sakakura M, Fukuda N, Takiya T, Shimotsuma Y, Miura K 2016 *Journal of Laser Applications* **28 (2)** 022603
[7] Vartapetov S, Ganin D, Lapshin K, Obidin A Z 2015 *Quantum Electronics* **45**(11), 1029-1036
[8] Morita N, Shimotsuma Y, Nishi M, Sakakura M, Miura K and Hirao K 2014 *Appl.Phys. Lett.* **105** 201104
[9] Dergachev A A , Ionin A A , Kandidov V P, Mokrousova D V , Seleznev L V, Sinitsyn D V, Sunchugasheva E S, Shlenov S A and Shustikova A P 2015 *Laser Phys. Lett.* **12 (1)** 015403

Journal of Physics: Conference Series **737** (2016) 012024 doi:10.1088/1742-6596/737/1/012024

Efficient material treatment by axi-symmetrically polarized laser radiation

V S Makin[1], Yu I Pestov[1], R S Makin[2]

[1]Public Limited Company Scientific Researxh Institute for Optoelectronic Device Instrumentation, Sosnovy Bor, 188540 Leningrad oblast, Russia
[2]National Research Nuclear University MEPhI (Moscow Engineering Physics Institute), Kashirskoe shosse 31, 115409 Moscow, Russia

E-mail: makin@sbor.net

Abstract. Recent years the increased interest is to the problem of interaction of nontraditionally polarized laser radiation with condensed media. The experiments with axisymmetrical polarization attract more attention. The peculiarities of interaction of axisymmetrical laser radiation with condensed matter are considered in framework of universal polariton model. It is shown that more effective is interaction of radially polarized laser radiation with surface active media. The optical schemes for efficient material treatment by radially polarized radiation are sketched.

For the increasing the efficiency of material treatment by laser radiation different approaches are known [1]. Conditionally they can be divided into two types: a) using different geometry of materials irradiation and b) using excitation of surface electromagnetic waves (waveguide modes, surface polaritons, localized surface plasmons, channel and wedge polaritons) at material interfaces and dissipation their energy in materials. Below the approach based on the surface polariton's excitation and their energy dissipation in materials will be considered. The consideration will be made on the basis of universal surface polariton model [2].

It is known that under the interaction of linear polarized laser radiation with materials having surface-active interfaces the linear gratings may be produced the grating (g) the periods of which are of the order of λ at near normal incidence, $d=\frac{\lambda}{\eta}$ and orientation $\mathbf{g} \parallel \mathbf{E}$, where λ is the wavelength of incident laser radiation and η is the real part of refractive index of surface polariton for considered boundary. For non normal incidence the angular dependence of the period arises.

The application of laser radiation with nontraditional polarization (for instance, axisymmetrical polarization) put forward the question about the resonant gratings parameters arising in framework of polariton model.

Let us consider the action of a radially polarized radiation (RPR) with selected direction of polarization φ_0, where $\varphi=\varphi_0$ is the angle characterizing (local) direction of the electric field vector of the incident wave. It is known that the SPP frequency is equal to that of the laser radiation, while its wave vector $\mathbf{k}_s$ is determined by the law of conservation of momentum (quasi-momentum) [3]:

$$k_s=\mathbf{k}_t+m\mathbf{g} \quad . \qquad (1)$$

where $\mathbf{k}_t$ is the tangent component of the wave vector of laser radiation, **g** is the vector of the resonance grating that allows conversion of laser radiation to the surface polaritons (SP), and *m* is the order of diffraction. At normal incidence of radiation, $\mathbf{k}_t=0$, $k_s=\mathbf{g}$, and

$$\mathbf{g} \parallel \mathbf{k}_s(\varphi_0) \ . \tag{2}$$

The induced grating is characterized by orientation (2) and period

$$d=\frac{2\pi}{|k_s|}=\frac{\lambda}{\eta} \ . \tag{3}$$

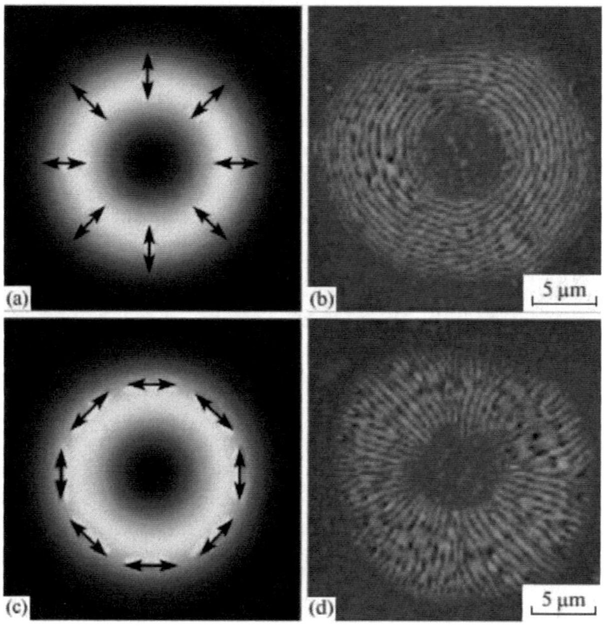

Figure 1. Distribution of (a, c) intensity and (b, d) corresponding microstructures formed under the influence of a train of femtosecond laser pulses with (a) radial and (c) azimuthal polarization on the surface of silicon [4].

If $\varphi \in [0, 2\pi]$, i.e., takes all possible values, there appears a grating in the form of axial ring structures with a period determined by (3). Hence, the resonance grating corresponding to interference of the radially polarized laser radiation with driven by it SPs represents axisymmetric ring structures with a period determined by (3). Based on similar arguments, it can readily be shown that the azimuthally polarized laser radiation causes formation of radial structures of damage.

The experimental results well support the suggested approach for ultrashort pulse duration radiation for semiconductors (monocrystalline silicon) [4, 5], dielectrics (quartz glass) [6] and metals (brass, stainless steel) [7- 8] under the action of pulses series of radial and azimuthal polarization, see Figures 1 and 2. In listed examples the effect was observed under the action of ultrashort laser pulses and for surface plasmon polaritons. The effect can also exists for longer pulse durations and for surface phonon polaritons. Really, the formation of circular gratings was observed under the interaction of series of chaotically polarized pulses of TEA CO_2 laser radiation (λ=10.6 µm) with quartz glass (frequency range of reststruhlen band). Possibly, in this case the radiation polarization direction chaotically changes its polarization direction from pulse to pulse. The period of the grating was near 9 µm.

Take attention to the peculiarity of dynamic structures with appear under the action of radially polarized laser radiation. The circular grating has a resonant period and directions of excited surface plasmon polaritons are perpendicular to the local grating direction. Hence part of the SPP are focused by surface grating into the spot diameter of $d=\frac{\lambda}{\eta}$ and another part is defocused and decays out of irradiated area, see Figure 3. So we have dynamic focusing of radiation which transfers SPP's energy to the spot center. In addition the radially polarized radiation itself is focused to the spot of smaller size [9] than linear polarized one. This results in effective reducing the size of overall other all energy dissipation and efficiently increases the power density of incident radiation.

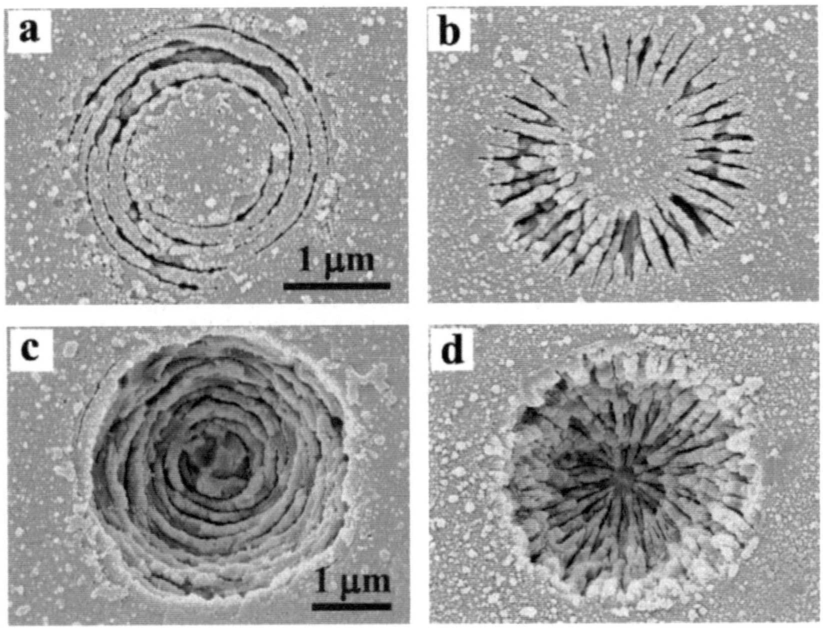

Figure 2. Ordered microstructures formed on the surface of fused silica by a train polarized radiation; (c, d) azimuthally polarized radiation. The power densities in Figs. 2b and 2d are 1.6 times higher than in Figs femtosecond laser pulses (λ = 775 nm, τ = 200 fs): (a, b) radially. 2a and 2b [6].

In universal polariton model the resonant grating **g** (for linear, circular and axisymmetrically polarized radiation) arises from small ones presented in the whole spectrum of initially rough surface during the series of laser pulses due to the positive feedback via grating amplitude (or value of dielectric permittivity module change). To increase the efficiency of material treatment we suggest the artificially creating resonant grating which will transfer incident radiation into SPP's. For *linear* polarized laser radiation the interference of two laser beams creates the artificial grating $\mathbf{g}_{inter.}$ which transforms incident radiation of one or two beams into SPP's [11]:

$$\mathbf{g}_{inter.} = \mathbf{g}_{res..} \tag{4}$$

One of possible examples of such grating, symmetrical incidence of laser beams and transformation of radiation from each beam is shown on Figure 4. In this case interference grating period is $d_{inter.} = \frac{\lambda}{2sin\theta}$ and $d_{res} = \frac{\lambda}{\eta - sin\theta}$. So using equation (4) for grating modules we obtain:

$$\theta = \arcsin\frac{\eta}{3}. \tag{5}$$

Here θ is the incidence angle for the case of symmetrical incidence, see Fig.4. In this casec the treated surface is placed in the place of maximal beams overlapping, see Fig.4. For the case $\eta \approx 1$ we have $\theta \approx 27°\text{-}28°$. The practical realization of symmetrical two beam scheme will be a simple use of

Fresnel bi-prism with appropriately chosen prism's angle. Note that it is well known that by optimizing the value of the resonant grating depth near 100% of incident laser radiation will be absorbed by the boundary with surface active media [12].

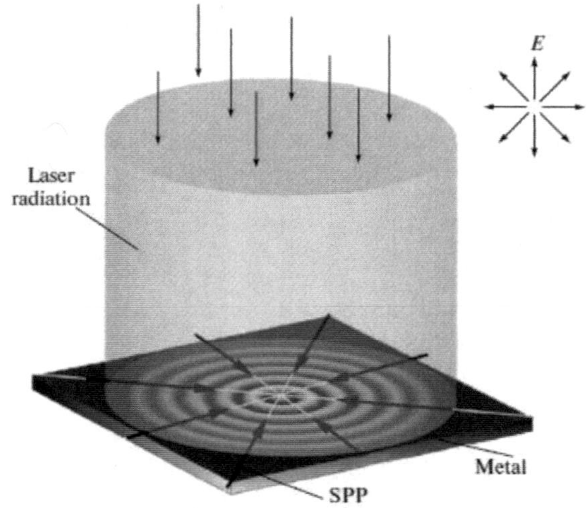

Figure 3. Schematic illustration of formation of an axisymmetric ring resonance grating on the surface of a condensed medium by radially polarized laser radiation; the grating converts incident radiation into surface polaritons that are focused onto the structure center. The inset shows polarization of laser radiation in the beam cross section [10].

This method can be used for resonant grating production of higher quality over large area via sample translation in the direction of SPP propagation.

Such approach can be applied to the case of radially polarized laser radiation. In this case we must create the circular resonant grating. The axicon optical element with proper chosen angle α (see equation (5)) can be used to create the circular grating, see Figure 5. Note that it will be desirable to take in mind the influence of the produced periodical surface modulations on the SPP's dispersion relation and fine alignment of axicon angle (β) in limits of half-width of resonant absorption line of SPP. The radially polarized laser radiation which illuminates the axicon can be obtained by different methods, for instance, with the help of the β-plate.

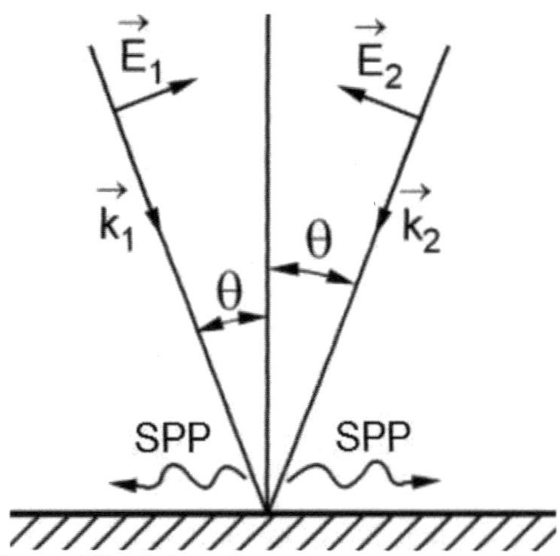

Figure 4. Optical scheme for artificial interference resonant grating formation by symmetrically incident p-polarized coherent laser beams and subsequent SPP excitation by each of beams via this dynamic grating.

Figure 5 shows a novel simple technique for generating radial polarization light from a linearly or circularly polarized laser light [14]. A birefringent plate of predetermined thickness is placed in between a telescope module. Light rays of a convergent beam or divergent beam travel different optical paths lengths when they transmit a birefringent plate. The light rays at the central part of the beam travel shorter distance than those at the edge through a birefringent material. This result in shift in electric field vector in the plane of the beam cross section is parallel to the radial direction.

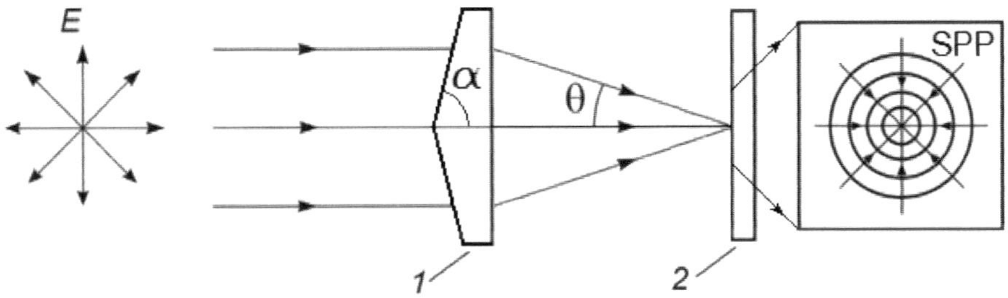

Figure 5. Optical scheme for materials treatment by radially polarized radiation. 1 – axicon, 2 – sample. The inset at right shows enlarged view of dynamic (residual) grating formation in irradiated zone. Note that axicon can be joint with β–plate. In this case the joint element must be irradiated by linear polarized radiation. When the possibility arises for material treatment by radially and azimutally polarized radiation.

Figure 6 shows a novel simple technique for generating radial polarization light from a linearly or circularly polarized laser light [13]. A birefringent plate of predetermined thickness is placed in between a telescope module. Light rays of a convergent beam or divergent beam travel different optical paths lengths when they transmit a birefringent plate. The light rays at the central part of the beam travel shorter distance than those at the edge through a birefringent material. This result in shift in electric field vector in the plane of the beam cross section is parallel to the radial direction.

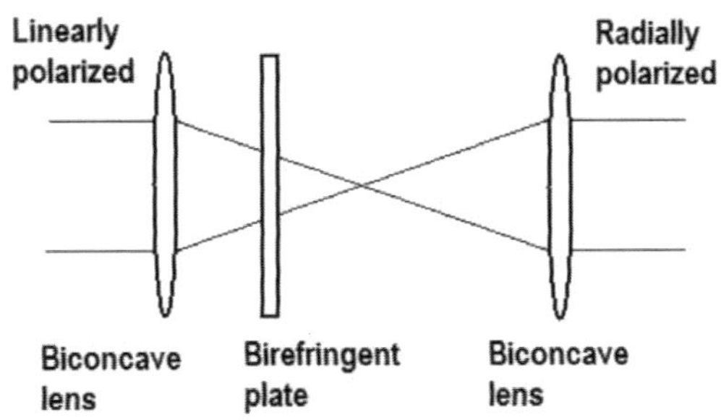

Figure 6. Polarization converter with birefringent plate in between a Keplerian telescope.

The influence of the (partially) radially polarized continuous wave CO_2 laser radiation on the metal cutting velocity has demonstrated an increase in cutting velocity up to 1.4 times in comparison with circular polarization [14, 15].

A diode pumped solid state Nd: YAG laser at 355 nm wavelength of 11 W average power (nanosecond range pulse duration) was used to provide the linearly polarized laser beam. The pulse repetition rate was from 50 kHz to 100 kHz. Then the scheme with birefringent plate was used it results in smaller ablation spot size on silicon plate and increased ablation depth due to shift the polarization state to radial. So radially polarized beam increases the machining efficiency by (40 ÷100) % and reduces the focal spot size by more than 30 % [13]. In the laser pulse silicon was melted and achieves the property of surface-active media.

In the regime of Q-switching the preliminary results of metal drilling by radially polarized radiation shave show a two-fold increase of the drilling speed [16]. Depending on the optical properties of the metal either radial or azimuthal polarization shows the best efficiency. For steel, a comparison to linear or circularly polarized laser radiation indicates that doughnut–shaped beam with azimuthal polarization is more energy efficient to produce holes of the same diameter and depth.

Note that one must take into account the further input to the effective material absorptivity from additional excitation by polarized laser radiation of channel and wedge surface plasmon polaritons, especially for ultrashort pulses [17].

In conclusion, the formation of circular grating and radial microstructures have been theoretically predicted in framework of universal polariton model and verified by experiments with series of ultrashort laser pulses and materials having different physical properties: semiconductors, dielectrics and metals. It was shown that radially polarized laser radiation more effectively interacts with surface active media. Number of examples has illustrated the efficiency effectiveness of interaction for continuous wave radiation, series of nanosecond and femtosecond pulses in cutting and drilling processes for metals, semiconductors and dielectrics. The optical schemes were suggested for further radially polarized laser treatment enhancement.

References

[1] Grigor'jantz A G, Shiganov I N and Misurov A I 2006 *Technological processes of laser treatment* (Moscow: MGTU N.E Bauman Publisher) 663
[2] Bonch-Bruevich A M, Libenson M N, Makin V S and Trubaev V V 1992 *Optical Engineering* **31** 718
[3] Akhmanov S A, Emel'yanov V I, Koroteev N I and Seminogov V N 1985 *Usp. Fiz. Nauk* **147** 675
[4] Lou K, Qian S -X, Wang X -L, Li Y, Gu B, Tu Ch and Wang X -T 2012 *Opt. Express* **20** 120
[5] Anoop K K, Rubano A, Fittipalti R, Wang X, Paparo D, Vechione A, Marucci L, Bruzzese R and Amoruso S 2014 *Appl. Phys. Lett.* **104** 241604
[6] Hnatovsky C, Shvedov V, Krolikovski W and Rode A 2011 *Phys. Rev. Lett.* **106** 123901
[7] Quyang J, Perrie W, Alligre Q J, Heil T, Jin Y, Fearon E, Eckford D, Edwardson S P and Dearden G 2015 *Optics Express* **23** 125662
[8] Anoop K K, Fittipalti R, Rubano A, Wang X, Paparo D, Vechione A, Marucci L, Bruzzese R and Amoruso S 2014 *J. Appl. Phys.* **116** 111302
[9] Dorn R, Quabis S and Leuchs G 2003 *Phys. Rev. Lett.* **91** 233901
[10] Makin V S, Makin R S 2013 *Optics and Spectroscopy* **115** 670
[11] Makin V S, Bazhenov V V and Libenson M N 1999 Patent RF No 1181413
[12] Hutley M S and Maystre D 1976 *Optics Communications* **19** 431
[13] Venkatakrishnan K and Tan B 2012 *JLMN Journal of Laser Micromachining* **7** 274
[14] Niziev V G and Nesterov A V 1999 *J. Phys. D: Appl. Phys.* **32** 1456
[15] Niziev V G and Nesterov A V 2000 *J. Phys. D: Appl. Phys.* **33** 1817
[16] Meier M, Romano V and Feurer T 2007 *Appl. Phys. A.* **86** 329
[17] Makin V S, Logacheva E I and Makin R S 2016 *Optics and Spectroscopy* **120** 118

Journal of Physics: Conference Series **737** (2016) 012025 doi:10.1088/1742-6596/737/1/012025

Interference filter with amorphous silicon layer and direct laser recording on it

A Kutanov[1], Nurbek Sydyk uluu[1], I Snimshikov[1] and Z Kazakbaeva[2]

[1]Institute of Physical-Technical Problems and Material Science, National Academy of Sciences, Chuy prospect 265-a, Bishkek 720071, Kyrgyz Republic
[2]Department of Computer Engineering, Kyrgyz-Turkish Manas University, Chyngyz Aytmatov Campus, Bishkek 720038, Kyrgyz Republic

E-mail:askarktnv@gmail.com

Abstract. The interference spectral filters with amorphous silicon layer deposited by magnetron sputtering on the reflective metal layer on a glass substrate are developed. Interference filter select from white light source components corresponding to quasi-monochromatic wavelength with a narrow bandwidth. The thickness of the amorphous silicon layer determines the center wavelength of the pass band of the filter. It proposed to use interference filter with amorphous silicon layer for direct laser recoding on it. Results on direct laser recording on amorphous silicon layer of the interference filter by single-mode Blu Ray laser ($\lambda = 405$ nm) with high contrast reflected image are demonstrated.

1. Interference spectral filters on the amorphous silicon films

The interference spectral filters obtained by magnetron sputtering of amorphous silicon films to the metal layer on the glass substrate are developed. Photo of interference filters with the amorphous silicon layer with different bands of light reflection is shown on figure 1. These filters have a very narrow bandwidth in the visible region of spectrum.

Figure 1. Photo of interference spectral filters with amorphous silicon layer with different color bands of reflected light.

Journal of Physics: Conference Series **737** (2016) 012025 doi:10.1088/1742-6596/737/1/012025

The structure of the developed interference filter with amorphous silicon layer is shown on figure 2. Amorphous silicon film thickness d determines the principal wavelength of the band pass filter.

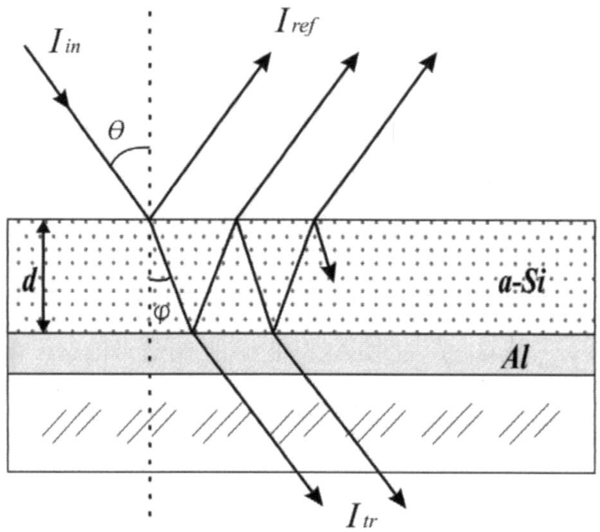

Figure 2. The structure of an interference filter with a-Si layer.

Interference filters with amorphous silicon layer reflect a part of the incident white light spectrum due to the phenomenon of multiray interference in thin dielectric films.

Multiray interference occurs when a light wave passing through the amorphous silicon layer with thickness d and refractive index n. θ and φ - angles of incidence and refraction rays at the upper boundary ($\sin\theta = n\sin\varphi$). The path difference for rays will be

$$\Delta = 2nd\cos\varphi \tag{1}$$

The reflection wave from the upper boundary of the layer in accordance with the Fresnel formulas its phase is changed to π. Therefore, the phase difference δ of folded waves is

$$\delta = (4\pi/\lambda_0)nd\cos\varphi \pm \pi = k\Delta \pm \pi \tag{2}$$

here λ_0 - wavelength in vacuum, $k = 2\pi/\lambda_0$

The multiray interference light intensity of the transmitted and reflected waves is described by the formulas Airy:

$$I_{tr} = I_{in}\frac{(1-R)^2}{1+R^2-2R\cos\delta} = I_{in}\frac{1}{1+(4R/(1-R^2)\sin^2(\delta/2)} \tag{3}$$

for reflected wave

$$I_{ref} = I_{in}\frac{2R(1-\cos\delta)}{1+R^2-2R\cos\delta} = I_{in}\frac{(4R/(1-R)^2\sin^2(\delta/2)}{1+(4R/(1-R)^2)\sin^2(\delta/2)} \tag{4}$$

The interference spectral filters with the amorphous silicon can be used in spectroscopy, fabrication of laser, for direct laser writing, laser marking and security applications.

2. Direct laser recording on a-Si layer of the interference filter

Method of direct laser recording on an amorphous silicon layer is based on the phenomenon of the direct formation of a micro relief by focused laser radiation. There is laser induced transition from amorphous to polycrystalline silicon by local laser heating and temperature rise. Previously, we demonstrated the possibility of direct laser writing on the a-Si films by laser radiation with different wavelengths [1]. Direct laser recording on amorphous silicon film is attractive for diffractive structures recording without wet chemical processing and it bring new possibilities of diffractive structures fabrication with the use of laser writing device [2]. The use of high-speed scanning system of interference lithography (SIL) [2] forming diffractive structures by consecutive writing of moderate-size fragments or cells is expecting to assist in solving the problem of deposition of antireflective diffractive coatings onto large-size surfaces of solar panels. Early, we demonstrated direct laser recording of dot holograms on amorphous silicon film by pulsed UV laser with λ= 355 nm[3].

In this paper we demonstrate possibility of direct laser recording on a-Si layer of the interference filter. Results of direct laser recording on amorphous silicon film of interference filter using single mode Blu Ray laser with λ= 405 nm is shown on figure 3. The Latin letter L is chooses for test direct laser recording on a-Si layer of interference filter. The experimental results demonstrate change in transmission of the amorphous silicon layer after its exposure by focused laser radiation. Registration layer becomes more transparent in place interacted by laser radiation. Early, measurements of absorption spectrum demonstrated that amorphous silicon layer's reflex index is changing after exposure by focused laser beam [1]. Thus, in the area of interference filter exposed by focused laser radiation multiray interference is not working and we could see contrast image of letter L in the reflected light on the blue background from the interference filter.

For direct recording on amorphous silicon the Mitsubishi 120mW single mode Blu Ray laser is used. Laser beam is collimated by aspheric lens after laser diode and then it focused by micro objective to recording media. Computer controls laser pulse duration and their frequency. For direct laser recording the amorphous silicon films with varying thickness deposited by magnetron sputtering on a glass substrate are used.

Figure 3. Direct laser recording on the a-Si layer of the interference filter by Blu Ray laser with a λ= 405nm.

As seen from photo on figure 3, direct laser writing on layer of a-Si interference filter provides a good contrast of the image in the reflected light due to the refractive index change and the transparency of the recording layer, the formation of the relief structure after interaction by laser radiation.

The relief formed by direct laser recording on a-Si layer of interference filter is studied on scanning electron microscope. A relief structure formed by a focused laser beam on a-Si layer is shown on photo taken on scanning electron microscope (figure 4).

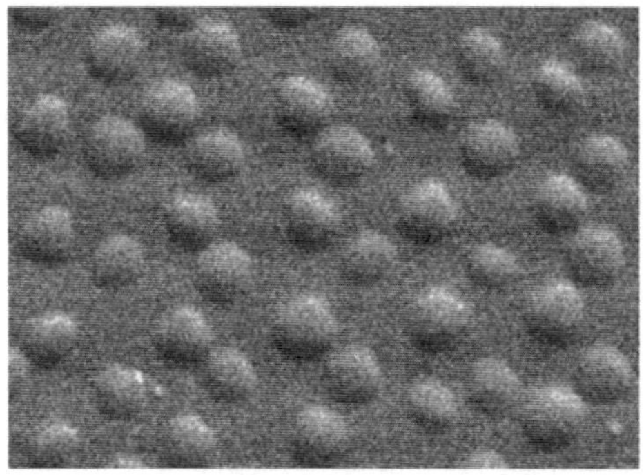

Figure 4. Photo of relief structures formed by a focused laser beam on a-Si layer taken on scanning electron microscope.

It's shown that under interaction of *a*-Si film by focused laser beam the local crystallization appears and relief is forms with height ~100 nm. Diameter of hemispheres induced by focused laser radiation of Blu Ray laser on a-Si layer is about 40 micron.

The drawback of single mode laser diode used for experiments was a short length of coherence and it dependence from output radiation power. Thus, it was difficult to record dot holograms by Blu Ray laser for fabrication of master matrix hologram for embossing holograms reproduction. Therefore, it will be useful to use Blu Ray laser with the greater length of coherence for dot holograms recording.

Direct laser writing on the amorphous silicon film could be used for fabrication of diffractive optical elements (DOE), security applications.

Summary

The interference spectral filters with amorphous silicon layer deposited by magnetron sputtering on the reflective metal layer on a glass substrate are developed. Direct laser recoding on amorphous silicon layer of interference filter is proposed. Results on direct laser recording on amorphous silicon layer of the interference filter by single-mode Blu Ray laser (λ = 405 nm) with high contrast reflected image are demonstrated. Laser radiation induced relief formation on a-Si layer is studied on scanning electronic microscope. It makes attractive use of direct laser recording on amorphous silicon for laser marking, DOE fabrication and security applications. Use of single mode Blu Ray laser opens opportunities to design compact laser device for micro structures fabrication for different applications.

References

[1] Kutanov A, Snimshikov I and Nurbek Sydyk uulu 2015 *Physics Procedia* **73** 82-86
[2] Poleshchuk A G, Kutanov A *et* 2010 *Optoelectronics, Instrumentation and Data Processing* **46** 171–180
[3] Kutanov A, Snimshikov I 2011 *Proceeding Optics Photonics Japan* 30aCS2

Journal of Physics: Conference Series **737** (2016) 012026 doi:10.1088/1742-6596/737/1/012026

Influence of electro-chemical etching parameters on the reflectance spectra of porous silicon rugate filters

R A Rakhimov, E V Osipov, D S Dovzhenko, I L Martynov, A A Chistyakov
National Research Nuclear University MEPhI (Moscow Engineering Physics Institute), 31 Kashirskoe shosse, 115409 Moscow, Russia

E-mail: dovzhenkods@gmail.com

Abstract. Today the porous silicon photonic crystals are widely used as components of the optical sensors. The structures with the resonance properties, e.g. a narrow reflection band, which provides high sensitivity of the device, generate considerable interest. One of the most interesting structures of this type are rugate filters. Rugate filter is a photonic crystal with a gradually varying under harmonic law index of refraction and narrow spectral reflection band. In this paper, we present our results of the optimization of silicon electrochemical etching parameters, in order to fabricate filters in a visible spectral range with reflection bandwidth of less than 30 nm and a reflectance higher than 80%.

1. Introduction

Currently, optical sensors with fast response to the change of external conditions and high sensitivity are of interest. Special attention is drawn to the sensors based on photonic crystals (PhC), which provide the opportunity for management of the optical radiation [1]. One of the PhC type, for use in optical sensors application, is rugate filter [2], which have relatively narrow reflection spectrum in a contrast to conventional Bragg mirror. PhCs with sorption properties are of special interest, particularly porous silicon (pSi) photonic crystals have a large specific surface area [3] and branchy pore system. Photonic crystals based on pSi, including rugate filters, could be obtained by electro-chemical etching of silicon in aqueous-alcoholic solution of hydrofluoric acid [4]. The key optical characteristics of the fabricated samples depend on the etching parameters such as current density, current profile, etc. Frequently these relations are non-linear, thus an experimental selection of optimal etching parameters is required to obtain rugate filters with a narrow band and high reflectance.

2. Experiment

We used p-type silicon wafers with a resistivity of 0.003-0.005 Ohm·cm and surface orientation (100), manufactured by Telecom-STV, for the fabrication of rugate filters. Pure isopropyl alcohol, methanol and hydrogen peroxide were used for auxiliary operations during the sample production. Rugate filters based on porous silicon were fabricated using the standard technology of electro-chemical etching of monocrystalline silicon wafers. We took a water-alcohol solution of hydrofluoric acid (HF: C_2H_5OH - 1: 2). The etching current density varied in the range of 4-47 mA / cm^2. Current profile was sine-shaped in the vicinity of a base value. All silicon samples were pre-cleaned using ultrasound in isopropyl alcohol before etching and sequentially washed in methanol and hexane after. The final stage of rugate filters fabrication was their oxidation in solution of hydrogen peroxide to stabilize the surface chemical properties.

The etching of monocrystalline silicon was produced using programmable DC power KEITHLEY 2600A. Measurements of the reflectance spectra were carried out at the facility, which contained spectrometer OceanOptics USB2000+ with reflection fiber probe as the main element.

3. Results and discussion

To predict the optical parameters of manufactured rugate filters, it is necessary to determine the relation between the porosity value, and thus the structure refractive index, and parameters of the etching current. In order to solve this problem we made a series of porous silicon monolayer samples obtained with the use of different etching time and currents. Then, we measured mass of material removed in the etching process and the thickness of the porous layer for all the samples. To measure the mass of removed agents, each sample was weighed twice before and after etching. Measurement of the mass of samples was carried out using scales with an accuracy of 1 mg. In order to make sure that the electrolyte was completely removed from the porous layer after the etching we made the test weighing of pSi samples 3 days after the etching procedure. Control experiment showed similar results. The thicknesses of the porous layers were determined using SEM images of sample cross-sections, fig. 1. The obtained thickness of the etched layers allowed us to calculate the speed of sample etching and porosity at the different current settings. The porosity was calculated using the following equation:

$$P = \frac{m_1 - m_2}{\rho \cdot l \cdot S} \cdot 100\%,$$

where m_1 and m_2 - sample weights before and after etching respectively, l - thickness of the porous layer, S - etching surface area. The results are presented at Fig. 2.

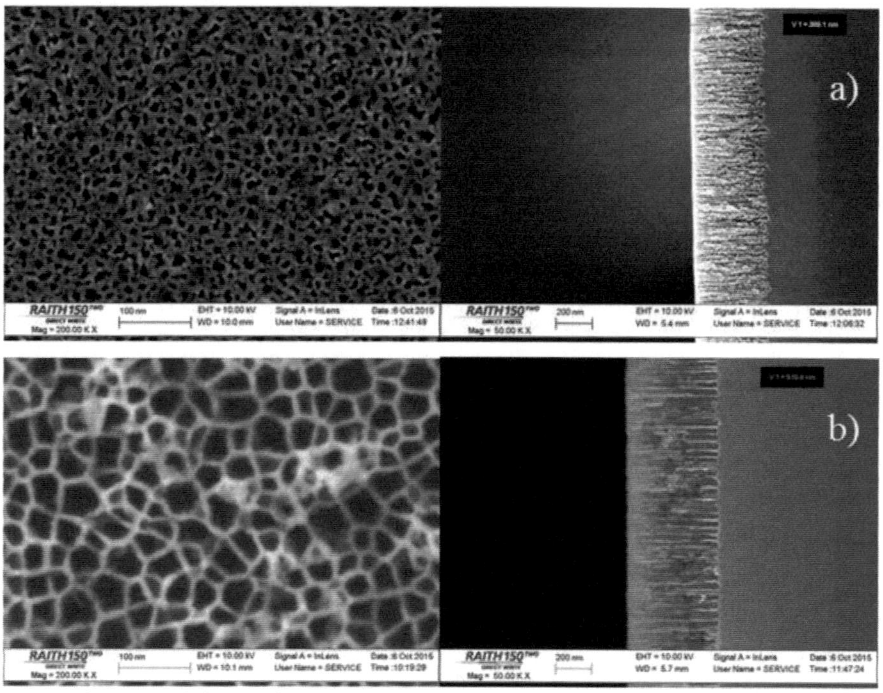

Figure 1. SEM images of the surface (left) and cross-section (right) of monolayer porous silicon obtained by etching current density of 4 (a) and 47 (b) mA/cm^2.

The experimental results allow one to investigate the dependence of the effective refractive index on the value of PC etching current and link the etching time with the thickness of the resulting porous layer using the Bruggeman model [5]. Hence, it becomes possible to compare the results of theoretical modeling of rugate filters optical properties with the parameters of the experimental profile of the etching current density. This principle has been realized with the use of transfer matrix method to calculate optical spectra of multilayer structures.

It is known [6], that the profile of the refractive index n in rugate filters varies sinusoidally. Therefore, each half-cycle of the n change was divided into 32 discrete elements in the theoretical modeling of reflectance spectra. Within each element n considered to be permanent.

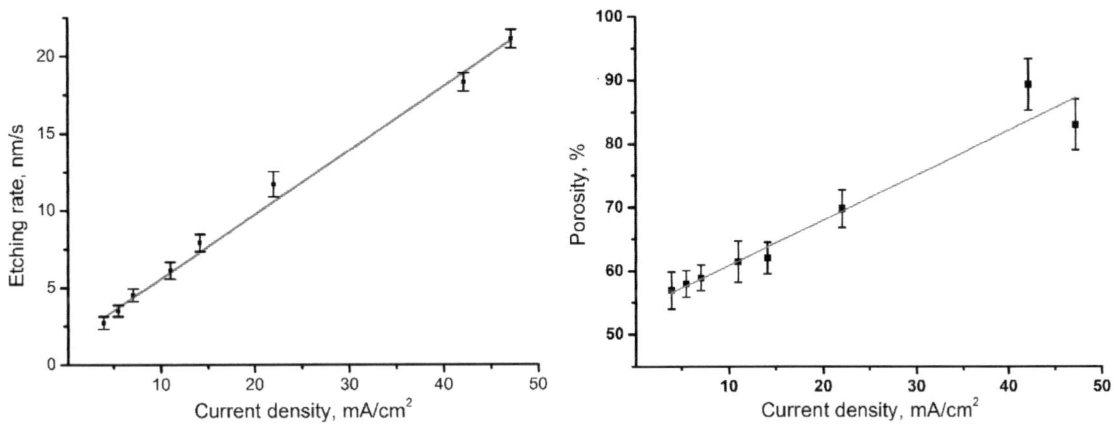

Figure 2. The dependence of the etch rate (left) and porosity (right) of the formed layer on the magnitude of the pSi etching current.

The resulting model helped in optimization of the parameters of rugate filters based on pSi. Detailed calculations have been carried out for a filter with a reflectance spectrum maximum in the range of 550-650 nm. The first stage was the examination of the dependence of the rugate filters optical properties on the number of layers (periods) of the refractive index harmonic variations. A typical single layer thickness in this case was 160 nm for the mean value n = 1.8. It was found that a sharp increase in the maximum reflectance and narrowing of rugate filter spectral reflection band occurs as the number of layers N grows from 0 to 50. The significant improvement of the reflection profile shape and decrease of the secondary peaks amplitude of more than 25% was observed in the range of 50-100 number of structure layers (periods). Further layers number growth is not advisable, as it is not accompanied by a significant change in rugate filter parameters. Next, we investigated the effect of a smooth change in current density at the beginning and at the end of etching on the reflection spectra. This method theoretically allows one to reduce the difference in refractive index at the pSi/air boundary, which should lead to a further decrease of the secondary peaks intensity. The calculation results have confirmed this assumption and demonstrated reduction of the secondary peak amplitude more than a factor of 2, which corresponds to the literature data [6]. We placed extra emphasis on the dependence of the rugate filter reflection bandwidth on the amplitude change of the refractive index (etching current). Reducing of the amplitude modulation of the refractive index Δn, as expected, resulted in the narrowing of the high-reflection bandwidth ($\Delta\lambda$), but the value of the reflection coefficient R decreased at the same time. However, as can be seen from Fig. 3 there is an optimal ratio between the values Δn, $\Delta\lambda$ and R. Thereby, in case of rugate filter with a number of layers (periods) of the structure N = 100 and the mean value $\langle n \rangle = 1{,}8$ (which corresponds to the porosity of 62%) the optimal value of Δn is 0.05.

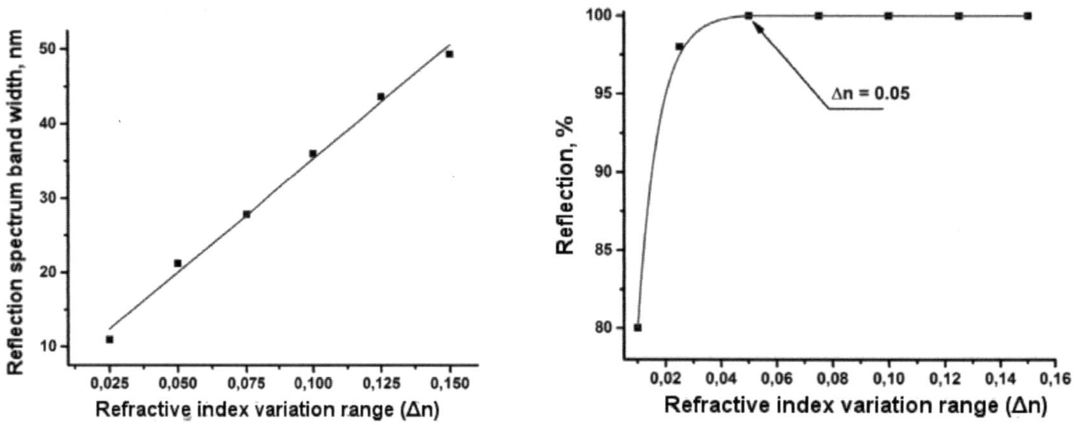

Figure 3. The dependence of rugate filter bandwidth (left) and its reflection (right) on the modulation of the refractive index values of n = <n> ± Δn. The calculation was performed for rugate filter with <n> = 1,8, single layer thickness (period) of 160 nm and the number of layers N = 100.

To examine our calculation we compared mathematical modeling results with experimentally measured reflection spectra for a series of rugate filter samples fabricated using different parameters for etching current. The calculated and experimentally measured reflection spectra of the rugate filter with a maximum reflectance at a wavelength of 615 nm represented as an example at Fig. 4.

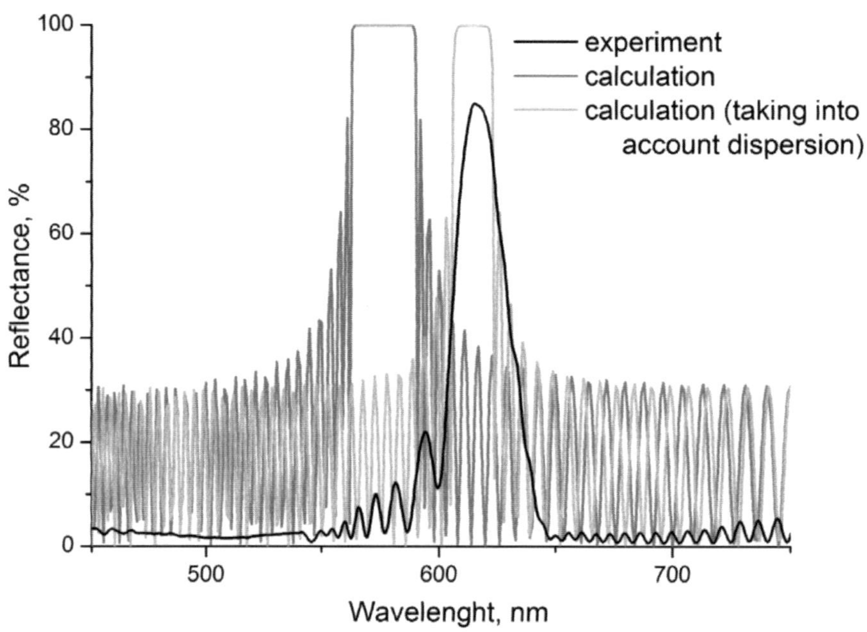

Figure 4. Comparison of the calculated and experimentally measured reflection spectra of porous silicon rugate filter with the reflection spectrum peak wavelength of 615 nm. Etching current - 7.5 mA, etching current modulation - 1.5 mA, the thickness of a single rugate filter layer - 160 nm, the number of layers - 100.

It can be seen that the measured width of rugate filter equals 30 nm and is in a good agreement with the calculated value. The maximum reflectivity of experimental spectrum is lower than calculated one, but this fact is obviously caused by the defects in porous silicon structure and its absorption in the visible range. Furthermore, there is a significant shift between high reflectivity band position for measuring and calculated spectra, but this difference disappears when dispersion of refractive index is taken into account in calculation [7].

4. Conclusion

In this work we have investigated the dependence of reflectance spectra of the porous silicon rugate filters on electrochemical etching parameters. For silicon p-type KDB monocrystalline wafers with resistivity of 0.003-0.005 Ohm·cm and a surface orientation (100), we measured the dependence of the porosity and the etching rate on the etching current density. We have demonstrated that the modeling of the optical properties of the pSi rugate filter in the visible range could be successfully performed using transfer matrices method and Bruggeman effective refractive index model. The calculation results allow us to optimize the etching process and fabricate rugate filters with maximum reflectance in a range of 550-650 nm, less than 30 nm bandwidth and a reflectance above 80%.

Acknowledgments
This study was supported by grant of Russian Federation President no. MK-4937.2016.2 and grant of Ministry of Education and Science of Russian Federation no 3.1718.2014/K (State Task no 1718).

References

[1] Dovzhenko D, Osipov E, Martynov I and Linkov P 2015 *Proceedings of International Conference NANOMEETING 2015* pp 144-147
[2] Bovard B G 1993 *Appl. Opt.* **32** 5427
[3] Halimaoui A 1994 *Surf. Sci.* **306** L550–4
[4] Lehmann V and Gösele U 1991 *Appl. Phys. Lett.* **58** 856
[5] Sailor M J 2011 *Porous Silicon in Practice* (Weinheim, Germany: Wiley-VCH Verlag GmbH & Co. KGaA)
[6] King B H 2011 *J. Nanophotonics* **5** 051510
[7] Green M A and Keevers M J 1995 *Prog. Photovoltaics Res. Appl.* **3** pp 189–192

Modification of properties of the rapidly quenched TiNiCu alloy underlaser irradiation

K A Borodako[1], A A Grigoriev[1], A V Shelyakov[1], N N Sitnikov[1,2], A A Ivanov[1], V Yu Fominski[1] and D V Shayfer[3,4]
[1]National Research Nuclear University MEPhI (Moscow Engineering Physics Institute), 31 Kashirskoe shosse, 115409 Moscow, Russia
[2]Federal State Unitary Enterprise "Keldysh Research Center", 8 Onezhskaya str., 125438 Moscow, Russia,
[3]Deutsches Elektronen-Synchrotron, Hamburg, Germany
[4]The Hamburg Centre For Ultrafast Imaging (CUI), Hamburg, Germany

E-mail: AVShelyakov@mephi.ru

Abstract. The work deals with the study of the influence of combined action of thermo-mechanical and laser treatments of rapidly quenched TiNiCu thin ribbons on the properties of two-way shape memory effect. It was shown that increasing of the energy density of the laser radiation and the external mechanical stress leads to growth of the reversible strain of the received amorphous-crystalline composite in the interval of martensitic transformation.

1. Introduction

Progress of knowledge intensive technologies in most industrial branches has been caused by miniaturization, rise of economy and response speed of actuator elements. In last decades micro-electromechanical systems (MEMS) find rapidly their ways into a wide range of applications, such as aviation, automobile, nano-technological or biomedical devices [1-3]. That leads to a necessity of development and investigation of new functional materials, being capable of operating at the micro- and nano level and creating new principles of design of microsystems.

Efficiency of the use of alloys with the shape memory effect (SME) for making of different-type microdevices was demonstrated recently [4-8]. Practical use of SME alloys in micromechanical devices assumes, as a rule, a reversible change of shape in a heating-cooling cycle. Our recent demonstration has revealed [9, 10], that laser treatment of surface of thin TiNiCu ribbons, produced by melt quenching, results in forming amorphous-crystalline composite, that is capable of doing cyclic work due to realization of the two-way SME. This work has been aimed at research of the effect of the combined action of thermal-mechanical and laser treatment of the quenched TiNiCu alloy on its thermal deformational characteristics on realization of two-way SME.

2. Materials and techniques

The work had to do with the alloy of quasi-binary TiNi-TiCu system at 25% of the copper atomic content, produced in the form of amorphous ribbons of 40 μm thick and 2 mm wide with using rapid quenching from the liquid state [11-16]. Crystallization of the amorphous ribbon and specific shape setting were conducted through isothermal annealing in a muffle furnace at temperature of 500°C in the course of 5 minutes. It has been revealed previously [17], that such process of heat treatment leads to the formation of submicrocrystalline structure with the average size of grain from 400 to 600 nm and manifestation of the pronounced SME.

The technique for forming the two-way shape memory effect (TWSME) through combined action of thermal-mechanical and laser treatment of the ribbon out of the rapidly quenched TiNiCu alloy is schematically shown in Fig.1.The ribbon is first endowed with memory of rectangular shape with using a special mandrel (Fig.1a). Then it is lengthened in the longitudinal direction under the mechanical load, σ, in the process of forward martensitic transformation (MT) up to a certain strain, ξ. Next the ribbon surface is subjected to the action of short pulse of laser radiation at the energy density, ample for heating of the sample surface layer above the melting temperature. As it was revealed in [9], as a result, a laminated structural composite with the interface, separating the amorphous and crystalline states into layers, is being formed. If this composite would be heated above the temperature A_s of the start of the reverse MT in alloy, then, due to SME realization, the crystalline layer would strive to compression, which gives rise to composite bending at a certain radius R_{min}, like a bimetallic plate (Fig. 1b). On cooling below the temperature M_f of the end of the forward MT the ribbon is partially straightened up to the bending radius R_{max} due to elasticity of the amorphous layer, acting as a restoring force. Owing to the transformation plasticity effect [18], with the cooling during the forward MT, monodomain formation of martensite variants takes place in the crystalline layer and large strains, oriented along the direction of the applied elastic force of the amorphous layer, are accumulated. So, significantly lesser forces are required to deform the crystalline layer, and in which connection the amorphous layer thickness can be much less (a few times or even one or two orders), than that of the crystalline layer. Later on in the heating-cooling cycle reversible bending strains of the sample happen, i.e., the amorphous-crystalline composite ribbon, produced by this technique, is capable of showing evidence of TWSME.

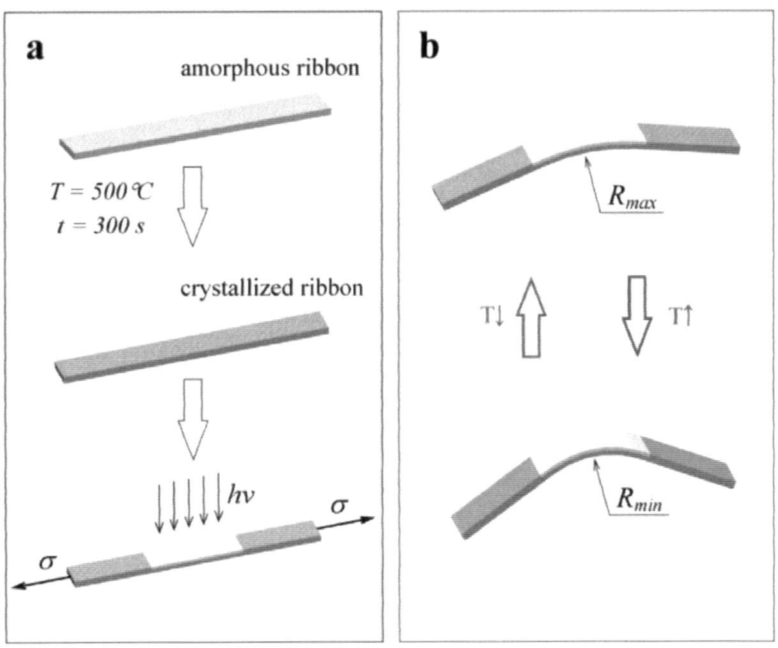

Figure 1. Diagrammatic representation of the technique for forming TWSME through the combined action of thermal-mechanical and laser treatment of the SME alloy ribbon: (a) – the treatment of the ribbon, (b) TWSME in the ribbon.

In accordance with the preliminary calculations the TiNiCu alloy ribbons were treated with using a single pulse (its duration – 20 ns) of radiation of the CL7000-series excimer laser with gas mixture KrF, with the wavelength - 248 nm, the laser energy density thereat was varied from 5 to 17 mJ/mm^2.

Metallographic cross-sections of the ribbons in hand were manufactured on BUEHLER® firm equipment: precise cutting-off machine «Isomet 1000», automatic hydraulic press for shrink fitting «Simplimet 1000», grinding-buffing machine «EcoMet 250+AutoMet 250».

Microstructure of the cross-section of the samples produced was studied by scanning electron microscope (SEM) «FEI Quanta 600 FEG» with field-emission cathode and integrated attachment for energy-dispersion X-ray microanalysis EDAX.

X-ray structural analysis was performed with the sliding X-ray beam method on diffractometer D8 Discover. X-ray lines profiles were obtained by survey in Cu Kα_1 radiation with pyrolytic graphite monochromator.

3. Results and discussion

According to the technique, proposed for the formation of TWSME, prior to the laser radiation action the ribbon with memory given to have rectangular shape, is extended lengthwise up to a certain strain ξ. To gain the maximum value of ξ the mechanical load σ is applied due to the suspension of a pertinent weight, P, to the ribbon, its heating above the temperature A_f, and its subsequent cooling below the temperature M_f. The process of the forward MT results in monodomain formation of the alloy (alignment of martensite domains along the load σ direction), which provides accumulation of the strain being maximum for the used magnitude of σ. By thermal cycling at the MT interval under loading the ribbon changes reversibly its length. If in the initial state it has the length l_0, on the cooling below M_f it is extended to l_1, but by the heating above A_f it decreases to l_2, then, the magnitude of the reversible strain makes up $\xi_r = (l_1-l_2)/l_0$. In so doing, at the great σ the ribbon can, when heating, not restore some strain $\xi_o = (l_2-l_0)/l_0$ because of plastic deformation.

Shown in Fig. 2 is the dependence of ξ_r and ξ_o on the magnitude of the applied mechanical stress σ by thermal cycling at the MT interval. As is seen, at small stresses (σ<60 MPa) the ribbon restores the strains practically fully, however, with increasing σ up to 120 MPa the magnitude of residual deformation ξ_o is getting noticeable, which has a negative effect on ribbon operating capacity with SME realization.

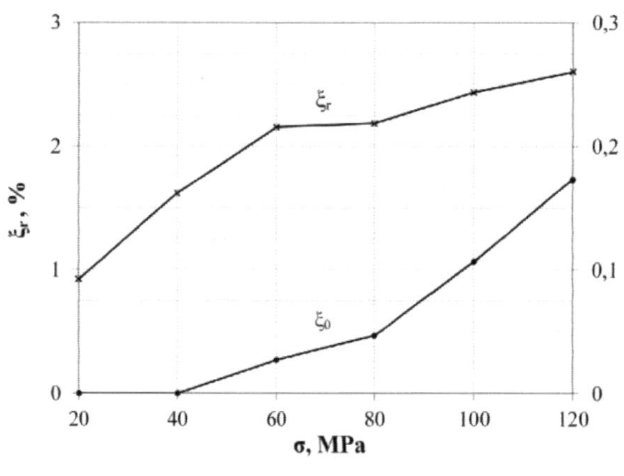

Figure 2. The dependence of the reversible ξ_r and the residual ξ_0 strains on the applied mechanical stress

The work dealt with the rapidly quenched ribbon of the $Ti_{50}Ni_{25}Cu_{25}$ alloy, treated by a single pulse of laser radiation with different energy density E from 5 to 17 mJ/mm^2, the value of tensile mechanical stress thereat varied in the range 20 to 120 MPa.

For the study on microstructure, metallographic cross-sections of the ribbons subjected to the action by the laser pulse of different power, were made. Representative electron microscope images of the cross-sections are given in Fig. 3 (for E = 5 mJ/mm^2 and E = 17 mJ/mm^2). It can be seen that at the ribbon surface the modified (amorphous) layer, separated by a sharp boundary from the rest of the crystalline ribbon bulk, is formed, as it was evident in previous works too [9, 10]. With the increase of the laser energy density the thickness of the amorphous layer grows from 0.4 to 1.4 μm.

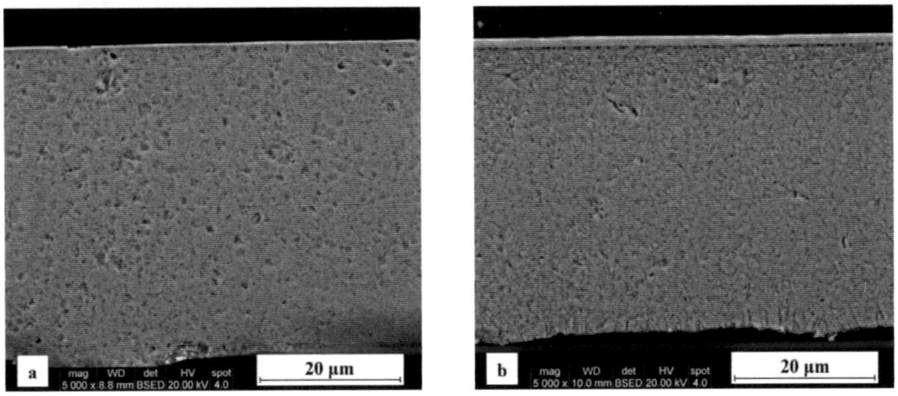

Figure 3. Metallographic cross-section of the ribbon treated by laser radiation at different energy density: 5 (a) and 17 (b) mJ/mm^2

In the course of X-ray structural analysis at the ambient temperature, a series of diffraction patterns was obtained both of the initial untreated ribbon (Fig. 4a), and samples after the laser action at different energy densities (Fig. 4b). On the diffraction pattern of the untreated ribbon characteristic peaks of reflection of the martensite phase B19 are evident [17]. It can be seen, that with a growth of energy of the laser pulse an appreciable decrease in most intense reflection peaks takes place near the angle 2θ=42°. This is indicative of the reduction of a part of the crystalline phase in the area of the laser beam action due to partial amorphisation of the alloy and of the formation of the surface amorphous layer.

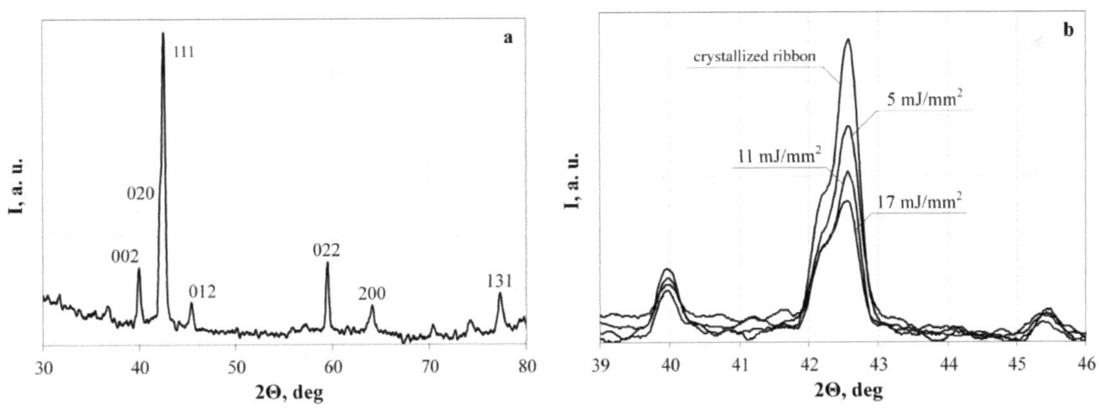

Figure 4. Diffraction patterns of the ribbon in the initial untreated state (a) and after the laser beam action with the different energy density (b).

Thus, structural investigations have revealed, that the$Ti_{50}Ni_{25}Cu_{25}$ alloy ribbons, subjected to the laser action, are the laminated amorphous-crystalline composite.

It was further established that the amorphous-crystalline composites produced have an ability to execute reversible angular displacements, as on Fig. 1б. For each composite, the bending radiuses R_{min} (in the "hot" state above the temperature A_f) and R_{max} (in the "cold" state below the temperature M_f) were determined, as well as the relevant strains, $\xi_{min} = D/2R_{max}$ and $\xi_{max} = D/2R_{min}$, where D is the thickness of the ribbon, were calculated. The reversible strain $\Delta\xi = \xi_{max} - \xi_{min}$ characterizes in essence the magnitude of the TWSME.

Presented in Fig. 5 are the bending strains ξ_{min} and ξ_{max}, as well as $\Delta\xi$, of the amorphous-crystalline composite as a function of the stress σ for the samples, treated at the three different values of the laser pulse energy density: 5, 11 and 17 mJ/mm^2.

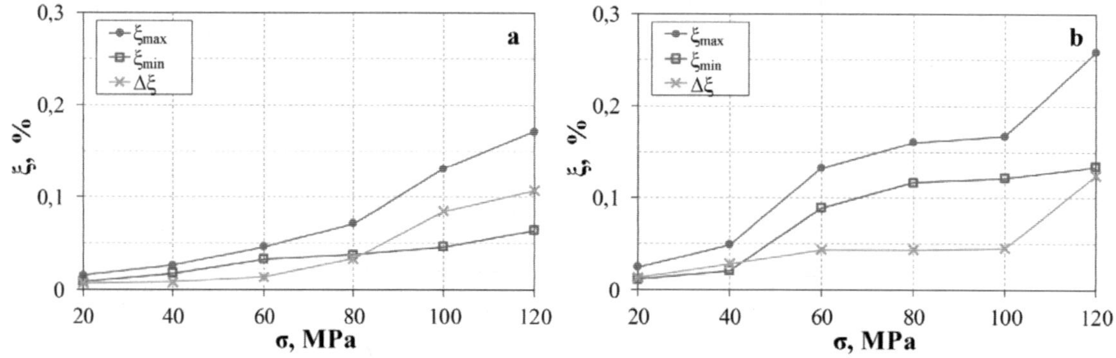

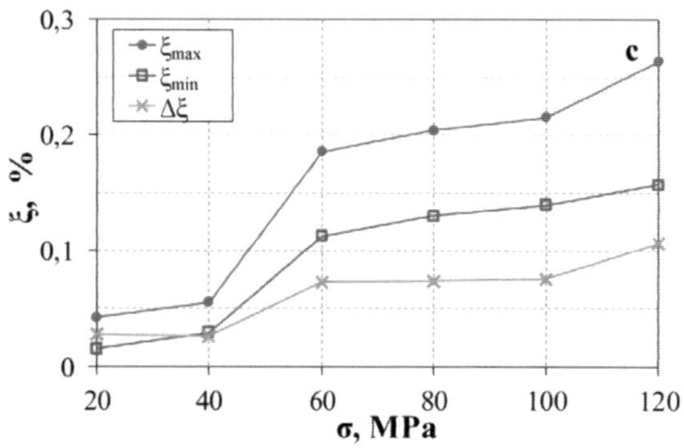

Figure 5. The bending strain of the amorphous-crystalline composite of TiNiCu alloy as a function of tensile stress after laser treatment at different energy density: 5 (a), 11 (b) and 17 (c) mJ/mm^2

It can be seen, with increasing the applied stress σ the bending of the composite grows. The reason is that at the sample tension with a growth of σ, a greater strain is accumulated in the crystalline layer (Fig. 3), which it strives to return with heating, which, in its turn, results in the increase in ξ_{max}. However, with increasing σ over 40 MPa, ξ_{min} increases appreciably too, which is caused by the accumulation of residual strain. As a result, at greater values of E (11 mJ/mm^2 and 17 mJ/mm^2) and, accordingly, the greater thickness of the amorphous layer, the reversible strain $\Delta\xi$ has somewhat greater values, but it does not essentially change with the growth of σ in the range 60 to 100 MPa.

Besides, the influence of the laser radiation energy density E on the composite strain was investigated at the fixed stress σ. The characteristic dependences ξ_{min}, ξ_{max} and $\Delta\xi$ are shown in Fig. 6 at σ = 80 MPa. It was established, that with increasing energy density from 5 to 17 mJ/mm^2 the reversible strain $\Delta\xi$ is nearly doubled. This may result from the fact, that, as was mentioned above, the increase in the laser beam power give rise to the formation of the thicker amorphous layer, that deforms the crystalline layer to a greater extent with cooling.

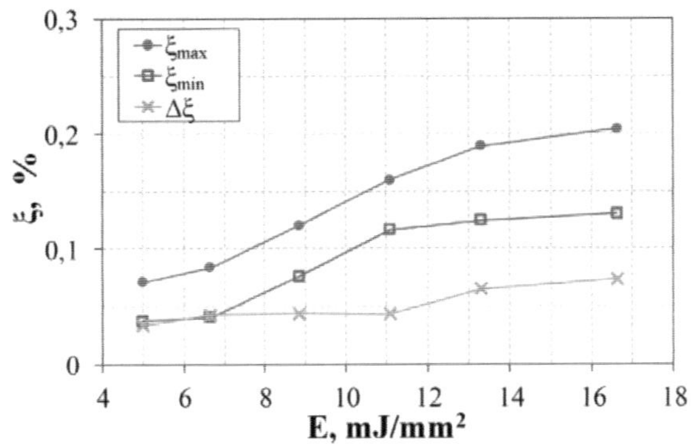

Figure 6. The bending strain of the amorphous-crystalline composite out of TiNiCu alloy as a function of laser radiation energy density (σ = 80 MPa)

Thus, the combined action of thermal-mechanical and laser treatments of the rapidly quenched TiNiCu alloy makes it possible to produce the laminated amorphous-crystalline composite having the TWSME properties. In so doing, combined variation of the tensile mechanical stress and the energy density of pulsed laser radiation enables to gain optimal parameters of the composite for a specific problem, both in the magnitude of reversible angular displacements, and in the range of angles, where these displacements take place.

4. Conclusion

The work has to do with investigation into the rapidly quenched ribbons of the $Ti_{50}Ni_{25}Cu_{25}$ (at%) alloy, subjected to the combined action of external mechanical stress and pulsed laser radiation (λ = 248 nm, τ = 20 ns). The interaction of laser radiation and surface of the alloy was simulated, which allowed choosing parameters of the combined action: the magnitude of the tensile stress was varied from 20 to 120 MPa, the laser radiation energy density – from 5 to 17 mJ/mm^2.

It has been revealed that after this treatment the laminated amorphous-crystalline composite with the distinct two-way SME is formed. It has been established, that greater amplitude of the reversible bending strain of the amorphous-crystalline composite at the MT interval can be gained both with the creation of high stresses in material, and the action of a high-power laser pulse.

Acknowledgements

The work was partly supported by the Ministry of Education and Science of the Russian Federation within the framework of the state assignment 3.1540.2014/K and by the Competitiveness Enhancement Program of National Research Nuclear University MEPhI.

References

[1] Jani J M, Leary M, Subic A and Gibson M A 2014 *Mater. Des.* **56** 1078
[2] Otsuka K and Ren X 2005 *Prog. Mater. Sci.* **50** 511
[3] Sun L, Huang W M, Ding Z, Zhao Y, Wang C C, Purnawali H and Tang C. 2012 *Mater. Des.* **33** 577
[4] Bellouard Y 2008 *Mater. Sci. Eng. A* **481-482** 582
[5] Kohl M 2010 *Shape memory microactuators (microtechnology and MEMS), first ed.* (Heidelberg: Springer-Verlag Berlin)
[6] Zhang H J and Qiu C J 2006 *Mater. Sci. Eng. A* **438–440** 1106
[7] Nespoli A, Besseghini S, Pittaccio S, Villa E and Viscuso S 2010 *Sens. Actuator A* **158** 149
[8] Irzhak A, ZakharovD, Koledov V, Lebedev G, Mashirov A, Afonina V, Akatyeva K, Kalashnikov V, Sitnikov N, Tabachkova N, Shelyakov A and Shavrov V 2014 *J. Alloy. Compd.* **586** S464
[9] Shelyakov A V, Sitnikov N N, Sheyfer D V, Borodako K A, Menushenkov A P and Fominski V Yu 2015 *Smart Mater. Struct.* **24** 115031
[10] Shelyakov A, Sitnikov N, Borodako K, Menushenkov A and Fominski V 2015 *Physics*

Procedia **73** 108

[11] Schlossmacher P, Rösner H, Shelyakov A V and Glezer A M 2000 *Mater. Sci. Forum* **327-328** 131

[12] Potapov P L, Kulkova S E, Shelyakov A V, Okutsu K, Miyazaki S and Schryvers D 2003 *J. Phys. IV France* **112** 727

[13] Chang S H, Wu S K and Kimura H 2007 *Intermetallics* **15** 233

[14] Morgiel J, Cesari E, Pons J, Pasko A and Dutkiewicz J 2002 *J. Mater. Sci.* **37** 5319

[15] Shelyakov A V, Sitnikov N N, Menushenkov A P, Koledov V V, Irjak A I 2011 *Thin Solid Films* **519** 5314-5317

[16] Park S, Oh J, Kim Y and Nam T 2006 *Mater. Sci. Eng. A* **438–440** 695

[17] Shelyakov A V, Matveeva N M and Larin S G 1999 *Shape Memory Alloys: Fundamentals, Modeling and Industrial Applications* (Quebec City, Canada: Canadian Inst. of Mining, Metallurgy and Petrolium) 295

[18] Fischer F D, Sun Q-P and Tanaka K. 1996 *Appl. Mech. Rev.* **49** 317

Journal of Physics: Conference Series **737** (2016) 012028 doi:10.1088/1742-6596/737/1/012028

Laser deposition of bimetallic island films

A O Kucherik[1], S M Arakelyan[1], S V Kutrovskaya[1], A V Osipov[1], A V, Istratov[1] T A Vartanyan[2] and T E Itina[3]

[1] A.G and N.G. Stoletov Vladimir State University (VSU), 87 Gorki st., Vladimir, Russia, 600000
[2] St. Petersburg National Research University of Information Technologies, Mechanics and Optics, St. Petersburg, 197101 Russia.
[3] Université de Lyon, UJM-Saint-Etienne, CNRS, UMR5516, Laboratoire Hubert Curien, F-42023, Saint-Etienne, France

E-mail: kucherik@vlsu.ru

Abstract. In this work the results of a bimetallic Au-Ag structure deposition from the colloidal system by nanosecond laser radiation are presented. The formation of the extended arrays of gold and silver nanoparticles with controlled morphology is examined. We report the results of formation bimetallic islands films with various electrical and optical properties. The changes in the optical properties of the obtained thin films are found to depend on their morphology.

Introduction

The formation of transparent metal coatings is one of the promising fields in the modern nanotechnology. Particular attention is attracted by thin nanostructured metallic films demonstrating non-linear optical effects in the visible part of spectrum attributed to the surface plasmon resonance [1]. The optical properties of the films strongly depend on the composition and periodicity of the surface structures [2]. Such effects as local field enhancement can be observed either for distinct frequencies or in a wide diapason depending, for instance, on the structure spacing and ordering. In the case when the gap between the deposited particles is comparable with their sizes, randomly distributed structure properties can considerably differ from the periodically ordered ones [3]. When, furthermore, bimetallic thin films are deposited, one can expect even more complicated optical properties behavior [4] since they depend not only on composition, order and spacing, but also on their shape and morphology.

Thus, a considerable change in the optical properties can be observed upon laser-assisted deposition of fractal clustering thin films with respect to the continuos ones with the same composition [5,6]. Since traditional methods of the ordered deposition of the individual particles have a very low productivity and require more time [7], such new laser-assisted methods are of a considerable interest
In this work, we investigated the influence of the film morphology (particle diameter in the colloid, the distance between the deposited particles, the number of layers, etc.) on the optical and elctrical properties of the deposited thin film of bimetallic clusters.

For the formation of a quasi-periodically spaced clusters of particles, we used a method of deposition of colloidal systems allowing nanoparticles deposition in colloids by the local laser heating and the following diffusion of particles towards the surface [8]. Changes in the optical properties of the deposited bimetallic films are examined and the observed optical phenomena are explained based on a performed electromagnetic simulation.

1. Deposition of bimetallic thin films

1.1. Nanoparticle generation

The initial colloidal systems are prepared by continuous laser ablation of silver and gold targets in liquid environment (deionized water). As a result, the average particle size of 10 nm is obtained at a concentration of ~1mkg/ml [9]. The particle size is controlled by dynamic light scattering.

1.2. Pulse laser deposition of nanoparticle

For deposition of bimetallic structures, colloids were mixed in equal concentration proportions. KV8 glass substrate was placed in the cuvette with the solution and laser deposition method was applied to deposit particles from colloidal systems on the substrate [10].

For deposition, we used ytterbium fiber laser (λ = 1.06 microns) with a pulse duration of 100 ns, pulse repetition rate of 20 kHz, and a pulse energy of up to 1mJ. The diameter of the laser beam at the focal plane was 5 microns. Nanoparticle array formation on the substrate surface was performed by scanning the laser beam along the same direction (from five to twenty-five times) at a speed of 0.6 mm/s to 1.5 mm/s. The film deposition was obtained as a result of a step-by-step process, which is typical of the island structures.

In the proposed method, particle settlement/deposition occurs due to the local particle heating, so that the energy absorbed by the particles can be determined from the relation:

$$E_{abs} = \pi r_0^2 Q_{abs} \int I(t)dt\,, \qquad (1)$$

where $I(t)$ is laser intensity distribution; r_0 is particle radius; and Q_{abs} is the effective parameter.

To calculate the Q_{abs} classical Mie theory is used [11,12] for spherical particles of 10 nm diameter as follows

$$Q_{abs} = \frac{8\pi r_0}{\lambda} Im\left(\frac{\varepsilon^2-1}{\varepsilon^2+2}\right), \qquad (2)$$

where $\varepsilon = n + ik$ and the calculation parameters of gold and silver at the laser wavelength λ = 1064 nm are given in [13]. Depending on the conditions of laser exposure for a single pulse, the value of the absorbed energy is 10^{-6}-10^{-9} J.

During laser irradiation of a colloidal system, it is necessary to take into consideration the amount of heat transferred to nanoparticles from the laser field. Taking into account small size of particles relative to laser wavelength (1064 nm), the temperature of a particle can be found from relation [14]:

$$T_{max} = \frac{I_p k_a d}{8\chi} \qquad (3)$$

where I_p is the peak laser power, x is the thermal conductivity of liquid phase of the colloid, coefficient k_a takes into account the small size of the particle relative to the wavelength, and

$$k_a = e^{(-0.2[\sqrt{n^2+k^2}-1])}\left(1-e^{\left(-\frac{4\pi kd}{\lambda}\right)}\right) \qquad (4)$$

Taking into account that coefficients n and k for gold and silver are close to each other for radiation with a wavelength of 1064 nm and that we are using particles of equal diameters, we can assume that they reach close temperatures. Substituting laser parameters typical of our experiments (average power 20W, pulse width 100 ns, pulse-repetition rate 20 kHz, and beam diameter 5 mkm), we find that the temperature of nanoparticles in a colloid calculated from (3) can reach 600 K.

However, under conditions of multi-pulse exposure local heating may occur, leading to the stability loss of the colloidal system [10, 15]. In addition, agglomerates can be formed due to particles heating, that will absorb a larger fraction of laser energy and will be able to move thermally in the colloid. The local heating of the liquid will, furthermore, result in the formation of convective currents also ensuring particle motion of to the substrate surface.

The proposed deposition method allows one to change the morphology of the deposited structures depending on the laser exposure conditions. In what follows, we describe the consecutive stages in the formation of bimetallic island films depending on the number of passes in the deposition area.

After 10 skans, islands significantly increase their area (Figure 1a), and the average height of the relief becomes 18nm.

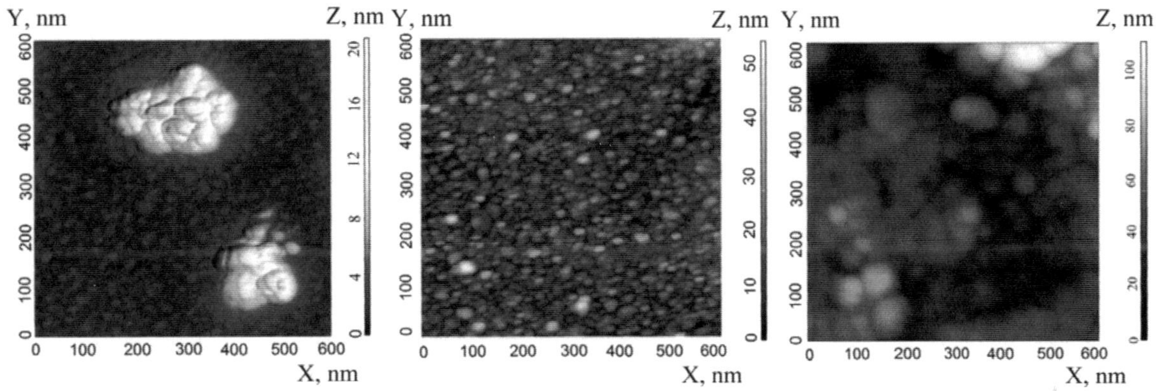

Figure 1. AFM images of bimetallic nanostructures deposited on a glass substrate. The power of the laser radiation formed is 2.5W, scanning speed is 1.2 mm/s for different number of laser pas passes: (a) - 10 scans, and (b) -25 scans, (c) – 35 scans.

With further scans the surface of the substrate islands completely fused and formed the structure of quasi-periodically arranged nanoparticles, the total height of the deposited topography increases monotonically with increasing number of passes of the laser beam in the same area. As a result, after 25 scans a granulate film is formed with an average height of the relief 52nm (Figure 1b), and after 35 scans the film with average height 112nm is formed (Figure 1c).

2. Measured optical properties of the depositing structures

The optical properties of the formed thin films are studied in the spectral range from 300 nm to 800nm by using SF -2000 spectrophotometer (Figure 2) .

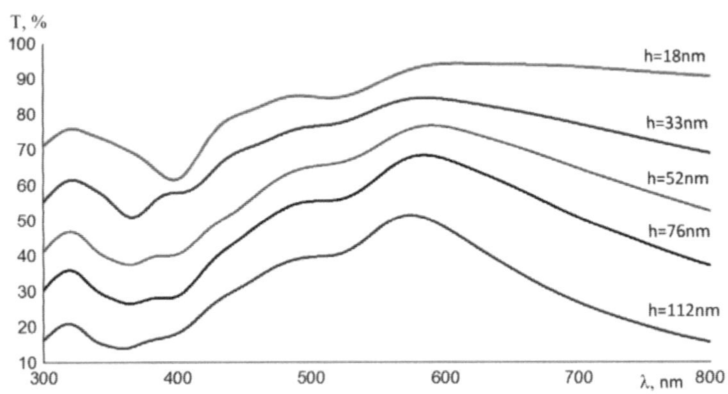

Figure 2. Transmission spectra of bimetallic films deposited with an average height of 18 to 112 nm .

Type transmission spectrum depends strongly on the structure of the deposited layer. The regions of plasmon resonance for silver (340-420nm) and gold (520-580nm) slightly stand out depending recorded. However, as the number of layers is increased, significant changes in the spectrum are observed. In the plasmon resonance region for silver, a system of local minima and laxima is observed, while the absorption maximum experiences the "red", the "blue" shifts. For all the films, an enhancement of transmission is observed between 550 and 600 nm. In all multilayer films, there is a periodic variation in transmittance between 400 and 500nm , which can be explained by the formation of the colloidal alloys of silver and gold nanoparticles [6], and by the interference –based antireflection of thin film.

3. Investigation of the electrical properties of the depositing structures

The current-voltage characteristics of the films were investigated using four contact setups. Two outer contacts are the pinch point contacts and are located at a distance of 1 mm from each other. The two other (central) are tips of an atomic force microscope coated with a conductive coating, and could move to a distance of up to 100 μm in the XY plane. The observed deviations from linearity (Figure 3) can be explained by the fact that the deposited layer was a nanoparticle conglomerate; the schema is not uniform and is an ensemble of ohmic contacts [10, 16].

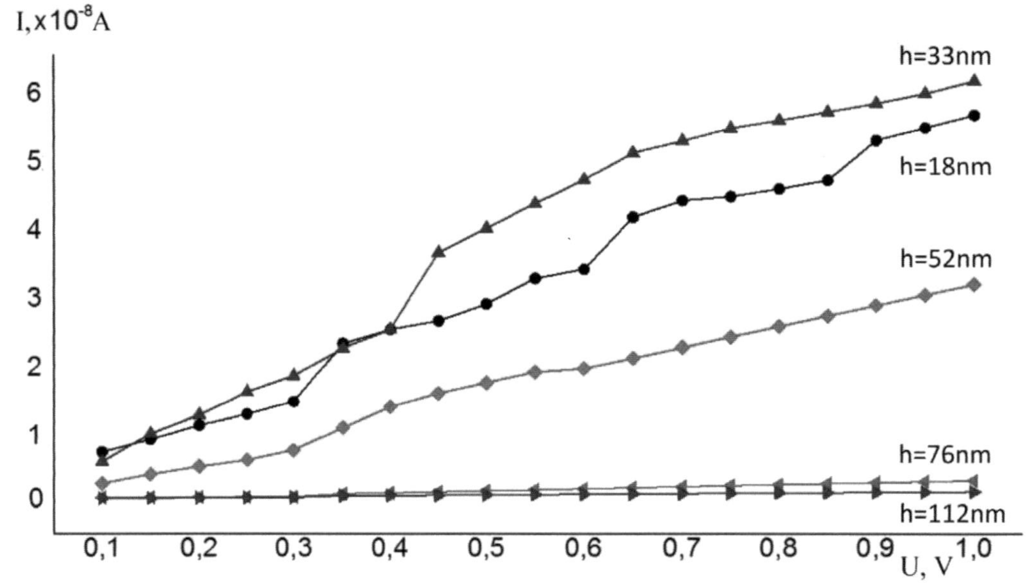

Figure 3. Current-voltage characteristics of deposited layers with different heights.

Changing current in the film can be described using the following equation:

$$j \propto U \sqrt{\frac{E_a}{kT + 2eUd/_L}} \exp\left(\frac{eUd/_L - E}{kT}\right) \tag{5}$$

U – potential difference between macrocontacts, E_a – ionization energy, e – electron charge, d – average distance between neighboring clusters, L – length of conductive layer between macrocontacts, k – Boltzmann constant, T - temperature

In deposited “islands” structures electrons move in unniform systems and the increasing of the voltage results in the opening of the new canal of the conductance. The obtained distribution in such system is shown in Figure 3. This system can be considered as a capacitor with capacitance C (tunneling between two metal contacts). "Coulomb blockade" is possible during the passing/tunneling like-charged particles across the border barrier/potential well, i.e., absence of the current in the structure when the external voltage owing to the Coulomb repulsion of the particles.

An estimation of the required minimum value of the energy (one-electron tunneling) for the overcoming of the "Coulomb barrier" can be done as follows:

$$\Delta E_K = hG/C, \tag{6}$$

where h - Planck's constant, G – the conductivity of tunnel transition.

The island films occurs electron scattering at the grain boundaries, this may change in the total resistance of the film depending on the ratio of the grains area (S_g) to the volume (V_f):

$$R = R_0 + \rho_g \left(\frac{S_g}{V_f}\right), \tag{7}$$

where R_0 – resistance of monocrystalline material , ρ_g – resistivity of grains border.

In our case, we have seen competition increase conductivity while opening new channels and increase the resistance by increasing the area of the grain. At the same time due to thermal activation also changes the current strength in the formed island films. Thus, the possibility of forming films of gold and silver nanoparticles for which the ability to control change in their electrical resistance by changing the morphology.

Conclusions

We demostrate the method of laser deposition of bimetal island films. The possibility of the formation of the surface array of gold and silver nanoparticles with controlled morphology is demonstrated.We have presented both the experimental data on the optical and electrical properties of nanostructured bimetallic films. We demosntrate what the properties of the films are found to depend on the film morphology.

Acknowledgments

The study was supported by the Ministry of Education and Science of the Russian Federation (state project no. 2014/13), RBFR grant number 16-32-60067 mol_a_dk and by the Government of Russian Federation (grant no. 074-U01).

References

[1] Genov D A, Sarychev A K, Shalaev V M and Wei A 2004 *Nano lett.* **4** 153-158
[2] Destouches N, Crespo-Monteiro N, Vitrant G, Lefkir Y, Reynaud S, Epicier T, Liu Y, Vocanson F and Pigeon F 2014 *J. Mater. Chem.* C **2** 6256
[3] Persson B N J. and Liebsch A 1983 *Phys. Rev. B* **28** 4247-54
[4] Baba K, Okuno T and Miyagi M 1993 *App. Phys. Lett.* **62** 437
[5] Pereira R M S, Borges J, Peres F C R, Pereira P A S, Smirnov G V, Vaz F, Cavaleiro A and Vasilevskiy M I 2015 *J. Nanophoton.* **9** 0937961-12
[6] Antipov A A, Arakelyan S M, Vartanyan T A, Itina T E, Kutrovskaya S V, Kucherik A O and Sapegina I V 2015 *Optics and Spectroscopy* **119** 119-123
[7] Bakhti S, Biswas S, Hubert C, Reynaud S, Vocanson F and Destouches N 2014 *J. Phys. Chem. C* **118** 7494-7500
[8] Antipov A A, Arakelyan S M, Kutrovskaya S V, Kucherik A O, Makarov A A, Nogtev D S and Prokoshev V.G. 2012 *Bulletin of the Russian Academy of Sciences: Physics* **76**611-617
[9] Arakelyan S M, Veiko V P, Kutrovskaya S V, Kucherik A O, Osipov A V, Vartanyan T A and Itina T E 2016 *J Nanopart Res* **18** 15501-12
[10] Abramov D V, Antipov A A, Arakelian S M, Khor'kov K S, Kucherik A O, Kutrovskaya S V and Prokoshev V G 2014 *Laser Phys* **24** 0740101-9
[11] Mie G 1908 *Ann. Phys. Leipzig* **25**, 377-452
[12] Xu Y-l and Gustafson B A S 2001 *JQSRT* **70** 419
[13] SCHOTT refractive Index Database, http://refractiveindex.info/
[14] Antipov A A, Arakelian S M, Kutrovskaya S V, Kucherik A O and Vartanian T 2014 *Optics and Spectroscopy* **116** 324-327
[15] Maximova K, Aristov A, Sentis M and Kabashin A 2015 *Nanotechnology* **26** 065601
[16] Antipov A A., Arakelyan S M., Kutrovskaya S V. and Kucherik A O 2014 *Technical Physics Letters* **6** 529–532

Reactive mesogen photoalignment on photopolymerizable composite layer

U V Mahilny, A I Stankevich, A V Trofimova

Belarusian State University, Minsk, Belarus

E-mail: mogilny@bsu.by

Abstract. The volume photoanisotropy (photoinduced birefringence) and surface photoanisotropy (LC photoalignment) of compositions of LC monomer – benzaldehyde polymer upon polarized UV radiation have been revealed and investigated. A high quality of photoalignment is confirmed by an extreme value of birefringence and low imperfection of phase plates fabricated on the basis of LC monomer on composite layer.

1. Introduction

Liquid crystal (LC) photoalighnment arises as a result of irradiation of aligning layer by linearly polarized light without any additional treatment of the material surface [1]. Such contact-free method makes possible to avoid typical disadvantages of rubbing method (dust, electrostatic charging, etc.), besides it allows to form multidomain aligning layer with different local orientation of LC director achieving by rotation of polarization plane of the activating light.

As aligning materials photocrosslinking polymers have the best prospects as the aligning molecular structures formed during the irradiation are appeared to be included in insoluble polymer network [2]. In previous works [3–5] we reported on LC photoalignment on the surface of photocrosslinking polymers, containing benzaldehyde groups in the side chain that upon linearly polarized UV irradiation create the molecular structures with a high anisotropic polarizability. But the azimuthal anchoring energy of the photoalignment for these materials is insufficient (~ 10^{-5} J/m^2) to form defectless multidomain phase LC films. The problem can be partially solved using an effect of photostimulated rubbing alignment [6]. Photocrosslinking makes the material insoluble, that allows to create polymer relieves with multidomain LC alignment [7]. Nevertheless this method just discrete changing of LC director orientation makes possible and the number of the orientations is limited.

In order to find the other opportunity to increase an azimuthal anchoring energy of the photoalignment in the present work we investigated photoanisotropy of composite material LC - monomer - benzaldehyde polymer (LC–B composition). Expected positive effect was connected with a partial alignment of LC monomer in the composite layer by photoproducts of benzaldehyde side groups of the polymer component.

2. Experiment

The experiments were performed with methacrylate polymers of M-series having different content of benzaldehyde fragments and LC monomer RMM-491 (Merck). The composition components were dissolved in the required ratio in toluene. Thin aligning layers (0.1 – 0.15 μm) were obtained by spin-coating without baking upon elevated temperature. Photoinduced birefringence of the composite layers

(0.5 – 0.6 μm), formed by exposing to linearly polarized radiation of a mercury lamp was measured by sample probing by laser beam (λ = 650 nm) as it is described in [8].

To study the photoalignment process LC monomer was coated on exposed composite layer and was polymerized with UV radiation to fix induced alignment. The transmittance coefficient T versus rotational angle of the sample placed between cross linear polarizers was measured and its birefringence was calculated according to:

$$|\delta n| == \frac{\lambda}{\pi l} \arcsin \sqrt{\Delta T} \quad , \qquad (1)$$

where $\Delta T = T_{max} - T_{min}$, l – the LC layer thickness, λ – wavelength of the probing laser beam (λ = 650 nm). The LC anisotropy axis direction (alignment direction) was determined by comparison of the experimental angle dependences for the examined sample and the reference one. The sample thickness was determined using interference microscope. The quality of the alignment of LC molecules was estimated by quality parameter that can be calculated as:

$$q = \frac{T_{max} - T_{min}}{T_{max}} . \qquad (2)$$

3. Photoinduced birefringence

The total flux of linearly polarized radiation (300 mW/cm^2) creates a negative photoinduced birefringence in the layers of the compositions (Fig. 1), it corresponds to lower value of the refractive index for the light with **E**-vector parallel to polarization of the activating radiation ($\delta n = n_{\parallel} - n_{\perp} < 0$).

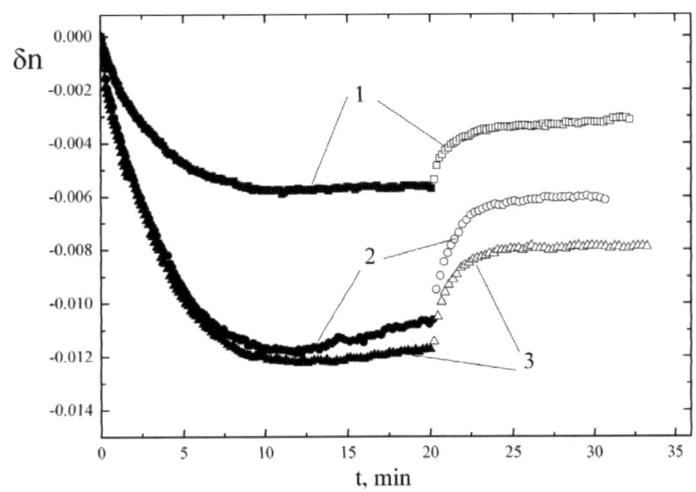

Figure 1 The kinetics of photoinduced birefringence recording and its dark behavior for the layers of LC–B composition (10:1 by weight) with different content of benzaldehyde groups in polymer: 10 (1), 20 (2), and 40 (3) mol.%

The positive (δn>0) birefringence of the composite layers is observed upon linearly polarized irradiation in the restricted spectral region and under less intensity (20 mW/cm^2), respectively, by using cut filter transmitting light with a wavelength more than 300 nm. And the time of exposure appeared to be sufficiently smaller (Fig. 2, curve 1). Upon dark relaxation the value of the negative birefringence decreases (Fig.1) whereas the positive one increases (curve 2).

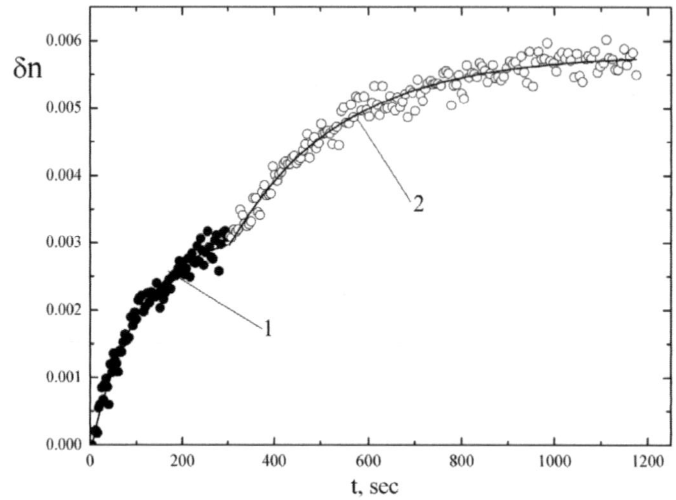

Figure 2 Kinetics of the photoinduced birefringence of the layer of LC–B composition with the components' ratio 6:1 (1) and its postexposure amplification (2)

Proper photoinduced birefringence of the benzaldehyde polymers consists of two basic components: non-stable negative birefringence and stable positive one [8]. The negative component of the birefringence foregoes positive one in the kinetics curves and lasts over a long period of time just upon high contents of benzaldehyde groups in macromolecules, i.e. upon their high concentration in polymer layer. As a rule values of both components are about 10^{-4} – 10^{-3}, and just upon high concentration of benzaldehyde fragments the negative birefringence exceeds value of 0,001 [9].

The positive component of photoinduced birefringence of the benzaldehyde polymers is formed by anisotropic products of photoreduction of benzaldehyde groups, included in crosslinks. The negative component arises due anisotropic (predominantely perpendicular to **E**) distribution of residual benzaldehyde fragments and the process of rotational chaotization makes the negative photoanisotropy rapidly decreasing [10].

The negative photoinduced birefringence of the composite layers (Fig. 1) arises during LC monomer photocuring and after its completing. According to the data of the wet development of irradiated composition layers, 5 minutes exposure time is quite enough to make the layers insoluble. It is hard to believe that the non-stable negative anisotropy of the ensemble of residual benzaldehyde fragments is able to align LC monomer, all the more after its curing. Therefore more probable reason of the negative birefringence forming in the composite materials is a phototransformation of mesogen centers of LC monomer and its polymer after curing that decrease their anisotropy.

The positive photoinduced birefringence of the composition layer (Fig. 2) might be considered as a result of LC monomer photoalignment effect. Actually the anisotropic axis direction in this case is determined by orientation of long axes of anisotropic products of benzaldehyde fragments along **E**, so the LC alignment does [4]. It is also proved by dark growth of the birefringence, that may be caused by finite duration of the alignment process, exposure time that is sufficiently less than curing time (5 min) and the absolute value of birefringence that exceeds considerably the values of the positive birefringence of pure benzaldehyde polymers.

Journal of Physics: Conference Series **737** (2016) 012029 doi:10.1088/1742-6596/737/1/012029

4. Photoalignment of LC monomer

The strong photoalignment effect was demonstrated by experiments of LC monomer coating onto thin (0.1 – 0.15 μm) composite layers, irradiated beforehand by linearly polarized light in the presence of the cutting filter. The effect appeared in the same time limits as for recording of the positive birefringence in composition layers (Fig. 2). It should be noted that there was no photoalignment effect upon total light flux exposure. Fig. 3 shows the dependences of birefringence of the cured LC monomer photoaligned on composite layers.

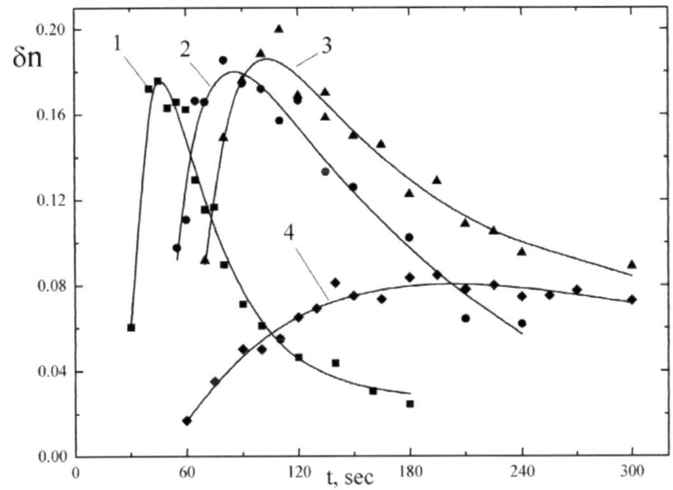

Figure 3 The kinetics of birefringence of LC layers aligned by composite layers upon different ratio LC – B: 10:1 (1), 7:1 (2). 5:1 (3), 3:1 (4). Polymer component contains 10 mol.% of benzaldehyde fragments

The dependences demonstrate dramatic growth of δn to the maximal value, and then relatively gradual decreasing of its value is observed. Maximal value of δn corresponds to the limit of anisotropy of cured LC monomer, it indicates to a low defectiveness of LC polymer film and high azimuthal anchoring energy. Low-intensity excitation $\lambda>300$ nm could make for creating of aligning benzaldehyde groups' photoproducts ahead of LC molecules' polymerization. In this case rotational mobility of LC and its alignment remains. At the same time the total light flux could polymerize LC quickly to prevent its alignment.

Applied potentialities of the photoalignment effect on the composite layers have been confirmed by polymeric retarder producing with radial director orientation based on linearly polarized irradiation of rotating photoalignment layer (Fig. 4).

Figure 4 The photo of LC film with radial director orientation in-between cross linear polarizers

Journal of Physics: Conference Series **737** (2016) 012029 doi:10.1088/1742-6596/737/1/012029

5. Conclusion

Photosensitive composite material consisting of LC monomer and benzaldehyde-containing polymer has demonstrated a photoalignment effect under linearly polarized irradiation λ>300 nm. Anchoring energy created by the material suffices for high quality alignment of the LC monomer layer. Co-axis photoinduced birefringence corresponds to assumption of partial alignment of LC molecules by benzaldehyde groups' photoproducts as a cause of the effective photoalignment.

References

[1] Cojocariu C and Rochon P 2004 *Pure Appl. Chem.* **76** 1479.
[2] Yaroshchuk O and Reznikov Y 2012 *J. Mater. Chem.* **22** 286.
[3] Kazak A *et al* 2008 *Techn. Phys. Lett.* **34** 861–864.
[4] Mahilny U *et al* 2013 *NPCS* **16** 79.
[5] Trofimova A and Mahilny U 2014 *JOSA B* **31** 948.
[6] Mahilny U V *et al* 2009 *J. Phys. D: Appl.Phys.* **42** 075303.
[7] Muravsky A *et al* 2010 *SID Int.Symp. Dig. Tech. Pap.* **41** 1727.
[8] Mahilny U V and Trofimova A V 2011 *J. Opt.* **13** 105601.
[9] Mahilny U V *et al* 2014 *J Appl. Spectr.* **80** 824.
[10] Mahilny V V and Trofimova AV 2008 *Bul. Russ. Acad. Sci. Phys.* **72** 1648.

Journal of Physics: Conference Series **737** (2016) 012030 doi:10.1088/1742-6596/737/1/012030

A low-symmetrical zinc phthalocyanine-based Langmuir-Blodgett thin films forNO_2 gas sensor applications

D M Krichevsky[1,2], A V Zasedatelev[1,2], A Yu Tolbin[3], Yu M Zelenskiy[1], V I Krasovskii[1,2], A B Karpo[2], L G Tomilova[3,4]

[1]National Research Nuclear University MEPhI (Moscow Engineering Physics Institute), 31 Kashirskoe shosse, 115409 Moscow, Russia
[2]Prokhorov General Physics Institute, Russian Academy of Sciences, Vavilov Str. 38, 119991 Moscow, Russia
[3]Institute of Physiologically Active Compounds, Russian Academy of Sciences, Chernogolovka, 142432 Moscow region, Russia
[4]Department of Chemistry M.V. Lomonosov Moscow State University, 119991 Moscow, Russia

e-mail: gomercheg@mail.ru

Abstract. For many years effective detection of hazardous substances such as nitrogen oxides has remained a crucial task for environmental safety. In this article, we demonstrate high promising NO_2–sensitive Langmuir-Blodgett monolayer structures based on 2-((2′-hydroxymethyl)-benzyloxy)-9(10),16(17),23(24)-tri-*tret*-butyl- substituted low symmetrical zinc phthalocyanine complex bearing hydroxyl group on the periphery (compound **1**). Amphiphilic arrangement of macrocycles was demonstrated to eliminate disordered molecular aggregation, resulting in a marked NO_2 gas sensing effect under real atmospheric conditions. The optical response of monolayers was at room temperature, with the significant spectral changes being caused by the specific charge transfer process in phthalocyanine π-conjugated electronic system.

1. Introduction

As a result of increased toxic gas production in chemical industry and its influence on human health, an effective detection of CO_2, NO, NO_2, NH_3, and other toxic gases has become a crucial task in environmental safety. Commonly used gas sensors are based on chemiresistive oxide thin films [1,2]; however they have some drawbacks, such as low selectivity and high power consumption [3]. In contrast, organic optical sensors are potentially more selective and have faster response times. Optical gas sensors based on organic thin films can operate at room temperatures [3] without external thermal stabilization. Moreover, manufacturing of organic sensing thin films is potentially cheaper due to utilizing only aqueous solution techniques, such as spin-coating, drop-casting and Langmuir-Blodgett (LB). However, the stability of sensors and signal recovery should be specifically improved.

Phthalocyanines (Pcs) are eligible sensing materials for NO_2 detection. Pcs are thermally and chemically stable, environmentally friendly and easy to produce. They exhibit strong light absorption in UV and visible regions.

It is known that interaction of NO_2 molecules with Pc macrocycles leads to reduction of Q-band absorption (600 – 700 nm) and an increase in the additional bands in the range of 500 – 550 nm as well as beyond 700 nm [4]. The similar changes can be also observed in absorption spectra during formation of charge-transfer (CT) complexes due to the interaction of Pcs with electron withdrawing molecules [4].

In this article, we demonstrate a recent study of NO_2 gas sensitivity towards monolayer structures based on low-symmetrical zinc phthalocyanine complexes.

2. Formation of monolayers

The synthesis of complex **1** was previously described [5]. For the preparation of LB films, KSV mini through system was applied. Ultrapure water with resistivity of 18 MOhm·cm was used as a subphase. The subphase temperature was regulated at 20°C by Julabo CD200F circulator. Complex **1** was uniformly spread from chloroform solution (100 μl of concentration $2 \cdot 10^{-4}$ M) onto the subphase. After evaporation of the solvent, the floating film was compressed at a constant rate of 5 mm/min until the surface pressure reached 20 mN/m (a value which corresponds to the linear region of the complex **1** compression isotherm). This value was held during the film transfer process onto hydrophilic glass substrate (KnittelGlaser 20x20 mm). The substrate was preliminary dipped into the water. The monolayer was deposited with constant transfer rate of 5 mm/min, with the monolayer transfer ratio being equal to 1±0.3.

3. Optical measurements and gas testing experiment

For spectral measurements of solution and thin film samples, a Perkin Elmer Lambda 1050 spectrometer was used. Experimental setup for gas testing is presented in Figure 1. The optical transmittance of the sample was measured during the gas exposure. In our experiments, NO_2 was diluted with N_2 as a carrier gas, with their ratio being carefully controlled to obtain the best response.

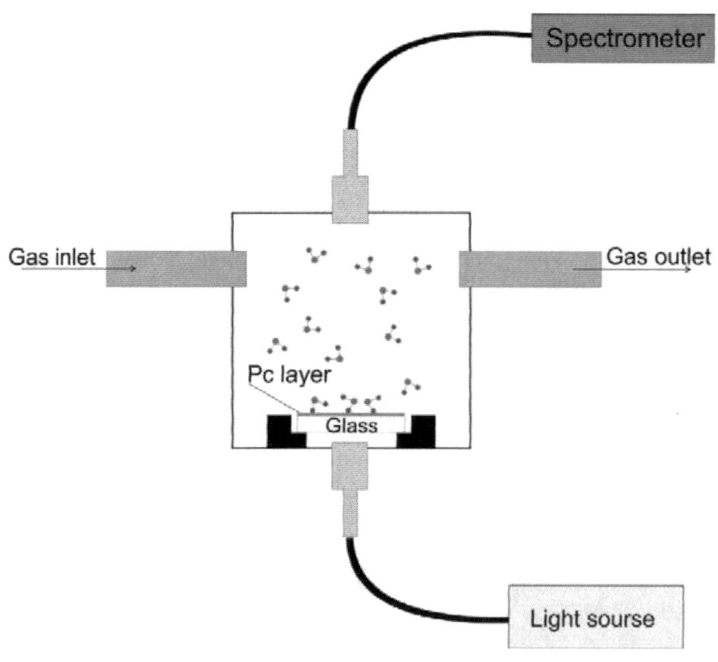

Figure 1. Scheme of experimental setup for gas sensing

Journal of Physics: Conference Series **737** (2016) 012030 doi:10.1088/1742-6596/737/1/012030

4. Results and discussions

*4.1. Chemical structure and optical properties of phthalocyanine complex **1**.*

Chemical structure of complex **1** and DFT optimized geometry of a model based on this compound are presented in Figure 2. In this work, quantum chemical calculations were performed using density functional theory (DFT). The Perdew–Burke–Ernzerhof (PBE) functional [6] and PRIRODA software package, supplied with the cc-pVDZ basis set [7], were used for optimization of the model structure corresponding to the low-symmetrical phthalocyanine complex **1** in the steady state. *Tert*-butyl substituents were replaced with hydrogen atoms to reduce a calculation time. The valence shells were described by basis sets with the following contraction schemes: {6s2p}/[2s1p] on H; {10s7p3d}/[3s2p1d] on C, N, O; and (17s13p8d)/[12s9p4d] on Zn atoms respectively. Systematic vibrational analysis was performed to confirm whether an optimized geometry corresponds to a local minimum without imaginary frequencies.

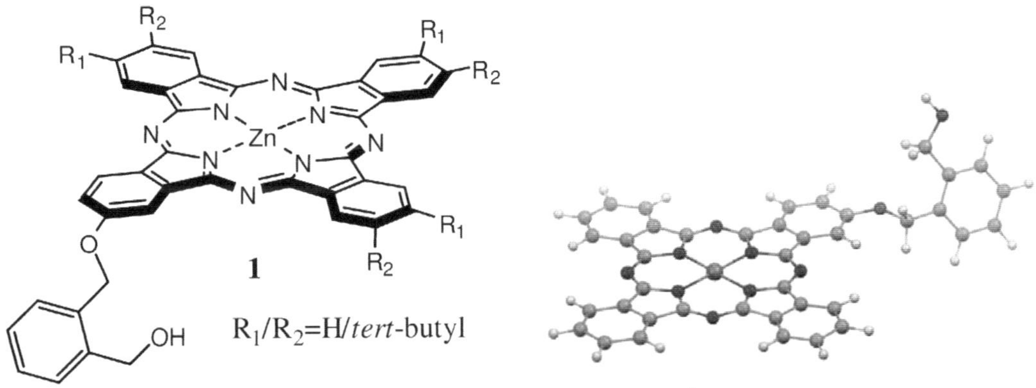

a b

Figure 2. Phthalocyanine complex **1** (a) and DFT optimized (b) geometry of a model, in which peripheral alkyl substituents were replaced with hydrogens to reduce a calculation time

UV/Vis spectrum of complex **1** in chloroform (Figure 3a) contains three typical well-resolved absorption bands (λ_1=351 nm – Soret band, λ_2=614 nm – vibrational satellite of Q-band, λ_3=680 nm – Q-band). UV/Vis spectrum of monolayer film (Figure 3b) also exhibits three basic bands which have nearly the same position as the complex **1**.

The majority of phthalocyanines have a tendency to form H- or J-type aggregates in thin films caused by strong π-π intermolecular interactions. The molecular aggregation in thin films alters UV/Vis spectra dramatically – new blue-shifted (for H-aggregates) and red-shifted (for J-aggregates) bands toward the Q-band as well as considerable spectral broadening are typically observed. However, significant suppression of intermolecular interaction in LB thin film of compound **1** was demonstrated. The UV/Vis spectra of Langmuir monolayers and unsymmetrically substituted complex **1** from which they were derived (Figure 3) clearly indicate the significant decreasing in the aggregation behavior compared to the symmetrical analogs [8].

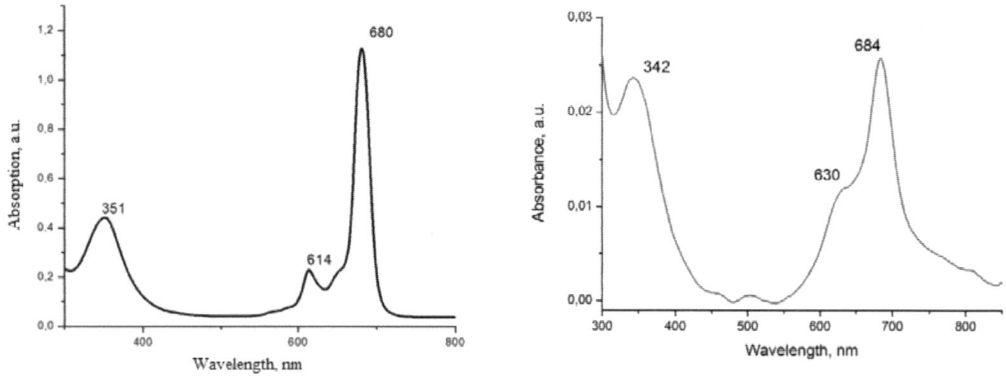

a b

Figure 3. UV/Vis spectra of complex **1** in chloroform (a) and the LB thin film (b)

4.2. Spectral changes induced by NO_2 exposure

The interaction of NO_2 with the monolayer structure leads to appearance of a new peak at 506 nm and a decreasing the Q-band (Figure 4a). This may indicate a formation of the charge transfer complex of **1** with NO_2 molecule, describing a new band as the CT-band. In order to recover sensitive properties, we heated the film up to 150°C and held it for 10 minutes. The CT-band has disappeared demonstrating a desorption of NO_2 (Figure 4b). At the same time, heating reveals spectral changes in Q-band region (Figure 4a) which can be attributed to the reorganization of the macrocycles providing significant strengthening the H-type molecular aggregation. The similar facts were previously reported [9], demonstrating a method to produce density-packed phthalocyanine thin films [10].

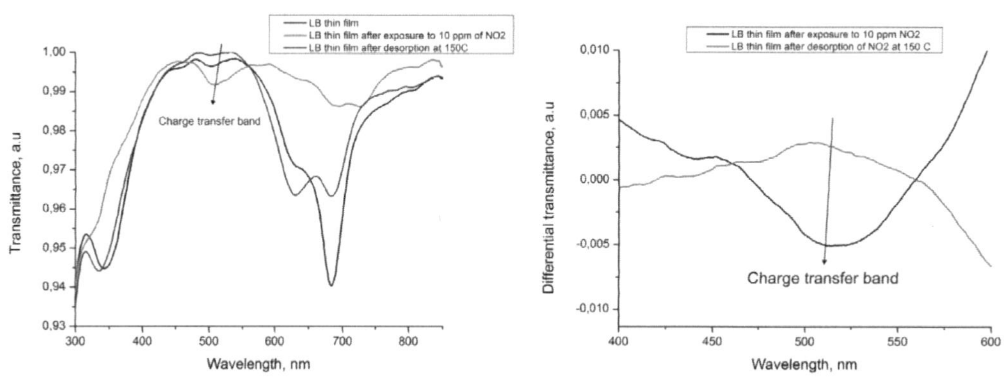

a b

Figure 4 Linear (a) and differential (b) transmittance spectra of the LB thin film after exposure and desorption of NO_2

5. Conclusion

A low-symmetrical zinc phthalocyanine complex **1** bearing a hydroxy group on the periphery was used for the preparation of LB thin films to estimate a possibility of their use in optical gas sensing applications. Owing to the unsymmetrical structure, complex **1** has a tendency to form ordered monolayer structure on glass substrates. The interaction of NO_2 with the monolayer structure leads to dramatic spectral changings induced by formation of charge-transfer complexes. We demonstrated recovery properties of CT-band by utilizing of heating. Moreover, significant aggregation features of the macrocycles were also demonstrated on heating the structures.

Acknowledgements
The research was supported by Russian Foundation for Basic Research (Grants No. 16-32-80032, 16-02-00694, 16-03-60031) and Program of fundamental researches of the Presidium of the Russian Academy of Sciences I.39P. The authors also thank Joint Supercomputer Center of RAS (www.jscc.ru) for providing computing resources.

References

[1] Collins G E, Armstrong N R, Pankow J W, Oden C, Brlna R, Arbour C and Dodelet J-P 1993 *J. Vac. Sci. Tech.* **11(4)** 1383
[2] Albert K J, Lewis N S, Schauer C, Sotzing G A, Stitzel S E, Vaid T P and Walt D R 2000 *Chem. Rev.* **100** 2595
[3] Dooling C, Worsfold O, Richardson T, Tregonning R, Vysotsky M O, Hunter C A, Kato K, Shinboc K and Kanekoc F 2001 *J. Mater. Chem.* **11** 392
[4] Pochekailov S, Nozar J, Nespurek S, Rakusan J, Karaskova 2012 *Sens. Actuator B-Chem.* **169** 1
[5] Tolbin A Yu, Pushkarev V E, Nikitin G F, Tomilova L G 2009 *Tetrahedron Lett.* **50 (34)** 4848
[6] Erznzerhof M, Scuseria G E 1999 *J. Chem. Phys.* **110** 5029
[7] Laikov D N 2005 *Chem. Phys. Lett.* **416** 116
[8] Nyokong T, Antunes E 2010 *Handbook of PorphyrinScience (*World Scientific Press) **7** 247
[9] Spano F C, Silva C 2014 *Annu. Rev. Phys. Chem.* **65** 477
[10] Roy D, Das N M, Shakti N, Gupta P S 2014 *RSC Adv.* **4** 42514

Enhancement of NO_2 gas detection in hybrid silver nanoparticles-phthalocyanine thin films

A V Zasedatelev[1,2], D M Krichevsky[1,2], Yu M Zelenskiy[1],A Yu Tolbin[3], V I Krasovskii[1,2], A B Karpo[2], L G Tomilova[3,4]

[1]National Research Nuclear University MEPhI (Moscow Engineering Physics Institute), 31 Kashirskoe shosse, 115409 Moscow, Russia
[2]Prokhorov General Physics Institute, Russian Academy of Sciences, Vavilov Str. 38, 119991 Moscow, Russia
[3]Institute of Physiologically Active Compounds, Russian Academy of Sciences, Chernogolovka, 142432 Moscow region, Russia
[4]Department of Chemistry M.V. Lomonosov Moscow State University, 119991 Moscow, Russia

e-mail: anton.zasedatelev@gmail.com

Abstract. Phthalocyanine-functionalized plasmonic sensing systems are typically based on Kretschmann configuration. Such scheme of detection utilizes spectral or angular modulation of reflected light, which is induced by surface plasmon's excitation in the metal film on prism. Phthalocyanine's layer plays a role of analyte adsorber. In present paper we offer another approach to phthalocyanine-plasmonic sensing, where both local surface plasmon resonance and optical absorption of phthalocyanines are simultaneously detected. Hybrid Ag nanoparticles (AgNps) – low symmetrical A3B zinc phthalocyanine (ZnPc) thin films were prepared, and their NO_2 gas sensitive properties were examined. Since the plasmon resonance of AgNps was properly tuned to charge-transfer band of ZnPc-NO_2 complex, we found out more than two-fold increase of the optical response to NO_2 exposure in AgNps-ZnPc thin films compared to ZnPc films without AgNps.

1. Introduction

Environmental safety and pollution prevention have always been a focus of leading science centers all over the world. As of today many different methods and techniques for detecting poisonous agents in gases and liquids have been developed. Particular attention among them is paid to methods based on excitation of surface and localized plasmons [1,2].

For detecting such gases as NO_2 and CO_2 plasmonic sensors, based on thin metal film covered by phthalocyanine layer, have been developed [3,4]. Phthalocyanine molecules serve as an adsorber of analyte molecules, providing changes in a permittivity of metal/phthalocyanine interface. As a consequence, conditions of surface plasmon resonance (SPR) are also changed. Kretschmann configuration is usually applied [5] to detect changes in SPR. Such scheme of surface plasmon excitation offers a possibility to detect SPR conditions by measuring angular or spectral distribution of reflected light. Significant response of SPR to alteration in the permittivity of metal/phthalocyanine interface results in high sensitivity of nitrogen dioxide detection close to 0.1 ppm. Moreover, metal/phthalocyanine SPR sensors also provide fast, real-time gas analysis without any additional analyte preparation. However, there are some drawbacks in this approach which are associated with relative complexity and overall size of the detection setup in comparison to classic sensors based on phthalocyanine thin films.

In contrast with surface plasmons, localized surface plasmons in nanoparticles can be easily excited by plane electromagnetic wave without utilizing any special excitation scheme. The light, captured by the metal nanoparticles, is confined to quite a small volume, not much larger than the particle itself, which concentrates the energy gained from the incident field. Therefore, the localized field is much stronger than the incident field. The strong local field is commonly used to increase various photo-induced processes in chromophores. Particularly, considerable increase of optical absorption and decay rate of excited state of zinc phthalocyanines which are placed on a surface of gold nanoparticles have been demonstrated [6,7]. In our recent paper we have shown many-fold plasmonic enhancement of nonlinear saturable absorption effect in subphthalocyanines [8].

In present research we found out a two-times improvement in sensitive properties of AgNps-ZnPc thin films to nitrogen dioxide in a gas phase. We suppose that plasmonic enhancement of ZnPc-NO_2 charge-transfer complex's absorption as well as red-shifting of LSPR are the most probable reasons of the improvement.

2. Formation of sensing structures

The synthesis of 2-((2′-hydroxymethyl)-benzyloxy)-9(10),16(17),23(24)-tri-tret-butyl- substituted low symmetrical A3B zinc phthalocyanine complex bearing hydroxyl group on the periphery (complex **1**) was previously described [9]. Langmuir-Blodgett films of complex **1** were deposited onto two types of substrates: 1 hydrophilic glass (KnittelGlasser 20x20 mm), 2 – plasmonic thin film based on silver nanoparticles. Plasmonic thin film was formed by a two-step process. At the first stage, 20 nm silver nanoparticles were deposited onto glass substrates through spin-coating technique. At the second stage, in order to stabilize and homogenize AgNps thin film, the structure was being annealed during 4 hours at 230°C. The annealing of AgNps has led to agglomeration of single nanoparticles to larger particles with an average size of 100-120 nm, as shown on Figure 1.

Figure 1. SEM image of the plasmonic thin film based on Ag nanoparticles

For preparation of Langmuir-Blodgett films KSV mini through system was applied. Ultrapure water with resistivity of 18 MOhm·cm, and temperature of 20°C was used as a subphase. Complex **1** was uniformly spread from chloroform solution (100 µl of concentration $2 \cdot 10^{-4}$M) onto the subphase. After evaporation of the solvent, the floating film was compressed to the surface pressure of 40 mN/m. Ten-monolayers films were deposed at the same surface pressure (40 mN/m) onto both hydrophilic glass and plasmonic thin film based on silver nanoparticles as well. For further convenience we introduced Sample 1 and Sample 2 abbreviations for 10-monolayers phthalocyanine films on the glass and on the plasmonic substrate respectively.

3. Optical properties of the samples

For spectral measurements Ocean Optics HR4000 spectrometer with 0.5 nm resolution and reflection/backscatter probe OceanOptics QR600-7-SR-125F were applied. Spectral measurements were carried out in reflectance mode with usage of the glass and the plasmonic substrate without phthalocyanine layers for reference measurements. Relative reflectance spectra are shown on Figure 2.

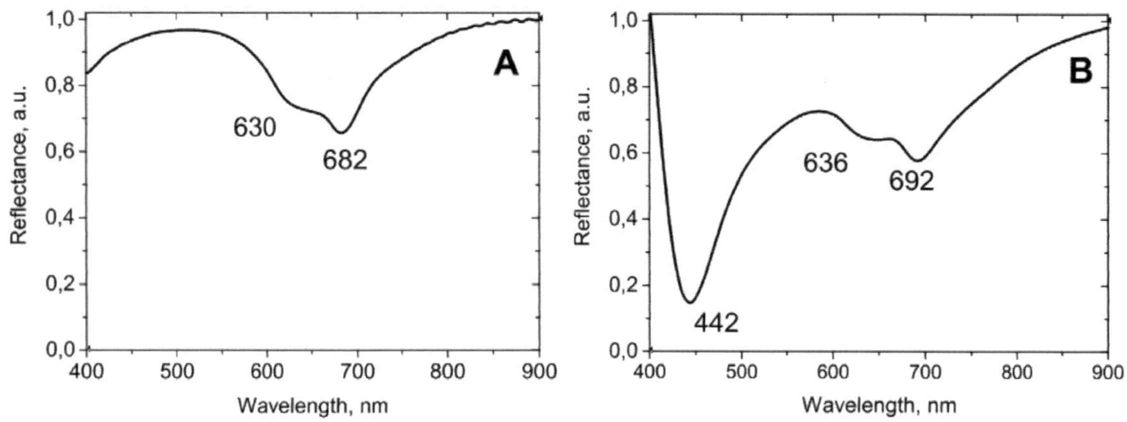

Figure 2. Reflectance spectra of Sample 1 – a, and Sample 2 – b.

Minimum in the reflectance spectrum of Sample 1 (Figure 2a) in a range of 600 – 700 nm is associated with strong Q-band absorption, where the dip at 682 nm is the main peak of the absorption band [9]. As can be seen from Figure 2b, Q-band absorption of Sample 2 is red-shifted in comparison with the band of Sample 1. Moreover, we obtained a new strong dip in the spectrum at 442 nm, which corresponds to LSPR of silver particles covered by phthalocyanine layers having a high real part of permittivity. Despite an overlap between Q-band and short-wavelength plasmonic band we see 10 nm red-shift of the Q-band absorption maximum. The reason of it is not clear. We suppose, that the charge transfer between ZnPc and AgNps, as well as some peculiarities in the molecular arrangement onto the plasmonic film, can be responsible for such non-additive behavior of the energy losses in the spectrum. New maximum of Q-band absorption of Sample 2 is located at 693 nm.

4. Results and discussions

4.1. Gas testing experiment

Experimental setup for gas testing is represented on Figure 3. Nitrogen dioxide was diluted with pure nitrogen (N_2) as inert carrier gas, the percentage of gases in amixture was regulated by mass flow controllers. In such way the concentration of nitrogen dioxide in the mixture was adjusted at the level of 10 ppm. Spectral measurements in relative reflectance mode were also performed.

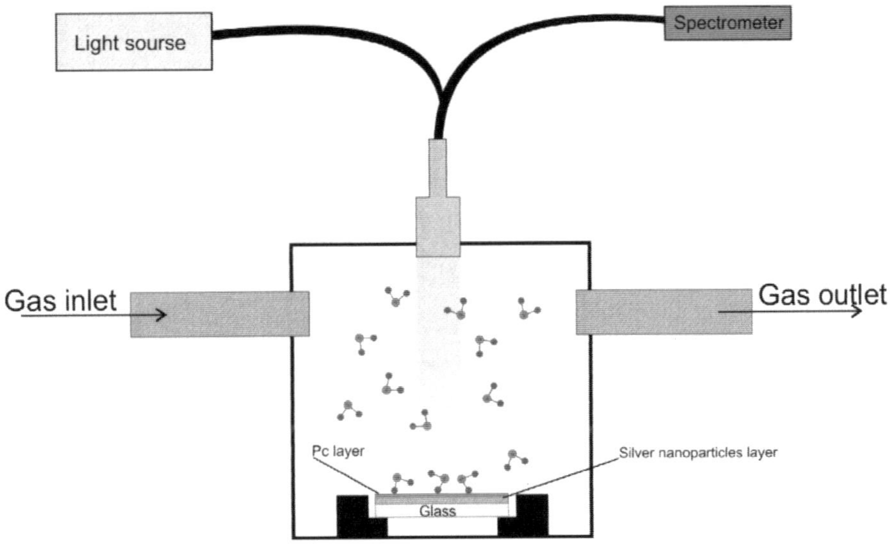

Figure 3. Scheme of experimental setup for gas sensing

4.2. Spectral changes induced by NO_2 exposure

The interaction of NO_2 with the Sample 1 results in considerable spectral modification of its reflectance properties. Figure 4A shows reflectance spectra of Sample 1 before and after NO_2 exposure (black and red line respectively). Additional losses in the spectrum at 470 – 580 nm after the gas exposure and increase of reflectance in the Q-band region clearly indicate a formation of the charge-transfer ZnPc-NO_2 complex, where ZnPc plays a role of electron-donating molecule. An absorption band at 470 – 580 nm is also well-known as a charge-transfer band (CT-band) [10]. In contrast to Sample 1, there is no well-resolved CT-band in reflectance spectrum of Sample 2, but we observe a red shift of LSPR as well as a considerable broadening of the dip (Figure 4B), which can be caused by superposition of CT-band and plasmonic shifting.

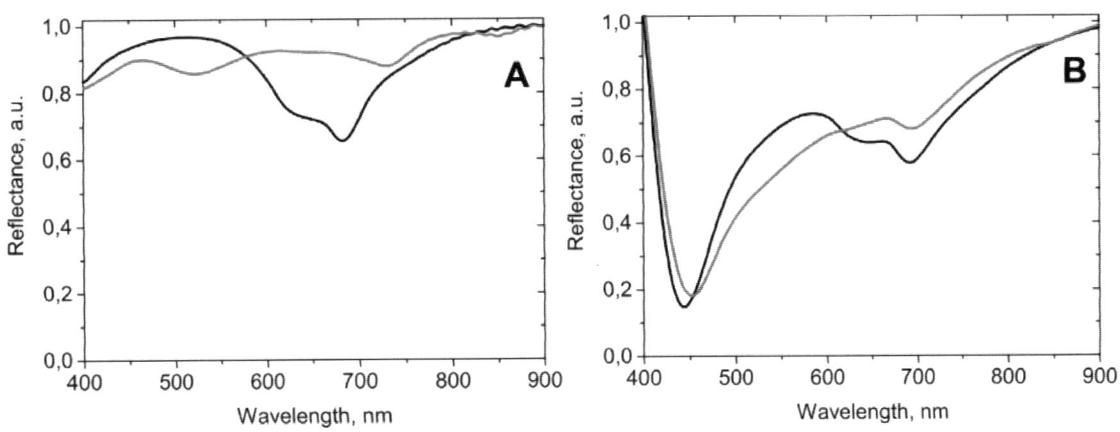

Figure 4. Reflectance spectra of Sample 1 – A, and Sample 2 – B. In both: black line – before NO_2 exposure, red line – after NO_2 exposure

Since the reflectance is not linear function of losses, it is more convenient to consider absorption spectra which can be easily obtained from our reflectance measurement assuming, that absorption losses are dominant. Figure 5 shows effective absorption spectra of Sample 1 and Sample 2 before and after exposure of nitrogen dioxide.

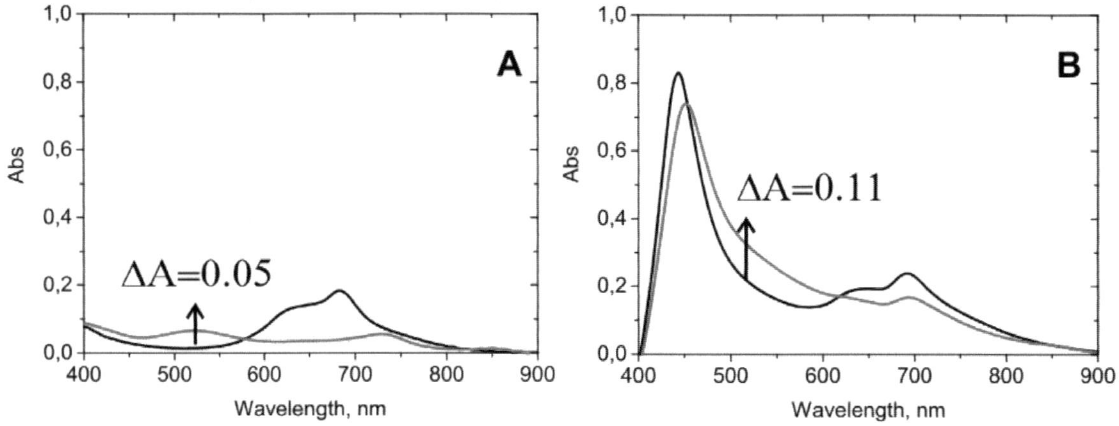

Figure 5. Absorption spectra of Sample 1 – A, and Sample 2 – B. In both: black line – before NO_2 exposure, red line – after NO_2 exposure

Absorbance is a linear function to the number of molecules and therefore it can be utilized to adequate comparing responses of the samples to influence of nitrogen dioxide. Here we considered only CT-band (the range of 470 – 580 nm) because Q-band is not specifically sensitive to analyte molecules and can be easily altered during different external processes, such as heating. To evaluate enhancement factor we compared maxima of differential absorption of both samples in CT-band. While the maximum value of differential absorption of Sample 1 equals to 0.05, the maximum of Sample 2 is 0.11, that is more than 2-times larger. We suppose that the effect is induced by 2 factors. The first factor is the shifting of LSPR caused by altering phthalocyanine's permittivity. Whereas the second one is plasmon-related enhancement of CT-band absorption. Of course, different chemical and photochemical processes, associated with interaction between silver particles and ZnPc molecules especially in the presence of nitrogen dioxide, cannot be excluded from consideration.

5. Conclusion

Here we proposed a new approach to designing of plasmonic phthalocyanine sensors which is based on simultaneous registration of LSPR spectral position and specific optical absorption of charge-transfer complex. Using Langmuir-Blodgett and spin-coating techniques, we fabricated hybrid thin-film sensitive structures based on novel A3B zinc phthalocyanine and silver nanoparticles, and found out more than two-fold improvement in the response of NO_2 exposure. As we are aware, this is the first time demonstration of plasmon-related enhancement in sensitive properties of phthalocyanines in similar hybrid structures. However, this is just a first step and it is required to conduct more research for further approach development.

Acknowledgements

The research was supported by Russian Foundation for Basic Research (Grants No. 16-32-80032, 16-02-00694, 16-03-60031) and Program of fundamental researches of the Presidium of the Russian Academy of Sciences I.39P.

References

[1] Homola J and Piliarik M 2006 *Surface Plasmon Resonance Based Sensors* (Springer Series on Chemical Sensors and Biosensors vol 4) ed J Homola (Springer Berlin Heidelberg) pp 45–67
[2] Mayer K M and Hafner J H 2011 *Chem. Rev.* **111** 3828
[3] Vukusic P S and Sambles J R 1992 *Thin Solid Films* **221** 311
[4] Wright J D, Cado A, Peacock S J, Rivalle V and Smith A M 1995 *Sens. Actuator B-Chem.* **29** 108

[5] Maier S A 2007 *Plasmonics: Fundamentals and Applications* (Springer US)
[6] Zasedatelev A, Krasovskii V, Dubinina T and Krichevsky D 2014 *Proc. SPIE* **9163** 91633E-1
[7] Zasedatelev A V, Dubinina T V, Krichevsky D M, Krasovskii V I, Gak V Yu, Pushkarev V E, Tomilova L G and Chistyakov A A 2016 *J. Phys. Chem. C* **120** 1816
[8] Zasedatelev A, Dubinina T, Krasovskii V, Suprunova O, Tomilova L and Chistyakov A 2015 *J. Phys.: Conf. Ser.* **643** 012049
[9] Tolbin A Yu, Pushkarev V E, Nikitin G F and Tomilova L G 2009 *Tetrahedron Lett.* **50** 4848
[10] Pochekailov S, Nozar J, Nespurek S, Rakusan J and Karaskova M 2012 *Sens. Actuator B-Chem.* **169** 1

Journal of Physics: Conference Series **737** (2016) 012032 doi:10.1088/1742-6596/737/1/012032

Superconducting detector of IR single-photons based on thin WSi films

V A Seleznev[1,2], A V Divochiy[1,2], Yu B Vakhtomin[1,2], P V Morozov[2], P I Zolotov[1,2], D D Vasil'ev[3], K M Moiseev[3], E I Malevannaya[3] and K V Smirnov[1,2,4]

[1] Moscow State Pedagogical University, 119435 Moscow, Russia
[2] CJSC "Superconducting nanotechnology" (Scontel), 119021 Moscow, Russia
[3] Bauman Moscow State Technical University, 105005 Moscow, Russia
[4] National Research University Higher School of Economics, 101000 Moscow, Russia

E-mail: seleznev@rplab.ru

Abstract. We have developed the deposition technology of WSi thin films 4 to 9 nm thick with high temperature values of superconducting transition (T_c~4 K). Based on deposed films there were produced nanostructures with indicative planar sizes ~100 nm, and the research revealed that even on nanoscale the films possess of high critical temperature values of the superconducting transition (T_c~3.3-3.7 K) which certifies high quality and homogeneity of the films created. The first experiments on creating superconducting single-photon detectors showed that the detectors' SDE (system detection efficiency) with increasing bias current (I_b) reaches a constant value of ~30% (for λ=1.55 micron) defined by infrared radiation absorption by the superconducting structure. To enhance radiation absorption by the superconductor there were created detectors with cavity structures which demonstrated a practically constant value of quantum efficiency >65% for bias currents $I_b \geq 0.6 \cdot I_c$. The minimal dark counts level (DC) made 1 s^{-1} limited with background noise. Hence WSi is the most promising material for creating single-photon detectors with record SDE/DC ratio and noise equivalent power (NEP).

1. Introduction

Development of thin films superconducting technologies has created multiple superconducting low-current electronics devices among which there can be marked out a group of electro-magnetic radiation sensors: mixer based on SIS Josephson junction (superconductor-insulator-superconductor)[1], direct bolometers and heterodyne receivers based on hot electron phenomena in superconductors (HEB, Hot Electron Bolometer)[2], bolometers based on superconducting transition (TES, transition edge sensor)[3], detectors based on kinetic inductance (KID, kinetic inductance detector)[4], superconducting single photon detector (SSPD)[5] etc. Progress in developing different detectors is as a rule ensured by development of the technologies of creating structures and by application of new superconducting materials. For example the initially discovered effect of superconductivity destruction at absorption of single infrared photon by a thin and narrow NbN strip biased by current close to critical value has allowed creating detectors of single infrared photons with quantum efficiency, or photons detection probability, at the level of several percent [5],[6]. Whilst usage of other superconducting materials such as NbTiN [7], MoRe [8], WSi [9], MoSi [10], NbSi [11], TaN [11], and improve of the technology of creating superconducting single photon detectors using resonator structures and antireflection coatings allowed creating a superconducting single photon detector with increased quantum efficiency value up to several dozen percent [13] and improved values of all main characteristics of the detectors: counting rate [14], Dark Count probability [15], timing resolution [16].

One of the problems in creating SSPD is to get detectors with simultaneous achievement of high level of detection efficiency and low level of dark counts. Both SDE and DC depend on bias current I_b, both indexes go down while displacement current goes down. As a rule, DC reduction along with I_b reduction is linked to fluctuation processes reductions (temperature, normal electrons and Cooper pairs concentricity) [17],[18]. While SDE reduction along with the detector's bias current reduction is linked to inhomogeneity of the superconducting strip (thickness, width, composition and structure of film) and, as a rule, difficulties in achieving the internal detection efficiency close to 100%; the situation when each photon absorbed by the superconducting strip leads to destruction of superconductivity across the whole cros-section of the superconducting strip. Creation of highly homogeneous superconducting films and films-based structures is one of the most important tasks of producing SSPD with the lowest NEP or a high SDE/DC ratio. We have to note that the records for different types of SSPD published until now show considerably falls of SDE when the bias current goes down.

This study is aimed to elaborate and create superconducting single photon detectors with saturation of current dependence of SDE(I_b) at the currents range of $0.6 \cdot I_c < Ib < I_c$. When choosing materials for the detectors we first took into account experience of creating SSPD with high SDE values. When comparing with NbN (or NbTiN as close to it) we chose WSi for several reasons. First of all, NbN films are polycrystalline with axial structure aligned with the film's thickness. At that, the size of crystallites depends much on the parameters of the film deposition and on the substrate type [19]. Achieving the highest homogeneity level of the NbN film on considerable areas comparable to typical SSPD active area is absolutely not a trivial task. In the meantime as testified by several research groups [20] WSi has amorphous structure which makes the aim of creating homogeneous superconducting nanostructures still more technological. Besides, WSi have a lower value of energy gap that also assumes possibility of producing detectors with highest values of internal DE. Hence in this study we present the results of elaboration of the technology of WSi thin films deposition and creation of superconducting single photon detectors for the near infrared range with internal detection efficiency close to 100% for bias current range of $0.6 \cdot I_c < I_b < I_c$, with high potential to realize record-breaking values of NEP and SDE/DC ratio.

2. Deposition of WSi thin films

For deposition tungsten silicide film we used the magnetron sputtering method realized on the basis of VUP-11M technological unit. W and Si were sputtered simultaneously from two magnetron sources. To sputter Si there was used a magnetron working in high frequency mode. To sputter W there was used a magnetron working in impulse mode at 15 kHz frequency. Magnetrons were located at the angle of 90° to each other and the substrate holder at the angle of 45° to axial lines of both magnetrons. As planar sizes of the substrate in our experiments did not exceed $20 \cdot 20$ mm^2 and the distance between each magnetron center and the substrate exceeded 15 cm, our method allowed insuring high level of the film homogeneity [21], confirmed by measurements of sheet resistance and temperature of transition into superconducting state taken at different areas of the substrate. Deviations of the above mentioned values within the substrates did not exceed 5%. We must also mention that the method employed and the technique realized allows to easily changing the ratio of W and Si materials in produced films by varying magnetron power. This fact is doubtless importance while optimizing the process of films deposition.

While optimization of ratio of W and Si as a criterion the film's optimal composition we used its critical temperature value for superconducting transition (T_c). These researches were made for the film thickness fixed at 4 nm. Normalized dependencies for films resistance on temperature are shown on figure 1. As a substrate we used high-resistivity Si-wafer of 400-micron thick with a thermal silicone dioxide layer of 250-nm thick. The chart we have obtained shows that when percentage of W in WSi films is changed within the range of 71÷81% the critical temperature of the film's changed within the range of 2.78 K to 3.35 K. The highest T_c value was achieved with atomic ratio $W_{0.76}Si_{0.24}$. It can be also noted that dependencies of R(T) for WSi films had a typical 'metallic like' rate marked by reduction of resistance upon changes in temperature from 300K to 4K . Electrical resistance of the films made about 200·μOhm·cm did not practically depend on W content within 71-81%, and complied with the data published earlier [9],[20],[21]. The parameters of WSi films deposition process corresponding to the optimal atomic ratio are shown in table 1. In further work we used WSi films of 5 nm, 7 nm and 9 nm thick with the highest T_c value corresponding to the atomic ratio $W_{0.76}Si_{0.24}$ which made T_c(5 nm)=3.65 K, T_c(7 nm)=3.75 K и T_c(9 nm)=3.8 K.

Table 1. Parameters of WSi films deposition process.

P (residual), mbar	$3.0 \cdot 10^{-5}$
P (operating), mbar	$2.4 \cdot 10^{-3}$
Argon flow, l/h	2.6
W magnetron power, watt	120
Si magnetron power, watt	109
W deposition rate, nm/sec	0.30
Si deposition rate, nm/sec	0.08

3. WSi single photon detectors and measuring methods

3.1. Technology WSi SSPD fabrication.

Using $W_{0.76}Si_{0.24}$ films 5, 7, 9-nm thick we produced single photon detectors which was in shape of a superconducting strip ~120 nm wide bended into meander with ~200 nm step and covering the area of 7·7 micron^2 (active area of the detector). The superconducting strip had contact pads set as a coplanar line. The technology of creating the detectors was based on direct e-beam lithography and following reactive plasma etching and was similar to NbN detectors technology [22]

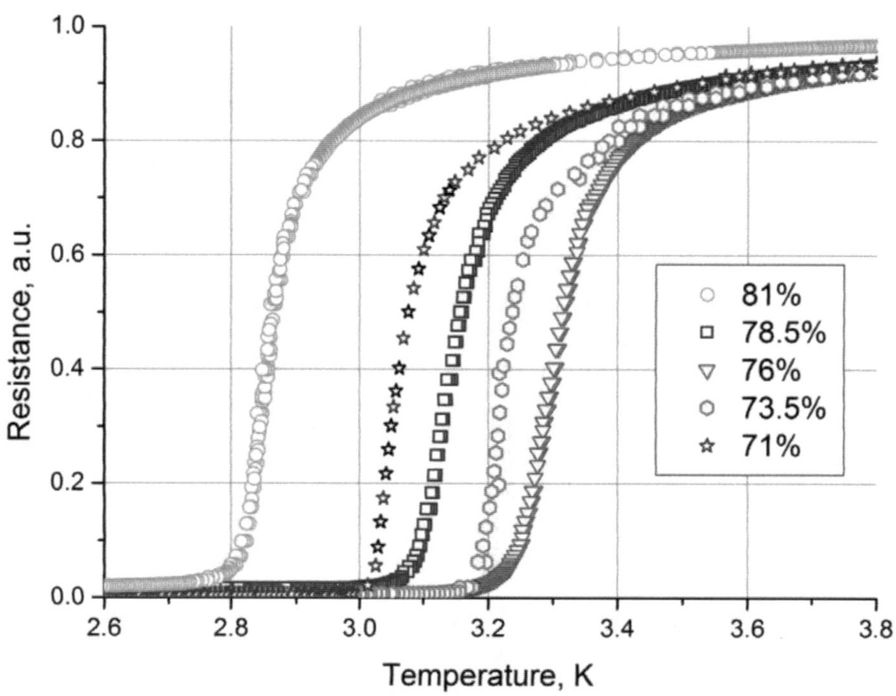

Figure 1. Temperature dependencies of resistance of W_xSi_{1-x} 4 nm thick films. x-values percentage shown on the insert to the Figure.

We produced WSi detectors on high-resistivity silicon wafers 400 microns thick with a thermal silicone dioxide layer 250-nm thick (detector type I). For the 2nd type of detector we apply a resonator structure before deposition of a WSi film. The resonator structure consists of a metal mirror and a quarter-wave dielectric layer (optimized for λ=1.55 micron). The metal mirror of 50·50 micron^2 was made of a 70-nm thick gold layer applied with e-beam evaporation method by using direct photo lithography and following wet etching. The metal mirror was located under the SSPD active area. A silicon nitride layer 180-nm thick applied by PECVD method onto the whole substrate surface acted as quarter-wave resonator layer (figure 2). Applying a resonator structure we solved the problem of increasing radiation absorption by a superconductor that does not exceeds 30% in case of simple Si/SiO_2 substrate.

3.2. Coupling the SSPD with radiation.

In our experiments we used direct coupling of the WSi detector with radiation through standard single-mode fiber Corning SMF 28e. Since the active area of the detector made 7·7 micron^2, mode field diameter for that fiber type at λ=1.55 micron made ~9 micron and in-fiber radiation have a Gaussian profile, then the best geometrical coupling in our experiments was ~85%.

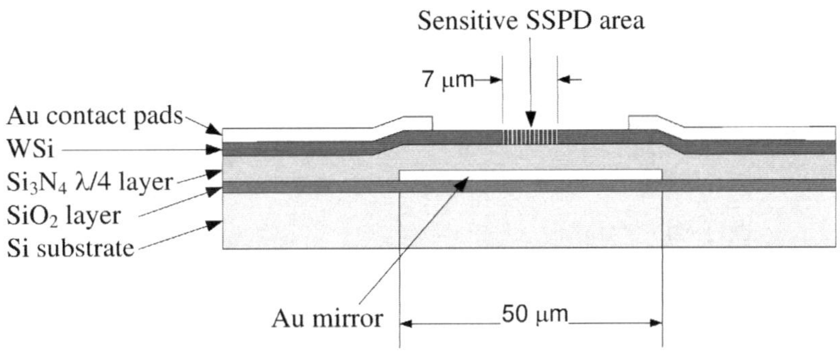

Figure 2. Schematic the WSi detector cross-section having resonator structure.

To accurate positioning the detector against the fiber we used two methods. The first one is based on applying a flip-chip positioning device Fineplacer Lambda which allows observing and positioning two space-divided objects against each other (the fiber core and the detector's active area in our case), and then automatically space-match them with accuracy no worse than 0,5 micron. The second method is based on possibility to observe the bolometric signal from the superconducting detector at room temperature at absorption of modulated optical radiation [14]. By bringing the optical radiation via single-mode fiber and under the condition the detector is biased by DC current it becomes possible to define how accurately the fiber is positioned against SSPD. After reaching the maximum signal the fiber was fixed against the detector.

3.3. Low-temperature technique of the experiment.

The experiments for defining WSi films superconducting transition temperature and measuring SDE and DC of WSi single photon detectors were made with a low-temperature technique that allowed temperatures down to 1.6K. The technique is based on dipstick into a standard liquid helium Dewar. Helium entered to the dipstick internal volume through capillary that connected to the helium Dewar bath. The double-wall dipstick with vacuum between the walls allowed for thermal insulation of the dipstick's internal volume from surrounding. Evacuation of the helium vapor from the dipstick's internal volume with a forevacuum pump to reach the pressure of ~10 mbar allowed reducing the temperature in the dipstick down to 1.6K.

3.4. SDE and DC measuring methods.

The block chart of the set-up we applied for measuring SDE and DC detectors is shown on figure 3. As a source of photons for detection efficiency measuring we employed a fiber laser diode with 1550 nm wavelength in a continuous wave generation mode. Radiation was going from the source via a calibrated fiber coupler into two channels: detector channel and powermeter channel. Radiation power was measured by Ophir IRG300 powermeter with a accuracy better than 1 picowatts. In the detector channel radiation was passing through a calibrated sensitive attenuator and further through the single-mode fiber to the detector. The radiation power level was fixed at 10^7 photons per second coming to the single-mode fiber connected to the WSi single photon detector. Voltage pulses from the detector passed through the bias-T adapter into the amplification chain consisting of two ultra wideband Mini-circuit amplifiers (bandwidth 0.1-1000 MHz, total gain ~46 dB), and further to the Agilent 53131A pulse counter. To control the shape and scape of the pulse the signal was also send to oscilloscope. The bias-T adapter, an amplification chain and DC biasing unit were integrated into a single unit.

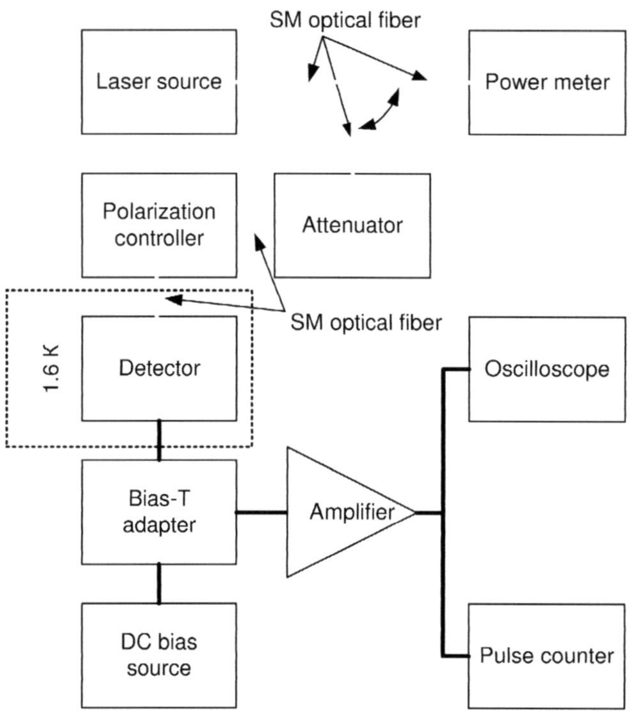

Figure 3. Block chart of the set-up for measurement SDE and DC of single photon detectors based on WSi films.

Detection efficiency of the detector was defined as a ratio between number of electrical pulses coming from the detector per second to the number of photons reaching the detector via single-mode fiber within the same time period. The detection efficiency defined above is the System Detection efficiency because it does not take into account radiation loss related to optical coupling between single mode optical fiber and detector as well as losses related to a certain radiation absorption by the superconductor.

For measuring the dark count rate we made use of the same experimental set-up but in case of screening from radiation of the standard single-mode optical fiber input connected to the detector. For measuring the lowest dark count rates the integration time of the pulse counter was extended to 10 seconds.

4. Results of measurements

To define an optimal thickness of superconducting WSi films in order to reach best SDE/DC ratio and NEP) we produced single-photon detectors based on 5-nm, 7-nm and 9-nm thick films. Figure 4 shows dependencies of SDE and DC for these detectors from relative bias current I_b/I_c, with I_b being the detectors current and I_c being the critical current. We must note that at maximum bias currents the single photon detectors based on 5-nm and 7-nm films showed SDE equal to ~20-30% for λ=1.55 μm, which is determined by the maximum radiation absorption and optical coupling between single-mode

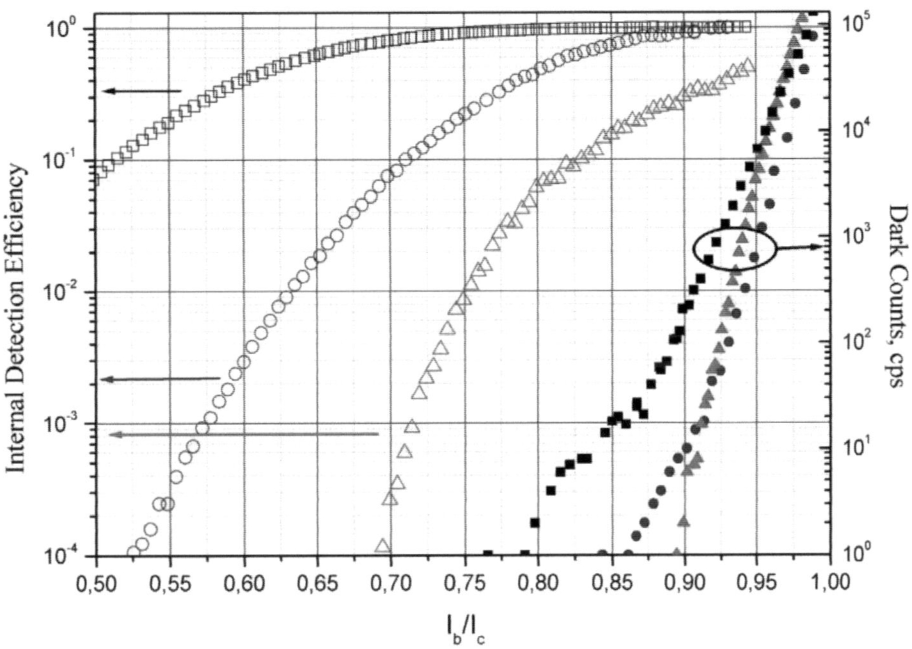

Figure 4. Standardized dependence of SDE (open symbols) and DC (closed symbols) for single photon detectors on Si/SiO_2 substrate from the modified displacement current for detectors based on WSi films of the following thickness: □, ■- 5 nm, ○, ●- 7 nm, △, ▲- 9 nm.

optical fiber and detector. Whilst the detector based on the 9-nm film had an absolute value of 15% for the same wavelength. For making it more convenient to compare and analyze dependencies of SDE (I_b/I_c) for different detectors the present dependencies normalized to the absorption coefficient. DC values are given in the chart in absolute units. The dependencies obtained certify the following: a) higher levels of SDE for the detectors based on 5-nm and 7-nm thick films are determined by the absorption coefficient but achieved at different I_b/I_c values; SDE values for the detectors based on 9-nm thick film do not reach saturation; b) enduring saturated dependence SDE (I_b/I_c) is only indicative for the detectors produced from 5-nm thick WSi films; saturation arise already at currents of ~0.75 I_b/I_c; c) DC dependencies (I_b/I_c) are similar for detectors based on films of different thickness.

Presented SDE dependencies (I_b/I_c) allow for the following conclusions. First of all as the detection efficiency for the 5-nm film detector reaches its constant value at I_b/I_c>0.75 and does not grow further together with the detector bias current. It allow to conclude that the internal quantum efficiency of this detector reaches its maximum level of 100% i.e. each photon absorbed by the superconducting structure generates a voltage pulse. Hence we specified the internal detection efficiency on normalized system detection efficiency chart, the first reaching its maximum value of 1 at I_b/I_c>0.75. Besides,

Journal of Physics: Conference Series **737** (2016) 012032 doi:10.1088/1742-6596/737/1/012032

reduction of the bias current within this currents range causes considerable reduction of the dark count rates down to 1 cps with the quantum efficiency preserved at a constant level.

The detector based on 7-nm thick film also reaches its highest value of the internal quantum efficiency but this takes place solely with maximum levels of bias currents. Despite the 5-nm film detector, this detector reacts onto reduction of dark count rates during the displacement current reduction by reducing considerably its quantum efficiency. The detector based on 9-nm thick film has a still stronger dependence of the detection efficiency on bias current and its internal quantum efficiency does not reach its maximum. We must note that saturated dependence of SDE vs. bias current was not earlier observed for single photon detector of NbN films. We believe that strong saturation of SDE dependence (I_b/I_c) is connected to the WSi films deposition and structuring technology when such parameters as thickness and width the superconducting strip, the superconductor composition and stoichiometric formula are homogeneous well. Undoubtedly the main advantage of the created composition is the possibility of reaching maximum values of SDE/DC and

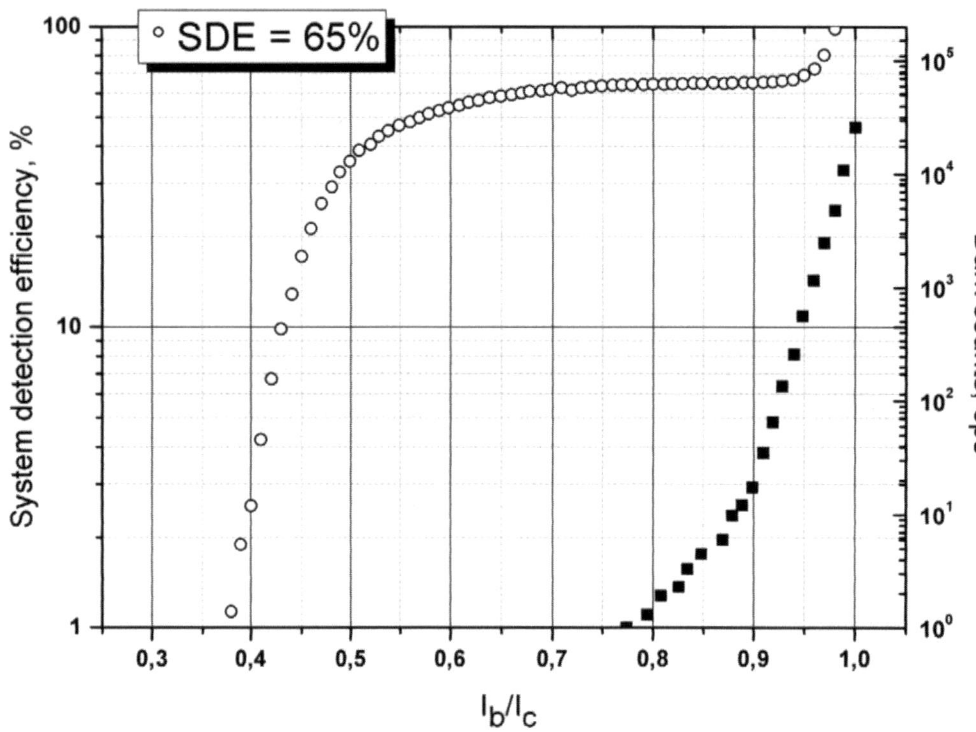

Figure 5. Dependence of SDE (open symbols) and DC (closed symbols) for single photon detectors on $Si/Au/Si_3N_4$ substrate with λ/4 resonator layer from the modified displacement current for 5-nm thick WSi films detectors.

NEP by reducing the detector displacement current and correspondingly reducing DC while preserving SDE value as unchanged. It is also evident that highest levels of SDE/DC and NEP can be in demand just in case high absolute level of SDE reached.

To increase SDE of the 5-nm WSi films based structures we created detector with λ/4 resonator structure located under the superconducting film. Technology and topology of these structures are described above. SDE (Ib/Ic) and DC (I_b/I_c) dependencies for such detectors are shown on Figure 5 and have the dependencies equal to detector based Si/SiO_2 substrate. The maximum absolute value of SDE measured for such detectors makes 65% for a 1.55 μm wavelength. Constancy of the system quantum efficiency is observed in a wider range of bias currents, that is in the area of $>0.6I_b/I_c$. DE dependence (I_b/I_c) of the created detector shows that when the displacement current is reduced the dark count rate falls down to 1 cps with a constant SDE value. Unfortunately with further reduction of the bias current it proved impossible to reduce the DC level because of the background radiation. The additional cooled filters for the background radiation [23] if used can allow for further reduction of DC in the area of $0.6I_b/I_c$ where the system quantum efficiency is still of constant value.

The best NEP value usually determined for quantum detectors as $NEP = \frac{h\nu}{SDE}\sqrt{2DC}$, where $h\nu$ is photon energy, made $4.4 \cdot 10^{-20}$ W·Hz$^{-1/2}$ (for λ=1.55 μm, SDE=65 %, DC=1 cps), and the ratio of SDE/DC=0.65 s. Despite the fact that the authors [13] have come to SDE=90% when using WSi single photon detectors, the best SDE/DC ratio obtained in this study made SDE/DC=$3 \cdot 10^{-3}$ s, with the value of NEP=$3.9 \cdot 10^{-19}$ W·Hz$^{-1/2}$, which is considerably worse than the results presented in this study owing to the lack of dependence of SDE (I_b/I_c) shows a constant value of the system detection efficiency in a wide range of bias currents.

5. Conclusion

Whereas this study presents technology of deposition and structuring of WSi ultrathin films and creating superconducting single photon detector for telecommunication wavelengths (λ=1.55 μm) which can simultaneously reach high values of the system quantum efficiency (65%) and low levels of the dark count rate (1 cps). Obtained records of NEP=$4.4 \cdot 10^{-20}$ W·Hz$^{-1/2}$ (for λ=1.55 μm) and SDE/DC=0.65 s are the best amongst superconducting single photon detectors and in comparison with other alternative types of single photon detectors.

Acknowledgments

The research was financially supported by the Ministry of Education and Science of the Russian Federation (Contract № 3.2655.2014/K). This work was partly supported by the Ministry of Education and Science of the Russian Federation, Contract No. 14.B25.31.0007.

References

[1] Tucker J R and Feldman M J 1985 *Reviews of Modern Physics* **57(4)** 1055
[2] Shurakov A, Lobanov Y and Goltsman G 2015 *Superconductor Science and Technology* **29(2)** 023001
[3] Lita A E, Miller A J and Nam S W 2008 *Optics Express* **16** 3032–3040
[4] Doyle S, Mauskopf P, Naylon J, Porch A and Duncombe C 2008 *Journal of Low Temperature Physics* **151(1-2)** 530-536
[5] Gol'tsman G N et al. 2001 *Appl. Phys. Lett.* **79** 705–707
[6] Semenov A D, Gol'tsman G N and Korneev A A 2001 *Physica C* **351** 349–356
[7] Dorenbos S N, Reiger E M, Perinetti U, Zwiller V, Zijlstra T and Klapwijk T M 2008 *Applied Physics Letters* **93** 131101
[8] Gol'tsman G N et al. 2007 *Infrared and Millimeter Waves 15th International Conference on Terahertz Electronics IRMMW* (Hong Kong) 115-116
[9] Baek B, Lita A E, Verma V and Nam S W 2011 *Appl. Phys. Lett.* **98** 251105
[10] Korneeva Y P, Mikhailov M Y, Pershin Y P, Manova N N, Divochiy A V, Vakhtomin Y B, and Goltsman G N 2014 *Superconductor Science and Technology* **27** 095012
[11] Dorenbos S N, Forn-Diaz P, Fuse T, Verbruggen A H, Zijlstra T, Klapwijk T M and Zwiller V 2011 *Appl.Phys. Lett.* **98** 251102
[12] Engel K A, Aeschbacher A, Inderbitzin K, Schilling A, Ilin K, Hofherr M, Siegel M, Semenov A and Hubers H-W 2012 *Appl. Phys. Lett.* **100** 062601
[13] Marsili F, Verma V B, Stern J A, Harrington S, Lita A E, Gerrits T and Nam S W 2013 *Nature Photonics* **7(3)** 210-214
[14] Sidorova M V, Divochiy A V, Vakhtomin Y B and Smirnov K V 2015 *Journal of Nanophotonics* **9(1)** 093051-093051
[15] Smirnov K, Vachtomin Y, Divochiy A, Antipov A and Goltsman G 2015 *Applied Physics Express* **8(2)** 022501

[16] Shcheslavskiy V, Morozov P, Divochiy A, Vakhtomin Yu, Smirnov K and Becker W 2016 *Review of Scientific Instruments* **87** 053117
[17] Bell M, Sergeev A, Mitin V, Bird J, Verevkin A and Gol'tsman G 2007 *Phys. Rev. B* **76** 094521
[18] Kitaygorsky J et al 2005 *IEEE Trans. Appl. Supercond.* **15** 545
[19] Wolf S and Lowrey W H 1977 *Phys. Rev. Lett.* **39** 1038
[20] Kondo S 1992 *Journal of materials research* **7(04)** pp 853-860
[21] Vasil'ev D D, Malevannaya E I and Moiseev K M. 2015 *Vacuum science and technics Materials of XXII scientific and technical* (Moscow) pp 18-22
[22] Gol'tsman G N et al. 2003 *IEEE Transactions on.* **13** (2) pp192-195
[23] Shibata H, Shimizu K, Takesue H and Tokura Y 2013 *Applied Physics Express* **6 (7)** 072801

Journal of Physics: Conference Series **737** (2016) 012033 doi:10.1088/1742-6596/737/1/012033

Numerical simulation of polariton Bose gas thermalization

P F Kartsev, I O Kuznetsov

National Research Nuclear University MEPhI (Moscow Engineering Physics Institute), 31 Kashirskoe shosse, 115409 Moscow, Russia

E-mail: `PFKartsev@mephi.ru`

Abstract. In this work, we present the numerical simulation of the process a Bose gas thermalization and the formation of the condensate. Our approach is based on kinetic equations and "Fermi's golden rule" in the incoherent approximation. Direct summation of terms is performed using GPGPU OpenCL parallel code using AMD Radeon HD 7970.

1. Introduction

The problem of describing the evolution of a weakly interacting Bose gas (such as gas thermalization, i.e. relaxation to thermal equilibruim after a perturbation, or the formation of a Bose-Einstein condensate at lowering temperature) has received new applications recent years. First of all, they are (i) possible Bose condensation at room temperature for a system of exciton polaritons in semiconductor sample and (ii) the gamma-laser scheme based on annihilation of the Bose-condensed the positronium gas.

The evolution of weakly interacting Bose gas is essential to achieve Bose condensation in the polariton system in the semiconductor sample in the optical resonator under optical pumping [1]. This system is interesting because it can lead to the development of new optoelectronic devices and optical converters. The rate of a Bose condensate formation must prevail over the rate of decay of polaritons, as well as other losses. Due to experimental success, the system of exciton polaritons has become the object of intense theoretical [2], [3] and experimental [1] studies. The decisive role in the establishment of thermal equilibrium in this system is played by the interparticle interaction.

The proposed [4, 5] scheme of gamma-laser making use the annihilation of positronium atoms in the state of BEC, raises question of calculating the rates of all needed processes in the system, including thermalization and Bose condensate formation.

However, there are not many reliable theoretical works applying to these systems[6, 7]. To determine the optimal experimental parameters, it is necessary to build a reliable theoretical model of the system. In this work, we employ simple approach based on the "Fermi's golden rule" and direct summation of kinetic terms.

2. Model

The basic model in the equilibrium statistical mechanics of the system is the case of low density and weak interaction. Let us consider the weakly interacting Bose gas with the contact interaction. The Hamiltonian of the system is

$$\hat{H} = \hat{H}_0 + \hat{V} = \sum_k \varepsilon_k \hat{a}_k^\dagger \hat{a}_k + \sum_{k_1 k_2 k_3 k_4}' U_0 a_{k_1}^\dagger \hat{a}_{k_2}^\dagger \hat{a}_{k_3} \hat{a}_{k_4} \tag{1}$$

The prime at the summation sign denotes the momentum conservation law ($k_1 + k_2 = k_3 + k_4$).

In the case of temperature high enough (i.e. no condensate) and incoherent limit (neglecting correlation of particle occupations) we can apply the "Fermi's golden rule" for the relaxation time:

$$\frac{1}{\tau} = \frac{2\pi}{\hbar} \sum_f |< i \mid \hat{V} \mid f >|^2 \, \delta(E_i - E_f) \tag{2}$$

Thermalization dynamics of the system can be described by the quantum Boltzmann equation (QBE) taking into account the law of conservation of energy [8]. Thus a change of the particle density n_p is given by the equation:

$$\frac{dn_{\mathbf{p}}}{dt} = \frac{2\pi}{\hbar U_0^2} \sum_{\mathbf{qmn}}' \delta(\varepsilon_{\mathbf{q}} + \varepsilon_{\mathbf{p}} - \varepsilon_{\mathbf{m}} - \varepsilon_{\mathbf{n}}) \times$$
$$\times [(n_{\mathbf{p}} + 1)(n_{\mathbf{q}} + 1 + \delta_{\mathbf{pq}}) n_{\mathbf{m}} (n_{\mathbf{n}} - \delta_{\mathbf{nm}}) - n_{\mathbf{p}} (n_{\mathbf{q}} - \delta_{\mathbf{pq}})(n_{\mathbf{m}} + 1)(n_{\mathbf{n}} + 1 + \delta_{\mathbf{nm}})] \tag{3}$$

In further calculations we assume $\frac{2\pi U_0^2}{\hbar} = 1$ and consider the dimensionless time.

3. Numerical calculation

Main obstacle to simulate the system evolution using Eq.(3) for reasonable system sizes ($L > 6..8$) is caused by enormous amount of needed calculation. In our experience, it scales roughly as L^8.

This required to apply the corresponding high-performance solution. In our work, we created a parallel GPGPU code using OpenCL language with acceptable resulting speed able to take as high as $L = 12..18$. Our calculations were performed on the AMD Radeon HD 7970 graphic accelerator. Further acceleration is also possible, with the help of parallel computing and task distribution on multiple accelerators.

In Figs. 1,2,3,4 we present the results of simulation for the system $L^3 = 12^3$ for several time slices in the time intervals of 10^{-11} (dimensionless units). We plot the 2D momentum distribution for slice $p_z = 0$, and additionally 1D slice for $p_z = p_y = 0$.

We take different starting distributions of particles which reflect different temperatures ($T < T_c$ and $T > T_c$, Figs. 1 to 3) and shifted initial distributions due to e.g. optical pumping (Fig. 4).

All starting distributions were chosen as rectangular ($n_p = n_{max}$ at $|p| < \sigma$) with small nonzero background ($n_p = n_{min}$ at $|p| \geq \sigma$), which proven needed to get equilibration due to *Umklapp* processes.

All figures show reasonable behaviour. The systems reach reach expected thermal distribution.

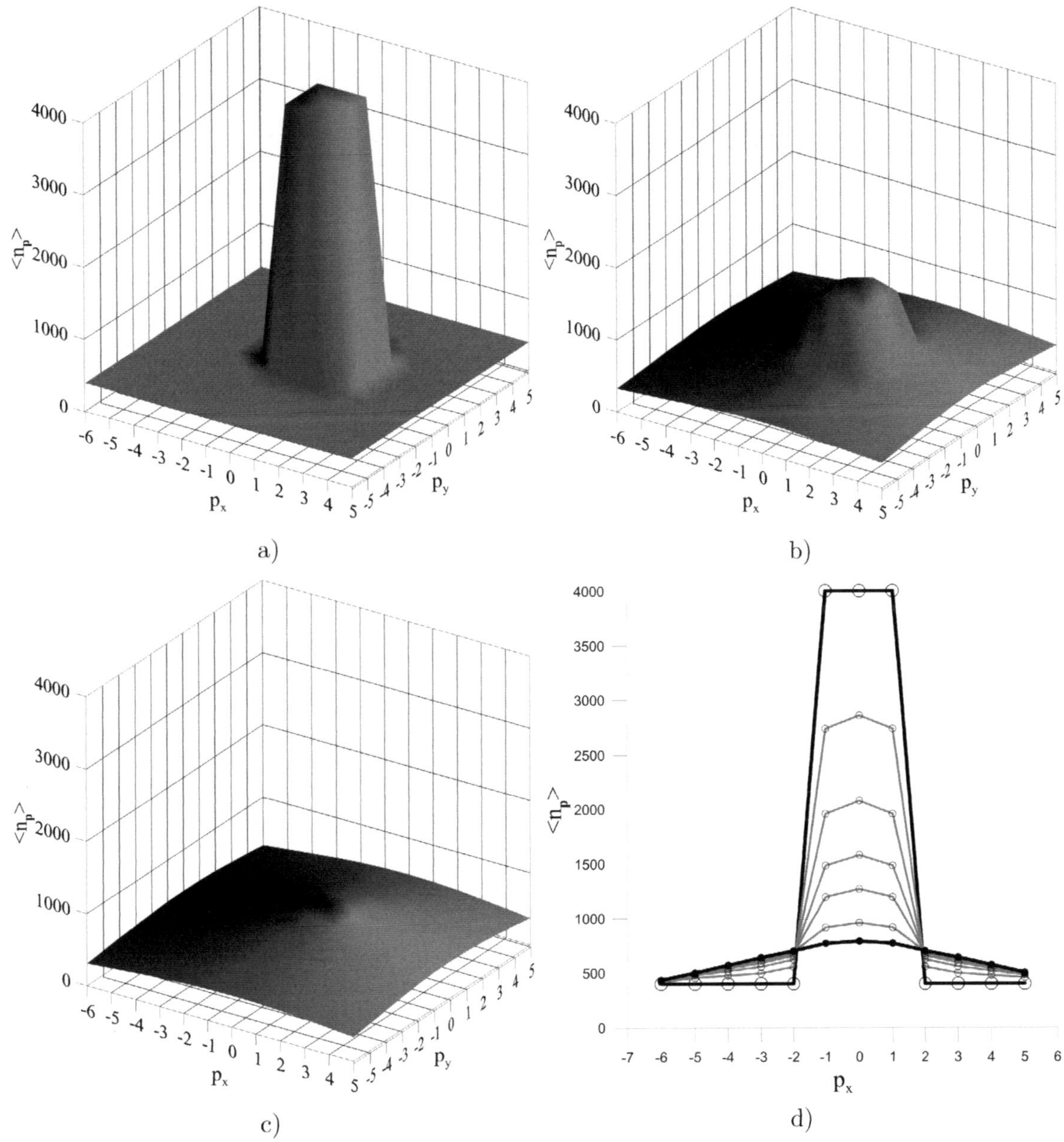

Figure 1. The initial distribution of the particles (a) in three dimensional space 12x12x12 size gradually redistributed (b, c). The 2D slices on $p_z = 0$. Part (d) represents 1D slice on $p_y = p_z = 0$. We see the gradual change of distribution to equilibrium

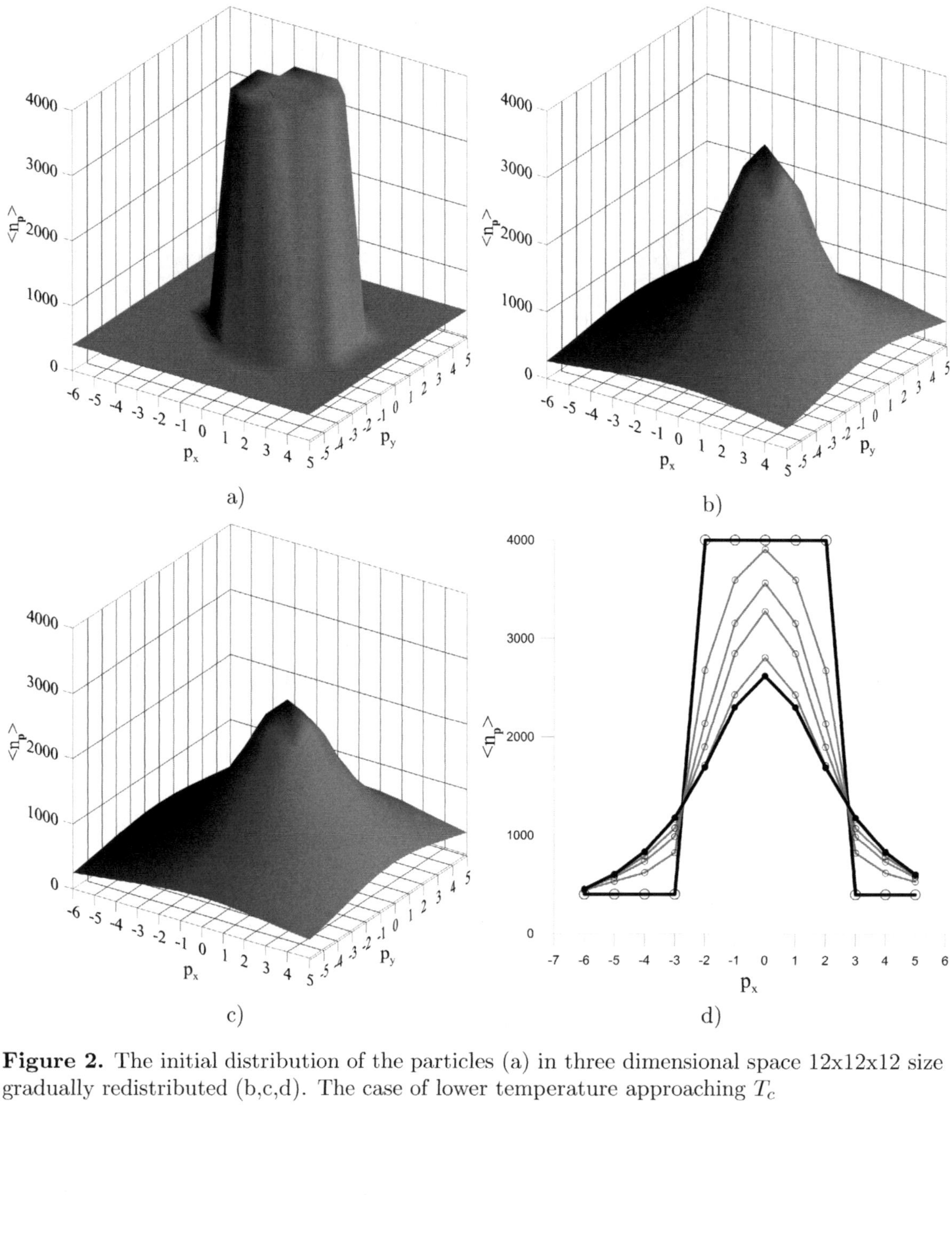

Figure 2. The initial distribution of the particles (a) in three dimensional space 12x12x12 size gradually redistributed (b,c,d). The case of lower temperature approaching T_c

Journal of Physics: Conference Series **737** (2016) 012033 doi:10.1088/1742-6596/737/1/012033

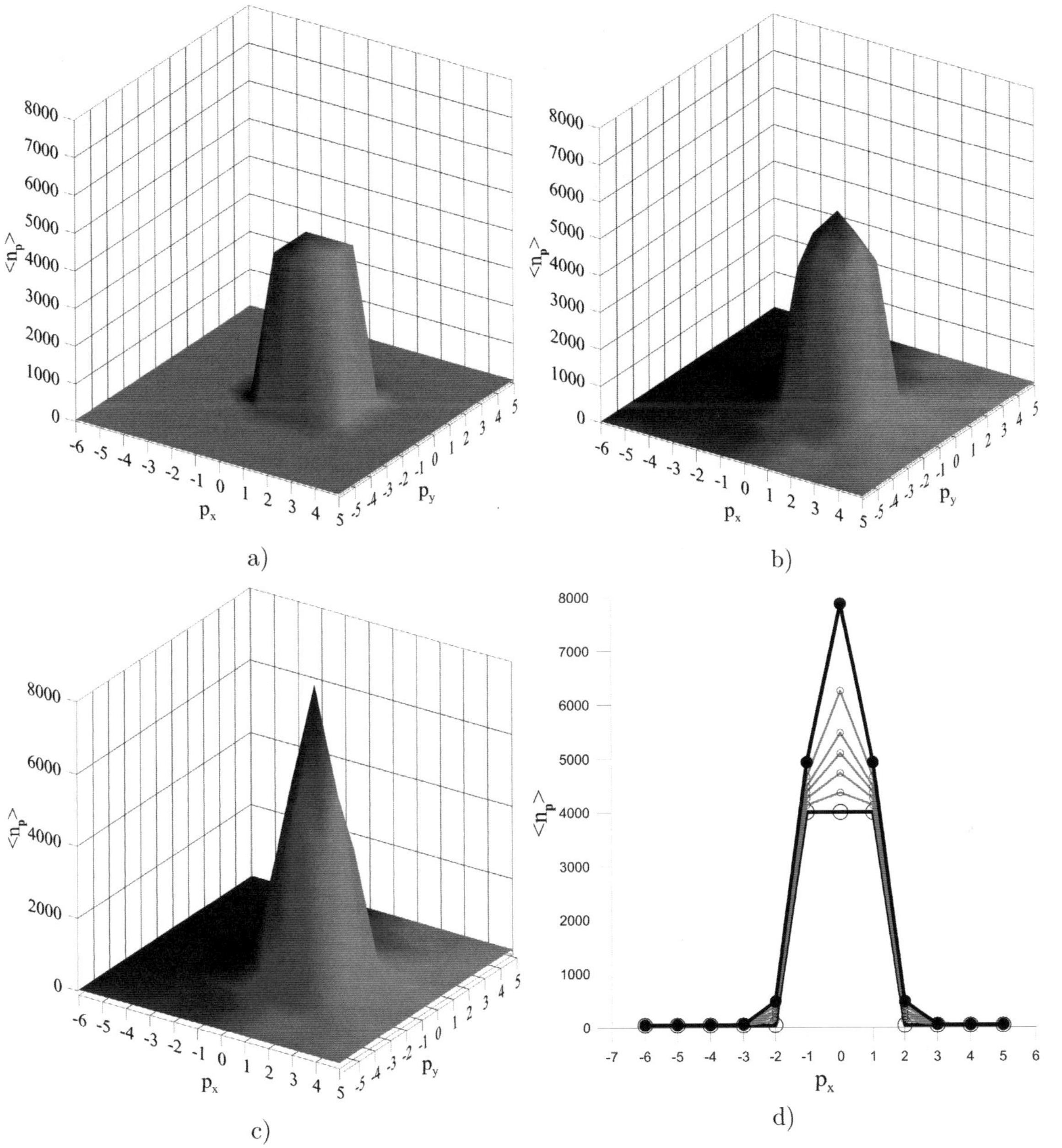

Figure 3. The initial distribution of the particles (a) in three dimensional space 12x12x12 size gradually redistributed (b,c,d). The case of temperature lower than T_c

Journal of Physics: Conference Series **737** (2016) 012033 doi:10.1088/1742-6596/737/1/012033

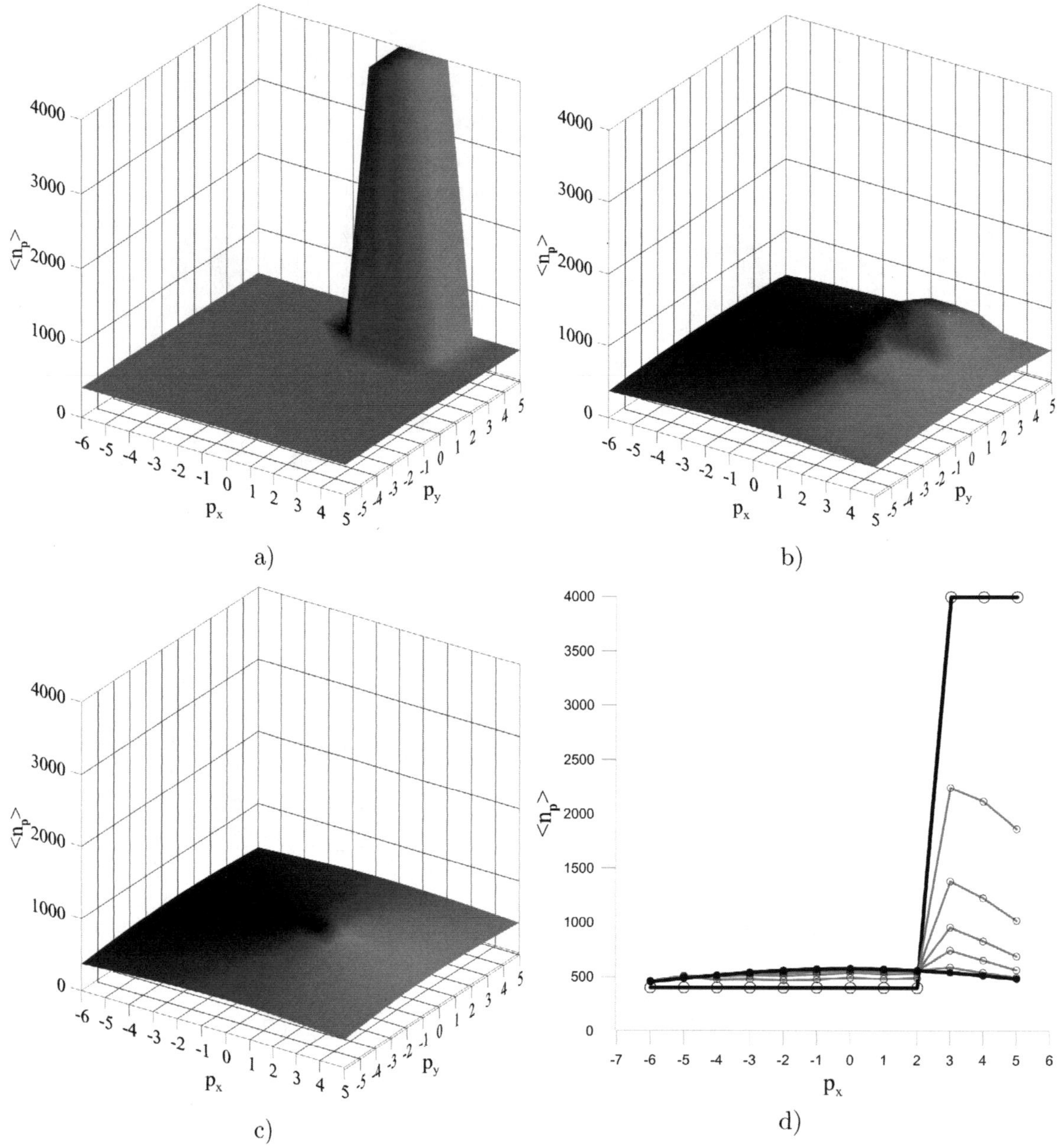

Figure 4. The initial distribution of the particles (a) in three dimensional space 12x12x12 size gradually redistributed (b,c,d). The case of shifted initial distribution modeling optical pumping of polaritons

The figures clearly show the process of redistribution of occupation numbers, giving relaxation to thermal function and the beginning of the Bose condensate formation. In figure (1) the spread of the particles takes place over the entire space with a uniform increase in the occupation on all momenta, in contrast to the given initial filling. The case of a larger initial density in figure ((2) corresponds to lower effective temperature of the system.

For figure (3) with even lower temperature we note the accumulation of particles in the center of the momentum space, reflecting beginning of the formation of Bose condensate.

4. Conclusion

We performed numerical simulation of the evolution of the particle distribution function in the incoherent approximation on the basis of kinetic equations and "Fermi golden rule" by direct calculation on a discrete momentum grid. The results of the calculations are presented as Figures 1 - 4, which show the evolution of particle distribution for different temperatures and starting occupations.

Such a direct numerical calculation of the Bose gas thermalization can be applied to determine optimal parameters of new BEC experiments with alkali gases and give new information on polaritonic condensation in the semiconductor sample under optical pumping.

The work was supported by the Russian Foundation for Basic Research, Grant No. 15-02-02764.

References

[1] van Vugt L K, Piccione B, Cho C-H, Nukala P and Agarwal R 2011 *PNAS* **108** 10050–5
[2] Voronova N S and Lozovik Yu E 2012 *Phys. Rev.* B **86** 193305
[3] Das A, Bhattacharya P, Banerjee A and Jankowski M 2012 *Phys. Rev* B **85** 195321
[4] Avetissian H K, Avetissian A K and Mkrtchian G F 2014 *Phys. Rev.* Lett **113** 023904
[5] Wang Y-H, Anderson B M and Clark Ch W 2014 *Phys.Rev.* B **89** 043624
[6] Kagan Yu, Muryshev A E and Shlyapnikov G V 1998 *Phys. Rev.* Lett **81** 933
[7] Wertz E, Amo A, Solnyshkov D D, Ferrier L, Liew T C H, Sanvitto D, Senellart P, Sagnes I, Lemaitre A, Kavokin A V, Malpuech G and Bloch J 2012 *Phys. Rev.* Lett **109** 216404
[8] Davis M J, Ballagh R J and Burnett K 2001 *J. Phys.* B **34** 4487–512

Band structure calculation of GaSe-based nanostructures using empirical pseudopotential method

A V Osadchy[1,2], S G Volotovskiy[3], E D Obraztsova[1,2], V V Savin[4] and D L Golovashkin[3]

[1]A.M. Prokhorov General Physics Institute, RAS, 38 Vavilov Str., 119991 Moscow, Russia
[2]National Research Nuclear University MEPhI (Moscow Engineering Physics Institute), 31 Kashirskoye Shosse, 115409 Moscow, Russia
[3]Image processing system institute of Russian academy of sciences, 151 Molodogvardeyskaya str., 443001 Samara, Russia
[4]Immanuel Kant Baltic Federal University, 14 Nevskogo str., 236041 Kaliningrad, Russia

E-mail: aosadchy@kapella.gpi.ru

Abstract. In this paper we present the results of band structure computer simulation of GaSe-based nanostructures using the empirical pseudopotential method. Calculations were performed using a specially developed software that allows performing simulations using cluster computing. Application of this method significantly reduces the demands on computing resources compared to traditional approaches based on ab-initio techniques and provides receiving the adequate comparable results. The use of cluster computing allows to obtain information for structures that require an explicit account of a significant number of atoms, such as quantum dots and quantum pillars.

1. Introduction

Layered semiconductors AIIIBVI groups (e.g., gallium selenide GaSe, gallium sulfide GaS, etc.) are widely used as nonlinear elements in laser technology. In this list, the crystalline GaSe is especially allocated because it is used as a second harmonic generator in CO2 laser [1,2]. Despite this, the applicability of this material, due to a limited spectral operation range, is not so wide, as it may be expected. In recent years, due to the significant advances in the study of another layered material - graphene, the interest to such material grows back. This is one of the reasons for a new direction of work on the synthesis, a comprehensive study and practical application of nanostructured materials based on GaSe. The layer struc-ture similar to graphite, makes it relatively easy to tear the individual layers for obtaining the samples consisting of one or few layers. Recently several new works dedicated to this subject have been published [3,4,5]. Nevertheless this area is still insufficiently investigated.

This work aims to study the band structure of nanomaterials based on gallium selenide using the method of empirical pseudopotential. The main idea is to demonstrate the size effects observed in this area for further experimental work on the fine tuning of optical properties and broaden the application range of GaSe in laser technology.

2. Computation methods

Computer modeling was carried out using the method of empirical pseudopotential, which belongs to a group of techniques in which the effective potential in the one-electron Schrodinger equation in part is parameterized so that the obtained values of the energy and electron wave functions of electron states reproduced the experimental data.

The method is the solution of the pseudo wave Schrödinger equation:

$$\left[-\frac{1}{2}\Delta + V_{ps}(\vec{r})\right]\varphi_{ps,j}(\vec{r}) = \epsilon_j \varphi_{ps,j}(\vec{r}), \quad (1)$$

where $V_{ps}(\vec{r})$ – pseudopotential of cristall, $\varphi_{ps,j}(\vec{r})$ – one-electron pseudo wave functions, ϵ_j – electron eigenvalue energies.

Pseudo wave function is represented as an expansion in plane waves:

$$\varphi_{ps}(\vec{r},\vec{k}) = \sum_{v=1}^{M} C_v e^{i(\vec{k}+\vec{G}_n)\vec{r}}, \quad (2)$$

where $\vec{k}$ – electron wave vector, M – plane wave number, C_v – coefficients of expansion, $\vec{G}_n$ - reciprocal lattice vectors. By substituting (2) at (1) and multiplying scalar on $e^{i(\vec{k}+\vec{G}_m)\vec{r}}$ we obtain secular equation:

$$\det\left\{\left[\left(\vec{k}+\vec{G}_n\right)^2 - \epsilon(\vec{k})\right]\delta_{m,n} + V_{ps}^{m,n}\right\} = 0, \quad (3)$$

where $V_{ps}^{m,n}$ - the matrix element of the pseudopotential operator.

$$V_{ps}^{m,n} \equiv V_{ps}(\vec{G}_m - \vec{G}_n) \equiv \frac{1}{\Omega}\int e^{-i\vec{q}\vec{r}} V_{ps}(\vec{r}) dr, \quad (4)$$

where Ω – crystal volume, $\vec{q} = \vec{G}_m - \vec{G}_n$. Crystal pseudopotential is represented as a superposition of atomic pseudopotentials and represented as:

$$V_{ps}^{m,n} = \frac{1}{N}\sum_{i=1}^{N} \exp(-i\vec{q}\vec{r}) FF_i(|\vec{q}|), \quad (5)$$

where N – amount of atoms in crystal basis, $FF_i(|\vec{q}|)$ - form factor of i-th atom:

$$FF_i(|\vec{q}|) = \frac{c_i^1(|\vec{q}|^2 - c_i^2)}{\exp(c_i^3(|\vec{q}|^2 - c_i^4)) + 1} \cdot \lambda, \quad (6)$$

where C^k_i –empirical coefficients of pseudopotential for the i-th atom, λ – normalization factor.

Thus, substituting (6), (5) and (4) in the secular equation (3) and solving it, it is possible to obtain electronic dispersion curves for investigated material.

The above described approach allows simulation of a wide range of materials, including various nanostructures, with the method of supercell [6]. At the same time, the usage of this approach for calculation of nanomaterial parameters, especially those, requiring an explicit consideration of significant numbers of atoms (quantum dots, disks, pillars, and layered materials such as gallium selenide) encounters significant problems in the large Hamiltonian matrices requiring huge amounts of computer memory. This problem is a major difficulty in the simulation of quantum dots and quantum pillars of gallium selenide and other layered semiconductors. To solve the problem the computer algorithm parallel computation was implemented where eigenvectors and eigenvalues of the Hamiltonian matrix were obtained by filling and diagonalizing matrix in a part using a large number of cluster supercomputer nodes [7]. Implement of this approach on computational center K100 which located in the M.V. Keldysh IPM made it possible to take into account up to 10^7 atoms in the system.

The next problem of the method applicability is a significant dependence of adequacy of the obtained results on the used pseudopotential parameters. In case of crystalline GaS, the pseudopotential parameters providing a good coincidence with the experimental data have been published [8,9]. However, the research carried out for the nanostructures based on these materials have shown that the quality of electron dispersion relations is not sufficient for the practical applicability of the results. The work for optimization of the pseudopotential parameters for Ga and Se was needed. The problem was solved by developing an algorithm of pseudopotential parameters C^k_i variation for performing the minimization of deviation function of electronic dispersion dependences for a bulk crystal and for a single layer of GaSe, obtained by the empirical pseudopotential method, and the reference data published in the literature [10]. A gradientless method of finding a local minimum has been used for optimization [11]. The obtained pseudopotential parameters are provided in Table 1.

Table 1. A comparison of parameters of atomic pseudopotentials presented in the literature and obtained by optimization.

Atom	Source	C1	C2	C3	C4
Ga	Schlüter [9]	0.21	2.75	0.6	3.0
	Optimization	0.545	2.870	1.867	0.255
Se	Schlüter [9]	0.27	3.45	0.42	1.86
	Optimization	0.332	2.305	1.655	2.553

The optimization results listed above have been used for calculations of band structure of nanomaterials based on GaSe. In the simulation, the number of plane waves of M used for the expansion of electronic wave vector was selected based on the cut-off energy of 14 Ry. The expansion involved the plane waves with energy lower than the cut-off energy.

3. Results and discussions

The main objective of this work was the modeling of band structure of different types of nanomaterials, based on layered semiconductor GaSe. Figure 1 shows the electronic dispersion relations for the bulk crystal of GaSe, calculated by the method based on density functional theory (Fig. 1) [12] and the empirical pseudopotential described above, using the pseudopotential parameters listed in Table 1 (Fig. 2).

From obtained dependence it can be seen that the width of the direct band gap for bulk β-GaSe, calculated by method based on the electron density functional theory, and obtained in this work are significantly different. The bandgap in Figure 1 is 0.8 eV, while in figure 2 it is 1.8 eV. At the same time, the experimental data published in literature demonstrate a band gap of 2.1 eV [13]. Thus, a method based on the electron density functional theory, significantly lowers the bandgap (by more than 1 eV). This prevents the use of this method for predicting the optical properties of the material. At the same time, the method used in this work shows the band gap value which is much closer to the experimental values.

The next step of the work was modeling various types of nanostructured materials based on gallium selenide. Figure 3 shows the band structure of one- and two-layered GaSe, calculated by the method of empirical pseudopotential.

From the obtained dependence it can be seen that the width of the indirect band gap for a single layer of gallium selenide is 2.11 eV. This value is in a good agreement with published data [12]. Also, a ring-shape dependence of the dispersion in valence band on k-vector can be seen from Figures. This is completely similar to data shown in [14].

Thus, a good agreement between results obtained with the abovementioned method, the published experimental and theoretical data, have been obtained. Thus, a promising possibility of this approach to simulate different nanostructures with the explicit account of a considerable number of atoms along with much less requirements for computational resources than the widely used methods (including those based on electron density functional theory and GW-approximation) and the possibility of performing calculations on cluster supercomputers have been demonstrated.

Journal of Physics: Conference Series **737** (2016) 012034 doi:10.1088/1742-6596/737/1/012034

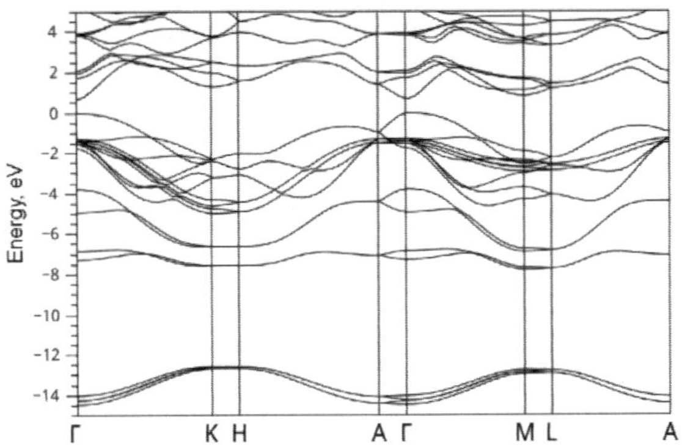

Figure 1. The band structure of β-GaSe, calculated by the electron density functional method [12]

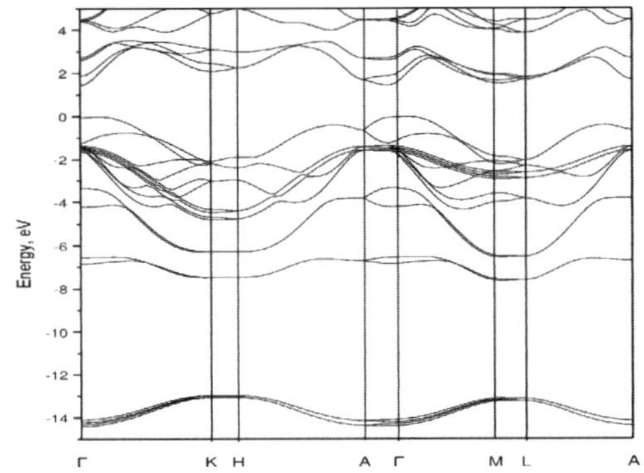

Figure 2. The electronic structure of β-GaSe, calculated by the method of empirical pseudopotential.

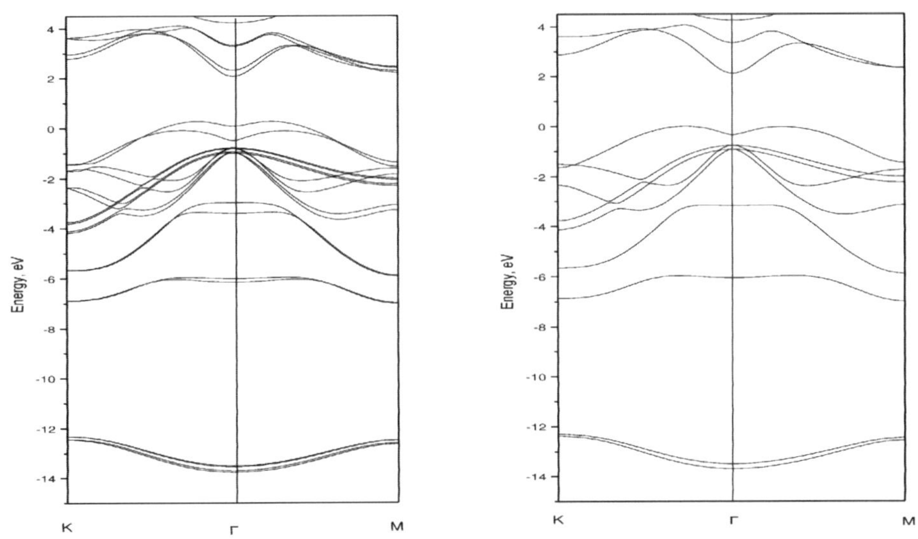

Figure 3. The band structure of a single-layer (right) and a double layer (left) of gallium selenide, calculated by the method of empirical pseudopotential.

4. Conclusions

In the work the numerical simulation of the band structure of different types of nanostructures based on layered gallium selenide has been performed. The calculation was performed using the method of empirical pseudopotential. The parameters of gallium and selenium atomic pseudopotentials have been optimized in this paper for the best reproduction of published experimental data. The results showed a good agreement with the experimental data for a variety of structures, along with the significantly lower requirements for computing resources than the widely used ab-initio methods, and made this method more prospective for such structures modeling.

During the work a computer program implementing this approach has been developed allowing using both personal computers and clusters with the distributed memory supercomputers. This significantly extends the application range of the method (allows one to explicitly take into account the significant number of atoms). Because the program is developed in a general way, one can use it for a wide range of different materials.

Acknowledgments
This work was supported by RFBR 15-02-09073 and the Competitiveness Programm of NRNU MEPhI.

References

[1] Auerhammer J M and Eliel E R 1996 *Opt. Lett.* **21 (11)** 773.
[2] Singh N B, Suhre D R, Balakrishna V, Marable M, Meyer R, Fernelius N, Hopkins F K and Zelmon D 1998 *Progress in Crystal Growth and Characterization of Materials* **37 (1)** 47–102
[3] Chikan V, Kelley D F 2002 *Nano Lett.* **2 (2)** 141-145
[4] Kohler Th, Frauenheim Th, Hajnal Z, Seifert G 2004 *Phys. Rev. B* **69 (19)** 193403-1 - 193403-4
[5] Rybkovskiy D V, Osadchy A V and Obraztsova E D 2013 *Journal of Nanoelectronics and Optoelectronics* **8** 110-113
[6] Payne M C, Teter M P, Allan D C, Arias T C and Joannopoulos J D 1992 *Reviews of Modern Physics* **64** 1045
[7] Blackford L S, Choi J, Cleary A, D'Azevedo E, Demmel J, Dhillon I, Dongarra J, Hammarling S, Henry G, Petitet A, Stanley K, Walker D, Whaley R C 1997 *ScaLAPACK Users' Guide* (Society for Industrial and Applied Mathematics)
[8] Cohen M L, Bergstresser T K 1966 *Phys. Rev.* **141 (2)** 789-796
[9] Schlüter M 1973 *Nuovo Ci-mento B.* **13 (2)** 313-360
[10] Rybkovskiy D V, Vorobyev I V, Osadchy A V, Obraztsova E D 2012 *Journal of Nanoelectronics and Optoelectronics* **7** 65–67
[11] Powell M J D 1998 *Acta Numerica* **7** 287-336
[12] Rybkovskiy D V 2014 *The electronic structure of two-dimensional and zero-dimensional nanoobjects of gallium selenide* (*Cand. of physical and mathematical sciences diss.*) (GPI RAS)
[13] Aulich E., Brebner J L, Mooser E 1969 *Phys. Stat. Sol. (b)* **31 (1)** 129-131
[14] Rybkovskiy D V, Osadchy A V and Obraztsova E D 2014 *Physical Review B* **90** 235302

Vibrational spectroscopy of $GdCr_3(BO_3)_4$: quantitative separation of crystalline phases

E A Dobretsova[1], K N Boldyrev[1], M N Popova[1], V A Chernyshev[2], E Yu Borovikova[3], V V Maltsev[3] and N I Leonyuk[3]

[1]Institute for Spectroscopy Russian Academy of Sciences, Moscow, Troitsk, 108480, Russia
[2]Ural Federal University, Ekaterinburg, 620002, Russia
[3]Department of Crystallography and Crystal Chemistry, Faculty of Geology, Moscow State University, Moscow, 119991, Russia.

E-mail: elena-dobrecova@yandex.ru

Abstract. This work is devoted to the investigation of $GdCr_3(BO_3)_4$ crystals by the method of infrared spectroscopy. Incongruently melting borate $GdCr_3(BO_3)_4$ was obtained as a result of spontaneous crystallization. Crystal structures were identified by the method of infrared spectroscopy. *Ab initio* calculations in the frame of density functional theory enabled us to separate modes belonging to the *R*32 and *C*2/*c* phases and to estimate the ratio of these phases in $GdCr_3(BO_3)_4$ crystals. We have found that the content of the rhombohedral *R*32 (non-centrosymmetric) modification is about 85 %.

1. Introduction

Rare-earth (RE) borates described by the general formula $RM_3(BO_3)_4$ (R = RE or yttrium, M = Al, Ga, Sc ,Fe, or Cr) crystallize in a number of space groups (SG) with structure type of the natural mineral huntite [1]. These compounds are of interest due to their optical, magnetic, and magnetoelectric properties [2, 3]. RE aluminum borates possess excellent luminescent, nonlinear optical, thermal, and mechanical properties, are chemically stable and are widely used in self-frequency doubling and self-frequency summing lasers [4-6]. RE iron [7-11], aluminum [12-15] and gallium [16] borates demonstrate attractive magnetoelectric properties. Materials with a large magnetoelectric effect have an application potential as magnetoelectric sensors, magneto-electric and electro-magnetic memory elements, etc. It is important to notice that nonlinear optical and magnetoelectric properties are characteristic only for non-centrosymmetric structures.

RE chromium borates have polytype nature and crystalize in two modifications, namely, the rhombohedral one with the non-centrosymmetric SG *R*32 or the monoclinic one with the centrosymmetric SG *C*2/*c*. The crystal type depends on the growth conditions. Moreover, a coexistence of differently ordered structural fragments in one single crystal is a characteristic feature of RE chromium borates. [17, 18].

The objective of our present study are to identify the crystal structure of $GdCr_3(BO_3)_4$ obtained by the method of spontaneous crystallization and to evaluate its potential for practical applications.

2. Crystal structure of huntite-like borates

As was already mentioned in Introduction, RE borates have the structure type of huntite $CaMg_3(CO_3)_4$. The latter one crystallizes in the SG *R*32 [19]. Three types of coordination polyhedra are present in this structure, namely, CaO_6 trigonal prisms, MgO_6 octahedra and two types of CO_3 plain triangles [20] The MgO_6 octahedra are connected by common edges forming chains along the *c* axis. The RO_6 trigonal prisms are isolated from each other. In the rare-earth double borates, Ca atoms are replaced by RE elements but instead of Mg atoms one finds *M* cations (which form a quasi-one-dimensional magnetic system in the case of Fe^{3+} and Cr^{3+} ions) [21]. The CO_3^{2-} groups are replaced by the BO_3^{3-} triangles (figure 1).

Besides the rhombohedral *R*32 structure there is a monoclinic one with SG *C*2/*c*. In each of these two structures it is possible to distinguish two different types of layers, L_{2n} and L_{2n+1}, identical for two modifications (figure 2). The first type layer (L_{2n}) contains pairs of edge-sharing MO_6 octahedra which are further connected into a layer via isolated BO_3 triangles. The second one (L_{2n+1}) is built by trigonal RO_6 prisms and MO_6 octahedra connected in columns by vertices and also connected via the BO_3 triangles. In the rhombohedral (SG *R*32) polytype, the L_{2n+1} layers which contain prisms are multiplied around the L_{2n} layers by two-fold axes, whereas in the monoclinic modification (SG *C*2/*c*) by centers of symmetry [22, 23].

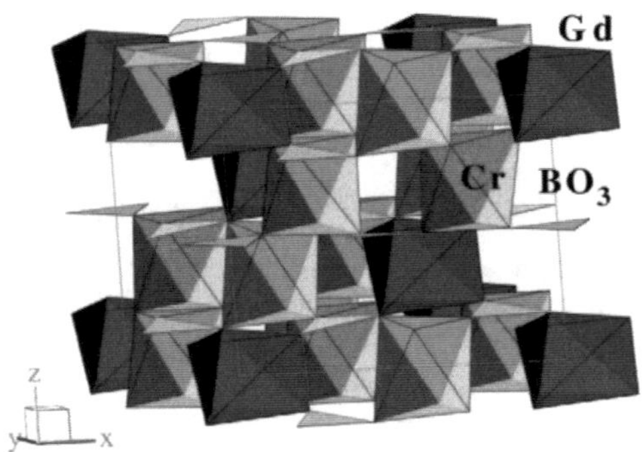

Figure 1. Crystal structure of $GdCr_3(BO_3)_4$ with SG *R*32

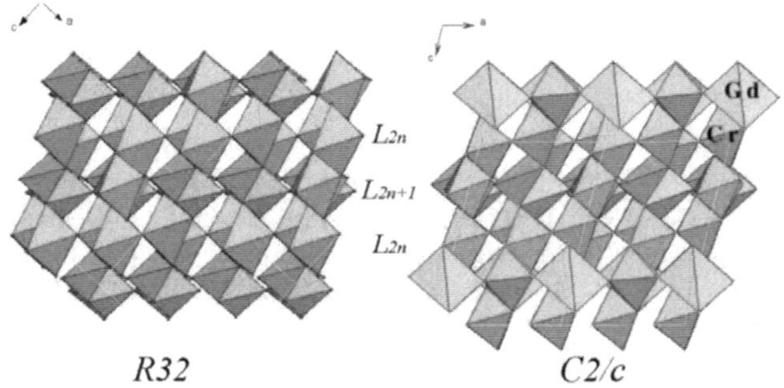

Figure 2. Schematic structures of $GdCr_3(BO_3)_4$ with: a) SG *R*32 (D_3^7) and b) SG *C*2/*c* (C_{2h}^6)

3. Experimental and computational details

3.1. Sample preparation

The incongruently melting borate $GdCr_3(BO_3)_4$ was obtained from a molten solution on the basis of K_2MoO_4 - $3MoO_3$ as a result of spontaneous crystallization. Experiments were conducted with a ratio of the borate to the solvent in the mixture 1:1. The crucible with the charge was maintained in a furnace until a complete homogenization of the melt for 5 hours at 1100 – 1150 °C. Thereafter, the temperature was lowered to 1000 °C at a rate of 2 °C/h, then at a rate of 1°C/h to 840 °C and to 300 °C at a rate of 10 °C/h. Crystallization of the borate occurred at 1050 - 840°C.

For infrared measurements, a single crystal was grinded into a powder in a pobedite mortar.

3.2. Measurements by the methods of infrared spectroscopy

Absorption spectra of the powder sample of $GdCr_3(BO_3)_4$ in the far-infrared (FIR) region 50–550 cm^{-1} were registered with the help of a Bruker IFS 125HR Fourier spectrometer (the spectral resolution was up to 2 cm^{-1}), using polyethylene disc technique. The infrared absorption spectra in the middle IR region (MIR) 550-1600 cm^{-1} were registered on the same spectrometer with an attenuated total reflection (ATR) module. The powder sample was pressed to a diamond crystal of the ATR module. The spectral resolution was 2 cm^{-1}. We applied methods of vibrational spectroscopy combined with the group-theoretical analysis [24] in order to understand the peculiarities of the structure of the $GdCr_3(BO_3)_4$ borate.

3.3. Ab initio calculations

Ab initio calculations of phonon frequencies and intensities of the infrared-active modes in $GdCr_3(BO_3)_4$ were performed for both *R32* and *C2/c* structures in a frame of the density functional theory (DFT) with hybrid functional B3LYP. The calculations were carried out within MO LCAO approach implemented in CRYSTAL14 code. Non-relativistic pseudopotential was used to describe the inner shells of the rare-earth ion. Details of the calculations can be found in Ref. [25].

4. Results of IR measurements and DFT calculations

Figure 3 shows the IR spectrum in the region 50-1600 cm^{-1}. It is typical for RE chromium borates with the SG *R*32. The majority of the observed modes find their counterparts in the calculated energy spectrum for this phase of $GdCr_3(BO_3)_4$. It is not possible to extract oscillator strengths from the measured spectra of powder samples. But we have to note that, in general, large calculated oscillator strengths correspond to strong lines in the measured spectrum, whereas for low ones one finds either weak lines or no line at all (see Table 1). Thus, the calculated data for the *R*32 phase are in good agreement with the experimental ones, which, in turn, confirms the assumption of predominantly rhombohedral structure of gadolinium chromium borate.

However, several features point to the presence of a monoclinic phase fragment. For internal vibrations of BO_3^{3-} groups, they are (i) a shoulder at 747 cm^{-1} in the region of the symmetric bending vibrations, (ii) a splitting of the band at 980 cm^{-1} in the region of the symmetric stretching vibrations, and (iii) shoulders at 1278 and 1372 cm^{-1} in the region of asymmetric stretching vibrations (Figure 3, Table 1). In the FIR region, translational vibrations of the Gd^{3+} and Cr^{3+} ions, as well as translational and librational motions of the BO_3^{3-} groups are observed. Signatures of monoclinic inclusions in this spectral region are (i) a weak band at 73 cm^{-1} of the Gd^{3+} vibrations, (ii) bands at 126, 148, and 275 cm^{-1}, (iii) a weak shoulder at 248 cm^{-1} due to the BO_3^{3-} vibrations, and (iv) a splitting of the band at 434 cm^{-1} in the region of Cr^{3+} vibrations. (Figures 3 and 4, Table 1).

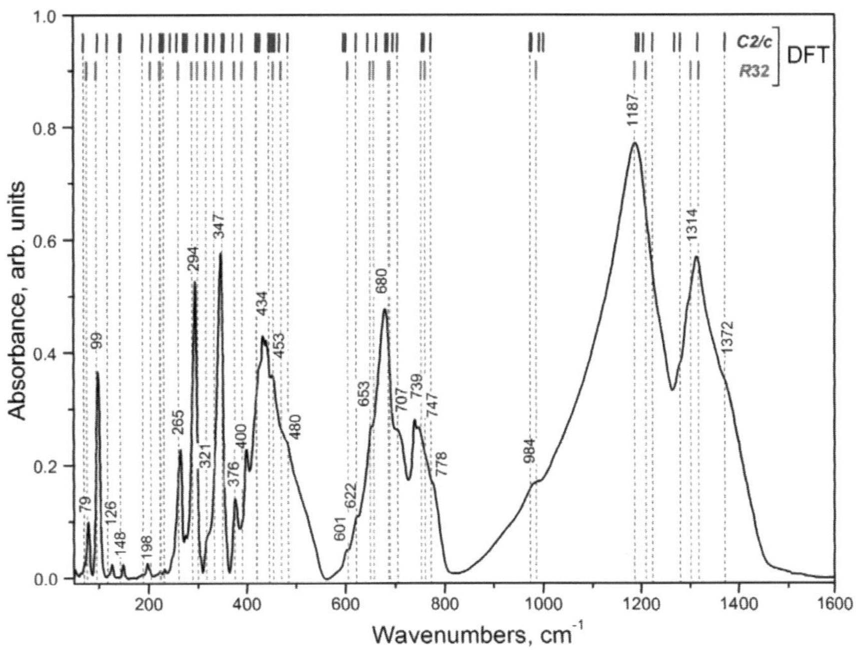

Figure 3. Infrared spectrum of $GdCr_3(BO_3)_4$. At the upper part of the figure, results of the *ab initio* calculations for the *R*32 (red bars) and *C2/c* (blue bars) phases are presented.

Journal of Physics: Conference Series **737** (2016) 012035 doi:10.1088/1742-6596/737/1/012035

Table 1. Comparison between calculated and experimentally observed vibrational modes of $GdCr_3(BO_3)_4$ in the *R*32 phase. The assignments of particular modes are also given in accordance with the calculated eigenvectors. Experimental extra peaks ascribed to the monoclinic C2/*c* phase are in bold (their symmetry is indicated in parentheses). The corresponding calculated frequencies are plotted in Figures 3 and 4.

Assignments		*Ab initio* calculations		Experiment
		Wavenumbers, cm^{-1}	Oscillator strength	Wavenumbers, cm^{-1}
A_1	L BO_3^{3-}	216		
	T (BO_3^{3-}) T (Gr^{3+})	342		
		495		
	ν_4	660		
	ν_1	975		
		1004		
	ν_3	1209		
A_2	T (Gd^{3+})	76	2,97	**73 sh (B_u)** 79 m
	L (BO_3^{3-})	205	0,11	**126 w (B_u)** **148 w (B_u)** 198 w
		225	0,001	233 w
		261	0,72	265 s **275 w (A_u)**
	T (BO_3^{3-}) T (Cr^{3+})	289	0,17	290 sh
		334	0,78	321 w
		391	0,32	400 s
		419	0,91	424 sh
	ν_4	657	0,0004	653 sh
	ν_2	690	0,73	707 sh
		760	0,12	778 sh
	ν_3	1302	0,02	**1278 sh (B_u)** 1296 sh
E	T (R^{3+})	95	2,08	99 s
	L (BO_3^{3-})	205	0,0005	202 w sh
		224	0,0016	223 w
		261	0,044	**248 sh (Bu)** 262 sh
		300	0,36	294 s
	T (BO_3^{3-}) T (Cr^{3+})	317	0,05	306 w
		351	0,15	347 s
		375	0,015	376 m
		420	0,33	434 s **440 s (Bu)**
		454	1,80	453 m
		470	0,00007	480 sh
	ν_4	605	0,0011	601 w
		650	0,0038	622 w
		687	0,053	680 s
	ν_2	752	0,017	739 s **747 sh (Au)**
	ν_1	986	0,005	980 w **990 w (Bu)**
	ν_3	1187	0,78	1187 s
		1210	0,001	-
		1318	0,30	1314 s **1372 sh (Au)**

sh – shoulder, w – weak, s – strong, m – medium

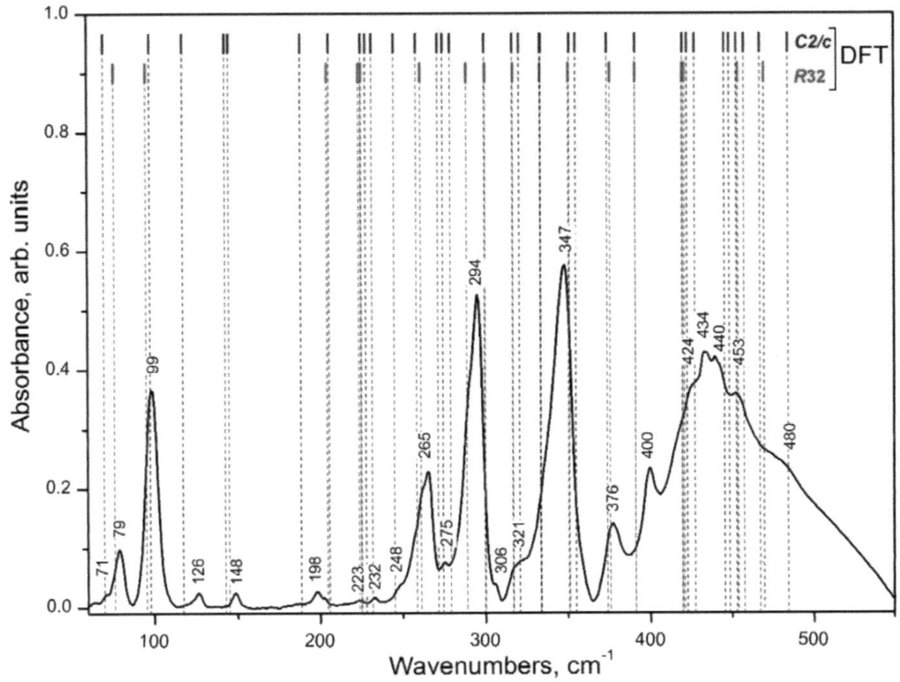

Figure 4. FIR part of the spectrum of $GdCr_3(BO_3)_4$.

5. Estimate of the relative quantity of the monoclinic phase in the sample

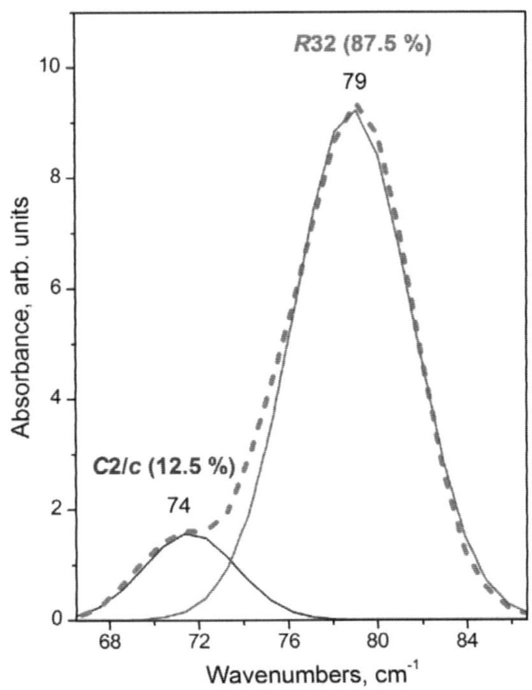

Figure 5. Long-wavelength part of the absorption spectrum of $GdCr_3(BO_3)_4$ (dashed grey line). Decomposition into two Gaussian peaks is shown by thin lines.

Figure 5 shows the experimental IR spectrum at long wavelengths where the translational vibrations of Gd^{3+} are observed. The band at 79 cm^{-1} has a shoulder 73 cm^{-1}, which is associated with a small admixture of the monoclinic phase. We have fitted the experimental spectrum by two Gaussians and found that the intensity of the strongest band constitutes 82.3%. According to calculations, the oscillator strength of the considered vibration in the monoclinic phase is 1.5 times greater than in the rhombohedral one. Taking into account this value, we could estimate that the content of the rhombohedral phase in the sample was about 87.5%.

6. Conclusions

Good quality single crystals of gadolinium chromium borate with the structure of the mineral huntite were obtained by spontaneous crystallization from the flux. IR spectroscopy delivered experimental data on the IR-active phonons. The data were compared with the results of *ab initio* DFT calculations. This comparison shows that the crystal has a polytype structure, containing mainly the rhombohedral phase *R32* and a small part of the monoclinic phase *C2/c*. The *ab initio* calculations allowed us to distinguish between vibrations belonging to the one and the other structural phases. From the relative intensities of the two low-frequency bands corresponding to translational motions of Gd in the *R32* and *C2/c* phases and taking into account the calculated oscillator strengths, we were able to evaluate the content of monoclinic phase in the predominantly rhombohedral single crystal of $GdCr_3(BO_3)_4$, which was around 15%. Thus, we have shown that the gadolinium chromium borate crystallizes, mainly, in the noncentrosymmetric phase *R*32. This indicates a potential of owning magnetoelectric properties, by analogy with noncentrosymmetric iron, aluminium and gallium RE borates. Currently, the work is underway to obtain large single-phase crystals of $GdCr_3(BO_3)_4$ for investigation of the magnetoelectric effect.

Acknowledments

This work was supported by the Russian Science Foundation under Grant No 14-12-01033 and by the President of Russian Federation (D.E.A. project СП-754.2015.1).

References

[1] Leonyuk N I and Leonyuk L I 1995 *Prog. Cryst. Growth and Charact.* **31** 179
[2] Dorozhkin L M, Kuratev I I, Leonyuk N I, Timchenko T I and Shestakov A V 1981 *Sov. Tech. Phys. Lett.* **7 (11)** 555
[3] Leonyuk N I, Maltsev V V, Volkova E A, Koporulina E V, Nekrasova N V, Tolstik N A and Kuleshov N V 2009 *J. Physics: Conf. Series* **176** 012010A
[4] Nikogosyan D N 2005 *Nonlinear Optical Crystals: A Complete Survey* Springer Berlin
[5] Brenier A, Tu C, Zhu Z and Wu B 2004 *Appl. Phys. Lett.* **84** 2034
[6] Jaque, D. J. 2001 *Alloys and Comp.* **323–324** 204
[7] Mukhin A A, Vorob'ev G P, Ivanov V Yu, Kadomtseva A M, Narizhnaya A S, Kuzmenko A M, Popov Yu F, Bezmaternykh L N and Gudim I A 2011 *JETP Lett.* **93(5)** 275
[8] Zvezdin A K, Krotov S S, Kadomtseva A M, Vorob'ev G P, Popov Yu F, Pyatakov A P, Bezmaternykh L N and Popova E A 2005 *JETP Lett.* **81(6)** 272
[9] Kadomtseva A M, Popov Yu F, G. P. Vorob'ev, Pyatakov A P, Krotov S S, Kamilov K I, Ivanov V Yu, Mukhin A A, Zvezdin A K, Kuz'menko A M, Bezmaternykh L N, Gudim I A and Temerov V L 2010 *Low Temp. Phys.* **36(6)** 511
[10] Zvezdin A K, Vorob'ov G P, Kadomtseva A M, Popov Yu F, Pyatakov A P, Bezmaternykh L N, Kuvardin A V and Popova E A 2006 *JETP Lett.* **83(11)** 509
[11] Popov Yu F, Pyatakov A P, Kadomtseva A M, Vorob'ov G P, Zvezdin A K, Mukhin A A and Ivanov V Yu 2010 *JETP* **138(2)** 226
[12] Liang K - C, Chaudhury R P , Lorenz B, Sun Y Y, Bezmaternykh L N, Temerov V L and Chu C W 2011 *Phys. Rev. B* **83** 180417
[13] Begunov A I, Demidov A A, Gudim I A and Eremin E V 2013 *JETP Lett.* **97(9)** 528

[14] Kadomtseva A M, Popov Yu F, Vorob'ev G P, Kostyuchenko N V, Popov A I, Mukhin A A, Ivanov V Yu, Bezmaternykh L N, Gudim I A, Temerov V L, Pyatakov A P and Zvezdin A K 2014 *Phys. Rev. B* **89(1)** 014418

[15] Chaudhury R P, Lorenz B, Sun Y Y, Bezmaternykh L N, Temerov V L and Chu C W 2010 *Phys. Rev. B* **81(22)** 220402

[16] Volkov N V, Gudim I A, Eremin E V, Begunov A I, Demidov A A and Boldyrev K N 2014 *JETP Lett.* **99(2)** 67

[17] Borovikova E Yu, Dobretsova E A, Boldyrev K N, Kurazhkovskaya V S, Maltsev V V and Leonyuk N I 2013 *Vibr. spectrosc.* **68** 82

[18] Kurazhkovskaya V S, Dobretsova E A, Borovikova E Yu, Mal'tsev V V and Leonyuk N I, 201 *J. Struct. Chem.* **52**(4) 699

[19] Mills A D 1962 *Inorg. Chem.* **1**(4) 960

[20] Dollase W A and Reeder R J 1986 *Amer. Miner.* **71(1-2)** 163

[21] Campá J A, Cascales C, Gutiérrez-Puebla E,Monge M A, Rasines I and Ruíz-Valero C 1997 *Chem. Mater.* **9**(1) 237-240.

[22] Belokoneva E L and Timchenko T I 1983 *Sov. Phys. Crystallogr.* **28** 658

[23] Plachinda P A and Belokoneva E L 2008 *Cryst. Res. Technol.* **43** 157

[24] Fausti D, Nugroho A A, van Loosdrecht P H M, Klimin S A, Popova M N, Bezmaternykh L N 2006 *Phys. Rev. B* **74** 024403

[25] Chernyshev V A, Nikiforov A E, Petrov V P, Serdtsev A V, Kaschenko M A and Klimin S. 2016 *Phys. Sol. St.* **58 (8)** 1587

Properties of TiO_2 films with gold nanoparticles

S A Aliev, N E Nikolaev, N S Trofimov and T K Chekhlova

Department of Applied Physics, Peoples' Friendship University of Russia, Moscow, 117198, Russia

E-mail: altersam@mail.ru

Abstract. The physicochemical and optical properties of titanium dioxide films, made by gel technology and doped with gold nanoparticles, were investigated. The structures of the titanium dioxide films synthesized by different techniques have been compared. Using methods of high-resolution microscopy and the results of X-ray diffraction analysis it was shown, that the developed gel technology allows getting almost 100% nanostructured anatase phase. Titanium dioxide was modified by nanoparticles of gold with different concentration and transmittance spectra of the samples were studied.

1. Introduction

Nowadays, nanocrystalline titanium dioxide (TiO_2) is the subject of active research due to several unique properties of this compound [1-3]. TiO_2 films are photosensitive, possess chemical and mechanical stability, good optical properties. Particular interest in titanium dioxide is caused by its photocatalytic properties, which makes it a promising material to solve many environmental problems, including the creation of highly efficient solar cells. In addition, the titanium dioxide films have good optical properties, which can be used to create photonic devices.

Titanium dioxide is an effective photocatalyst for a variety of chemical reactions [4-5]. Its photocatalytic properties are caused by the formation on the surface during the illumination of a number of radicals or paramagnetic centers (PC), which are capable of entering into the secondary reaction. This TiO_2 property is widely used to create a filter for water and air treatment from toxic organic impurities. Free radicals, such as O^{2-} and OH^- [6], formed on the surface of TiO_2 under the oxygen and water influence, can destroy environmental pollutants, because they are strong oxidants. Nano-titanium dioxide in anatase form structure has increased specific surface area, which allows to increase the yield of photo oxidation reaction by several orders.

Large specific surface area of titanium dioxide in the anatase form structure makes it a promising material for the creation of efficient solar cells, since the absorption capability of the material is increased by nanostructuring. Currently, there appeared studies where nanoparticle net was replaced by single crystals of titanium dioxide having pores with diameter of about 10 nm. Solar cells with such a structure have a record rate of conversion of light energy into electrical energy (7.4%) [7]. Moreover, since titanium dioxide is a porous material, it can be activated using dye molecules such as rhodamine 6G. The development of the injection type solar cells is possible [8], because energy position of the conduction band of TiO_2 coincides with the energy of excitation of many dyes. The perspective of improving device characteristics on the base of titanium dioxide is associated with increasing catalytic activity and widening its absorption spectrum.

Recently, new technologies for the synthesis of titanium dioxide are being actively developed, allowing to obtain samples with enhanced photoactivity which is associated with an increase in the content of TiO_2 films proportion of fine-grained structure in the form of anatase.

Titanium dioxide is doped with nanoparticles of metals [9], in order to expand the spectral absorption band and raise the amplitude of the absorption of the material, for increasing photocatalytic activity. Moreover, the injection of such particles into a composite film leads to an appearance of qualitatively new physical properties [10-12]. The substance that is in a nanoscale modification differs significantly by many features from the volume materials. For example, gold nanoparticles display ferromagnetic and catalytic properties, specific optical properties which consist in the appearance of absorption bands in the visible spectrum, caused by resonance events on plasmons. The intensity of the resonance absorption and spectral position depends on the amount of metal nanoparticles included in the dielectric matrix. Titanium dioxide can be used as the matrix. Such templates can be used to improve sensitivity of the spectral methods for the analysis of substances composition, to create a variety of sensors and new metamaterials, superlattices synthesis, including photonic crystals, and others. Of special interest are optical and nonlinear optical properties of such structures.

The present study investigates the properties of titanium dioxide films, synthesized using a new method – gel method [13]. The results are compared with the research results obtained for films produced by sol-gel technology. In addition, the influence of modification by gold nanoparticles on the characteristics of the TiO_2 gel-films is investigated.

2. Formation of titanium dioxide films by gel method and modifying by gold nanoparticles

Sol-gel method is most often used for the production of titanium dioxide. It is known that titanium dioxide can exist in three configurations: amorphous, nanocrystalline (anatase) and rutile forms. As for the above applications large specific surface area is of importance, the most interesting is the highly structured anatase modification. In practice, a mixture of these crystal modifications is usually used. Relative content of structural modifications depends on the synthesis method and on the temperature of heat treatment conditions.

Amorphous TiO_2 is transformed into anatase at a temperature greater than 300°C. Furthermore, the anatase content of the amorphous film depends on the film thickness. Films of thickness 100-200 nm are strictly amorphous, and at the thickness of 500 nm and more they contain anatase [14]. This is due to the fact that the initial degree of crystallization decreases with decreasing film thickness.

Manufacturing films by the sol-gel method leads to the formation of titanium hydroxide $TiO_2 \times H_2O$, which, depending on the deposition conditions may contain a variable number of OH groups connected to titanium. During annealing of titanium dioxide in the amorphous state, first anatase is formed (thus OH-groups are partially removed), and then – rutile. Complete removal of water occurs at temperature above 600°C.

In the present work, TiO_2 films were obtained not only by the sol-gel method described in [15], but also by the gel method [16], which allows to get nearly 100% of anatase form of titanium dioxide. Unlike the sol-gel method, in this case gelation takes place in the true solutions. Film formation occurs from the solution of gel, which is the result of reaction of titanium tetrabutoxide (TBT) compounding with triethylene glycol (TEG), in butanol-1 in the air and subsequent transformation during annealing. The chemical structure is shown on the following scheme:

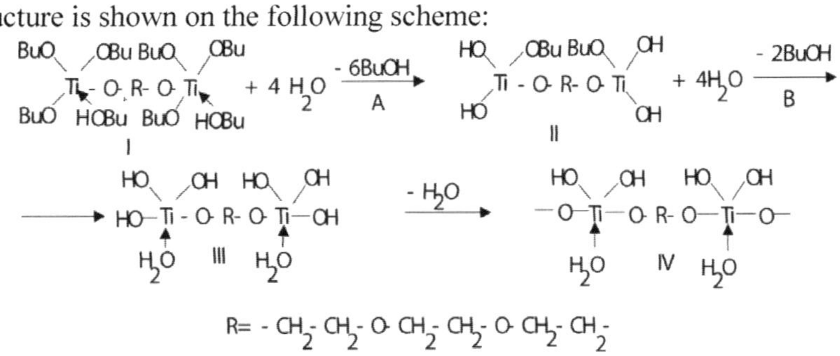

Unlike the sol-gel process, in this case, sol-formation does not occur, films are drawn directly from solution, rather than from the suspension, and there is no direct hydrolysis reaction. Hydrolysis takes place without the addition of water to the solution, water condenses out of the atmosphere, moreover the OH groups for hydrolysis split out while removing the alcohol from the solution.

The process of making gel films samples of the titanium dioxide was similar to manufacturing process of sol-gel films described in [15]. It consisted of several phases: preparing a solution, applying it to the prepared substrate, drying and subsequent annealing. As a basic material for producing titanium dioxide films the titanium tetrabutoxide $Ti(OCH_2C_3H_7)_4$ was used, which was mixed in the appropriate proportions with triethylene glycol $C_6H_{14}O_4$, then butanol $C_6H_{12}O_3$ was added to the predetermined volume and stirred again.

Substrate layers with applied titanium dioxide were dried at ~ 100°C for 10-20 min. Porous film-frame was formed on a substrate due to the evaporation of the solvent. Subsequent annealing at temperature ~ 300-800°C led to the formation of a continuous film, where the porosity does not exceed 10-15%. Some samples were doped with gold.

A series of titanium dioxide films samples was produced by the described technology (table 1), the parameters of which, namely the refractive index and thickness, were varied by changing the ratio of the solution components, the drawing speed of the substrate from the solution, the solution temperature and annealing time and temperature. Several samples were modified by gold nanoparticles with different concentration.

Table 1. The ratio of components of the solution, annealing temperature and relative content of gold in the samples.

TEG:TBT		**1:2**	**1:1**	**2:1**	**1:1 (Au:TiO_2 = 1:100)**	**1:1 (Au:TiO_2 = 1:50)**
Annealing temperature	450°C	K1	K2	K3	KZ1	KZ2
	700°C	K4	K5	K6	KZ3	KZ4

Experimental samples were studied by integrated optics methods, optical spectrophotometry and EPR spectroscopy, thermal analysis, high-resolution microscopy and Raman scattering.

3. Experimental studies

3.1. Study of dependence of the gel films refractive index on the parameters of technological regime and the introduction of gold nanoparticles

The thickness and refractive index of the manufactured samples were determined using the method of integrated optics based on the waveguide propagation of radiation through the film [17].

Calculations of the refractive index and film thickness were carried out using the dispersion equations for the optical waveguide composed of the studied film deposited on quartz substrate. Dispersion equations determine the relationship between the effective refractive index (ERI) for the modes propagating in the optical waveguide, the film thickness, the refractive indices of the film and the substrate, and the wavelength of the radiation injected into the film. For determining the thickness and refractive index of the material of the film using the dispersion equations, ERI of two modes (TE_1 and TM_1) were measured at wavelength of light injected into the waveguide. The film thickness in this case must exceed the critical thickness h_{cr} for each mode, which has been calculated for the given waveguide. The refractive index of the quartz substrate was equal to 1.4585, and the wavelength of the radiation – 632.8 nm.

The He-Ne laser radiation was injected into the waveguide through the prism at certain resonant angles. After measuring angles of the excitation of the waveguide, the ERI were determined using the relationship between the angles of the excitation, ERI and the refractive index of the input prism.

Table 2. The results of calculation for the samples of titanium dioxide films produced with different ratios of the components of stock solutions, and different values of the annealing temperatures.

Sample	γ_{TE}	γ_{TM}	h_{crTE}, nm	h_{crTM}, nm	h, nm	n_2
K1	2.0339	1.6424	24	64	120	2.48
K2	1.9419	1.5064	24	64	89	2.54
KZ1	1.8494	1.5769	32	71	130	2.24
KZ4	1.7911	1.4817	28	69	91	2.35

The table 2 shows that the refractive index of the films of titanium dioxide, synthesized by gel method, has a rather high value of ~ 2.4, which is much higher than the refractive index of the sol-gel films [15], and is almost independent of the ratio of stock solution components and annealing temperature.

The introduction of gold nanoparticles decreases the refractive index, and the higher is the concentration of gold, the smaller is the refractive index.

3.2. Determining of the phase transition temperature and study of the films morphology

In order to choose the proper annealing temperature for the films, the thermal analysis was conducted, which resulted in finding the temperatures of phase transitions for gel material.

Determination of the phase transitions temperatures was carried out by the thermal analysis on SDTQ600. By simultaneous thermal analysis, thermogravimetric analysis (TGA) (figure 1) of samples was performed.

As seen in figure 1 the addition of gold does not change the phase transition temperature.

The morphology of the manufactured films was investigated using Raman spectroscopy (figures 2, 3).

Raman spectra were obtained at the facility NTegra Spectra. In the Raman spectrum of the sample annealed at the temperature 450°C, 4 peaks were observed, which correspond to the position of anatase [18], and at the temperature 700°C peaks characteristic for rutile appeared. It was found that increasing the proportion of TEG in the stock solution, as well as adding the gold in the solution, both lead to the rise of the temperature of phase transition anatase-rutile. Raman spectra of the commercial film fabricated by company "Aldrich" and of sample (solution 1) annealed at 450°C confirm the presence of the anatase only.

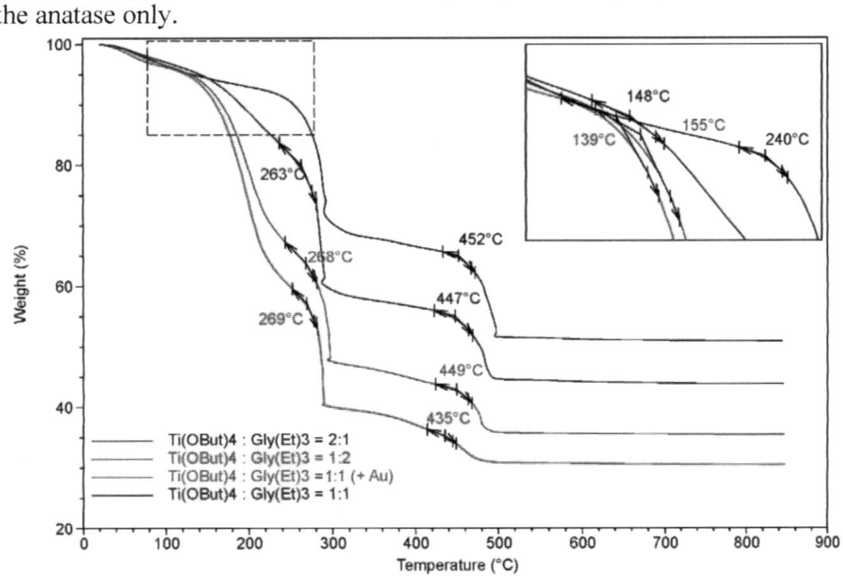

Figure 1. TGA of samples.

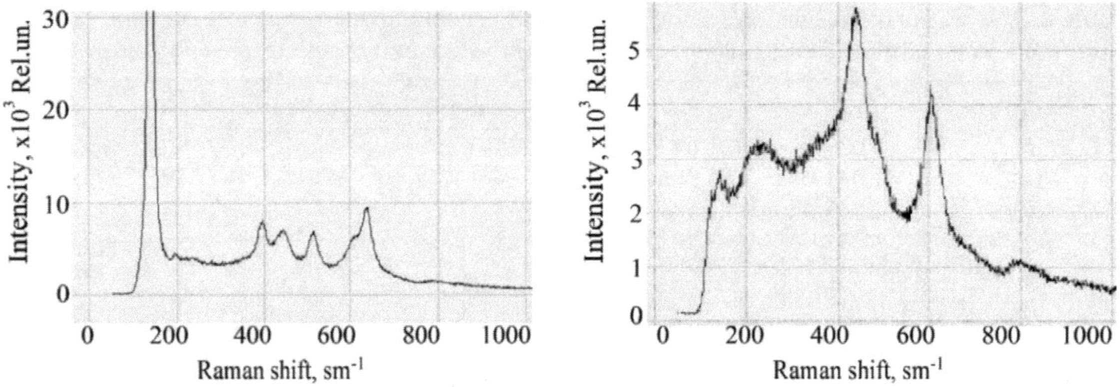

Figure 2. Raman spectra of the films confirm the presence of the anatase only (left – solution 2, 450°C) or of mixed (anatase + rutile) phase (right – solution 3, 700°C).

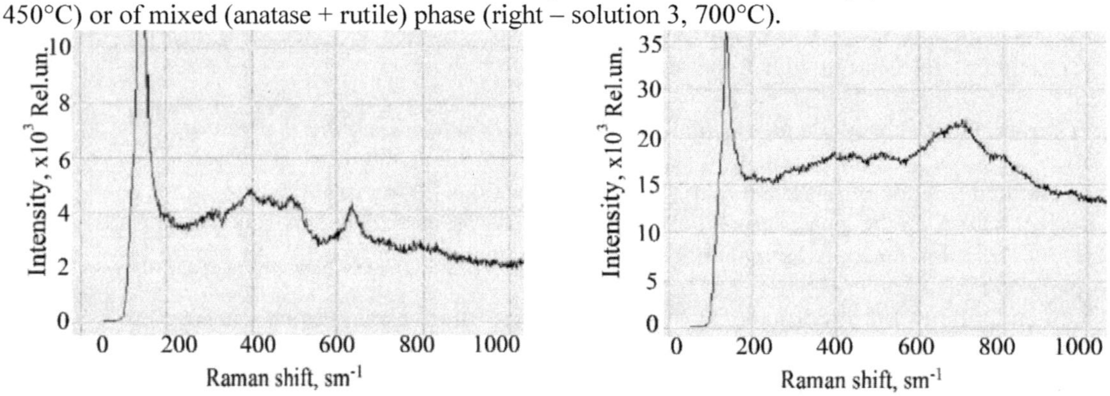

Figure 3. Raman spectra of the films of TiO_2 with Au (TiO_2:Au = 50:1) annealed at 450°C (left) and at 700°C (right).

Studies of the optical properties of TiO_2 films were carried out at room temperature using a monochromator-spectrograph MS 3504I and photometric lamp SIRSH 6-100, grating 1200 grooves/mm, the measurement step 1 nm, lamp color temperature 2840 K. The data show changes of photosensitivity depending on changing the parameters of the technological regime, the ratio of components of solutions, as well as the addition of the gold nanoparticles.

3.3. Study of optical characteristics of gel films with gold nanoparticles

Analysis of transmission spectra of samples without the addition of gold (K1, K2, K3) with the annealing temperature of 450°C (figure 4a) showed that the absorption edge corresponds to $\lambda = 330$ nm. The maximum absorption of K1 sample was observed at $\lambda = 500$ nm with an amplitude of about 40% and about 150 nm wide. For sample K2 absorption peak (50%) shifted 50 nm to the shorter wavelengths $\lambda = 450$ nm with width of 80 nm. Transmittance was 97% at a wavelength of 600 nm, and at $\lambda = 850$ nm a second absorption peak appeared with width ~ 200 nm and amplitude of about 30%. The long-wavelength peak was similar to the peak of K1.

The edge of the absorption band shifted to 350 nm for all samples (figure 4b) by increasing annealing temperature (700°C). Absorption maximum of sample K4 corresponded to $\lambda = 450$ nm (60% amplitude, width 80 nm). For sample K5 slow growth of absorption intensity was observed in the range from 470 nm to 870 nm and the maximum absorption was 50%.

Samples K3 and K6 had a greater scattering associated with the ratio of solution components. Custom solution contained a large amount of solvent and alcohol, which prevented the formation of a strong structural frame of components, and as a result the fabricated film cracked during annealing and was not optically homogeneous. That is why K3 and K6 samples had about the same ~ 40-60% transmittance in the wavelength range from 500 nm to 870 nm.

Journal of Physics: Conference Series **737** (2016) 012036 doi:10.1088/1742-6596/737/1/012036

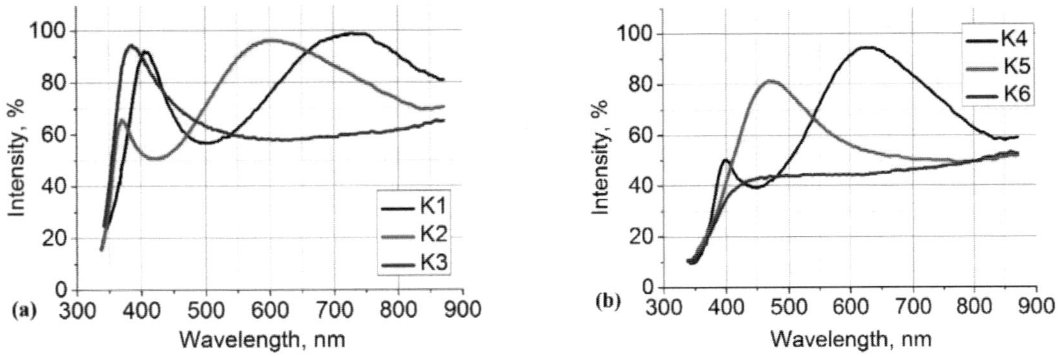

Figure 4. Transmission spectra of samples without gold after annealing at temperature of 450°C (a) and 700°C (b).

Analysis of the results allows to conclude that the transmission spectra is changed, after modifying film with gold nanoparticles (figure 5a). Introduction of nanoparticles changes the intensity of the absorption bands and their spectral position. A visual representation of spectra changes can be seen on the figure 5b, which is a difference spectrum obtained by subtracting the spectra of the sample with and without gold. Introduction of gold increases the absorption over the entire spectral range. As shown in figure 5b with annealing temperature of 450°C there is characteristic peak at the wavelength of 600 nm. Increasing the annealing temperature changes the spectral peak position by sliding it into the short-wavelength region by 130 nm. The shift of the absorption peak is associated with the change in nanoparticle size [19]. Similarly to silver nanoparticles, gold absorbs at different wavelengths depending on the diameter, which can be varied by changing the parameters of manufacturing process, specifically, the annealing temperature.

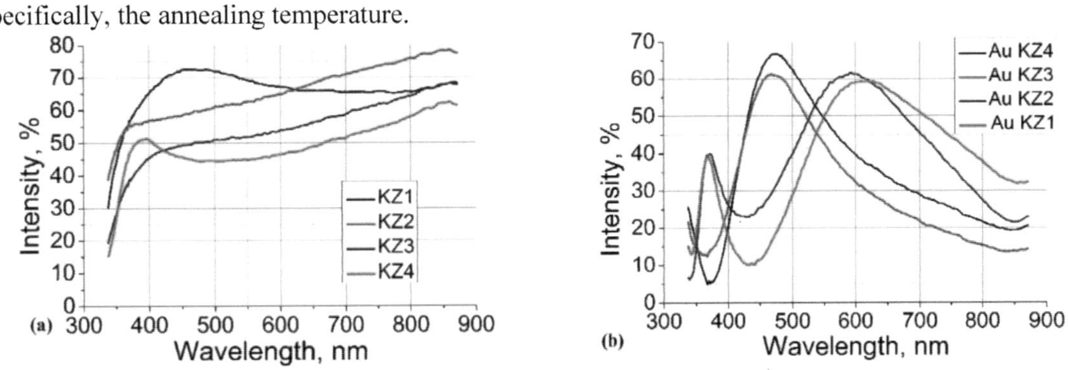

Figure 5. The transmittance spectra of samples with gold (a) and difference spectra of Au absorption in the films (b).

Table 3. The short-wavelength absorption edge λ of the samples in visible range and the corresponding band gap are presented.

Sample	Temperature	λ, nm	E, eV
K1	450°C	505	2.456
K2		421	2.947
K4	700°C	452	2.744
K5		342	3.627
Au KZ1	450°C	369/619	3.362/2
Au KZ2		372/593	3.335/2.092
Au KZ3	700°C	468	2.651
Au KZ4		475	2.612

3.4. Research of photosensitivity of titanium dioxide films by ESR spectroscopy

TiO_2 photoactivity is based on the semiconducting properties of the material, the electronic structure of which is due to the existence of defects on the surface of the crystalline structure. The absorption of a photon of light causes the transfer of an electron from the valence band to the conduction band of the semiconductor to form an electron and a hole, which leads to the formation of the paramagnetic centers (PC). The quantum yield of the process is largely determined by the size and geometry of the material owing to the exclusively superficial phenomena. Increasing the specific surface area can increase the degree of material photoactivity. Therefore, anatase, which has high specific surface area, exhibits a higher photocatalytic activity than the rutile. Increasing of photoactivity is explained, in particular, by a higher position of the Fermi level in anatase (3.3-3.4 eV), compared with rutile (3.1-3.2 eV) [19].

The intensity of the absorption of photons and the efficiency of photocatalysis are determined by the form of its electronic spectrum in the band gap. The band gap of titanium dioxide varies within 3.2-3.6 eV, depending on the method of synthesis. The band gap of the investigated material made by gel method, was estimated to be 2.456-3.362 eV. The photosensitivity of titanium dioxide films was investigated by electron paramagnetic resonance spectroscopy (ESR), which allows registering the paramagnetic centers, such as free radicals, ion radicals and other formations, possessing an unpaired electron. The study of paramagnetic centers was conducted on the spectrometer "SENS ESR 70-03 XD/2" [20] whose operation is controlled by a personal computer. The sample was placed inside the cavity, the sample irradiation was carried out through the cavity transparent window.

A high-pressure mercury lamp was used as a source of UV radiation. Spectrum of lamp (PRK-4) contained a number of intense lines. The most intense lines are at wavelengths of 315 nm, 367 nm and 407 nm, which correspond to radiation energies of 3.936 eV, 3.397 eV and 3.024 eV, respectively.

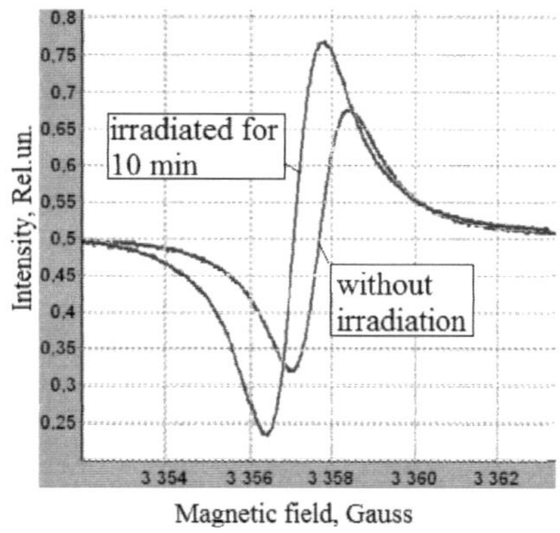

Figure 6. ESR spectra of the sample TEG:TBT = 1:2 annealed at 450°C irradiated with a mercury lamp.

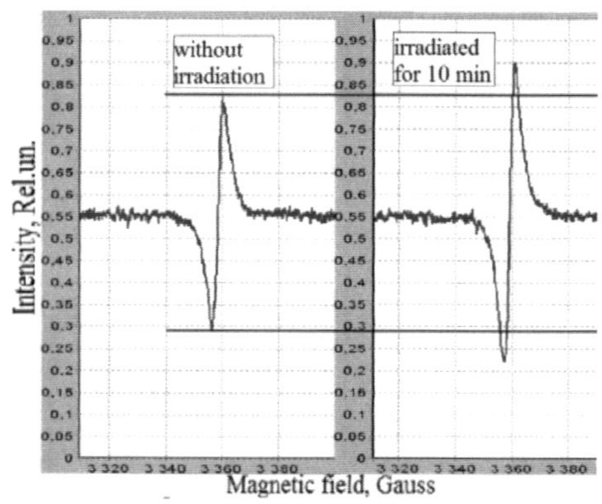

Figure 7. ESR spectra of the sample TEG:TBT = 1:1 annealed at 450°C and irradiated with a mercury lamp.

We can trace the dynamics of change in intensity of the lines after irradiation according to these graphs (figures 6, 7). Total relaxation was 2.5 hours. The difference in the ESR spectra of samples, apparently is due to the different ratios of the components of stock solutions.

Irradiation by UV light of the films annealed at 450°C showed increasing the number of paramagnetic centers.

4. Conclusion

Films samples, with TiO_2 in anatase form, were prepared using the developed gel technology. The mass fraction of anatase is close to 100%, this was confirmed by high-resolution microscopy techniques and the results of X-ray diffraction analysis. The physicochemical properties of titanium dioxide films were studied. The presented results of topography and morphology studies of the obtained film samples also confirmed the existence of anatase phase of material in films.

It is found that the refractive index of the film of titanium dioxide, synthesized by the gel method, has a quite high value of about ~2.4, which is much higher than the refractive index of the sol-gel films and almost independent of the ratio of stock solution components, and the annealing temperature. The modification by gold nanoparticles decreases the refractive index and the higher the concentration of gold, the smaller the refractive index.

The photoactivity research of the synthesized films by ESR spectroscopy showed an increase of films photoactivity under UV irradiation from mercury lamp. It is revealed that films of titanium dioxide in anatase form demonstrate greater photoactivity than the rutile and mixed modifications, while free radicals are not generated on the surface of titanium dioxide in the amorphous phase during the irradiation and the films do not demonstrate photosensitivity in UV range.

Transmission spectra research has shown their dependence on the ratio of the solution components while manufacturing the gel film, and the annealing temperature during their formation.

Raman spectra showed that for the sample annealed at the temperature 450°C, four peaks were observed, which correspond to the position of anatase [18], and at the temperature 700°C there appeared peaks characteristic of rutile. It can be concluded that with increasing the proportion of TEG in the stock solution increases the temperature of phase transition anatase-rutile.

In addition, the absorption spectra are significantly dependent on the parameters of the technological regime. Research of the absorption spectra of titanium dioxide films, containing gold nanoparticles, showed significant changes in the spectra, specifically, additional absorption peaks of various intensity and the shift of the absorption band edge were observed. These spectra features are caused, presumably, by changes in the film structure, specifically, by changes of the size and shape of the particles, while changing the parameters of the technological regime. An important role is played by the aggregation of gold nanoparticles, i.e., occurrence of agglomerates, resulting in an increase of absorption over the entire spectral range. The research of the absorption spectra of titanium dioxide films, modified with other elements such as cobalt and carbon, is of interest.

References

[1] Konstantinov E A, Gayvoronskiy V Y, Timoshenko V Y, Kashkarov P K 2010 *PTP* **44** 1093
[2] Zainullina V M and Zhukov V P 2013 *Solid State Physics* **55** 534–541
[3] Ollis D F and Al-Ekabi H 1993 *Photocatalytic Purification and Treatment of Water and Air* (Amsterdam: Elsevier)
[4] Fujishima A, Hashimoto K and Watanabe T 1999 *TiO_2 Photocatalysis. Fundamentals and Applications* (Tokyo: BKC Inc.)
[5] Hoffmann M R, Martin S T, Choi W and Bahnemann D W 1995 *Chem. Rev.* **95** 69
[6] Chen X and Mao S S 2007 *Chem. Rev.* **107** 2891
[7] Deb S, et al 2007 *Journal of Optics A: Pure and Applied Optics* **9** 555–59
[8] Gr¨atzel M J and Howe R F 1991 *Nature* **91** 3906
[9] Karpov S V and Slabko V V 2003 *Optical and photophysical properties of fractal-structured metal sols* (Novosibirsk: Publishing House of the SB RAS)
[10] Rupa A V, Divakar D and Sivakumar T 2009 *Catalysis Letters* **132** 259–67
[11] Papp J, Shen H S, Kershaw R, Dwight K and Wold A 1993 *Chemistry of Materials* **5** 284–88
[12] Thampi K R, Kiwi J and Grätzel M 1987 *Nature* **327** 506–08
[13] Evtushenko Yu, Romashkin S, Trofimov N, Chekhlova T 2015 *Physics Procedia* **73** 100–107
[14] Guo B, Liu Zh, Hong L. et al. 2005 *Thin Solid Films* **479** 310–15
[15] Zhai J, Yao X and Zhang L 2000 *J. Phys. D. Appl. Phys* **33** 3013–17
[16] Trofimov N S 2014 *Bulletin of PFUR: Mathematics. Information Sciences. Physics* **3** 194–201
[17] Nikolaev N, Pavlov S, Trofimov N, Chekhlova T 2012 *J. Comm. Technol. Electron.* **57** 15–21
[18] Shulga Y M, Courgettes E N et al. 2011 *Technical Physics* **81** chapter 1
[19] Artemyev Yu. and Ryabchuk V K 1999 *Introduction to heterogeneous photocatalysis* (St. Petersburg: SPb. Publishing House, University Press) p 304
[20] Instruction Manual 2008 *Automated compact ESR spectrometer ESR70-03XD/2* (Minsk) p 15-46

Journal of Physics: Conference Series **737** (2016) 012037 doi:10.1088/1742-6596/737/1/012037

Plasmon enhancement of Raman scattering and fluorescence for rhodamine 6G molecules in the porous glass and PVA films with nanoparticles of silver citrate hydrosol

E I Konstantinova[1,4], A U Zyubin[2,4], V A Slezhkin[3,4], I G Samusev[2,4] and V V Bryukhanov[4]

[1] – Department of Physics, Kaliningrad State Technical University, 1 Sovietsky prospect, 236022, Kaliningrad, Russia
[2] – Institute of Physics and Technology, Immanuel Kant Baltic Federal University, 14 A, Nevskogo street, 236041 Kaliningrad, Russia
[3] – Department of Chemistry, Kaliningrad State Technical University, 1 Sovietsky prospect, 236022, Kaliningrad, Russia
[4] – Centre for Laser Nanotechnologies and Information Biophysics, Immanuel Kant Baltic Federal University, 14 A, Nevskogo street, 236041 Kaliningrad, Russia

E-mail: konstantinovaeliz@gmail.com,

Abstract. The study of Raman and fluorescence spectra for Rhodamine 6G molecules in a film of polyvinyl alcohol on the modified by silver nanoparticles (NPs) porous glass and without the porous glass has been done. The gain of the scattering intensity and fluorescence emission has been obtained in the presence of silver nanoparticles. The gain order was obtained as ~ 10^{11}

1. Introduction

The nanoparticles of noble metals due to the surface plasmon resonance [1], arising on the surface of NPs under photoexcitation, are promising to create a variety of devices and biophysical sensors for diagnosis and destruction of cancer cells. Today, gold and silver NPs are investigated for the possibility of sensor elements creation based on nonradiative dipole-dipole energy transfer of plasmon excitation in molecular systems [2-4], and for detecting various compounds using fluorescent markers [5].

Raman spectroscopy has a special place in the study of biological structures and organometallic compounds, in case of its possibility to identify distinct groups of biological molecules and the interactions between them in a condensed media and on a solid surfaces. Therefore, the attention of researchers is attracted to use various nano- and microstructures like porous and mesoporous silica [5] ZnO nanorods [6], gold nanorods and nanoholes [5, 8], porous glass [9], etc. The porous glass - the unique material having both high biocompatibility and abundance, making it a very economical biosensor application. In addition, the production technology of porous glass allows us to control the size and the degree of porosity.

In the present study of the Raman scattering and fluorescence of rhodamine 6G (R6G) molecules, amplified by generation of surface plasmons of silver NP in a film of polyvinyl alcohol (PVA) and on porous glass were investigated. The main objective of the paper was to study non-radiative processes of electronic energy transfer between molecule adsorbates R6G and Ag NPs on porous glass in order to create an optical sensor.

2. Method of the sample preparation

2.1. Synthesis of Ag NPs

The Turkevich citrate method was used for the synthesis of silver NP [10-12]. The sodium citrate $Na_3C_6H_5O_7$ was added dropwise, up to $5{,}8\cdot10^{-4}$ mol/l concentration, into the boiling aqueous solution of silver nitrate at the concentration $C(AgNO_3) = 5{,}3\cdot10^{-4}$ mol/l. The resulting solution was boiled during one hour to obtain a yellow-green color. Citrate-ion acts both as a stabilizer and a reducing agent, therefore the ions concentration determines both the rate of recovery and the growing processes of metal particles. Total recovery of silver citrate reaction can be expressed by the following equation:

$$6AgNO_3+3Na_3C_6H_5O_7 \rightarrow 6Ag+3Na_2C_5H_4O_5+3CO_2+3NaNO_3+3HNO_3$$

Borohydride silver sol was synthesized by reduction of silver nitrate by sodium tetraborhydride (borohydride) [13]. The solution of silver nitrate at the concentration C(AgNO3) = 10^{-3} mol/l was added dropwise, with vigorous stirring, into the cooled to 0°C aqueous solution of sodium borohydride at the concentration of $C(NaBH_4) = 2\cdot10^{-3}$ mol/l. The process of recovery of silver can be expressed by the following equation:

$$AgNO_3 + NaBH_4 + H_2O \rightarrow NaNO_3 + B(OH)_3 + H_2 + Ag.$$

2.2. Preparation of PVA films with R6G on a surface of porous glass modified by silver NP

Glass used for the experiments has the following chemical composition (%): SiO_2 - 72,2; Na_2O - 14,3; K_2O - 1,2; CaO - 6,4; MgO - 4,3; Al_2O_3 - 1,2.

The porous surface was created by mechanical rubbing. The degree of porosity was controlled by the transmission spectra of porous glasses. Random error, calculated from the intensity, was 7%. Samples were washed thoroughly with distilled water and dried at 60° C. Silver NPs of borohydride sol at Ag concentrations (cm^{-2}): $0{,}23\cdot10^{12}$, $0{,}6\cdot10^{12}$, $1{,}1\cdot10^{12}$ were added to surface. After sol drying samples were thermally treated (T = 350°C) in a muffle furnace. During the heat treatment on the glass surface chemical processes were following:

$$2NaNO_3 \rightarrow 2NaNO_2+O_2,$$
$$2B(OH)_3 \rightarrow B_2O_3+3H_2O.$$

PVA films with R6G was applied to the porous glass with Ag NPs by adding aqueous solution containing R6G ($C = 5\cdot10^{-4}$ M) and PVA (ω= 4%) on the surface of porous glass. After drying at T = 60° C on the porous surface of the glass the film of R6G $d\approx3$ microns thick was formed.

2.3. Preparation of films with R6G and NPs of silver citrate sol on the polished surface of the glass

Films on a smooth glass was made by adding of an aqueous solution containing R6G, Ag NPs and PVA onto the surface. The concentration of components in the solution: R6G - $5\cdot10^{-4}$ M, Ag NPs - $0{,}8\cdot10^{12}$ см$^{-2}$, and PVA - ω=6%. The film thickness after drying was 12 micron.

3. Experimental results

The synthesized Ag NPs were investigated by photocorrelation spectroscopy with Photocor Complex unit. The dimensions of NPs were 1.0 nm and 16.0 nm (for borohydride sol), and 15,0 nm and 64,0 nm (for citrate sol). Measurement error was 5%.

The morphology of Ag NPs on the glass surface and inside the film was studied using a fluorescence microscope Olympus of M319/12-OG BX43F model. The agglomeration nature of NPs distribution appeared as dark areas on porous glass (Figure 1a) and in PVA films (Figure 1b).

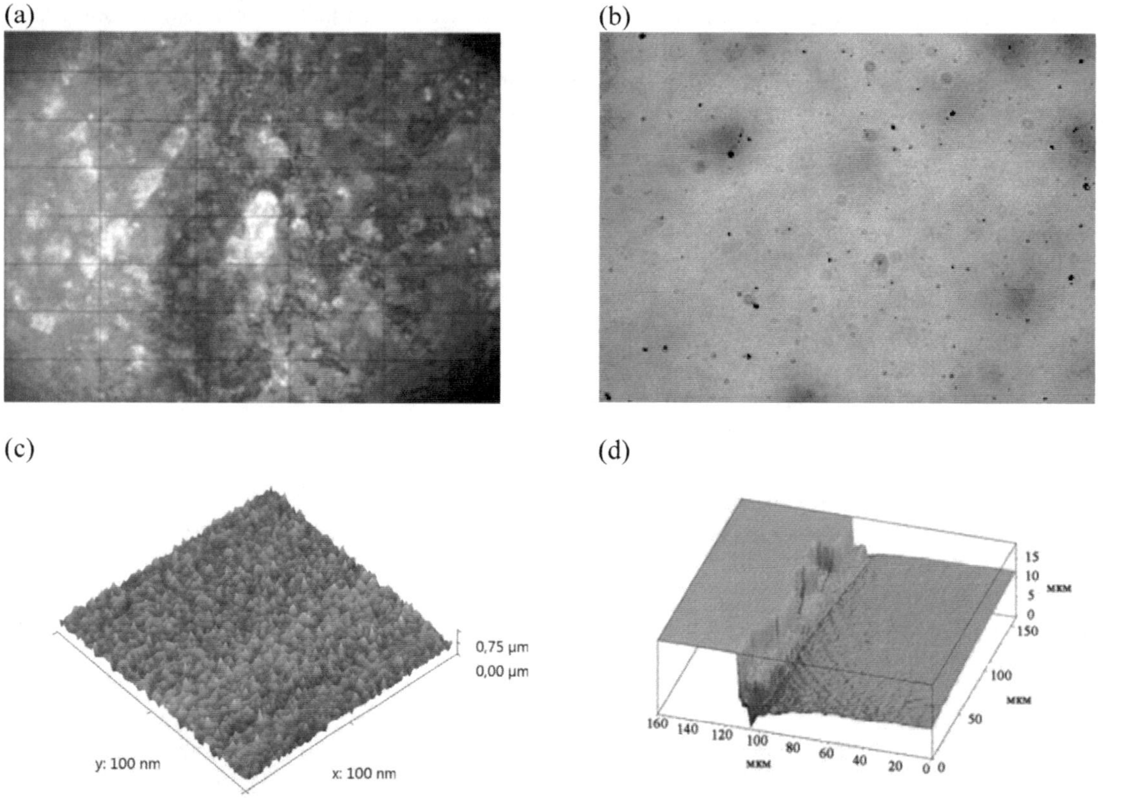

Figure 1. The morphology of the porous glass surface (a) and for PVA film with Ag NPs (b), the topology porous glass surface (c), and for the PVA film cut (d).

Raman spectra were obtained by CENTAUR U unit with diode-pumped solid-state laser (DPSS) excitation (λ = 632 nm, radiation power of 50 mW). Raman spectra were recorded using a CCD detector matrix with 5 s signal accumulation.

The porosity of the glass was investigated by using the SPM Certus Standart module with optical registration system (Russia), using tapping mode atomic force microscopy (AFM). MikroMasch cantilever probes with a curvature radius less than 10 nm were used for scanning. The obtained surface images (Figure 1c) showed the average pore size equal ~ 0,5 μm. However, it had been shown that the surface had nanometer sized pores.

Also, by using techniques of AFM with tapping mode scanning, crosscut image of PVA film was taken (Figure 1d). Determined crosscut thickness of the film was equal 12 microns and it is fully correlates to the theoretical calculation.

IR spectra were obtained on Shimadzu IRPrestige-21 (Japan) in the range of 500-4000 cm^{-1} with a resolution of 4 cm^{-1}. Background noise was detected in the range of 1300-2000 cm^{-1}, therefore, this area has been removed from PVA result spectra.

The absorption spectra were recorded on a double-beam spectrophotometer Shimadzu UV-2600 in the range of 200-800 nm. Fluorescence spectra were recorded on an optical module unit Fluorolog-22 Horiba (France).

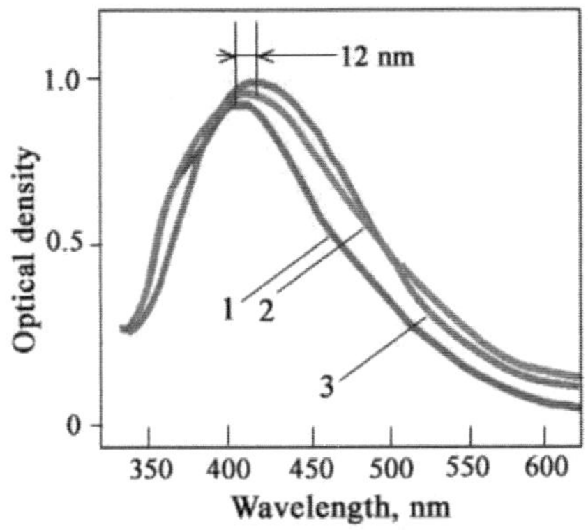

Figure 2. Normalized optical density of Ag NPs on the porous glass surface before (1) and after heat treatment (2) and for NPs inside the PVA (3).

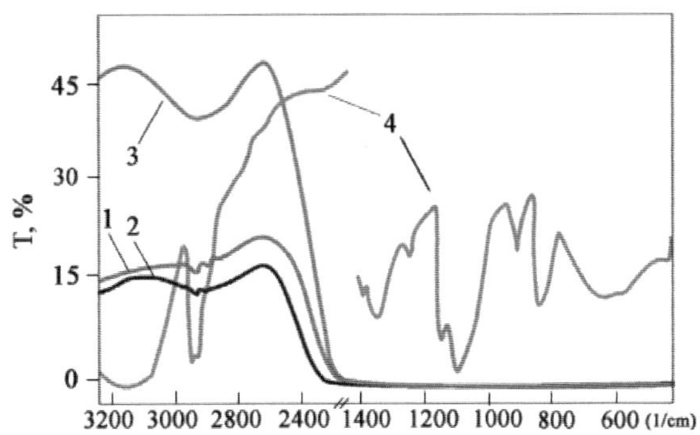

Figure 3. Transmittance spectra of porous glass, with surface, modified with Ag NPs, and PVA films: 1 - porous glass, 2 - porous glass with Ag NPs after heat treatment (350 °C), 3 - the polished glass, 4 - PVA film.

3. Results and discussion

Ag NPs melting temperature depends on their size [14], decreasing from 950 °C to ~ 450 °C for 40 nm sized NPs or less. Presence of 1nm and 15 nm sized NPs in hydrosols, gives us the reason to believe that the heat treatment of porous glasses with NPs leads to NPs caking with the glass matrix. The absorption spectrum of the porous glass with Ag NPs after heat treatment (Figure 2) shows the spectral broadening and long-wavelength shift of surface resonance peak, which indicates the increasing of NPs size.

The transmission spectra of porous glass (Figure 3) in the range of 2500-2800 cm^{-1}, show both reducing the intensity from 40% to 16% and a frequency split of SiH deformation vibrations groups at 2900 cm^{-1} into two: 1) - 2920^{-1} and 2) - 2980 cm^{-1}. The presence of two bands in this range of the spectrum is typical for silicon nanoclusters [15, 16]. Ag NPs, fixed on the porous surface of the glass reduce transmission rate on another 3.5%, as a result of the heat treatment.

NP on the porous glasses Raman spectra (Figure 4a) display 1333 cm^{-1} and 1577 cm^{-1} maxima which are characteristic of bridge-oxygen in B-O-Si group asymmetric vibrations and bridge hydrogen bonds in B-Si-H gropus [16]. Intense wide band (70 – 1000 cm^{-1}) with the 127 cm^{-1} peak, which appears at the maximum NP concentration (curve 1), can be associated with Ag_2O (96 cm^{-1} and 146 cm^{-1} [17]), Na_2O (298 cm^{-1} [18]) and $NaNO_2$ (186 cm^{-1} [19]) groups vibrations. The presence of the above mentioned groups' wavenumbers in confirmed by Gaussian spectra analysis (Figure 4b). The good fitting of experimental spectrum with Gaussian spectra allows to identify a 434 cm^{-1} band, which can be associated with Si stretching vibrations in SiO_2 group (484 cm^{-1} [15]). The analysis of the middle-range of the experimental spectrum (Figure 4c) displays the following bands: 1153 cm^{-1} (Si stretching vibrations in -Si-O-Si- group), 1346 cm^{-1} (B stretching vibrations in B-O group), 1234 cm^{-1}, 1517 cm^{-1} and 1602 cm^{-1} (NO_2 group symmetrical stretching vibrations) [20, 21]. All the identified vibrational bands are shifted by 5 – 25 cm^{-1} to the low-wavenumber region compared with literature data for "clean" compounds. The observed shift was assumed to be caused by phase decomposition in the system "porous glass – borohydride sol Ag NP". The shift decrease to the $NaNO_2$ and SiO_2 groups evidence of the strengthening bonding between the phases.

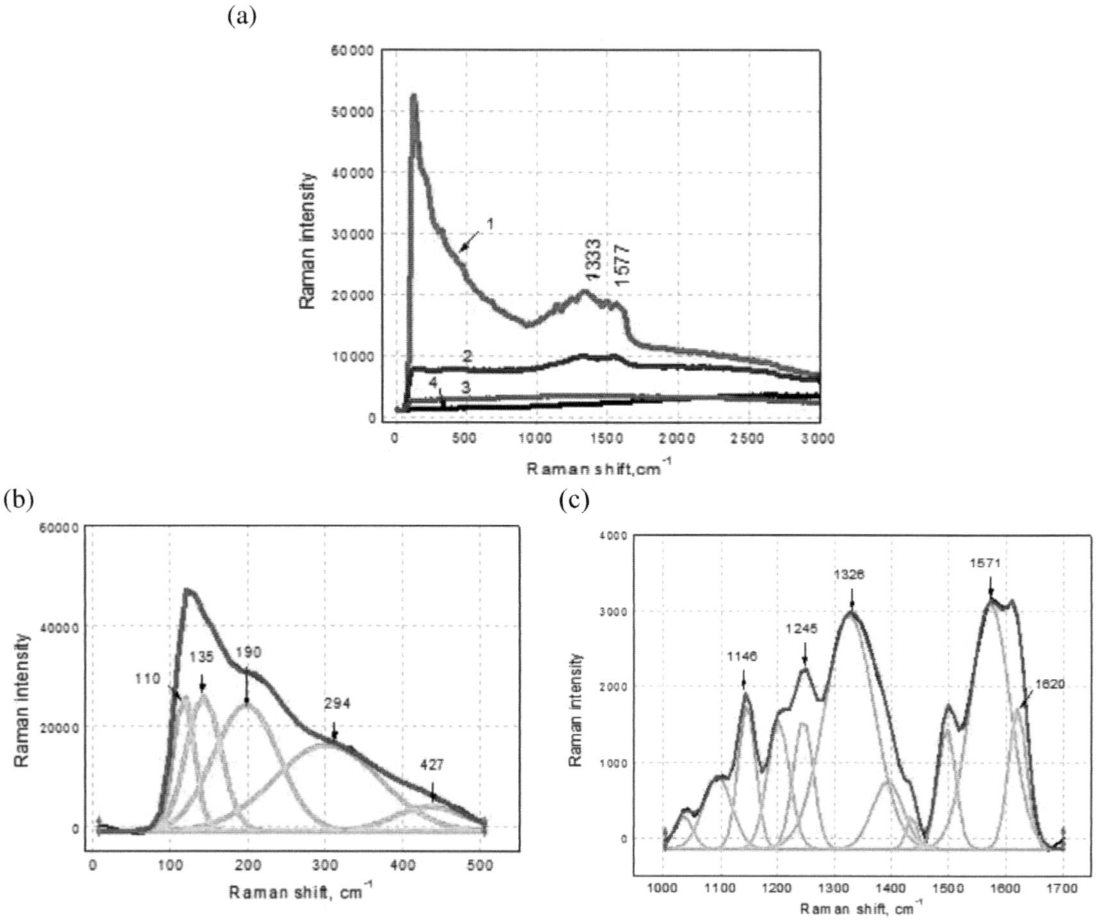

Figure 4. Ag NP on the porous glass experimental Raman spectra at various NP concentrations (cm^{-2}): 1 – 1,1 · 10^{12}, 2 – 0,6 · 10^{12}, 3 – 0,23 · 10^{12}, 4 – without Ag NP (a); Raman spectra Gaussian analysis at low- (b) and middle-wavenumbers (c) spectral ranges.

NP on the porous glasses Raman spectra (Figure 4a) display 1333 cm^{-1} and 1577 cm^{-1} maxima which are characteristic of bridge-oxygen in B-O-Si group asymmetric vibrations and bridge hydrogen bonds in B-Si-H gropus [16]. Intense wide band (70 – 1000 cm^{-1}) with the 127 cm^{-1} peak, which appears at the maximum NP concentration (curve 1), can be associated with Ag_2O (96 cm^{-1} and 146 cm^{-1} [17]), Na_2O (298 cm^{-1} [18]) and $NaNO_2$ (186 cm^{-1} [19]) groups vibrations. The presence of the above mentioned groups' wavenumbers in confirmed by Gaussian spectra analysis (Figure 4b). The good fitting of experimental spectrum with Gaussian spectra allows to identify a 434 cm^{-1} band, which can be associated with Si stretching vibrations in SiO_2 group (484 cm^{-1} [15]). The analysis of the middle-range of the experimental spectrum (Figure 4c) displays the following bands: 1146 cm^{-1} (Si stretching vibrations in -Si-O-Si- group), 1245 cm^{-1}, 1326 cm^{-1} (NO_2 group symmetrical stretching vibrations [19]) and 1620 cm^{-1} (H-O-H group deformation vibrations [20]). All the identified vibrational bands are shifted by 5 – 25 cm^{-1} to the low-wavenumber region compared with literature data for "clean" compounds. The observed shift was assumed to be caused by phase decomposition in the system "porous glass – borohydride sol Ag NP". The shift decrease to the NO_2 and SiO_2 groups evidence of the strengthening bonding between the phases.

Shift and vibrational bands intensity selective increase in Raman spectra indicates of light scattering plasmon enhancement by borohydride sol products on the porous glass surface. In this regard, it was interesting to study Raman spectra of R6G doped to PVA film on the porous glass surface.

According to the experimental spectra (Figure 5), the bands defying R6G [21], appeared to be the following: 602 cm^{-1}, 770 cm^{-1}, 1180 cm^{-1}, 1349 cm^{-1}, 1507 cm^{-1} and 1644 cm^{-1}.

Gain coefficient can be obtained by the formula [21]

$$SE = \frac{I_{SERS} \cdot C_{RS}}{I_{RS} \cdot C_{SERS}},$$

where I_{SERS} and I_{RS} denote SERS and Raman scattering intensities on the defined wavenumber, respectively; C_{SERS} and C_{RS} denote agent concentrations in SERS- and Raman scattering experiments, respectively. *SE* estimated value was equal to 10^{11}. Light scattering enhancement by R6G molecules in PVA film doped with Ag NP constituted approximately 10^3.

(a) (b)

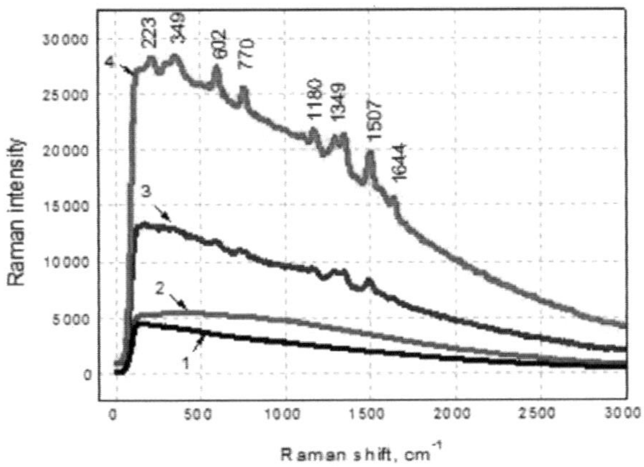

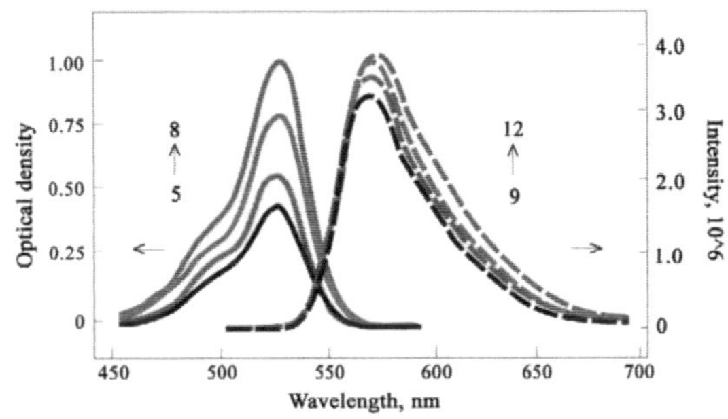

Figure 5. Raman (a), absorption (curves 5 – 8) and fluorescence (curves 9 – 12) spectra of R6G molecules in PVA film on the porous glass modified by Ag NP at various concentrations of Ag NP (cm^{-2}): 1, 5, 9 – without Ag NP; 2, 6, 10 – 0,23 · 10^{12}; 3, 7, 11 – 0,6 · 10^{12}; 4, 8, 12 – 1,1 · 10^{12}. Curves 3 and 4 are weighted by the factor of 10^2.

Significant increase of light scattering by R6G molecules with Ag NP on the porous glasses is caused by the presence of scattering centers which participate in both scattering and fluorescence. The spatial restriction that takes place due to surface pores leads to the molecular complexes formation. With Ag NP concentration increase, due to nonradiative energy transfer, R6G molecules absorbance increase, as shown in their absorption spectra (Figure 5b). However, in the time range between molecule excitation and deactivation there may be relaxation, recombination and inter-conversion processes leading to fluorescent molecules number decrease. For R6G molecular clusters localized in glass pores, due to their proximity to each other, the transfer to the triplet state probability can change, resulting in delayed fluorescence and phosphorescence emission along with that the number of singlet molecules decreases. This statement will lead to the further investigations of photoprocesses occurring in porous media with the presence of NP.

Ag NP doped into the polymer matrix tend to form centers which enhance light scattering of nearby dye molecules. As shown on Figure 4, Raman spectra of these matrices allow to obtain detected R6G molecules. The main bands (612 cm^{-1}, 1370 cm^{-1}, 1518 cm^{-1} and 1664 cm^{-1}) shift of R6G in PVA Raman spectra to the "clean" compounds range is minimal, which explains strong bonding of the polymer and dye in the CO- and OH-groups deformational vibrations range. The polymer-porous glass bonding formation occurs at these groups wavenumbers (Figure 5a).

Scattering intensity drastic increase bands occurs in Raman spectra of dye molecules adsorbed on NP surface, which is not observed for R6G molecules doped into polymer (Figure 5a and Figure 6a). This can be associated with energy redistribution while bonding formation with polymer.

R6G in polymer film with the presence of Ag NP absorption and fluorescence increase (Figure 6b) which is caused by local surface plasmon resonance in NP occurring on excitation [22, 23].

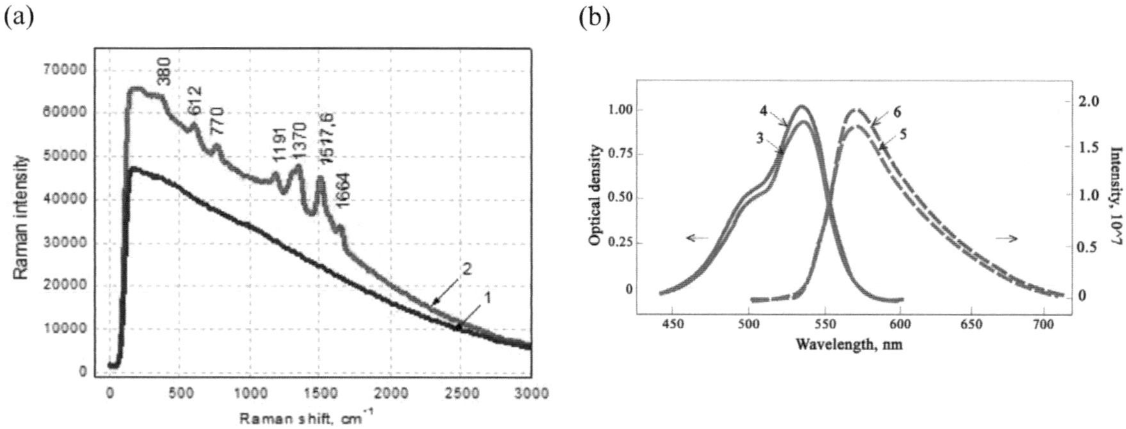

Figure 6. Raman (a) and absorption (curves 3, 4) and fluorescence (b) spectra of R6G molecules in PVA film with the presence of Ag NP: curves 1, 3, 5 – R6G in PVA film; 2, 4, 6 – R6G with Ag NP in PVA film

5. Conclusion

It was shown that the optical spectroscopy of dye molecules adsorbates and Ag NP in PVA films on the porous glasses surface allows to obtain and evaluate the efficiency of the energy exchange between electron-excited and unexcited molecules. It was found that electron and vibrational energy absorbance and scattering centers are formed in the presence of Ag NP and dye molecules which is shown in SERS and Raman spectra. FRET, plasmon excitation in Ag NP and R6G fluorescence enhancement was found to occur in these conditions. It is difficult to conclude which process is predominant due to the investigated processes complexity: the additional experiments are needed. In addition, porous glass modifying by Ag NP results in appearance of new properties (molecular fluorescence enhancement), which allows to use these systems as optical sensors. R6G fluorescence intensity detection in PVA film showed that fluorescence weak red shift was due to surface plasmons generation in Ag NP, while Ag NP Raman spectra Stokes shift was well detected. We suggest that Raman spectra of R6G molecules in PVA near glass surface is due to interaction between polymer chains and dye on the CO- and OH-groups vibrations wavenumbers range.

Thus, modified porous glasses allow to enhance Raman scattering on R6G molecules by Ag NP surface plasmons and detect polymer-dye system features.

The results obtained in the framework of the State task of Russian Ministry of Science and Education. Nr. 3.809.2014/K.

References

[1] Klimov V V 2010 Nanoplasmonics (Moscow: Fizmatlit)
[2] Demchenko A P 2013 Methods Appl. Fluoresc. **1** 1-29
[3] Dai H, Shi Y, Wang Y, Sun Y, Hu J, Ni P, Li X 2014 Sensor and Actuators B **202** 201-208
[4] Li H, Wang M, Wang C, Li W, Qiang W, and Xu D 2013 Anal. Chem. **85** 4492-4499
[5] Escobedo C, Chou Y, Rahman M, Duan X, Gordon R, Sinton D, Brolo A G and Ferreira J 2013 Analyst **138** 1450-1458
[6] Zhang H, Feng L, Liu B, Tong C, Lu C 2014 Dyes and Pigments **101** 122-129
[7] Lin Y, Xu C, Li J, Zhu G, Xu X, Dai J, and Wang B 2013 Adv.Optical Mater. **1** 887-990

Journal of Physics: Conference Series **737** (2016) 012037 doi:10.1088/1742-6596/737/1/012037

[8] Mahmoud M A, El-Sayed M A 2013 J.Phys.Chem.Lett. **4** 1541-1545
[9] Evstratov A A, Esikova N A, Rudninskaya G E, Antropova T V, Anfimova I N 2010 Scientific Instrumentation **20** 52-58
[10] Turkevich J, Stevenson P S, Hiller 1955 J. Discuss. Faraday Soc. **11** 55-75
[11] Lee P S, Meisel D 1982 J. Phys.Chem. **86** 3391-3395
[12] Schmid G 1992 Chem. Rev. **92** 1709-1727
[13] Creighton J A, Blatchford C G, Albrecht M G 1979 J. Chem. Soc. **75** 790
[14] Castro T, Reifenberger R, Choi E, Andres R P 1990 Phys. Rev. B **42** 8548–8557
[15] Lenshin A S, Kashkarov V M, Seredin P V, Spivak Yu M, Moshnikov V A 2011 Semiconductors 45 1223-1228
[16] Mikhaylov A V, Grigoryev L V, Konorov P P 2012 Journal of Optical Technology **79** 54-57
[17] Strelchuk V V, Kolomys O F, Golichenko B O, Boyko M I, Kaganovich E B, Krishchenko I M, Kravchenko S O, Lytvyn O S, Manoilov E G, Nasieka Iu M 2015 Semiconductor Physics, Quantum Electronics & Optoelectronics 18 46-52
[18] Zhuravlyov Yu N, Korabelnikov D V, Aleynikova M V 2012 Physics of the Solid State 54 1427-1434
[19] Gorelik V S, Pyatyshev A Yu, Krylov A S 2016 Physics of the Solid State **58** 163-169
[20] Averin I A, Karmanov A A, Moshnikov V A, Pronin I A, Igoshina S E, Sigaev A P, Terukov E I 2015 Physics of the Solid State **57** 16-24
[21] Le Ru E C, Blackie E, Meyer M and Etchegoin P G 2007 J.Phys.Chem. **111** 13794-13803
[22] Bryukhanov V V, Tsibulnikova A V, Samusev I G, Slezhkin V A 2014 Journal of Applied Spectroscopy **81** 515-521
[23] Bryukhanov V V, Konstantinova E I, Borkunov R Yu, Tsarkov M V, V A Slezhkin 2015 Quantum Electronics **45** 908-913

Journal of Physics: Conference Series **737** (2016) 012038 doi:10.1088/1742-6596/737/1/012038

Research of plasmon resonance in developed holographic photomaterials

N V Andreeva, T B Kuzmina and O V Andreeva

ITMO University, Saint Petersburg 197101, Russia

E-mail: andreevanv_3@mail.ru

Abstract. The research of aqueous suspensions of colloidal particles of silver obtained by photographic method was carried out, two optical methods were used: the standard photometric method and the method of dynamic light scattering. It is shown that in the researched preparations, which attenuation spectra in the visible region have a close form, distribution of the particles by sizes could vary considerably.

1. Introduction

Great interest to optics of metal nanoparticles that have arisen in recent decades is associated with the development of nanotechnologies and nanodiagnostics. This interest supports a rapidly evolving scientific field, called "Nanoplasmonics" [1], or "Plasmonics" [2]. Volume holograms on plasmon particles of silver represent the photoinduced periodic structures that demonstrate the practical use of plasmon resonance. The research of such structures is important for creation on their basis of new optical media with functional properties that surpass the capabilities of natural materials and traditional elements of optical devices and systems.

In the work the results of research of aqueous suspensions of colloidal particles of metallic silver obtained from the developed samples of high-resolution holographic photomaterials are presented. Silver particles in such suspensions could be considered isolated and during research we could evaluate their properties as properties of ensemble of isolated particles, which is impossible during research of patterns of the initial photomaterial.

2. Object and methods

The objects of research are holographic silver-halide photomaterials after exposure and development: holographic photoplates [3] and nanoporous silicate silver containing matrices Ag-NPM [4]-[5]. Developed photoplates (figure 1a) – silver particles distributed in gelatin matrice with a thickness of $\approx$ 10 μm, which is formed on a glass substrate. Developed samples AgNPM (figure 1b) – silver particles distributed in free volume of porous matrice with a thickness of $\approx$ 1 mm with the presence of gelatin (not more than 10% of sample weight).

Developed particles of metallic silver of colloidal structure, obtained in high-resolution photomaterials for holography, represent plasmon particles, which range of attenuation (plasmon resonance) has a maximum in the short-wave region of the spectrum (figure 2c). Maximums of spectra of attenuation of developed samples tend to have high values of optical density which are resistant to standard measurements on standard optical devices, especially in the case of samples on the basis of nanoporous silicate matrices even during production of sufficiently thin samples with a thickness of about 300 μm (see figure 2c, curve 2).

To obtain the spectral characteristics of ensemble of isolated particles of developed silver were made preparations that have the ability to adjust the concentration of researched particles. On figure 2d the spectral characteristics of preparations of samples of photoplates and nanoporous recording media are shown (data obtained by the author taken from the report of SOI 1994).

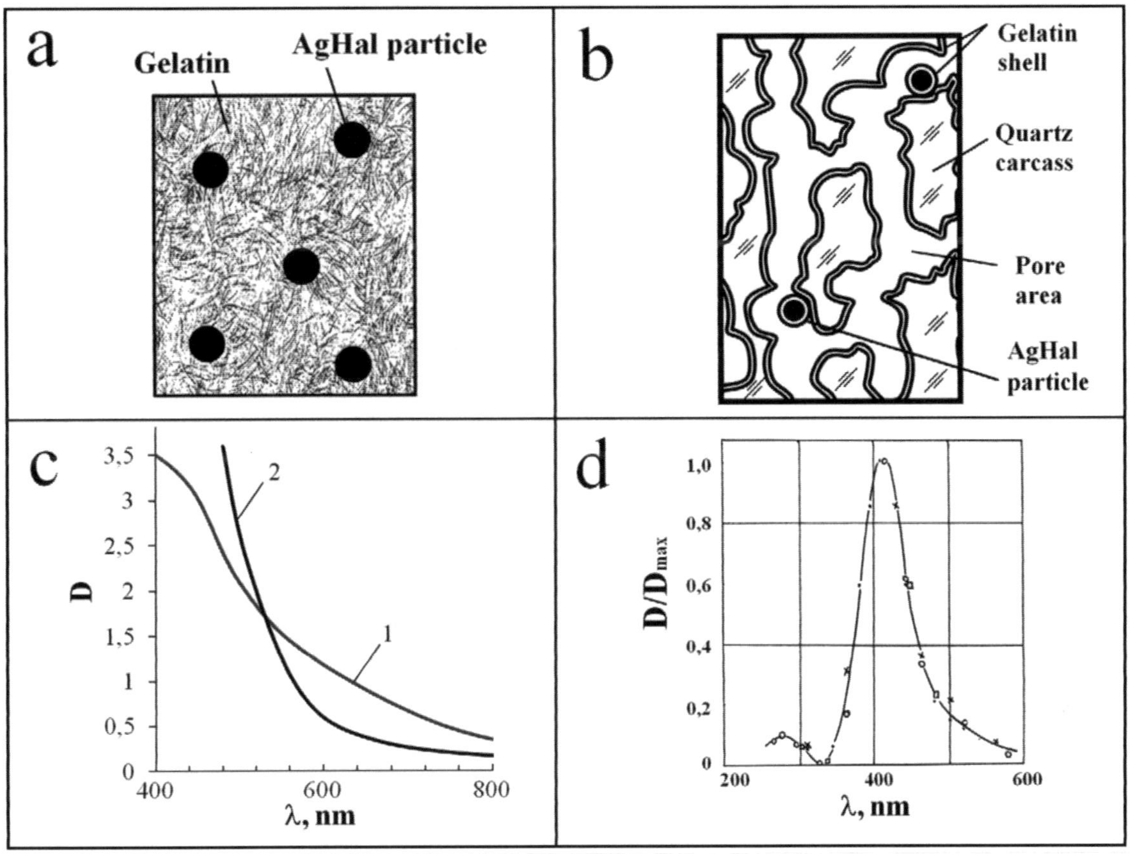

Figure 1. Schematic image of cross section of silver-halide photomaterial (a) and nanoporous silver-halide medium (b). c - spectra of attenuation of developed photoplate (curve 1) and AgNPM (curve 2), d - normalized spectra of attenuation of suspensions of particles of developed silver of these samples.

It should be noted that the average particle size of the prepared preparation could be higher than in initial samples due to the agglomeration of the smallest nanoparticles in the developement process of preparations as well as due to possible shell, which could be present on the particles of silver due to absorption of components of the mixture. For estimation the distribution of particles by sizes in the samples and for control of the state of preparations in parallel with the spectral measurements were carried out measurements of function of particles distribution by sizes using method of dynamic light scattering. Measurements of spectra of absorption were carried out on spectrophotometer Evolution 300 in the wavelength interval 300-800 nm. Estimation of the distribution of particles by sizes was carried out on device Horiba LB-550.

3. Experimental results

On figure 2 and in table 1 the results of measurements of parameters of preparations are shown. We could clearly see the difference in the distribution of particles by sizes of samples AgNPM and AgPP, as well as changing of state of preparation AgPP during storage, while their spectra of attenuation differ not so much. For example, preparations (curves 1 and 2), which for D_{ch} (Q_{max}) differ in 7 times, in spectra of attenuation have not such big differences.

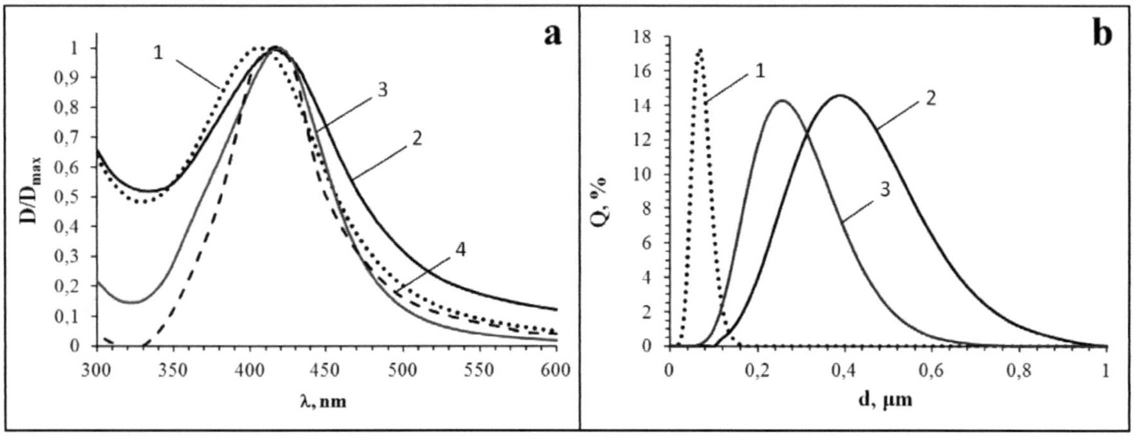

Figure 2. Spectra of attenuation (a) and the distribution of particles by sizes (b) of aqueous suspensions: AgNPM after 24 days (1), AgPP after 8 days (2) and 28 days (3), preparations of 1994 year (4, see figure 1d).

Table 1. Results of measurements of parameters of preparations.

№ of curve	Material	Time, days.	Spectral measurements		Distribution of particle sizes	
			$\lambda(D_{max})$, nm	$\Delta\lambda$, nm	D_{ch}, nm	ΔD, nm
1	AgNPM	24	407	116	50,7	60
2	AgPP	8	415	140	388,6	224
3	AgPP	28	417	88	258,6	340

4. Conclusions

In the work it was discovered that despite the coincidence of maximum values of optical density of researched suspensions, functions of distribution of silver particles by sizes obtained by method of dynamic scattering vary significantly. The size of developed silver particles in preparations AgNPM is several times less, than in preparations of photoplates PPH-03 – such assessments couldn't be done using only spectral measurements.

Thus, during research of developed silver nanoparticles with the plasmon resonance, the method of dynamic light scattering allows to do more accurately estimation in difference between parameters than spectrophotometric measurements do it.

References

[1] Klimov V V 2010 *Nanoplasmonics* (Moscow: Fizmatlit) 480 p
[2] Novotny L, Hecht B 2006 *Principles of nano-optics* (New York: Cambridge University Press) 541 p
[3] "Company SLAVICH", http://www.slavich.ru (accessed September 10, 2015).
[4] Andreeva O V, Obyknovennaya I E, Gavrilyuk E R, Paramonov A A, Kushnarenko A P 2005 *Journal of optical technology* **72 (12)** pp 916-922
[5] Andreeva O V, Bandyuk O V 2011 *Holograms - Recording Materials and Applications. Book 1* (INTECH Open Access Publisher) pp 45-70

Journal of Physics: Conference Series **737** (2016) 012039 doi:10.1088/1742-6596/737/1/012039

Laser desorption of traces of low volatile explosives

A E Akmalov, A A Chistyakov, G E Kotkovskii
National Research Nuclear University MEPhI (Moscow Engineering Physics Institute), 31 Kashirskoe shosse, 115409 Moscow, Russia

E-mail: geko@mail.ru

Abstract. Comparison of desorption effectiveness of two laser sources: continuous diode laser ($\lambda = 440$ nm, P = 0.5 W) and laser based on YAG: Nd^{3+} ($\lambda = 266$ nm, $\tau_{pulse} = 6$ ns, $E_{pulse} = 1.0$ mJ, f = 10 Hz, $I_{pulse} = 10^8$ W/cm^2) was carried out for the explosive cyclotrimethylenetrinitramine (RDX). The results of mass spectrometric studies of RDX laser desorption from a quartz substrate are discussed. It is shown that a continuous diode laser ($\lambda = 440$ nm) effectively desorbs the analyte without accompanying dissociation. The typical desorbed mass is 70 ng for RDX at 1.1 minutes exposure. However a laser based on YAG: Nd^{3+} seems to be more efficient for the same time exposure.

1. Introduction

A problem of detecting explosives is particularly acute due to the worldwide growth of terrorist acts. There are various devices based on principle of ion mobility spectrometry (IMS) and field asymmetric ion mobility spectrometry (FAIMS) that can solve the problem. However a lot of explosives have an extremely low pressure of saturated vapor ($\sim 10^{-14}$ g/cm^3 for RDX, $<10^{-16}$ g/cm^3 for 1,3,5,7-tetranitro-1,3,5,7 tetraazacyclooctane (HMX)), which greatly complicates the searching for such devices. A pre-concentration method based on processing of a surface with a special tissue, followed by heating of the tissue in a thermal desorption unit of IMS (FAIMS)-based detectors, is widely used to increase the sensitivity of gas phase analyzers. It should be noted that the "tissue" method may be dangerous because of the contact with a suspicious object. The difficulties arising from a handling of rough and relief surfaces could be also observed.

The method of laser desorption of explosives is elaborated [1] for solving these problems. The ultraviolet (UV) YAG: Nd^{3+} laser is one of the most effective sources for laser desorption having an opportunity of portable implementation [2-3]. Although the current progress allows construct such laser sources sufficiently miniature, the weight of them exceeds usually 1.5 kg. Therefore, a much more lightweight and compact (0,3 kg with power supply) continuous diode laser with wavelength of $\lambda = 440$ nm and laser power of 0.5 W has been used in this work to initiate desorption.

The purpose of this study is to compare the desorption efficiency of pulsed and continuous laser sources operating in UV and visible ranges.

2. Theoretical basis

The initiation of laser desorption may occur through various mechanisms, such as direct photo desorption, non-thermal desorption, desorption at a local heating of a surface. Direct photo desorption is observed in the course of resonance absorption of a quantum in a spectral range of excitation of adsorbed molecules [4]. Non-thermal desorption is similar to the direct photo desorption but is realized after electron excitation of a solid in the course of energy transmittance to an adsorbed molecule [4]. Local heating mechanism is a point localization of heat energy that provides desorption of molecules at low thermal conductivity of a substrate.

The most common process is thermal desorption - heating of a surface due to thermal relaxation of the excited states after absorption of a quantum of light [4-6].

3. Experimental setup and technique

A semiconductor laser diode operating at λ=440 nm and power of 0.5 W was used in this work as a source of radiation. The desorption efficiency of this source was compared with the same one demonstrated by YAG: Nd^{3+} (λ = 266 nm, τ_{pulse} = 6 ns, E_{pulse} = 1.0 mJ, f = 10 Hz, I_{pulse} = 10^8 W/cm^2) - [2,3].

The desorbed mass was determined by the quadrupole calibrated mass spectrometer Shimadzu GCMS-GP2010 with electron impact ionization and the ability of quick direct introduction of a sample into vacuum for immediate analysis.

The laser-vacuum module providing an ability of laser desorption near the ion source of the mass spectrometer was specially designed and used for this research (Figure 1).

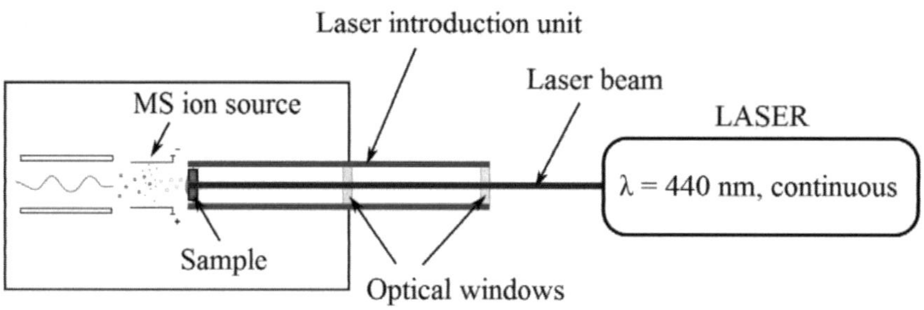

Figure 1. Experimental setup.

The sample was a quartz glass substrate covered by a solution of 10^{-5}g RDX in acetonitrile. After acetonitrile drying within 3 minutes the RDX molecules formed on the surface an optically thin layer for the wavelength used. Then the sample was fixed on the laser-vacuum rod after that within 2 seconds was entered into the mass spectrometer for analysis.

4. Results and discussion

Figure 2 shows the ion current time dependence characterizing desorption of sample molecules under both evaporation into vacuum and heating by an ion source of the mass spectrometer.

This dependence is a smooth curve with a maximum corresponding 5 minutes delay from a placement of the sample into mass spectrometer. In the case of laser exposure (Figure 2b) the time of maximum of the dependence is significantly less: one minute instead of five.

Mass spectra (Figure 3) in both cases correspond to RDX with an accuracy of 90% compared with the NIST database.

The dependence of RDX desorption time from a quartz substrate on the laser intensity was obtained (Figure 4).

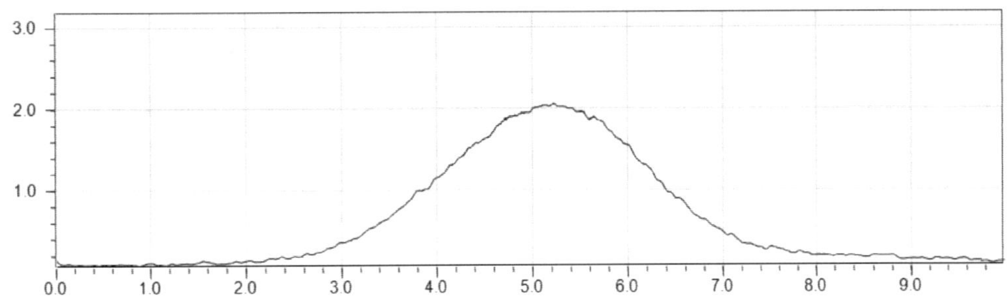

a

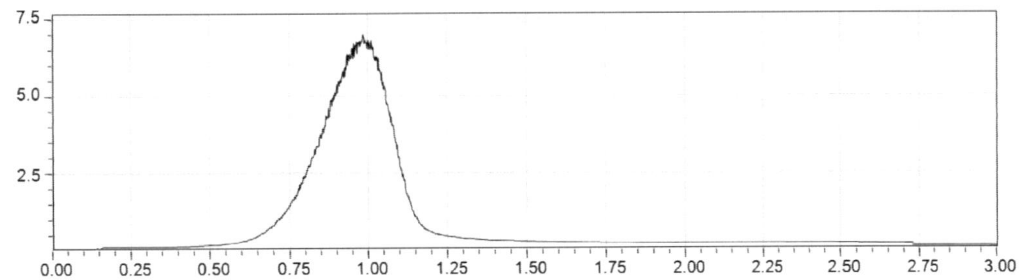

б

Figure 2. Ion current time dependence: a) without laser exposure; b) under laser irradiation (λ=440 nm, W=0.5 W, q=24 W/cm^2).

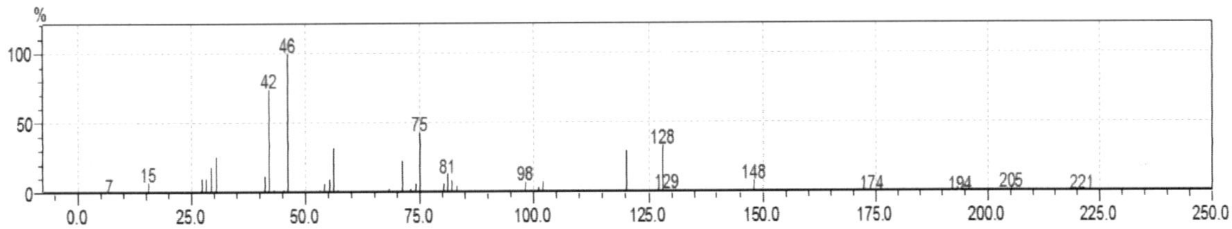

Figure 3. Mass spectrum of RDX under desorption of a sample without laser exposure. Coincides with the mass spectrum at laser exposure.

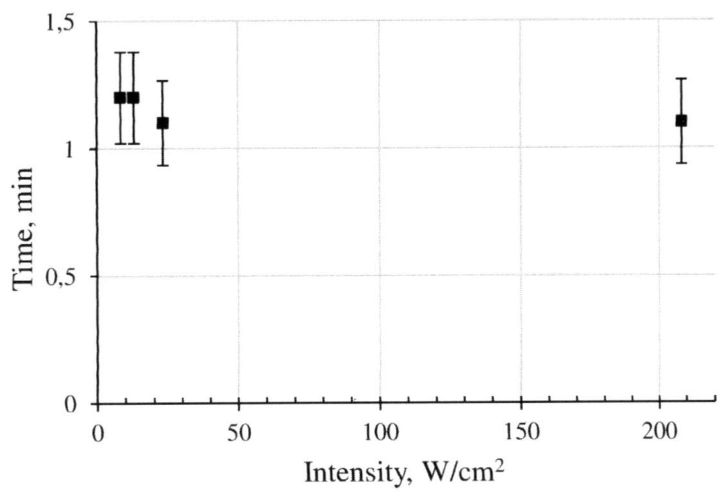

Figure 4. The dependence of RDX desorption time on laser intensity from a quartz substrate under laser exposure (λ=440 nm, P=0,5W). The desorbed mass is 7.0*10^{-8} g).

The intensity was varied by changing area of radiation spot on a sample surface. The radiation power was the same for all intensities. The time corresponding to desorption of about 90% total mass registered was chosen as desorption time. As seen in Figure 4, the amount of the desorbed substance (7.0*10-8 g) and the laser exposure time required for desorption are almost independent on the radiation intensity. It presumably means that the spreading of heat over the surface and in depth from irradiated region of the sample is rapid for all intensities used. So, desorption of RDX is possible for this laser source only through heating of bulk of quartz (and, of course, metal) substrate which exclude possibility of rather effective local heating.

We compared the efficiency of a nanosecond UV laser source with wavelength of $\lambda = 266$ nm, and the continuous laser source used in this research at comparable time intervals. The maximal quantity of desorbed RDX under $\lambda=266$ nm laser radiation was 2 ng per pulse [2,3]. So, the desorbed quantity can be estimated as 2ng* 66sek *10 Hz = 1320 ng for 1.1 minutes irradiation when operating in frequency mode at 10 Hz. Thus, the difference in desorbed quantities with continuous laser reaches 20 times in favor of the pulsed source (the spot area while processing the surface by the pulsed laser must be equal to the continuous source spot area). It should be noted that exposure of nitro molecules under UV nanosecond laser radiation should carry out with the pulse intensity no more than 10^9 W / cm^2 to reduce the probability of photo dissociation of these molecules with forming of NO and NO_2 fragments.

Thus, it can be concluded that initiating desorption process of RDX traces from a surface using a nanosecond pulsed UV source is preferable than using the continuous laser diode ($\lambda = 440$ nm), that expressed in the less time of registration and (or) bigger amount of the substance desorbed.

5. Conclusion

The objective of this work is to compare the efficiency of laser desorption of RDX molecules from a quartz glass substrate for two laser sources: continuous diode laser with wavelength of $\lambda = 440$ nm and laser power of 0.5 W and laser based on YAG: Nd^{3+} ($\lambda = 266$ nm, $\tau_{pulse} = 6$ ns, $E_{pulse} = 1.0$ mJ, f = 10 Hz, $I_{pulse} = 10^8$ W/cm^2). It is shown that a pulse laser is approximately 20 times more effective than the investigated continuous laser source. However, due to the small size and weight of the continuous laser it is also suitable to create the non-contact portable sampling device.

6. References

[1] Akmalov A, Bogdanov A, Kotkovskii G and et al 2013 *Instruments and experimental techniques* **56** pp 309-316
[2] Akmalov A, Chistyakov A, Kotkovskii G 2015 *Physics Procedia* **71** pp 207-211
[3] Akmalov A, Chistyakov A, Kotkovskii G 2015 *Proceedings of SPIE* **9652** p 965206
[4] Lazneva E 1990 *Laser desorption* (Leningrad: Publishing House of Leningrad University)
[5] Bechtel J H 1975 *J.Appl.Phes.* **46** pp 1585-1593
[6] de Boer J 1962 *The dynamic nature of the adsorption* (Moscow: Izd. Inostr. Lit.)

Detection of weak seismic waves in sea-ground interface by fiber-optic interferometric seismometer

O T Kamenev, Yu S Petrov, R V Khiznyak, R V Romashko

Institute of Automation and Control Processes, Far Eastern Branch of Russian Academy of Sciences, 5 Radio Street, Vladivostok, 690041 Russia

E-mail: okamenev@iacp.dvo.ru

Abstract. The possibility of detection of weak seismic waves in sea-ground interface by fiber-optic interferometric seismometer is experimentally demonstrated. High sensitivity of the seismometer to acceleration 10^{-7} m/s^2 is provided by application of original multiturn fiber-optical sensing element. Long term stability of the seismometer operation is provided by feed-back control of the interferometer working point.

Introduction

Detection of weak seismic waves in sea-ground interface is important but challenging task which solution can bring a practical benefit to investigation of a sea shelf. Seismometers which are commonly used for seismic detection in fact measure a displacement of mass caused by seismic wave. In its turn optical interferometry has always been considered as one of the most sensitive techniques for measuring mechanical displacements [1]. The sensitivity of classical homodyne interferometer is limited by shot noise of the photoelectrons in a photodetector and can be extremely high: the theoretical minimal detectable displacement is 1.1×10^{-16} m/Hz$^{1/2}$ for 10 mW of detected laser power at the wavelength of 500 nm [2]. Use of optical fiber in interferometers can provide development of compact and immune to EMI seismometer with low intrinsic noises. Moreover sensitivity of fiber-optic (FO) sensors can be additionally increased if multiturn sensing elements are used. As a result a combination of high interferometric sensitivity with long optical path length can provide a measurement of displacements with resolution up to 10^{-13} m in a wide frequency band. It opens opportunities for detection of weak seismic waves in sea-ground interface by fiber-optic interferometric seismometer.

Experiment

In our work the fiber-optic accelerometer on the basis of a Mach-Zehnder Interferometer was used as the seismometer. Among existing interferometers the Mach-Zehnder Interferometer (MZI) has one of the simplest schemes and can be easily used as a base for fiber-optic sensors.

The scheme of fiber-optic seismometer is shown in Fig. 1. Its base is a fiber-optic Mach-Zehnder interferometer. CW optical radiation from a laser (1) with power 1 mW is coupled by means of fiber-optic Y-splitter (3) into two single-mode optical fibers which work as signal and reference arms of the interferometer. Afterward the signal and reference waves are combined at the second Y-splitter (7). The result of an interference is registered by photodetector (8).

In order to maintain the interferometer output signal in the quadrature we used an approach based on active phase control. Here, the optical fiber in the interferometer reference arm is coiled to a piezoelectric cylinder (4) operated as a phase modulator (PM). PM is driven by feed-back voltage which, in its turn, is determined by DC level of the interferometer output signal. The latest is directly related to the interferometer working point position. So the working point drift being occurred is compensated by additional phase shift produced by PM.

When the interferometer is in quadrature, its output signal U_0(V) is linearly proportional to acceleration amplitude a_0 (m/s^2). For the particular scheme of the seismometer represented in Fig.1 the sensitivity to acceleration has reached the level 10^{-7} m/s^2. Such high sensitivity was provided by using a multiturn sensing element in the interferometer's signal arm [3, 4].

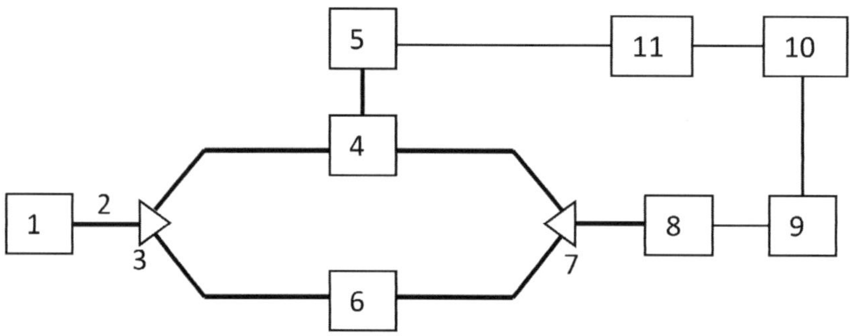

Figure 1. Scheme of the seismometer on the basis of a fiber-optic Mach-Zehnder interferometer. 1 – DFB-laser; 2 – optical fiber; 3, 7 – fiber-optic Y-splitters; 4 – phase modulator; 5 – amplifier; 6 – multiturn sensing element; 8 – photodetector; 9 – ADC; 10 – personal computer; 11 – DAC.

Figure 2. The fiber-optic interferometric seismometer (a) and place of its installation on a shore rock – shown by arrow (b)

For the experimental testing, the described fiber-optic seismometer was installed in sea-ground interface, in particular on the rock at 75 m above sea level (Fig.2). The seismometer was mounted at concrete base in a wind-proof room at a depth of 2 m.

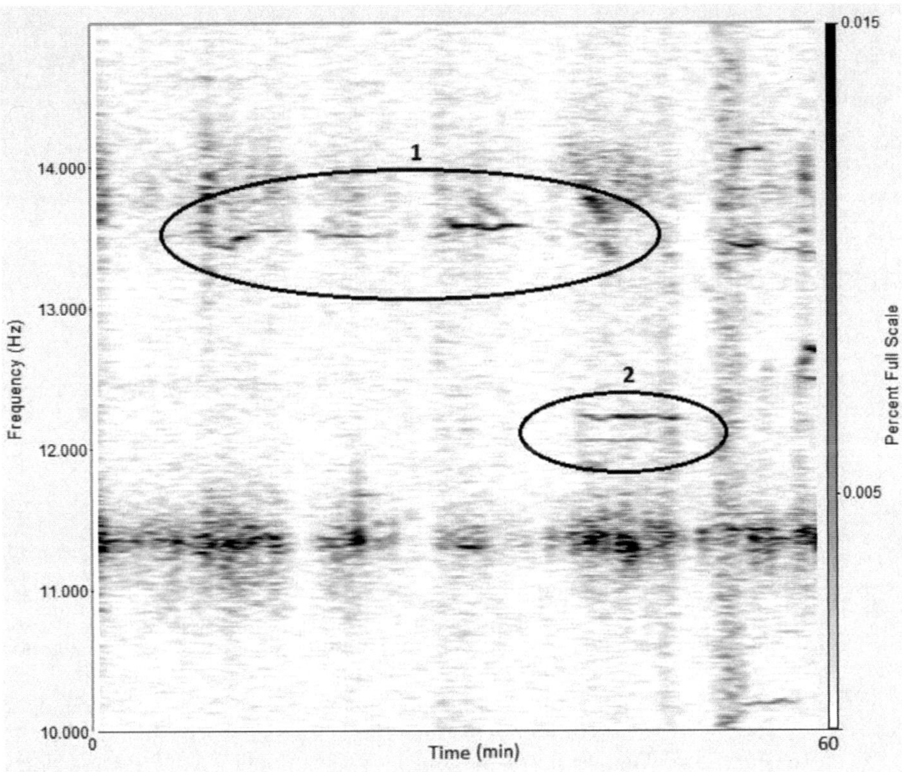

Figure 3. Part of time evolution of Fourier spectrum of the seismic signals produced by heavy-tonnage vessel (1) and small-size high-speed boat (2).

During the measurements (1 hour) two types of the marine vessels – heavy-tonnage vessel (HTV) and the small-size high-speed boat (HSB) – were running at the distance between 3 and 8 km apart from the rock. Fig.3 shows the time evolution of Fourier spectrum of the seismic signals detected by the seismometer. As seen there are noticeable signals which correspond to serial passing of both vessels: HTV – from 5 to 45 min, and HSB – from 35 to 52 min.

Thus, in this work we have experimentally demonstrated possibility of detection of weak seismic waves coming through the sea-ground interface by fiber-optic interferometric seismometer mounted at shore rock.

Acknowledgments
The research was financially supported by Russian Science Foundation (grant # 14-12-01122).

References

[1] Hariharan P 1990 *Rep. Prog. Physics* **54** 339
[2] Wagner J W, Spicer J B 1987 *J. Opt. Soc. Am. B.* **4** 1316
[3] Kamenev O T, Kulchin Yu N, Petrov Yu S, Khiznyak R V, Romashko R V 2016 *Sensors and Actuators A* **244** 133
[4] Kamenev O T, Kulchin Yu N, Petrov Yu S, Khizhnyak R V 2014 *Technical Physics Letters* **40** 119

Application of ultrashort laser pulses for timing characterization of silicon photomultipliers

E V Popova, P Zh Buzhan, A A Stifutkin, A L Ilyin, O B Mavritskii, A N Egorov, A A Nastulyavichius

National Research Nuclear University MEPhI (Moscow Engineering Physics Institute), 31 Kashirskoe shosse, 115409 Moscow, Russia

E-mail: OBMavritskij@mephi.ru

Abstract. The application of femtosecond laser irradiation for the investigation of Geiger discharge process in silicon photomultiplier (SiPM) is discussed. It is shown experimentally that sub-picosecond pulses of laser beam focused to micron spot sizes allow studying the dynamics of Geiger discharge process in single cell of silicon photomultiplier. These studies are aimed at identifying the factors limiting the timing resolution of this class of devices.

1. Silicon photomultiplier

Solid state Silicon Photo Multiplier (SiPM) – a fast-paced class of photodetectors intended to register the low intensity pulse of optical radiation. One of the most important SiPM applications is to use it as a part of medical facilities, such as positron emission tomography using the Time-of-Flight Positron Emission Tomography (TOF-PET) [1, 2].

SiPM is a matrix of silicon avalanche photodiodes (cells), combined on a common substrate (see. Figure 1a).

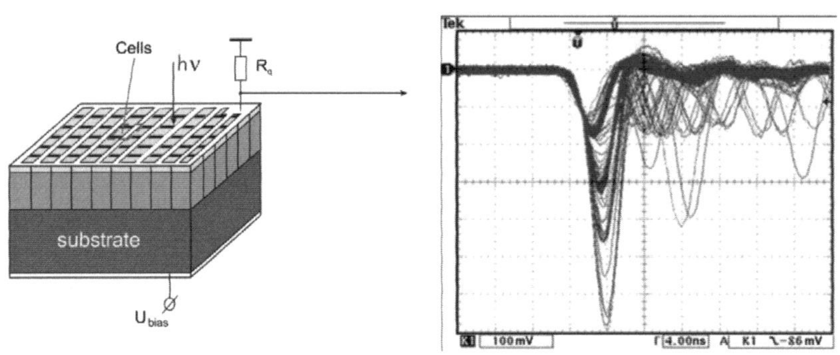

a b

Figure 1. Design (a) and typical SiPM matrix signal oscillogram (b).

Figure 1b shows typical example of SiPM matrix signal oscillogram. General view of its photosensitive surface is presented in Figure 2.

All cells operate in Geiger discharge mode (i.e. biased by voltage, slightly exceeding electrical breakdown threshold), in which photon-generated charge carrier initiates avalanche cell discharge (see. Figure 3).

Figure 2. General view and the magnified image of SiPM cells.

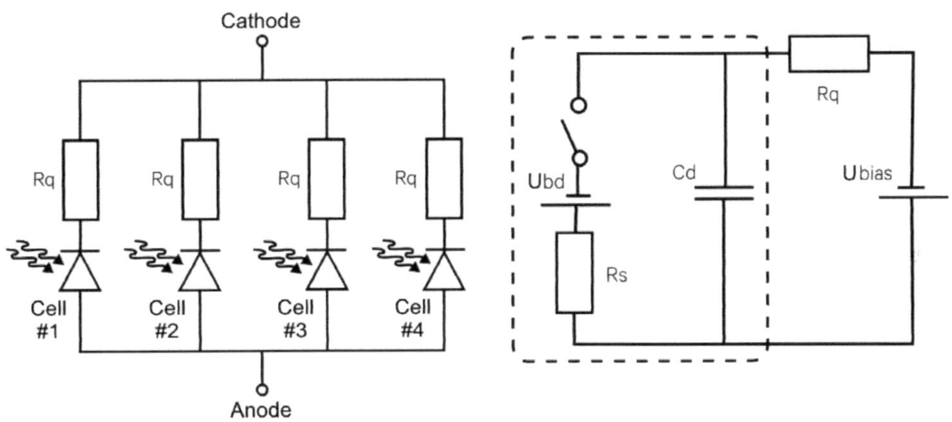

Figure 3. The SiPM equivalent circuit (parallel connection of avalanche photodiodes with quenching resistor Rq). The switch simulates the turn-on of photodiode under single photon excitation.

The most important characteristic of SiPM cell is its extremely high speed of avalanche discharge development. To ensure the best SiPM timing parameters a detailed study of the Geiger discharge process dynamics within a single cell is required. The speed of the avalanche development depends on the applied voltage and the thickness of the depletion region. For example, in [3] it was shown that the time required for the onset of Geiger discharge mode (the discharge signal front duration) may be of the order of 10 ps. To investigate these processes a light source with a light flash duration of 1 ps or less is required for the Geiger discharge initiation.

The research on emission microscopy [4] showed that the area of the avalanche discharge occupies only a very small part of the whole cell area (see Figure 4). So, to guarantee the consistency of the results, the sharp focusing of light radiation to a spot size of about 1 ... 3 μm is needed.

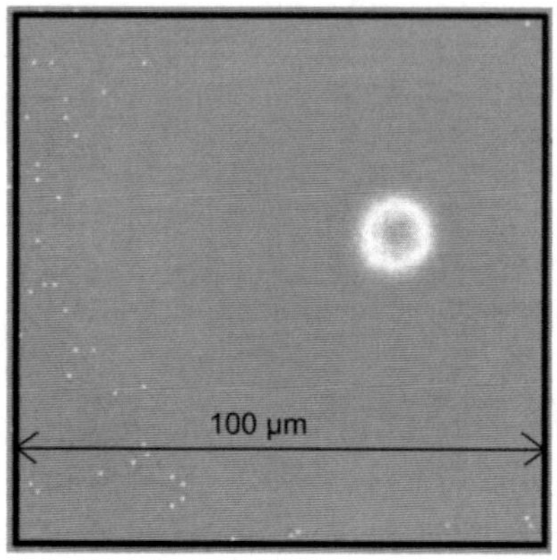

Figure 4. SiPM avalanche emission microscopy image (from [4]).

2. Experiment

In this work we have investigated experimentally the timing characteristics of the single-cell SiPM samples under the ultrashort pulsed optical irradiation. We used the experimental setup, based on femtosecond laser with 870 nm wavelength and 150 fs pulse duration combined with high resolution optical microscope with laser port (figure 5). This setup allowed to focus laser beam to a minimal spot of about 2 μm in diameter or obtain slightly defocused spot up to 200 μm by shifting the device under test away from the focal plane. For precise optical synchronization of the registration process high speed PIN photodiode was used. More detailed description of the setup can be found in [5].

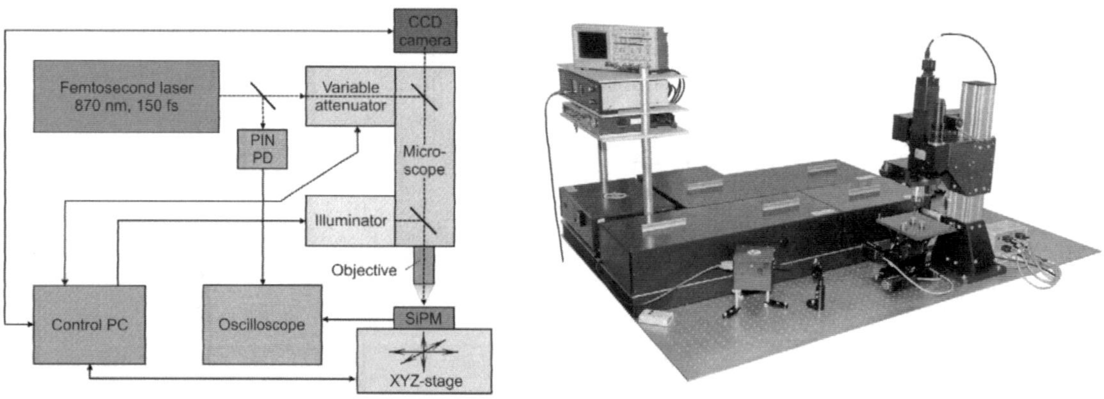

Figure 5. Block diagram and general view of the femtosecond laser experimental setup.

The SiPM electrical response was captured and analyzed statistically using the LeCroy WaveRunner 620Zi oscilloscope with the 2 GHz vertical channel bandwidth and 20 GHz sample rate.

In the experiment we have measured the delay and the jitter of the SiPM single cell discharge relative to the optical pulses registered by PIN photodiode, as a function of laser pulse energy E incident to the cell (see figure 6) and cell voltage for both focused and defocused irradiation with small number of photons.

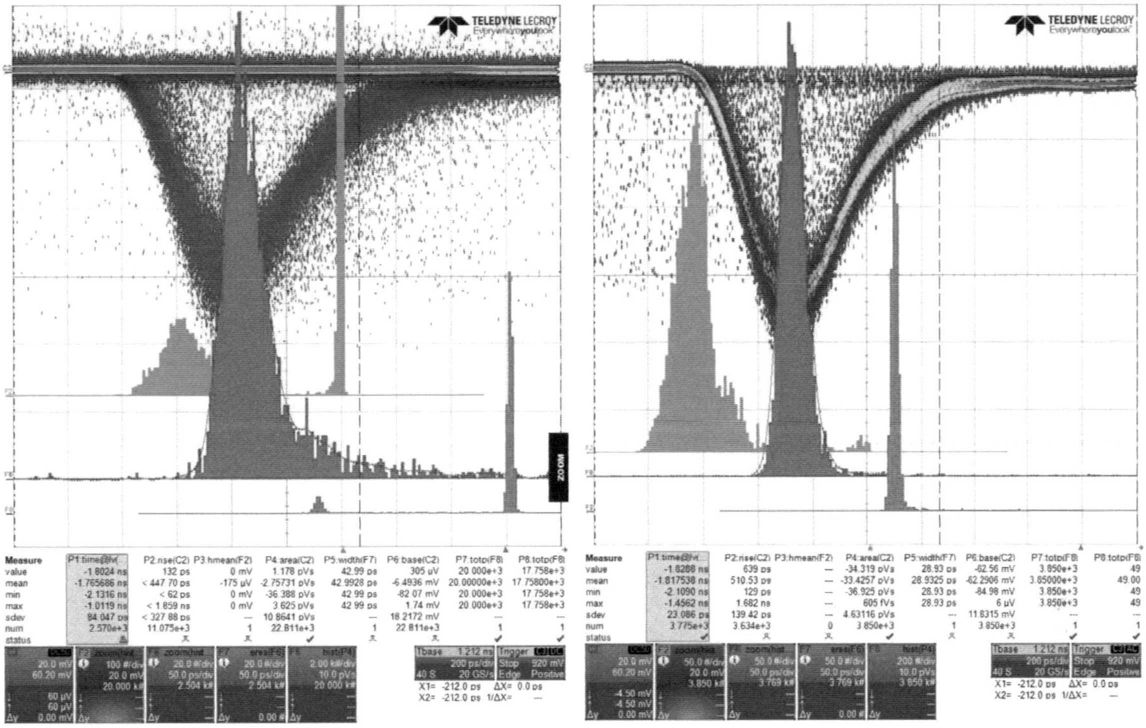

Figure 6. SiPM output electrical pulses at various femtosecond laser pulse energies: 0.1 nJ (left) and 10 nJ (right). Figure also shows the spectra measured by the oscilloscope: red – the registered SiPM electrical pulse time distribution at fixed threshold; green – the total charge (integral of SiPM signal) distribution; pink – the amplitude of SiPM signal distribution.

Figure 7 shows the result of the SiPM output signal oscillograms processing at various energies of laser excitation for 2 μm and 200 μm beam diameters and 35 V and 41 V SiPM bias voltages. Figures 8a and 8b present the dependences of the discharge time fluctuation, i.e. the time resolution (FWHM of the fitting curve), and discharge delay time, respectively, vs. the number of photoelectrons N_{phel} participating in the SiPM cell Geiger discharge initiation. In figure 8a one can see the noticeable decrease of time resolution at low irradiation energies, producing very few photoelectrons. Also, the time resolution for the defocused laser beam of about 10 photons is worse than for focused beam at the same SiPM bias voltage. Figure 8b shows that for the defocused laser beam the SiPM cell discharge happens earlier and takes less time due to the formation of several Geiger avalanches over the irradiated area. The value U_{th} stands for SiPM avalanche breakdown threshold voltage.

We have also investigated the influence of the SiPM bias voltage in the range from 35 – 41 V on the discharge time fluctuation and the delay under femtosecond pulsed laser irradiation. The experimental distributions of output pulse front delay relative to PIN photodiode trigger at various SiPM bias voltages are presented in figure 9 for focused (spot size 2 μm) and defocused (spot size 200 μm) laser beam.

One can see, that with the increase of bias voltage the SiPM cell discharges earlier and the avalanche develops faster. Also, the increase of overvoltage improves SiPM time resolution. Corresponding results of processing data presented in figure 9 are depicted in figure 10.

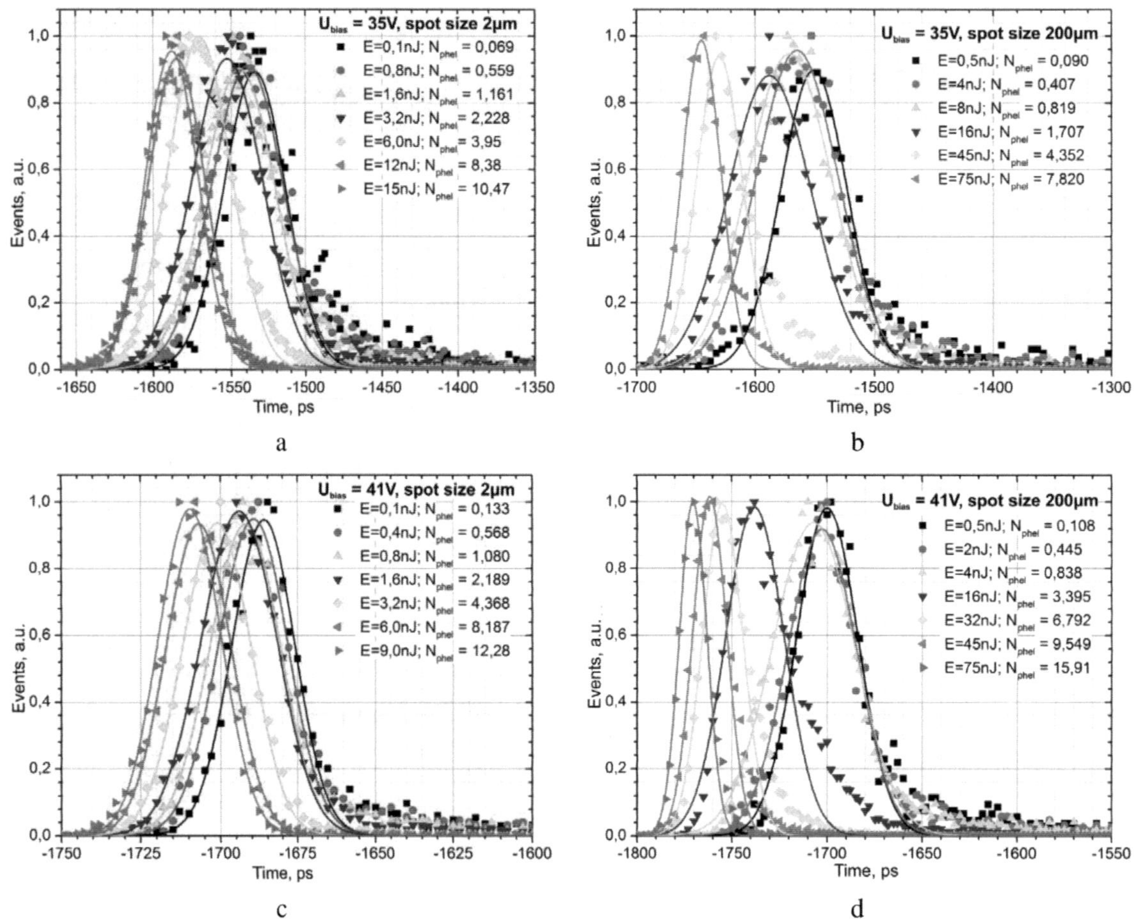

Figure 7. The distributions of output pulse front delay relative to the trigger signal from PIN photodiode for various laser excitation energies: (a) and (c) – beam diameter 2 µm, bias voltage 35 V and 41 V respectively; (b) and (d) – beam diameter 200 µm, bias voltage 35 V and 41 V, respectively. Dots correspond to the experimental data, solid lines – to Gaussian fit.

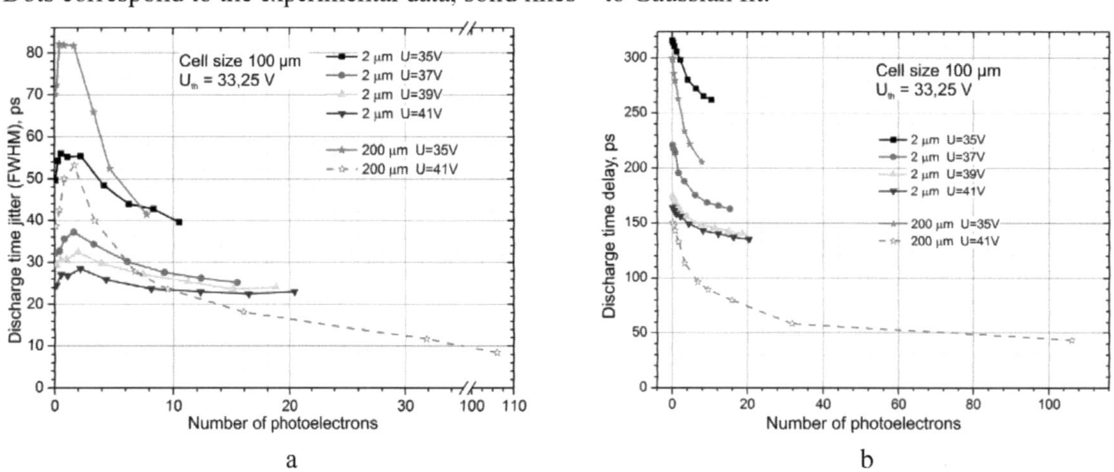

Figure 8. Photoelectrons number dependence of: (a) discharge time fluctuation (time resolution, FWHM of fitting curve); (b) discharge time delay.

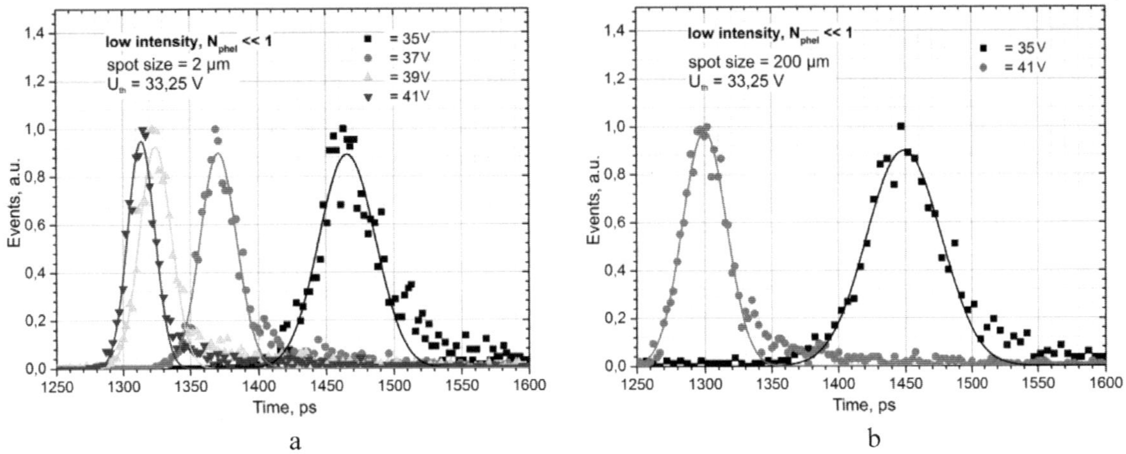

Figure 9. Distribution of output pulse front delay relative to PIN photodiode trigger at various SiPM bias voltages: (a) beam focused to 2 μm; (b) beam defocused to 200 μm.

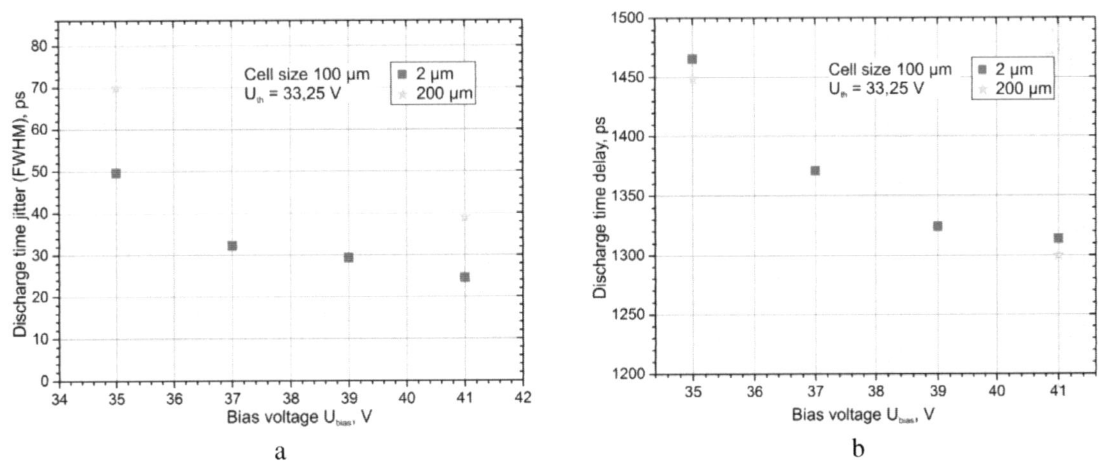

Figure 10. The bias voltage U_{bias} dependence of: (a) discharge time fluctuation; (b) discharge time delay.

3. Conclusion

This work has demonstrated the successful application of femtosecond laser excitation with focused and defocused beam of various intensity for the investigation of timing characteristics of SiPM cells. It was shown, that the usage of the focused laser beam allows to increase the accuracy of SiPM cell time resolution measurements. Such investigations are possible to perform only using the ultrafast light source.

Acknowledgements

The authors recognize and greatly appreciate the help of A.A. Pechenkin and D.V. Savchenkov, participated in the experimental setup creation and automation.

The work of E.V. Popova, P.Zh. Buzhan, A.A. Stifutkin and A.L. Ilyin has been supported by Megagrant 2013 program of Russia Agreement 14.A12.31.0006 from 24.06.2013.

The work of O.B. Mavritskii and A.N. Egorov has been supported by the Government Contract No. 3.1540.2014/K of the Ministry of Education and Science of the Russian Federation.

References

[1] Conti M 2011 *Nucl. Instr. Meth. Phys. Res., Section A* **648** pp S194-S198
[2] Yeom J Y, Vinke R and Levin C S 2013 *Physics in Medicine and Biology* **58 (4)** pp 1207-1220
[3] Marano D. et al. 2014 *IEEE Trans. Nucl. Sci.* **61 (1)** pp 23-34
[4] Popova E. et al. 2013 *IEEE NSS/MIC/RTSD (*Seoul)
[5] Egorov A N, Chumakov A I, Mavritskiy O B, Pechenkin A A, Savchenkov D V, Telets V A, Yanenko A V 2014 *IEEE Radiation Effects Data Workshop (REDW)* pp 247-250

Journal of Physics: Conference Series **737** (2016) 012042 doi:10.1088/1742-6596/737/1/012042

Correction of optic spectra on a absorption substrate

E. Kotlikov and E. Iurkovetc

State university of aerospace instrumentation, Saint-Petersburg, Russia

E-mail: iurkovetc@gmail.com

Abstract. Analysis of possibility of determination optical constants of films by numerical and analytical methods is done. It's shown, that universal method of minimization quality function does not get a unique solution. Methods of consistent approximation is proposed to use for determining optical constants. These methods are: correction spectra for absorption in the substrate; correction spectra of films for its absorption; using additional numerical and analytical methods. Method of correction substrate's absorption for determining spectra of films on substrate without absorption is proposed

Spectra of films are show important information about optic constants of film's material: refraction coefficient $n(\lambda)$ and extinction coefficient $k(\lambda)$ (or absorption coefficient ($\alpha(\lambda)$)). Spectrophotometric methods are widely used. This methods based on calculation optic constants of films from refraction and transmission spectra. Numerical [1] and analytical[2] methods for calculation optic constants on substrate without absorption are known.

The first group involves finding convenient analytical expressions for the direct calculation of optical constants in various particular cases. Examples of this approach are works [3-5]. In these works, values of optical constants are found at the points, which coincide to extremes of transmission and reflection of absorption films.

Let's analyze reflection spectra of films on the substrate without absorption. In this case, reflection coefficient of film is

$$R_{13} = \frac{r_{12}^2 + r_{23}^2 + 2r_{12}r_{23}\cos(-\Delta_{12} + \Delta_{23} - 4\pi n_2 h_2/\lambda)}{1 + r_{12}^2 r_{23}^2 + 2r_{12}r_{23}cos(\Delta_{12} - \Delta_{23} + 4\pi n_2 h_2/\lambda)} \quad (1)$$

Δ_{lm} – phase shift of reflection on the border *lm*. *r* – Frenel's amplitude coefficients of reflection on the borders film (2) – air (1) and film (2) – substrate (3)

$$r_{12} = (n_1 - n_2)/(n_1 + n_2) = (1 - n_2)\ /(1 + n_2) \quad (2)$$

$$r_{23} = (n_2 - n_3)/(n_2 + n_3) \quad (3)$$

Reflection of film R_0, transmission T_0, at absence of absorption R_0+T_0=1.

Let's analyze possibility of finding solution of equation (1) concerning n_2. Let's use calculated spectra of films for analysis. Transmission spectra of films with different refraction coefficients (n_2=1.30-1.50) and the same geometric thickness on the ZnSe substrate $n_3 \approx 2.42$ are shown on the figure 1.

As shown at the figure, big amount of solutions coincide for the same wavelength at the one geometric thickness. We showed graphic with rarefied n spectra. But series of ranges have several solutions. For example, value of refraction coefficient coincides for 3 lines better 0.003 at the wavelength 6.67 μm. It's showed impossibility finding the unique solution of equation (1) concerning refraction coefficient n_2. From this analysis clear, the unique solution is could be not found.

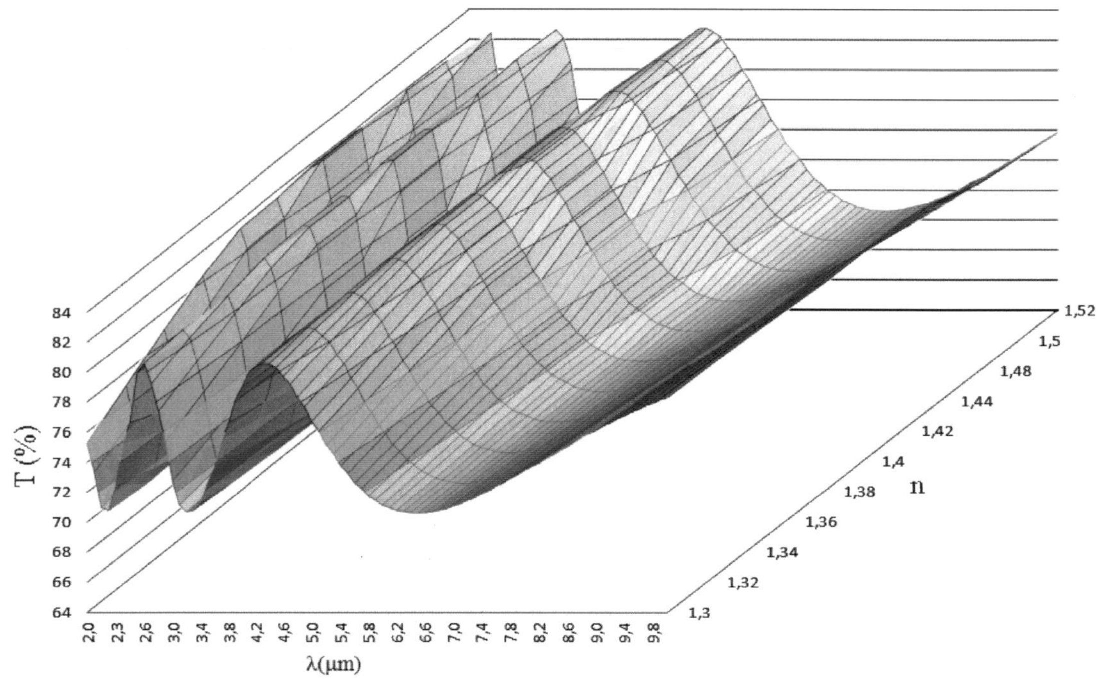

Figure 1. The dependence of the transmittance spectra on the wavelength and refraction coefficient. Solid lines on the 3d graphic are lines of the equal transmittance

The second group is based on different numerical methods. In the work[6] method of determining optical constants is proposed. This method based on solution system of nonlinear equations for energetic coefficients of reflection and transmittance of the system air-substrate at the wavelength λ by minimization of quality function. Quality function F is could be specified for T as:

$$F_T(n_2, h_2, k_2, \lambda_i) = \frac{1}{L}\sum_{i=1}^{L} |T(\lambda_i) - T_0(n_2, h_2, k_2, \lambda_i)|^2 \quad (4)$$

where $T(\lambda_i)$ – experimental value of transmittance coefficient, $T_0(n_2,k_2,\lambda_i)$ – calculated value of transmittance coefficient with absorption, L – amount points of spectrum. This method is universal, but, as rule, finding optical constants has big difficults by reason of absence of the unique solution, because equation (1) is transcendental and has a lot of solutions for the same values reflection and transmittance coefficients.

Third group of methods [7-8] based on correction spectra of films for absorption, which allows to ignore absorption in films. This simplifies determining optical constants. Idea of method is calculating reflection and transmittance spectra for ignoring effect absorption in film. So, obtained spectra depends on only one variable and this simplifies calculating. Besides, in work[2] is proposed to determine thickness of film in range without absorption. Method of determining k is described in works [1]. However, in this case, the unique solution does not exist, as shown at the figure 1. Therefore, we have to do some additional steps. For example, we can set dispersion dependence refraction coefficient on the wavelength. The other method is determining refraction coefficient at ranges without absorption. Determined value of n is used as initial approximation at neighboring ranges of spectrum.

Absorption in the substrate complicates determining optical constants, so method of determining spectra of films on absorption substrate with correction this is proposed. In the beginning, measurement of reflection, transmittance and absorption spectra of the substrate is made. Absorption coefficient α and refraction coefficient n_2 [3] may be determine from these spectra. This correction allows to calculate spectra of films (reflection R_0 and transmittance T_0) on the substrate without absorption. Let's consider this method in detail.

Recurrent method was used for calculating refraction and transmission spectra on absorption substrate. Refraction R and transmission T are calculated by:

$$R = R_0 + \frac{T_0^2 \rho_0 e^{-2\alpha d}}{1-\rho_0 R_0 e^{-2\alpha d}},\ T = \frac{T_0 \tau_0 e^{-\alpha d}}{1-\rho_0 R_0 e^{-2\alpha d}} \tag{5}$$

ρ_0 – refraction, τ_0 – transmission on border film-substrate. R_0 – refraction, T_0 – transmission of film on substrate without absorption, α – absorption coefficient.
Decision of system (1) for R_0 and T_0 is:

$$R_0 = \frac{-B \pm \sqrt{B^2-4AC}}{2A}, \tag{6}$$

$$C = R - \left(\frac{T}{\tau_0}\right)^2 \rho_0,\ B = 1 - R\rho_0 e^{-2\alpha d} + 2\left(\frac{T}{\tau_0}\right)^2 \rho_0^2 e^{-2\alpha d},\ A == \rho_0 e^{-2\alpha d} - 2\left(\frac{T}{\tau_0}\right)^2 \rho_0^3 e^{-4\alpha d}.$$

$$T_0 = \frac{T}{\tau_0} e^{\alpha d} - \frac{T}{\tau_0} R_0 \rho_0 e^{-\alpha d} \tag{7}$$

Obtained equations allow calculate spectra T_0 of films on the substrate, which does not have absorption.

Analysis of possibility using proposed method was done by using model, which includes film on the absorption substrate. As model, approximated values of real silicon plate were used. The following spectra are showed on the figure 2: transmittance spectra on substrate without absorption (line 1); absorption spectra on substrate with absorption (line 2); transmission of substrate (line 3); absorption of substrate (line 4).

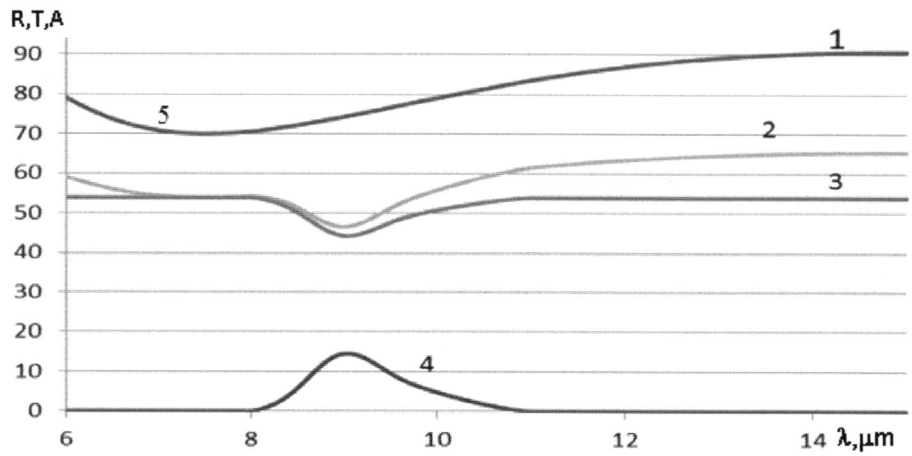

Figure 2. Transmittance and absorption spectra of model

Transmittance spectrum of film, which corrected by equations (6,7) is shown at the same figure (line 5). This spectrum coincides with spectra on substrate without absorption up to 0.001. This demonstrates possibility of correction spectra of films for absorption by this method.

In the future, line 5 is could be used for determining optical constants by one of the described method: analytical or numerical.

Conclusions

Analysis of possibility of determination optical constants of films by numerical and analytical methods is done. It's shown, that universal method of minimization quality function does not get a unique solution. Methods of consistent approximation is proposed to use for determining optical constants. These methods are: correction spectra for absorption in the substrate; correction spectra of films for its absorption; using additional numerical and analytical methods.

For applying this group of methods, reflection and transmittance spectra on the absorption substrate are recalculated for obtaining spectra of films on the substrate without absorption.[]

References

1. Kotlikov E, Varfolomeef G, Lavrovskaya N and Tropin A 2010 *Designing, mahing and studying of interference coating,* SUAI, tutorial, 185
2. Poelman D, Smet P F 2003 *J. Phys. D: Appl. Phys.* **36**. pp 1850–1857.
3. Andrievsiy B, Vakhulovich V, Kurlyak V, Romaniuk N 1988 *Determinatining dispersion of refraction coefficient and thickness of thin films by reflectance and transmittance spectra, V.65, №1, p.136-140*
4. Gao L, Lemarchand F, Lequime M 2011 *Thin Solid Films* **520** pp 501-509
5. Valiev A 1988 *Determining optical constants of weakly absorption dielectric layers on the transparent substrate,* OMP, №8, p.39-41
6. Kotlikov E, Tereschenko G, 1997 *Optics and spectroscopy* **82 (4)** pp 653-659
7. Kotlikov E, Novikova J 2014 *Optics and spectroscopy* **117 (3)** p.381-385
8. Kotlikov E, Kotlikov A, Novikova J, Iurkovetc E 2016 *Optics and spectroscopy*

Journal of Physics: Conference Series **737** (2016) 012043 doi:10.1088/1742-6596/737/1/012043

Analysis of the effect of the interline CCD-sensor dark signal increasing during gamma-irradiation

M E Cherniak[1,2], A V Ulanova[1,2], A Yu Nikiforov[1]

[1]National Research Nuclear University MEPhI (Moscow Engineering Physics Institute), 31 Kashirskoe shosse, 115409 Moscow, Russia
[2]Specialized Electronic Systems (SPELS), 31 Kashirskoe shosse, 115409 Moscow, Russia

E-mail: mecher@spels.ru, avulan@spels.ru, aynik@spels.ru

Abstract. This paper discusses results of the study of a dark signal degradation of visible range interline transfer CCD during gamma-irradiation.

1. Introduction
A spacecraft requires precise information about its location. On board Focal Plane Arrays (FPA), based on CCD-sensors are used for capturing current position relative to the preset star map. Star trackers have to operate in severe radiation environments of protons and electrons that can lead to Total Ionizing Dose (TID) effects [1-12].

2. Device under test
The operation of charge-coupled devices (CCD) is based on the formation, storage and transfer of charge packets in the potential wells formed in the surface layer of the semiconductor by application of external voltage to the electrodes.

The device under test (DUT) was a high-performance multimegapixel image sensor KAI-2093. The key features of this sensor, such as 7.4 μm square pixels with microlenses and the large full well capacity, result in high dynamic range. The split horizontal register offers a choice of single or dual output allowing either 15 or 30 frame per second (fps). The architecture allows either progressive scan or interlaced readout (fig.1).

Journal of Physics: Conference Series **737** (2016) 012043 doi:10.1088/1742-6596/737/1/012043

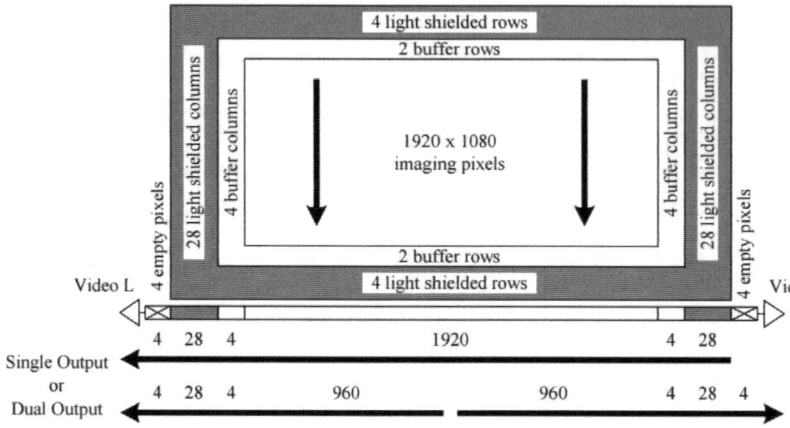

Figure 1. Structure of CCD image sensor KAI-2093

3. DUT parameters and characteristics

The frame capture was carried out between irradiation cycles. Dark current, dark signal non-uniformity, fixed pattern noise were calculated from the received frames. Supply currents were monitored within the test process.

4. Experimental technique

CCD-sensor irradiations were carried out at the gamma irradiation facility Panorama- MEPhI with Co^{60} source (NRNU MEPhI-SPELS) [13].

The device was biased during irradiation.

5. Results and discussion

One of the effects of gamma-irradiation is the increase of the dark signal of the CCD. The main contributor to the resulting dark signal is the surface dark current caused by electron generation at the Si/Si02 interface [14]. Figure 2 shows dark images obtained before and after irradiation.

Figure 2. Dark images before and after irradiation (100 krad[Si])

Due to the thermal generation in the depletion region, photosensitive pixel parts produce electron-hole pairs. Electrons and holes are separated and moved by the space charge region (SCR) field in the p and n region, respectively (Fig. 2).

When exposed to radiation, surface states are formed at the Si/SiO_2 interface, leading to increase of thermal generation due to additional levels in the SCR (Shockley-Read-Hall process). As a result, even in the absence of light, additional "dark" electrons are thermally generated and the resulting increase of the average dark signal is observed.

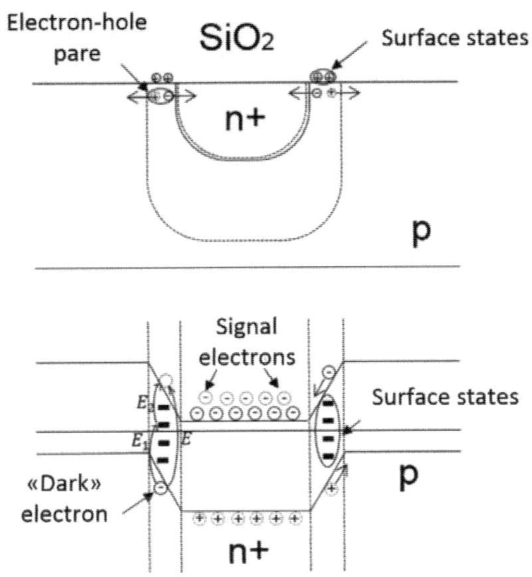

Figure 3. Band diagram of a photosensitive part of a pixel of the CCD

At the 40 krad[Si] TID level nonuniform degradation of a corner columns was observed. Dark signal averaged over all rows is presented in Fig.4. Region with dark reference pixels degraded more than photosensitive area. This effect can be explained by the influence of hydrogen incorporated into surface layers during processing of light shielded area [15].

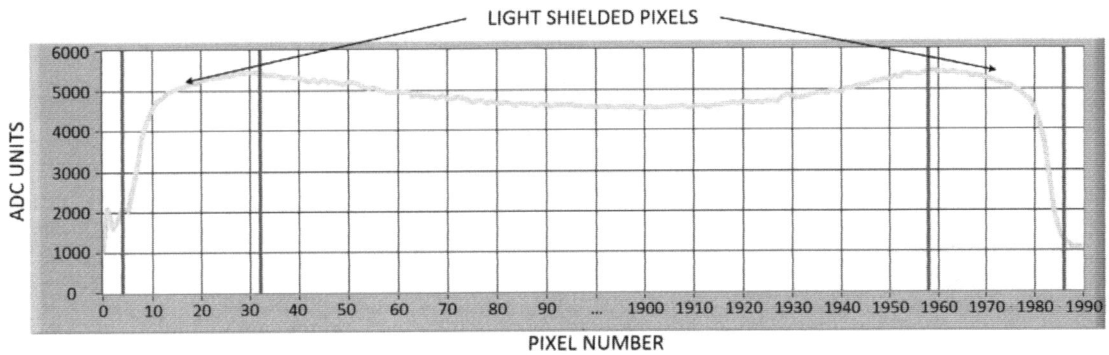

Figure 4. Averaged dark signal after 40 krad[Si]

After 100 krad[Si], dark signal of light sensitive area constitutes one-sixth of the ADC range. Also sequence of columns with different dark signal level was observed (Fig.5). Presumably, this effect is concerned with redistribution of radiation induced charge during transfer of signal charge.

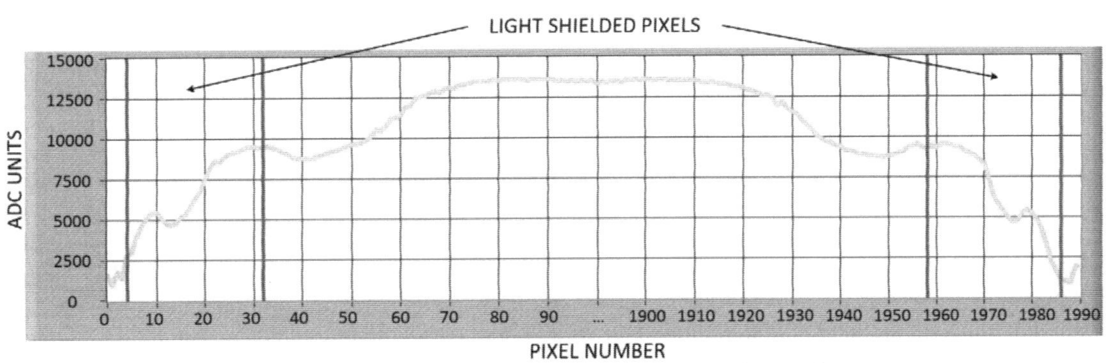

Figure 5. Averaged dark signal after 100 krad[Si]

6. Conclusion

The paper discusses TID degradation of interline transfer CCD image sensor. The main reason of degradation is the formation of surface states at the Si/SiO_2 interface leading to increase of thermal generation. That, in turn, results in the increase of the dark current of CCD image sensor. Nonunifrom degradation of corner regions was observed. It is therefore recommended not to use light shielded area as a dark level reference during image processing.

References

[1] Akhmetov A O, Boychenko D V, Bobrovskiy D V, Chumakov A I, Kalashnikov O A, Nikiforov A Y and Nekrasov P V 2014 *Proc. 29th Int. Conf. on Microelectronics* (*MIEL 2014*) (Belgrade, Serbia) pp 329-331

[2] Gilard O, Boatella-Polo C, Dolado-Perez J-C, Auvergne M, Quadri G, Boutillier M 2010 *IEEE Trans. Nucl. Sci.* **57(3)**

[3] Cherniak M E, Ulanova A V and Nikiforov A Y 2014 *Proc. 29th Int. Conf. on Microelectronics* (*MIEL 2014*) (Belgrade, Serbia) pp 195-198

[4] Skorobogatov P K., Sogoyan A V, Davydov G G, Egorov A N and Savchenkov D V 2015 *IEEE Trans. Nucl. Sci.* **62 (1)** pp 195-201

[5] Hardy T D, Deen M J and Murowinski R 1999 *Adv. Imaging and Electron Phys.* **106** pp 1-96

[6] Boychenko D V, Kalashnikov O A., Karakozov A B and Nikiforov A Y. 2015 *Russian Microelectronics* **44 (1)** pp 1-7

[7] Schwank J R *2002 Total-Dose Effects in MOS Device* (*IEEE NSREC Short Course Notes)* pp III-1-III-123

[8] Karakozov A B., Korneev O V., Nekrasov P V., Sokolov M N. and Zagryadsky D A 2013 *Proc. 14 th European Conf. on Radiation and its Effects on Components and Systems (RADECS-2013)* (Oxford; United Kingdom) article number 6937407

[9] Boudenot J C. and Augier P. 1990 *Proc ESA Electronic Components Conference (ESTEC)* (ESA SP-313) pp 319-324

[10] Boruzdina A B, Orlov A A, Ulanova A V, Grigor'ev N G and Nikiforov A Y 2015 *Proc.2015 International Siberian Conference on Control and Communications* (SIBCON 2015) (Omsk; Russian Federation) article number 7147007

[11] Sogoyan A V and Davydov G G 2011 *Russian Microelectronics* **40 (3)** pp 194-208

[12] Martin E, Nuns T, David J-P, Gilard O, Boutillier M and Penquer A 2011 *IEEE Trans. Nucl. Sci.* **58 (3)**

[13] Artamonov A S, Sangalov A A, Nikiforov A Y, Telets V A and Boychenko D V 2014 *IEEE Radiation Effects Data Workshop (REDW 2014)* (Paris, France) pp 258-261

[14] Li F, Nathan A 2005 *Image Sensors in Deep-Ultraviolet: Degradation Behavior and Damage Mechanisms* (Berlin-Heidelberg: Springer-Verlag,)

[15] Hopkinson G R 1993 *Proc. RADECS 93* (St Malo, France) pp 401-408

Journal of Physics: Conference Series **737** (2016) 012044 doi:10.1088/1742-6596/737/1/012044

Application of surface plasmon resonance method for visualization of processes in a near-wall liquid layer

I N Pavlov, B S Rinkevichyus and A V Vedyashkina

V.A. Fabrikant Physics Department, National Research University "MPEI", Krasnokazarmennaya st., 14, Moscow, Russia

E-mail: inpavlov@bk.ru

Abstract. The surface plasmon resonance method for investigation of different processes in near-wall liquid layers is described in the paper. Possibilities of application of this method for visualization and quantitative diagnostics (measurement of refractive index distribution in a near-wall layer) of evaporation, cooling, heating, crystallization, mixing of droplets are shown. Influence of the experimental parameters on sensitivity to the refractive index changing and the measurement range of this method were investigated. The scheme of experimental setup and obtained by it experimental results of visualization of mixing and crystallization of droplets are given.

1. Introduction

Investigation of physical processes in a near-wall liquid layer has fundamental and practical importance in many areas of science and technology. It plays an important role in the processes of heat and mass transfer, mixing and in the biological and chemical reactions. As is known, properties of a substance on surface (in thin boundary layer) and in volume can vary widely [1]. Therefore, the nature and results of processes in these layers and in the volume of medium can also be different. Thus, it is necessary to develop research methods to study the dynamics of processes in a thin boundary layer. There are several such methods (see, e.g., [2]), and the number of scientific and technical conferences separates special sections for discussion of it. Among these methods there is the method of frustrated total internal reflection of a wide collimated laser beam which was developed in dissertation [3]. It allows to measure the changing of refractive index in thin boundary layer with thickness of few hundred nanometers in the interface between the liquid and the glass prism. The method was successfully used for investigation of near-wall flows of inhomogeneous liquid in a cell [4], in flat and T-shaped microchannels [5], for investigation of evaporation, cooling, crystallization, spreading, mixing of droplets on a horizontal substrate [6-8]. A logical extension of research in this area is the use of the surface plasmon resonance method, described in detail below, to the specified range of tasks, as well as the expansion of possible areas of its application.

2. Principles of surface plasmon resonance method

The phenomenon of surface plasmon resonance (SPR) in optical wavelength range, appear to have been first described by Otto [9], and then Kretschmann [10]. They proposed two schemes excitation of surface plasmons – waves of variable density of free electrons in thin metal films – via total internal reflection (TIR). Formed over the glass-metal interface evanescent wave 6 [11] (Figure 1 [12]) penetrates through thin metal layer 4 and on the metal-dielectric interface (4-5) thereby causing excitation of surface plasmons 7. The efficiency of this excitation depends on the thickness of the metal film 4. By selecting of radiation incidence angle θ on the glass-metal interface, it is possible to reach the resonance in which a large part or even all of the energy of incident radiation will be spent on the excitation of surface plasmons, and intensity of the reflected light will tend to zero, despite the implementation of TIR condition.

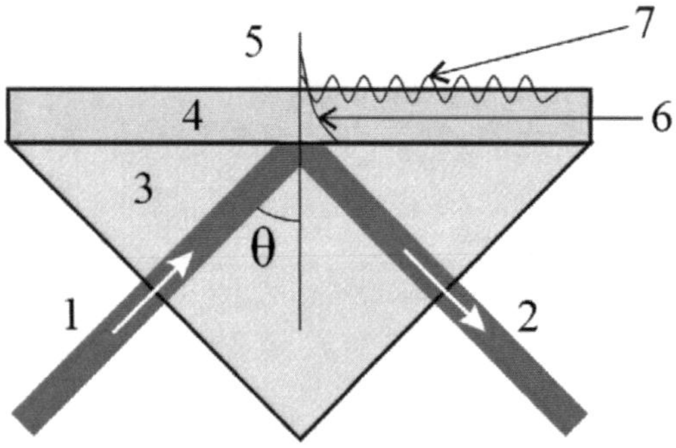

Figure 1. Kretschmann scheme for the excitation of surface plasmons via TIR of light: 1 – incident light beam, 2 – reflected light beam, 3 – glass prism, 4 – metal film, 5 – investigated medium (dielectric), 6 – evanescent wave, 7 – surface plasmons.

Figure 2 [13] shows how the reflectance $\rho_{\|}$ of the incident light beam depends on the angle of incidence θ in case when the investigated medium 5 is distilled water and wavelength of incident light equal to 650 nm.

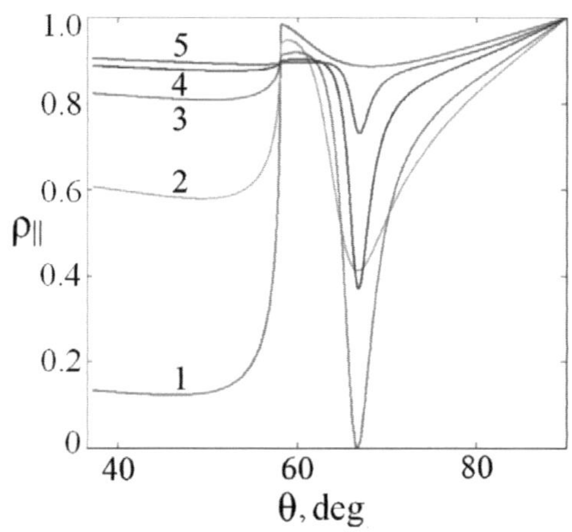

Figure 2. The graphs of the reflectance dependence on the angle of incidence on a glass-metal interface in when metal contacts with water for different thicknesses of a gold film: 1 – 10 nm; 2 – 30 nm; 3 – 50 nm; 4 – 70 nm; 5 – 90 nm.

It was calculated by the equations given in [14] obtained from the well-established Fresnel theory:

$$\rho_{\|} = \left| \frac{r_{12} + r_{23}\exp(2ik_{z2}d_2)}{1 + r_{12}r_{23}\exp(2ik_{z2}d_2)} \right|^2, \tag{1}$$

$$r_{12} = \frac{(\varepsilon_2 / k_{z2}) - (\varepsilon_1 / k_{z1})}{(\varepsilon_2 / k_{z2}) + (\varepsilon_1 / k_{z1})}, \quad r_{23} = \frac{(\varepsilon_3 / k_{z3}) - (\varepsilon_2 / k_{z2})}{(\varepsilon_3 / k_{z3}) + (\varepsilon_2 / k_{z2})}, \tag{2}$$

$$k_{zi} = \frac{\omega}{c}\left(\varepsilon_i - \frac{1}{\varepsilon_1 \sin^2\theta} \right)^{1/2}, \tag{3}$$

where $\rho_{\|}$ is the reflectance for p-polarized light, subscripts i = 1, 2 and 3 refer to glass of prism, metal film and investigated medium, respectively, d_2 is the thickness of metal film, k is the wave number which equal to the magnitude of the wave vector in the corresponding medium, ε is the dielectric constant which related to the refractive index by ratio $\varepsilon = n^2$, ω is the angular frequency, and c is the speed of light in vacuum.

It can be seen from Figure 2 that SPR angle (in this case, it is 66.86°) does not depend on the film thickness. It is mainly determined by the ratio of the refractive indices of the glass prism, metal film and investigated medium. Also Figure 2 shows that there is a certain optimum value of the film thickness (different for different metals) which corresponds to zero reflectance under SPR condition.

According to the above formulas (1)-(3) can also plotted the reflectance from the refractive index of investigated medium in defined conditions which shown in Figure 3.

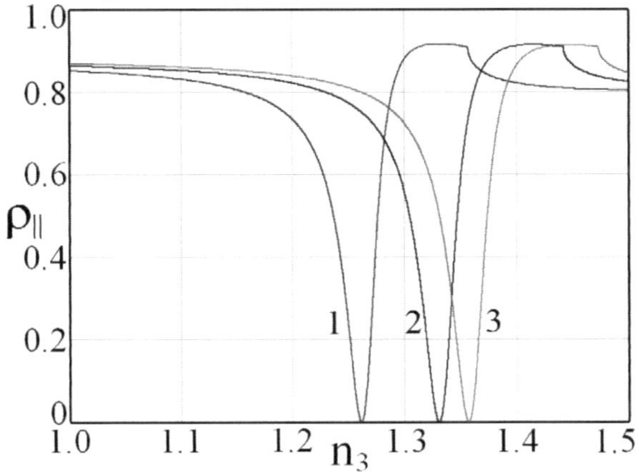

Figure 3. The graphs of the reflectance dependence on refractive index of investigated medium for wavelength 650 nm, thickness of a gold film 50 nm and three values of incidence angle: 1 – 60°; 2 – 67°; 3 – 70°

Figure 3 allows us to estimate the measurement range of the refractive index, as well as makes it possible to see that at a constant angle of incidence the reflectance is strongly dependent on the refractive index in the resonance area, and the highest sensitivity is achieved on the steep linear sections of curves.

Also it can be seen that SPR method has good sensitivity to a change in the experimental parameters, such as the angle of incidence or properties of investigated medium in contact with the metal film, therefore it allows to determine sufficiently small deviations of these parameters from the resonant values. Most often in the experiments used modification of this method in which the curve of dependence of reflectance on angle of incidence is measured, for example, by scanning [15]. This makes it possible to determine the refractive index of the investigated medium with an accuracy up to 10^{-6}, and in the case of interference SPR method – to 10^{-8} [16, 17]. Generally, this technique is used in biomedical research, which is required to determine the binding constants of antibodies to the antigen [18]. In this case, antigen molecules covalently attached to the surface of metal film, and then the interactions of various concentrations of antibody solutions to the surface is investigated.

Paper [14] describes the possibility of using the SPR method for investigation of properties of liquids, in particular, in processes of mixing, diffusion, and convection, as well as measuring of temperature distribution in the boundary layer of liquid and visualization of self-assembly of nanoparticles from solutions. It is shown that the spatial resolution which the method provides in plane of media interface is several micrometers, and in the perpendicular direction it may be up to one nanometer. The relative error of measurement of concentration, salinity or temperature is usually within 5%.

The authors of this paper proposed to use the SPR method for visualization of dynamics and quantitative diagnosis of heat and mass transfer and phase transitions (evaporation and crystallization) in liquid droplets. For this purpose, the modification of the SPR method was developed in which the wide collimated laser beam is used for excitation of SPR. In this case, you need not to measure the spectrum of the reflected radiation or dependence of reflection coefficient on angle of incidence, but you need to measure the intensity distribution over the cross-section of the reflected beam. This makes it possible to determine not only the refractive index distribution over the near-wall layer, but also visualize the size and shape of inhomogeneities associated with changes in physical parameters, and to visualize the dynamics of the processes. To this end the experimental setup was designed and constructed which described below.

3. Experimental setup

The scheme of the setup is shown in Figure 4. Principle of its operation is based on the excitation of SPR by Kretschmann's method with the help of a wide collimated laser beam.

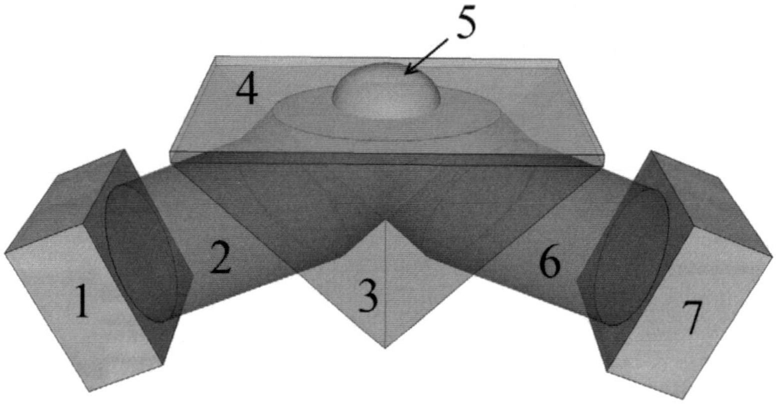

Figure 4. Experimental setup: 1 – system of the probe beam formation, 2 – incident beam, 3 – glass prism, 4 – plate with a gold film, 5 – investigated droplet, 6 – reflected beam, 7 – receiving system.

System of the probe beam formation 1 consists of a radiation source – compact laser with wavelength in the visible range, a beam expander of telescope type with a spatial filter in the form of a pinhole and a polarizer, which allocates emission polarization component parallel to the plane of its incidence on the boundary between two media. The experimental setup also comprises a prism 3, whereby TIR conditions are created, glass plate 4 on the outer side of which there is the gold film 50 nm thick and a receiving system 7 for image registration of reflected beam, which consists of a screen, a digital video camera and a computer.

A droplet of test liquid 5 is placed on the surface of the gold film, illuminated from below by wide collimated beam of laser radiation. The angle of beam incidence is chosen so that the SPR condition on the boundary film-liquid holds. When this is seen in the reflected beam the dark image of droplet on a light background of the rest of the beam. When changing the refractive index of the droplets boundary layer that can occur under the influence of any physical processes (for example, cooling, crystallization, mixing, etc.) corresponding reflection coefficient changes. Thus, by measuring the change of the intensity at a particular area of the reflected beam, it is possible to determine change of the refractive index in the corresponding near-wall layer area of liquid. Since the resulting image on screen of the reflected beam is recorded by the digital video camera, then we can trace the dynamics of change in the refractive index distribution of liquid in the boundary layer, and thus the dynamics of the process, which it is due.

4. Experimental results

Experiments on visualization of mixing, cooling and crystallization of droplets of different liquids were carry out by experimental setup which was assembled according to above scheme. Figure 5 shows a sequence of images obtained at visualization of mixing of glycerol and water droplets for different points in time.

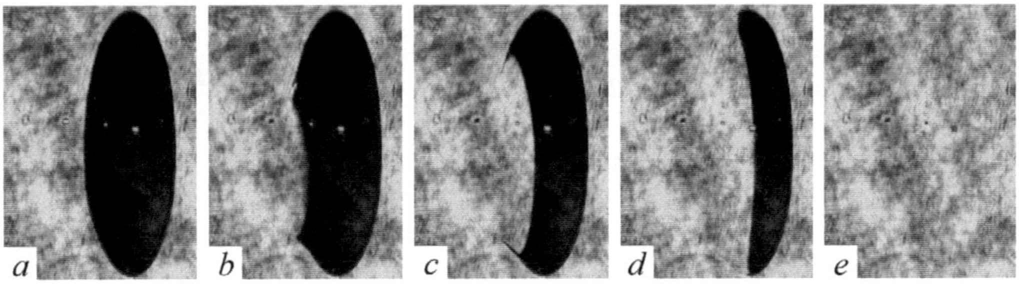

Figure 5. Experimental results on visualization of mixing of water and glycerol droplets for the different moments of time after placing a glycerol droplet on the substrate: a – 0 sec, b – 1 sec, c – 2 sec, d – 3 sec, e – 20 sec.

From the first image in Figure 5 it is possible to see the dark image of the water droplet, as the SPR angle set for this drop. The glycerol droplet is invisible in these images, as the SPR condition is not hold for it and it hardly changes the reflection coefficient (at least, it is imperceptible with the naked eye). As far as mixing of droplets the refractive index of mixture takes an intermediate value between the refractive index of water and glycerol, the mixture is also invisible on the above images. Therefore, it is possible to see how invisible drop absorbs dark drop, though, of course, in fact both of them are transparent. It is possible to select such experimental conditions in which both liquids change reflection coefficient, but in different ways, and therefore, both drops look different colors of gray. This case is more informative in terms of obtaining the quantitative characteristics of the experimental process, but less obvious, therefore, it is not given in this paper.

Figure 6 shows the examples of images obtained by experimental visualization of cooling and subsequent crystallization of distilled water droplet of 10 microliters. For carrying out of this experiment setup was equipped with a Peltier element, the cooling surface of it is contacted with the test droplet. Thus the droplet was sandwiched between the prism surface with the gold film and the Peltier element surface, forming a meniscus.

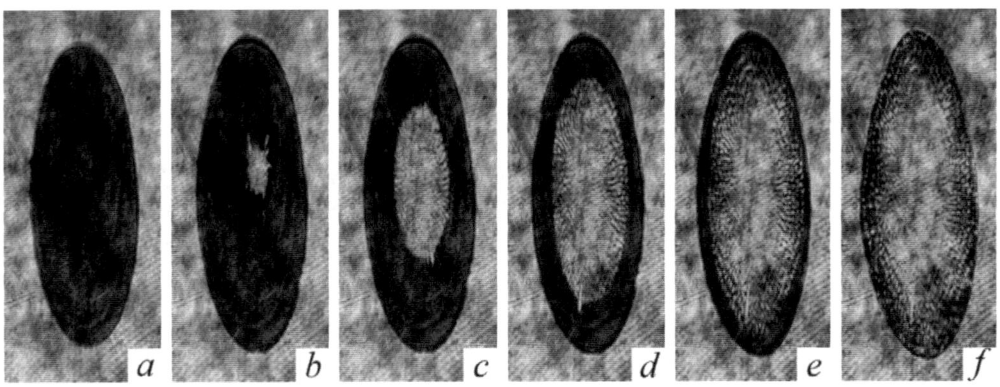

Figure 6. Experimental results on visualization of crystallization of a water droplet for the different moments of time after cooling start: a – 2.0 sec, b – 9.7 sec, c – 9.9 sec, d – 10.5 sec, e – 12.1 sec, f – 18.5 sec.

Figure 6 shows that the dark image of drop becomes brighter as it cools, and after crystallization is almost merges with the background color. Inhomogeneities in the image of crystallized drops are explained by the presence of air bubbles in the boundary layer structure which obtained by the process of crystallization. Moreover, this structure is different for different values of the gap between the prism surface and the Peltier element surface.

5. Conclusion

From the literature review in this paper it is possible to conclude that the SPR method is actively developed and used for research and diagnostic of various parameters of condensed media. The most widespread use it has received in biomedical researches, which are required to determine parameters of certain chemical and biological processes accompanied by changes in the refractive index of the medium under investigation. The authors believe that development and application of promising SPR method to research of heat and mass transfer and phase transitions will provide new original and useful scientific results in terms of fundamental knowledge and from the point of view of applied sciences.

Acknowledgments

This work was partially supported by the Russian Foundation for Basic Research (grants No. 14-08-00948-a, № 16-32-00530-mol_a) and Russian Federation President scholarships to young scientists and graduate students engaged in advanced research and development in priority areas of modernization of the Russian economy on the 2015-2017 years.

References

[1] Roldugin V I 2011 *Physical chemistry of surfaces* (Dolgoprudniy: Intellekt) p 564
[2] Repik E U and Sosedko Yu P 2007 *Turbulent boundary layer* (Moscow: PHYSMATLIT) p 312
[3] Pavlov I N 2013 *Opto-electronic complex for visualization of physical processes in near-wall layer of liquid* (Moscow: MPEI) Abstract of dissertation for the PhD degree
[4] Pavlov I N and Rinkevichyus B S 2009 *Optical Memory and Neural Networks (Information Optics)* **18(4)** pp 322-327
[5] Pavlov I N, Rinkevichyus B S and Tolkachev A V 2011 *Measurement Techniques* **53(10)** pp 1130-1134
[6] Pavlov I N, Rinkevichyus B S and Tolkachev A V 2013 *Instruments and Experimental Techniques* **56(2)** pp 242-246
[7] Pavlov I N, Rinkevichyus B S and Tolkachev A V 2013 *Measurement Techniques* **56(3)** pp 271-274
[8] Pavlov I N, Rinkevichyus B S and Tolkachev A V 2014 *Scientific Visualization* **6(3)** pp 1-13
[9] Otto A 1968 *Zeitschrift für Physik A Hadrons and Nuclei* **216(4)** pp 398-410
[10] Kretschmann E 1971 *Zeitschrift fur Physik* **241(4)** pp 313-324
[11] Novotny L and Hecht B 2006 *Principles of Nano-Optics* (Cambridge) 541 p
[12] Pavlov I N, Rinkevichyus B S, Tarasov A E and Tolkachev A V 2015 *Proc. of 13th Int. Sci. and Tech. Conf. "Optical Methods of Flow Investigations" (*Moscow*)* pp 281-287
[13] Pavlov I N, Rinkevichyus B S, Tolkachev A V and Vedyashkina A V 2016 *Proc. of V Int. Conf. "Photonics and Information Optics" (*Moscow: MEPhI Press*)* pp 169-170
[14] Kihm K D 2010 *Experiments in fluids* **48(4)** 547-564
[15] Wong C L and Olivio M 2014 *Plasmonics* **9** 809-824
[16] Kabashin A V and Nikitin P I 1997 *Quantum Electronics* **27(7)** 653-654
[17] Kashif M, Bakar A A, Arsad N and Shaari S 2014 *Sensors* **14** 15914-15938
[18] Homola J Surface 2008 *Chem. Rev.* **108** 462-493

Static and dynamic characteristics of angular velocity and acceleration transducers based on optical tunneling effect

V I Busurin, V V Korobkov, Naing Htoo Lwin, Phan Anh Tuan

Moscow Aviation Institute (National Research University), Volokolamskoe Road, 4, Moscow, A-80, GPS-3, 125993, Russia

e-mail: vbusurin@mai.ru

Abstract. Theoretical and experimental analysis of quasi-linear conversion function of angular velocity and acceleration microoptoelectromechnical (MOEM) transducers based on optical tunneling effect (OTE) are conducted. Equivalent oscillating circuit is developed and dynamic characteristics of angular velocity and acceleration MOEM–transducers are investigated.

1. Introduction

Angular velocity and acceleration transducer, constructed with structure «optical components of total internal reflection (TIF) – medium – sensing element (SE)» are suggested to apply for improving effectiveness of navigation and mobile objects control system with satisfied modern requirements [1-2]. Thus optical power, reflected from the structure «TIF – medium –SE», is received by photodetector and supports information about measuring result [3].

While developing such MOEM–transducers it is needed the complete description of actual conversion function of the measuring value (angular velocity and acceleration) in the change of optical output power in static and dynamic modes. The validity of the derived characteristics of MOEM–transducers should be based on the comparison results of theoretical calculations and experimental studies.

2. Correction of static characteristic of angular velocity and acceleration MOEM–transducers based on OTE according to experimental studies results

The operating principle of angular velocity and acceleration MOEM–transducers is based on dependence of optical source reflective coefficient on the value of operating gap between SE and TIF base. Therefore, the reflected part of optical power, reached to photodetector, is an informative value [4].

Experimental studies by using semi-natural setup has been conducted to verify the conversion function of MOEM–transducers based on OTE and determine the correction type of theoretical calculations (figure 1).

Semi–natural setup is comprised of FOT–930 Multifunction Loss Tester *1*, fiber optic cables *2*, TIF – prism made up of quartz *3*, piezoceramic SE in form of round plate *4*, load transferring ball *5*, aluminum beam *7* with hanging cable *6*. FOT-903 is used as optical source in generation mode with wavelengths λ_1=1.3μm and λ_2=0.85μm while supplying constant optical power source P_{OS}=2.045μW. Beam 7, fixed at one end, can change load F_H by changing mass m_H *8* and its hanging position. This varying load F_H, acting on SE center, leads to change of operating gap *d* between SE and TIF base.

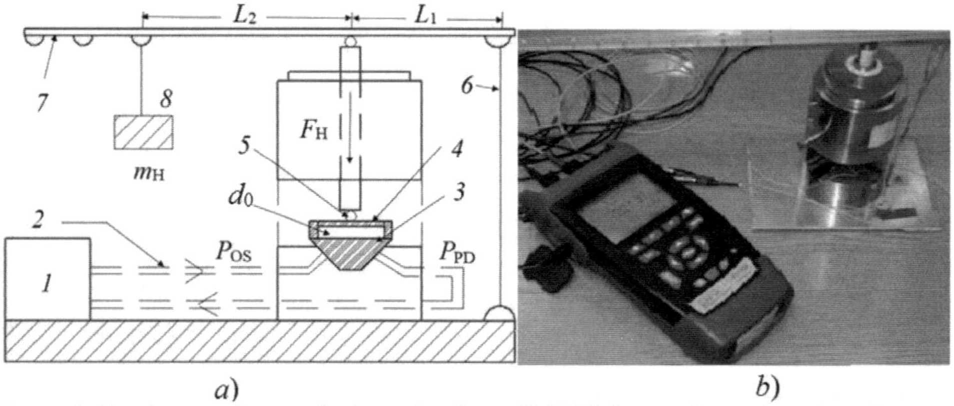

Figure 1. Semi–natural setup for investigation of MOEM–transducers based on OTE: *a* – structure, *b* – external view

From optical source *1*, signal enters to the structure «TIF–medium–SE» via optical fiber cables *2*. While reducing operating gap *d* from the influence of load F_H most of the optical energy penetrates to SE and is absorbed, which leads to decrease the optical input signal of photodetector P_{PD}.

While neglecting the mass of beam *7* (light compared to the suspended mass *8*) load F_H acting on the SE of MOEM–transducer is calculated by formula:

$$F_H = m_H \cdot g \cdot \left(\frac{L_1 + L_2}{L_1} \right). \tag{1}$$

SE of those MOEM–transducers based on OTE is clamped on the perimeter of circular end, and the largest deflection is observed at its center. Under the influence of the central load F_H magnitude of SE deflection $y_F(F_H)$ is function of parameters $y_F(F_H)=f\ (R_{SE}, h_{SE}, v, E_{SE})$ [5],where R_{SE} , h_{SE} – radius and thickness of SE; v – Poisson's ratio; E_{SE} – Young's modulus of SE material.

Magnitude of operating gap $d(F_H)$ between SE and TIF base is defined by initial gap d_0 and deflection of SE $y_F(F_H)$:

$$d_F(F_H) = d_0 - y_F(F_H). \tag{2}$$

The output optical power with the conversion function of such MOEM–transducer based on OTE in account of optical loss is determined by the following formula [6]:

$$P_{PD}(F_H) = P_{OS} \cdot k_{opt} \cdot R\left[d_F(F_H), \theta, \lambda\right], \tag{3}$$

where k_{opt} – optical loss; λ – wavelength of optical source; θ – angle of incidence, $R[d_F(F_H),\theta,\lambda]$ – reflectivity of structure «TIF – medium –SE».

Experimental studies of conversion function is conducted with following parameters: R_{SE} = 5mm, h_{SE} = 1mm, v = 0.17, E_{SE} = 72.5GPa (figure 2).

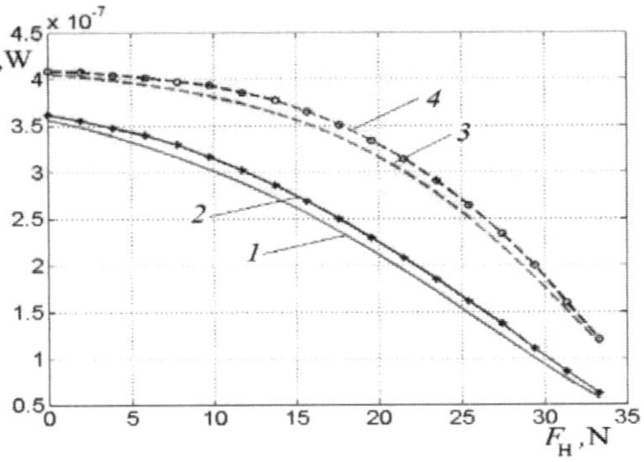

Figure 2. Dependence of conversion function of MOEM–transducer based on OTE on load F_H: *1, 3* – theoretical calculation results at λ_1=1.3μm, λ_2=0.85 μm; *2, 4* – experimental results at λ_1=1.3μm, λ_2=0.85 μm accordingly.

From comparing of theoretical calculation and experimental results it should follow that the analytical formula to calculate the conversion function can support to be quarsilinear and can be considered as acceptable for dependent description of output signal on the external influence to MOEM–transducer based on OTE, while ensuring not more than 15% of error.

To improve the accuracy of MOEM–transducer model is necessary to select the correction coefficient of theoretical calculations based on the experimental results, e.g., least squares method. The correction coefficient of conversion function must be satisfied in the condition with n experiments:

$$\sum_{i=1}^{n}\left[P_{\text{PD EXP}\,i}(F_{\text{H}\,i})-K_{\text{CORR}}(F_{\text{H}\,i})\cdot P_{\text{PD THEO}\,i}(F_{\text{H}\,i})\right]^2\rightarrow\min\,, \quad (4)$$

where $P_{\text{PD EXP}\,i}(F_{\text{H}\,i})$, $P_{\text{PD THEO}\,i}(F_{\text{H}\,i})$ – magnitude of optical power by means of experiments and theoretical calculation, corresponding value of load $F_{\text{H}\,i}$; $K_{\text{CORR}}(F_{\text{H}\,i})$ – correction coefficient of conversion function, $K_{\text{CORR}}(F_{\text{H}\,i})=f(F_{\text{H}\,i},f_1,f_2,\ldots,f_{k+1})$, $f_1,f_2,\ldots,f_{k+1}$ – variables of polynomial coefficient in k times.

Theoretical calculation of conversion function uses correction coefficient, described polynomial with 3[rd]order: $K_{\text{CORR}}(F_{\text{H}}) = (1.0137+0.00424\cdot F_{\text{H}}-0.00017\cdot F_{\text{H}}^2+0.00001\cdot F_{\text{H}}^3)\cdot 10^{-8}$ to provide the error not more than 2%.

3. Dynamic behaviors of angular velocity and acceleration MOEM–transducers based on OTE

Block diagram is composed to study the dynamic mode of angular velocity and acceleration MOEM–transducers based on OTE (figure 3.), which includes: a movable part - SE, which converts the measuring value Ψ (angular velocity Ω_z or acceleration a_z) according to varying of operating gap d_Ψ; modulator (M), which converts optical power source P_{OS} to output optical power of photodetector P_{PD}; converter "current-voltage" (I-U), which converts current I_{PD} into voltage, supplied to low pass filter (LPF) and generate output voltage; У- amplifier in feedback circuit (OC); PA - piezoactuator that converts the voltage U_{OC} with movement of d_{U}.

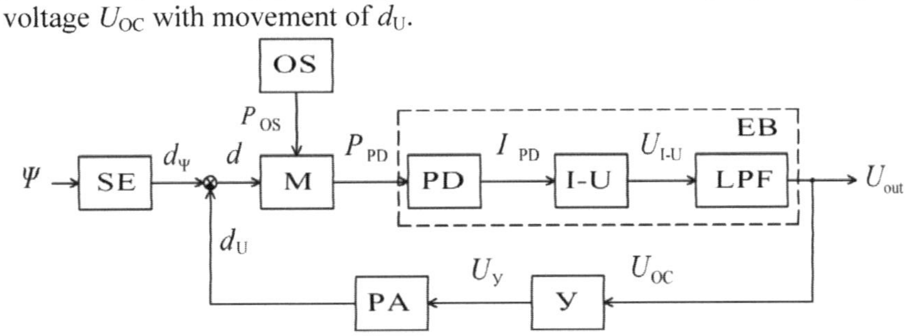

Figure 3. Block diagram of angular velocity and acceleration MOEM–transducers based on OTE with feedback circuit

Using feedback circuit can reduce the transfer ratio of conversion, which allows to increase the maximum measuring value to desired measuring range by adjusting operating gap of transducers [7]. In contravention of this part sensitivity and linearity of the MOEM–transducers conversion function can be decreased.

According to the block diagram of MOEM–transducers a transfer function $W_\Sigma(p)$ in form of operators is:

$$W_\Sigma(p) = W_{SE}(p) \cdot \frac{W_M(p) \cdot W_{EB}(p)}{1 + W_M(p) \cdot W_{EB}(p) \cdot W_Y(p) \cdot W_{PA}(p)}, \quad (5)$$

where $W_{SE}(p)$ – transfer function of SE; $W_M(p)=P_{OS}\cdot k_M(p)$ – transfer function of M; $W_{EB}(p)=W_{PD}(p)\cdot W_{I\text{-}U}(p)\cdot W_F(p)$ – transfer function of electronic block (EB); $W_{PD}(p)=k_{PD}/(T_{PD}\cdot s+1)$, $W_{I\text{-}U}(p)=k_{I\text{-}U}/(T_{PD}\cdot s+1)$, $W_F(p)=k_F\cdot \omega_F/(s+\omega_F)$, $W_Y(p)=k_Y/(T_Y\cdot s+1)$, $W_{PA}(p)=k_{PA}/(T_{PA}\cdot s+1)$ – transfer function of PD, I-U, LPF, Y, PA respectively; $k_M(p)$ – transfer ratio of M [1/m]; T_{PD}, k_{PD} – time constant, transfer ratio of PD[A/W]; $T_{I\text{-}U}$, $k_{I\text{-}U}$ – time constant, transfer ratio of I-U [Ohms]; $k_F= 1/[1+(\omega/\omega_F)^2]$, ω_F – transfer ratio, frequency pole of LPF; T_Y, $k_Y=R_2/R_1$ – time constant, gain of Y; T_{PA}, k_{PA} – time constant, transfer ratio of PA[m/V].

SE of MOEM–transducers based on OTE can be considered as a mechanical oscillating system which can be described with second order differential equation [8]:

$$J \cdot \frac{d^2 y}{dt^2} + K_{GD} \cdot \frac{dy}{dt} + G_{stiff} \cdot y = Q(t), \quad (6)$$

where J, G_{stiff} – inertial magnitude and stiffness of SE, K_{GD} – coefficient of gas dynamic damping due to viscous friction of the medium of structure «TIF – medium –SE», y – oscillating velocity.

Note that an arbitrary external factor $Q(t)$, acting on the SE of MOEM–transducers, can be expanded in Fourier series [9]:

$$Q(t) = \frac{1}{2} a_0 + \sum_{k=1}^{+\infty} A_k \cos(k\omega t + \theta_k) \ . \quad (7)$$

where A_k, $k\omega$, θ_k – amplitude, circular frequency, initial phase k times harmonics oscillation, a_0 – coefficient of Fourier series.

Let us assume that the investigated system is satisfied by filter condition, which do not effect with higher harmonics significantly the basic, i.e. can be neglected. Thus, the equation (7) can be written as:

$$Q(t) = \frac{1}{2} a_0 + A_1 \cos(\omega t + \theta_1) \, . \quad (8)$$

By following equation (8), the external factor $Q(t)$ according to the principle of superposition can be separately considered as a constant and harmonious action. Then, the external factor $Q(t)$, acting on the acceleration MOEM–transducer can be assumed constant, and for angular velocity MOEM–transducer - harmonic.

Angular velocity and acceleration MOEM–transducers are acoustomechanical oscillating system with distributed parameters, consisting of SE and the air gap between SE and prism base. To analyze the dynamic characteristics of MOEM–transducers, which are defined by properties of acoustomechanical oscillatory system, conveniently and efficiently by using well known methods of electrical engineering and circuit theory. The criteria of equivalence are governed by the laws of energy conservation. In addition, each factor of the differential equation (6) is replaced by combination of equivalent elements of electrical oscillating RLC–circuit, in which the value of inertial J and, gas dynamic damping coefficient K_{GD} and stiffness of SE G_{stiff} are corresponded to inductive L_E, capacitive C_E and active electrical resistance R_E.

Equivalence, acting on external factor of SE corresponds to EMF $U_E(t)$, and oscillating velocity of system y - current $I_E(t)$ (figure 4).

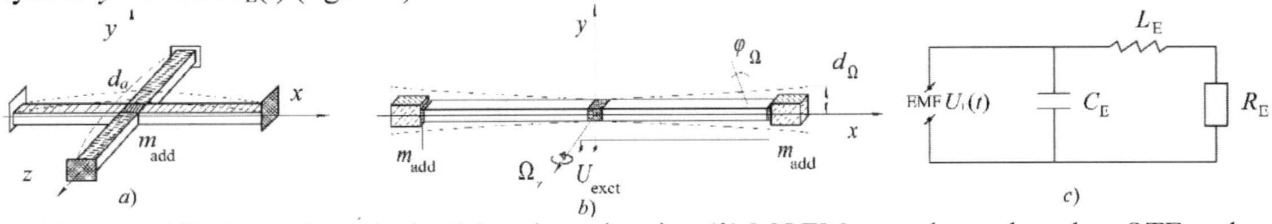

Figure 4. SE of angular velocity (*a*) and acceleration (*b*) MOEM–transducers based on OTE and equivalent schematic (*c*) of oscillating RLC-circuit

Under the influence of acceleration a_y acceleration MOEM–transducer based on OTE (Figure 4*a*) is deflected along the *OY* axis [2], which is corresponding to oscillating RLC-circuit (Figure 4*c*) with equivalent parameters: $U_E(t)=U_a=(m_a+m_{add})\cdot a_y$; $L_E=b_a\cdot h_a^3/12$; $R_E=K_{GD}=2\cdot\mu\cdot h_a^3\cdot b_a^3/[l_a^3\cdot[(h_a^2+b_a^2)]$; $1/C_E=G_{stiff}=(128\cdot E_a\cdot b_a\cdot h_a^3)/l_a^4$; where $\mu=f(p_c,T_0,\rho_c,v_c,k_B)$ – coefficient of dynamic viscosity, depend on density ρ_c, pressure p_c, average flow rate v_c, ambient temperature T_c and Boltzmann constant k_B (at temperature T_0=20°C, here $\mu\approx1{,}83\cdot10^{-6}$kg/m·s can be selected for air); l_a, h_a, b_a, m_a, E_a – length, thickness, width, mass of SE and Young's modulus of four-beam fused silicon SE, m_{add} – additional mass accordingly; and, electrical current $I_E(t)=dd_a(t)/dt$ – rate of central part deflection of SE. By the Kirchhoff's voltage law:

$$L_{E\,a}\cdot\frac{dI_{E\,a}(t)}{dt}+R_{E\,a}\cdot I_{E\,a}(t)+\frac{1}{C_{E\,a}}\cdot\int I_{E\,a}(t)\cdot dt=U_a\,. \tag{9}$$

Differentiating equation (9) with respect to time and multiplying by $1/L_E$, obtain:

$$\frac{d^2I_{E\,a}(t)}{dt^2}+2\alpha\frac{dI_{E\,a}(t)}{dt}+\omega_0^2I_{E\,a}(t)=0\,, \tag{10}$$

where $\alpha=R_{E\,a}\,/\,(2\cdot L_{E\,a})$ – attenuation coefficient of oscillation in RLC-circuit, $\omega_0=1/\sqrt{(L_{E\,a}\cdot C_{E\,a})}$ – natural oscillating frequency in RLC-circuit.

At the initial condition: current $I_{E\,a|t=0}=U_a/R_{E\,a}$ and the rate of change of current is different from zero $L_{E\,a}\cdot dI_{E\,a}(t)/dt|_{t=0}=-U_a$, the total current $I_{E\,a}(t)$ in an oscillatory RLC-circuit varies according to formula [10]:

$$I_{E\,a}(t)=\frac{U_a}{R_{E\,a}}-\frac{U_a}{L_{E\,a}\cdot\sqrt{\alpha^2-\omega_0^2}}e^{-\alpha t}\mathrm{sh}\left(\sqrt{\alpha^2-\omega_0^2}\cdot t\right). \tag{11}$$

Consequently, the transfer function $W_{SE\,a}(p)$ of SE of acceleration transducer based on OTE may be defined by formula:

$$W_{SE\,a}(p)=L\left\{\frac{\partial d_a(t)}{\partial a}\right\}=L\left\{\frac{\partial}{\partial a}\left[\frac{U_a}{R_{E\,a}}-\frac{U_a}{L_{E\,a}\cdot\sqrt{\alpha^2-\omega_0^2}}e^{-\alpha t}\mathrm{sh}\left(\sqrt{\alpha^2-\omega_0^2}\cdot t\right)\right]\right\}, \tag{12}$$

where L – Laplace transform.

Under the influence of measuring angular velocity Ω_z SE of angular velocity MOEM–transducers based on OTE (Figure 4*b*) occurs small deviations $d_\Omega(t)$ due to Coriolis effect [11]. This schematic can be seen as an oscillatory RLC-circuit (Figure 4*c*) with sinusoidal action at equivalent parameters: $Q(t)=U_0\sin(\omega t)=[l_\Omega\cdot F_0(\Omega_z)/(2\omega)]\cdot\sin(\omega t)$, where $F_0(\Omega_z)=\pi\cdot d_b\cdot U_{exct}\cdot\Omega_z\cdot[(\rho_\Omega\cdot l_\Omega\cdot h_\Omega\cdot b_\Omega\cdot\omega)/2+(4\cdot\pi\cdot m_{add})]$ – amplitude of Coriolis force); $R_E=\mu\cdot l_\Omega^4\cdot h_\Omega^3/[8\cdot b_\Omega^2\cdot(l_\Omega^2+4\cdot h_\Omega^2)]$; $1/C_E=2\cdot E_\Omega\cdot h_\Omega\cdot b_\Omega^3/2\cdot l_\Omega^3$, $L_E=J=(\rho_\Omega\cdot h_\Omega\cdot b_\Omega\cdot l_\Omega^3/24)+(m_{add}\cdot l_\Omega^2)/4$; where l_Ω, h_Ω, b_Ω, m_Ω, E_Ω, d_b – length, thickness, width, mass of SE, Young's modulus и piezomodule with feedback effect of piezoceramic SE, U_{exct} – excitation voltage with frequency ω accordingly; and electrical current $I_E(t)=d\varphi_\Omega(t)/dt$ – rate of change of angular displacement of SE along OY axis (for small value $\varphi_\Omega(t)\approx\sin[\varphi_\Omega(t)]$ is accepted, that $\varphi_\Omega(t)=2\cdot d_\Omega(t)/l_\Omega$). Then second order differential equation for such oscillatory RLC-circuit is:

$$\frac{d^2 I_{\mathrm{E}\,\Omega}(t)}{dt^2}+2\alpha\frac{dI_{\mathrm{E}\,\Omega}(t)}{dt}+\omega_0^2 I_{\mathrm{E}\,\Omega}(t)=\frac{\omega U_0}{L_{\mathrm{E}\,\Omega}}\sin(\omega t)\,. \quad (13)$$

Similarly, the general solution of equation (13) can be written as

$$I_{\mathrm{E}\,\Omega}(t)=\frac{U_0\cdot\omega\cdot\sin(\omega t-\Delta\varphi)}{L_{\mathrm{E}\,\Omega}\cdot\sqrt{(\omega_0^2-\omega^2)+4\alpha^2\omega^2}}-\frac{U_0}{L_{\mathrm{E}\,\Omega}\cdot\sqrt{\alpha^2-\omega_0^2}}\cdot e^{-\alpha t}\cdot\mathrm{sh}\left(\sqrt{\alpha^2-\omega_0^2}\cdot t\right), \quad (14)$$

where $\Delta\varphi=\arctan[(\omega^2-\omega_0^2)/(2\alpha\omega)]$ – phase shift.

Transfer function $W_{\mathrm{SE}\,\Omega}(p)$ of SE of angular velocity MOEM–transducers based on OTE is:

$$W_{\mathrm{SE}\,\Omega}(p)=L\left\{\frac{1}{2\cdot l_\Omega}\cdot\frac{\partial\varphi_\Omega(t)}{\partial\Omega_Z}\right\}=\frac{1}{2\cdot l_\Omega}L\left\{\frac{\partial}{\partial\Omega_Z}\left[\frac{U_0\cdot\omega\cdot\sin(\omega t-\Delta\varphi)}{L_{\mathrm{E}\,\Omega}\cdot\sqrt{(\omega_0^2-\omega^2)+4\alpha^2\omega^2}}-\frac{U_0\cdot e^{-\alpha t}}{L_{\mathrm{E}\,\Omega}\cdot\sqrt{\alpha^2-\omega_0^2}}\cdot\mathrm{sh}\left(\sqrt{\alpha^2-\omega_0^2}\cdot t\right)\right]\right\}. \quad (15)$$

Transient process modelling of angular velocity and acceleration MOEM–transducers based on OTE is shown in figure 5 with the following parameters of functional elements: PD T_{PD}=1µs, k_{PD}=0.4A/W; $T_{\mathrm{I\text{-}U}}$=10µs, $k_{\mathrm{I\text{-}U}}$=4kOhm; T_{Y}=10µs, k_{Y}=100; $k_{\mathrm{F}}\approx1$; T_{F}=10µs, T_{PA}=3µs, P_{OS}=0.5mW, λ=1.3µm, θ=47°.

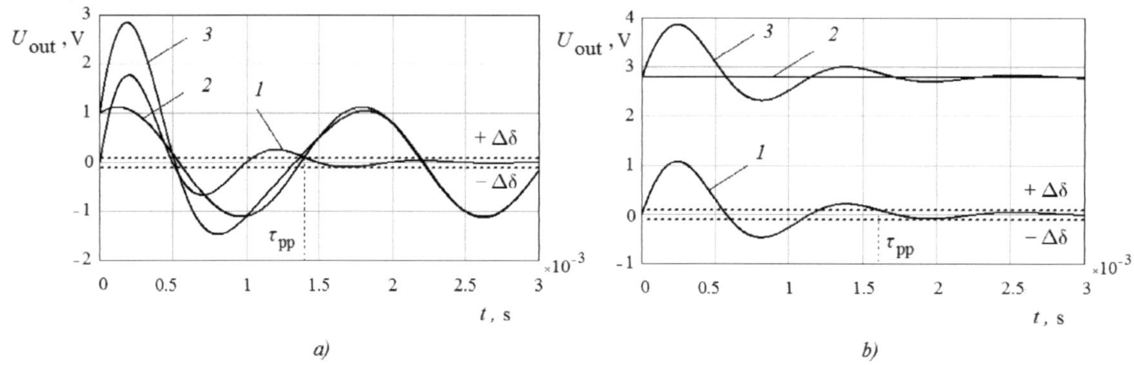

Figure 5. Transient processes of angular velocity (*a*) and acceleration (*b*) MOEM–transducers based on OTE *1*- composed damping, *2*- composed setting, *3* - resultant output voltage

In dynamic mode the considered transducers are stable and possess only small oscillatory transients (line *1*). When the permissible dynamic error $\Delta\delta$ is not exceed 5%, output characteristics (line *3*) of angular velocity and acceleration MOEM–transducers are included in the steady state (lines *2*) with a short duration transient $\tau_{\mathrm{PP}}\sim$ 1ms.

4. Conclusion

In this study angular velocity and acceleration MOEM–transducers based on OTE, providing quasi-linearity conversion function is investigated. An experimental study of this transducer conversion function is described. Correction coefficient of conversion functions based on comparing of theoretical calculations and experimental studies is defined and it supports to improve the accuracy of the calculation of the described model.

The block diagram is determined and equivalent oscillatory RLC-circuit is composed for the study of dynamic characteristics angular velocity and acceleration MOEM–transducers based on OTE. It is shown that the transition process of described transducers have oscillated character with permissible magnitude of transition process duration τ_{PP} is around 1ms, which allows to apply those transducers in mobile objects control systems.

Ackowledgment
This article was prepared with the financial support of the RFFR (*grant №16-08-00447-a*).

References

[1] Eric U, William B and Spillman Jr 2011 *Fiber Optic Sensors: An Introduction for Engineers and Scientists* (New Jersey: John Wiley & Sons, Inc) 512 p
[2] Busurin V I, Naing H L and Tuan P A 2015 *Physics procedia* **73** pp 198-204
[3] Born M and Volf E 1983 Osnovi optiki (Moscow: Nauka) 721 p
[4] Busurin V I, Zhelgov M A, Korobkov V V and Tuan P A 2015 *Mikro-optoelectro-mechanicheski dvuchocevoi datchik uglovoi i lineinovo uckorenir* (Patent No 2566384)
[5] Raspapov V Ya 2007 *Mikromechanicheskie pribori* (Moscow: Mashinostroenie) 400 p
[6] Busurin V I, Gorshkov B G and Korobkov V V 2012 *Volokonno – opticheskie informasionno-izmeritelnie sistemi* (Mosow: Moscow Aviation Institute Press) p 168
[7] Min H B 2000 *Handbook of sensors and actuators: micro mehcanical transducers pressure sensors, accelerometers and gyroscopes* (Elsevier Press) 392 p
[8] Fraden J 2004 *Handbook of modern sensors. Physics designs and application* (NewYork: Springer-Verlag) 579 p
[9] Kim D P 2010 *Teoriya avtomaticheskovo upravlenir:Linianie sistemi* (Moscow: Fizmalit) 312 p
[10] Dorf R C and Bishop R H 2008 *Modern Control Systems* (Upper Saddle River: Prentice Hall) 1018 p
[11] Busurin V I and Tuan P A 2016 *Avtometriya* **52 (2)** pp 124-30

Features of structures formation on the basis of chitosan derivatives by a prototype of 263 nm laser stereolithograph

D S Dudova[1,2], K N Bardakova[1], T A Akopova[3], P S Timashev[1], N V Minaev[1]

[1]Institute on Laser and Information Technologies, Russian Academy of Sciences, 142190 Troitsk, Moscow, Russia
[2]National Research Nuclear University MEPhI (Moscow Engineering Physics Institute), 31 Kashirskoe shosse, 115409 Moscow, Russia
[3]N S Enikolopov Institute of Synthetic Polymeric Materials, Russian Academy of Sciences, 117393 Moscow, Russia

E-mail: dudovadasha@mail.ru

Abstract. We have developed technology of polysaccharides based matrices formation by laser stereolithography (SLA) method using UV range laser radiation. Experimental data on a laser parameters selection of single-layer structures polymerization for compositions based on unsaturated chitosan derivatives with different degree of substitution and with the addition of polyethylene glycol diacrylate is shown.

1. Introduction

Experimental development of 3D structures printing which compatible with the human body began in the early 21st century [1]. Currently, the technology of biological printing is experiencing of exponential growth. The first experiments in this direction showed the possibility of fragments creating of different tissues and organs. A defining stage in the development of technology for the recovery of the bodies was the opportunity to create biocompatible porous matrix "scaffold" that can be created a given three-dimensional model with the specified parameters of the structure at the micro level. The scaffolds are three-dimensional porous or fibrous matrix, one of the main functions of which is to provide a mechanical frame for cells [2]. The scaffolds should have the following properties: biocompatibility, absence of immunological rejection, intoxicity, biodegradation and some others.

Previously it was shown that on the basis of reactive chitosan by the method of two-photon femtosecond microstereolithography it is possible to form a three-dimensional matrix, which can be used as carriers for primary neuronal cultures [3]. The advantage of this method is the possibility of creating complex structures with high resolution, however, the performance of the method is quite low and the cost of the installation exploitation is quite high. Moreover, experience shows that such installations have a number of advantages, but also they have significant disadvantages: low printing speed due to high resolution, which greatly limits the possibility of rapid prototyping; high cost and large sizes of installations, and the complexity of their configuration and operation.

Developed stereolithograph will considerably simplify the process of printing, increase the speed at the several times without significant loss of quality. In addition, a large variety of biocompatible printing materials provides wide opportunities for its application. Furthermore, a significant advantage is the simple design and low cost of optics, mechanics, laser and other components.

The aim of this work was to develop a prototype a 3D printer based on SLA technology, using the source of laser radiation at a wavelength of 263 nm for compositions based on chitosan, as well as conducting experiments on selection of optimal parameters for these compositions.

2. Scheme of experimental setup, experimental technique.

Experimental setup is shown on figure 1. We used diode pumped ultraviolet solid state laser (263 nm, 25 mW, 263 TECH-Basic, the pulse duration 4 ns and frequency up to 4 kHz). We focused laser pulses by spherical lens (f=8 cm), which made it possible to focus light to a spot about 100 microns in diameter. To move the laser radiation across the sample, was used the two-mirror galvanoscanner LScanH-10 (Ateko, Russia) with the basic parameters: the scan range of ± 6°; maximum scanning speed – up to 480°/s; field processing (in the current experiment) 11 x 11 mm. The positioning accuracy of the focused beam in the current configuration was about 15 μm. The laser light was got from the top perpendicular to the surface of the photopolymer. As software we used the program LDesiner 5.0.

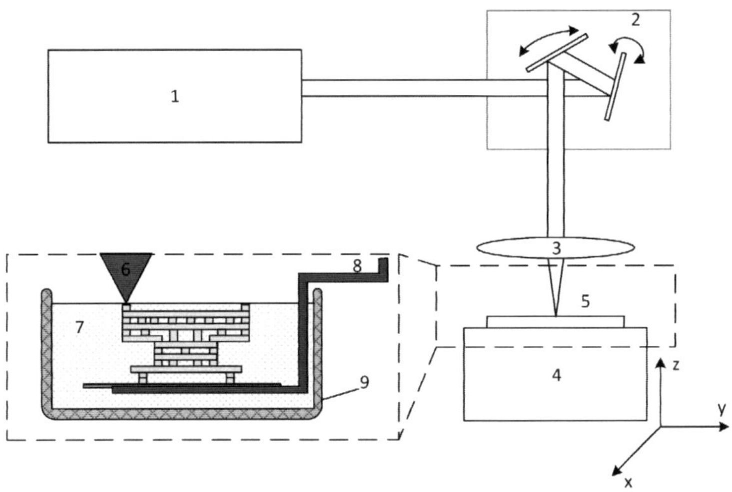

Figure 1. A schematic diagram of a 3D printer running on a single-photon photopolymerization: 1 – laser; 2 – two-mirror galvanoscanner; 3 – lens; 4 – three-coordinate table; 5 – substrate; 6 – focused laser beam; 7 - the photosensitive composition;8 – z-coordinate axis translator; 9 - quartz cuvette.

The following laser parameters were chosen for the experiment: pulse repetition rate 1.6 kHz, the spot diameter on the sample of 100 microns, the intensity of 0.2 W/cm^2. As a test figure was selected, a circle with a diameter of 6 mm, the area of which is filled by individual lines in laser scanning process.

3. Materials

One of the materials used to create tissue-engineering scaffolds is a natural polysaccharide chitosan and its derivatives. The main advantage of the selected polysaccharide is its non-toxicity, availability, and ability to form hydrogel system similar in mechanical characteristics to the soft body tissues [3]. Cleavage of the chitin N-acetyl-D-glucosamine an acetyl group or carrying deacetylation reaction underlies of chitosan forming. Chitosan and derivatives as a cellulose possess fibre- and film-forming [4–6] properties.

It was to perform a solid-state synthesis of unsaturated ether of chitosan – allyl chitosan – for the construction of 3D structures [7,8]. Number of allyl groups per chitosan's unit varies depending conditions and the ratio of reactants in which the synthesis was conducted. The greater the degree of substitution, the more unsaturated groups per unit volume of the composition, and hence more likely to be formed three-dimensional crosslinking during the photopolymerization process. Three type polymer compositions with different percentages of unsaturated groups were synthesized: CT-A1 (0.08-0.10)[7]., CT-A2 (0.25), CT-A3 (0.50).

To increase the number of unsaturated groups (double bonds) per unit volume, which can be the spatial crosslinking of the polymer composition and for varying the mechanical properties of the obtained matrix, swelling properties and biodegradation of [9] in the composition administered polyethylene glycol diacrylate (PEG-DA) with a molecular weight 2000 Da.

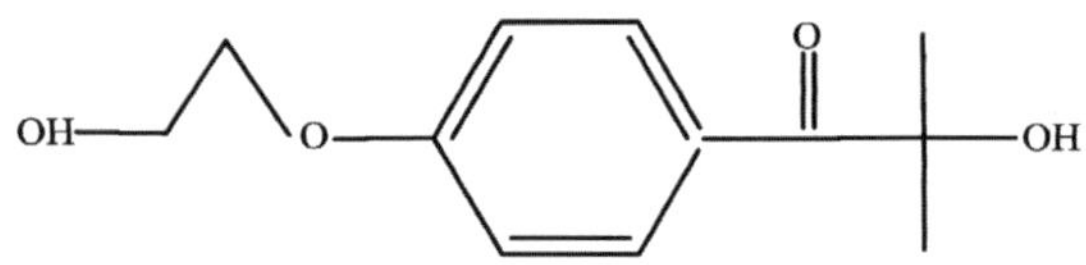

Figure 2. Irgacure 2959

In this study we used a biocompatible [10] photoinitiator Irgacure 2959 (Ciba), which upon irradiation of short-wave UV range (~ 263nm) decomposes to form benzoyl and ketyl fragments [11], which in turn all start the process of three-dimensional cross-linking.

4. Experimental results and Discussion

For polymer compositions with different degree of substitution and the addition of PEG-DA performed selection of optimal parameters. Fill density and the scan speed of the laser beam was varied. According to the results of test experiments, the optimal density of laser scan lines per millimeter was found - 15 lines/mm. At the test experiments the width of a represented single polymerized line is 60 μm, and it above parameters of the laser radiation. Most likely this is caused by the Gaussian profile at the laser light spot. This number of lines is provided complete filling of the area of the test figure in the form of polymerized areas with little lines overlap.

Figure 3. Photo of the test structure of the composition CT-A3.
Printed with a different speed in the silicone spacer. Diameter of cylinders 3 mm.

The most useful parameter for varying scan was chosen - speed of the laser radiation, because it affects the amount of energy that absorbs the sample. Scan range of printed speed was selected - from 2 to 20 mm/sec, from the results of test experiments.

4.1 A series of experiments for the samples CT-A1, CT-A2, CT-A3 *with 1% Irgacure 2959.*
After printing the sample were observed bright white spots. Then, the samples were placed in a petri dish filled with distilled water, the purpose of laundering the samples obtained from the glass substrate and unpolymerized material which dissolves in water completely. Best results were obtained with the composition CT-A2 and CT-A3 at speeds of 2, 4, 6 mm / sec. Samples polymerized films do not dissolve in water, retains its shape, the size of the increase was not observed.

For CT-A1 compositions was observed sharp "disintegration" of the polymerized film into fragments in the form of filaments, immediately after placing a sample in water. It is assumed that this is due to an insufficient number of double bonds which affect the degree of crosslinking and the strength of the sample.

4.2 A series of experiments for the samples CT-A1, CT-A2, CT-A3 *with 1% Irgacure 2959 and addition of PEG-DA.*
For all samples, the transparent areas of polymerized material were observed with a clear boundary. After 30 minutes, after placing all the samples in the CT-A3 water, we observed increase of a film size approximately 2-fold. After 60 minutes, samples printed with speed of 2, 4, 6, 8, 10 mm /sec is still observed in the water. For printing speed of 20 mm/sec were observed samples only in the form of separate fragments of the individual fibers.

After placing irradiated CT-A2 samples in the water, there was a gradual increase size of the film approximately 2-fold. After 60 minutes, samples irradiated at 2 and 4 mm / sec was observed in water as very transparencies film. Samples, printed with speed of 6, 8, 10, 20 mm/sec after 20 minutes in distilled water were not detected.

To the photosensitive compositions of CT-A1 observed much more rapid increase of sample size (relatively CT-A2 and CT-A3) when placed in water, that can be linked to less count of chemical links thru a less count of unsaturated groups at the sample. After 60 minutes in the water were observed samples printed at 2 and 4 mm/sec. The sample printed at a speed of 4 mm/sec significantly deformed and curled in the "roll". The samples were irradiated at higher speeds, dissolved after 10 minutes.

5. Conclusion

3D printer prototype was assembled using a UV laser radiation and galvo scanner. We investigated the original biocompatible compositions based on chitosan with different number of unsaturated groups per unit volume (with different degree of substitution) and with the addition of a copolymerizable component PEG-DA. Selection of parameters for of polymer compositions based on allyl chitosan with different degree of substitution was done - the density filling lines and scan speed.

In the first series of experiments without addition copolymerizable component PEG-DA for CT-A3 resistant film samples were obtained at speeds less than 6 mm / sec. In the case of compositions with less unsaturated bonds (CT-A2) resistant film were obtained at speeds of up to 4 mm/sec. For compositions CT-A1 resistant films failed.

In a series of experiments with addition copolymerizable component PEG-DA was observed increase size of samples approximately 2-fold by placing the samples in a petri dish filled with distilled water. Stable film for CT-A3 were obtained at speeds of 10 mm/sec. For CT-A2 and CT-A1 films were obtained at speeds up to 4 mm/sec, and a day later dissolved.

It should be noted that the CT-A3 compositions without addition copolymerizable component PEG-DA and with it resistant films were obtained which remained intact after soaking in distilled water for one week.

Acknowledgements
The work is executed at financial support of Russian scientific Foundation (project No. 14-25-00055).

6. References

[1] Murphy S V and Atala A 2014 *Nat. Biotechnol.* **32** 773–785

[2] Haycock J 2011 **695** 261–80

[3] Akopova T A, Rogovina S Z, Vikhoreva G A, Zelenetskiy S N, Gal'braykh L S and Enikolopov N S 1991 *Vysokomol. Soedin. Seriya B* **32** 735–737

[4] Stepnova E A, Tihanov V E, Babak V G, Krau Бабак В Г, Kraujina M A, Babievsky K K and Yamskov I A 2005 *Chemical fiber* **57** 8

[5] Kamada T and Takemaru T 1983 Coprinus cinereus *J. Gen. Microbiol.* **129** 703

[6] Seng J-M 1988 *Biofutur* **71** 40–44

[7] Akopova T A, Timashev P S, Demina T S, Bardakova K N, Minaev N V., Burdukovskii V F, Cherkaev G V., Vladimirov L V., Istomin A V., Svidchenko E A, Surin N M and Bagratashvili V N 2015 *Mendeleev Commun.* **25** 280–282

[8] Timashev P S, Demina T S, Minaev N V., Bardakova K N, Koroleva a. V., Kufelt O a., Chichkov B N, Panchenko V Y, Akopova T a. and Bagratashvili V N 2015 *High Energy Chem.* **49** 300–303

[9] Timashev P S, Bardakova K N, Demina T S, Pudovkina G I, Novikov M M, Markov M A, Asyutin D S, Pimenova L F, Svidchenko E M, Ermakov A M, Selezneva I I, Popov V K, Konovalov N A, Akopova T T, Solovieva A B, Panchenko V Y and Bagratashvili V N 2015 *Sovrem. Tehnol. v Med.* **7** 20–29

[10] Williams C G, Malik A N, Kim T K, Manson P N and Elisseeff J H 2005 *Biomaterials* **26** 1211–1218

[11] Scaiano J C, Stamplecoskie K G and Hallett-Tapley G L 2012 *Chem. Commun.* **48** 4798

Investigation of interaction femtosecond laser pulses with skin and eyes mathematical model

P U Rogov, S V Smirnov, V A Semenova, M V Melnik and V G Bespalov
ITMO University, Kronverkskiy pr. 49, St. Petersburg 197101, Russia

E-mail: rogovpu@niuifmo.ru

Abstract. We present a mathematical model of linear and nonlinear processes that takes place under the action of femtosecond laser radiation on the cutaneous covering. The study is carried out and the analytical solution of the set of equations describing the dynamics of the electron and atomic subsystems and investigated the processes of linear and nonlinear interaction of femtosecond laser pulses in the vitreous of the human eye, revealed the dependence of the pulse duration on the retina of the duration of the input pulse and found the value of the radiation power density, in which there is a self-focusing is obtained. The results of the work can be used to determine the maximum acceptable energy, generated by femtosecond laser systems, and to develop Russian laser safety standards for femtosecond laser systems.

Introduction

Femtosecond laser systems (FLSs) are widely used in science, technical applications and medical technologies [1, 2]. Despite the fast advancement of femtosecond technologies and their application in biology and medicine, currently, in the Russian Federation, there are no standards of safe energy levels of high-intensity femtosecond laser radiation. Note that the mechanism of interaction of intense femtosecond laser radiation with matter essentially differs from that for the pulses of longer duration, since the interaction time is smaller than the time, necessary to excite the phonon subsystem [3]. Besides, the high power density that arises when using the FLSs leads to various nonlinear processes, such as self-focusing, two- and three-photon absorption, multiphoton and impact ionization, and for the high energy density to the optical breakdown. [4]. This process is of particular importance in the case of femtosecond radiation incident on biological tissues, since the high-intensity femtosecond radiation may be essentially unsafe for skin, vision, and other human organs [5].

In the present work, we consider the linear and nonlinear interaction of femtosecond laser radiation with skin and eye and propose a mathematical model approximately describing particular interaction processes. In future, the model can be used to determine the maximal acceptable levels of FLS energy.

Interaction mechanism

1.1. Specific features of ultrashort-pulse lasers and the skin

The human skin is an optically turbid opaque medium (both the absorption and the scattering are present). At the same time, the skin is a heterogeneous structure containing the inclusions of different type and dimension (blood vessels, hair follicles, etc.), which essentially makes it difficult in the understanding of the processes that occur under the action of laser radiations on the cutaneous covering. The main constituents of the skin are water (70%) and proteins (27%), and the main structure protein is collagen (nearly 70% of dry skin weight). The water molecules can be divided into two groups, the free ones and the ones included in protein compounds (bound state, e.g., the three-fold screw group of collagen binds nearly 500 water molecules [6]). Thus, in the first approximation the skin may be considered as water with protein inclusions close to dielectric materials in their electrodynamic properties.

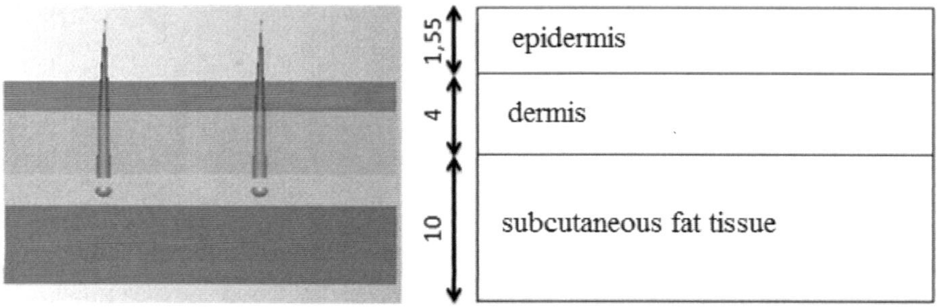

Figure 1. Model of human skin.

The mean optical (λ=800 nm) and thermal physical parameters used in the skin model for calculating was taken from [7]. For estimation, we have chosen the wavelength 800 nm typical for the most commonly used titanium-sapphire FLSs. The pulse duration in such systems can vary from 25 to 550 fs. The irradiation with ultrashort laser pulses leads to damages that cannot be explained by the thermal melting mechanism [8] and are due to the excitation of valence electrons by the laser pulses [9, 10]. The water molecules serve as the transmitting element that absorb the laser pulse energy and excite the vibrational (phonon) modes of collagen with the relaxation time of the order of 3 ps [11, 12].

1.2. Mathematical model

In the calculations, we utilized the set of balance equations that describe the dynamics of electron-atom subsystem. In this model, the atoms interact via the semi empirical potentials and the electron degrees of freedom are not taken into account explicitly. The laser radiation generates non-equilibrium charge carriers described by the integral concentration [13]. In this case one can ignore the processes of Auger recombination and impact ionization, since for the chosen parameters of radiation their semi empirical contribution is negligibly small.

E.B. Yakovlev et al. [11] proposed the idea of determining the dependence between the potentials of the atoms and the state of the electronic sub-system, thus making it possible to simulate the effects of heat transfer under the action of femtosecond pulses on dielectric materials. The differential Bouguer-Lambert law determines the distribution of the laser radiation intensity J (z, t) inside a solid. The system of heat transfer equation that describe the dynamics of electron and atom subsystems in the one-dimensional approximation has the form:

$$\frac{\partial T_e}{\partial t} = \alpha_e \frac{\partial^2 T_e}{\partial z^2} - \frac{1}{C_e \tau_{ep}}(T_e - T_a) + \frac{\alpha_e h\nu}{C_e} J(t,z) \quad (1.1)$$

$$\frac{\partial T_a}{\partial t} = \alpha_a \frac{\partial^2 T_a}{\partial z^2} + \frac{1}{c_a \tau_{ep}}(T_e - T_a) \quad (1.2)$$

where $J(t,z)$ is the intensity distribution in the solid, $C_e = \frac{\pi^2 k_B^2 N_e}{2E_F}$ is the heat capacity of the electron gas, C_a is the heat capacity of atoms, τ_{ep} is the time of electron-phonon relaxation, T_e is the temperature of electrons, T_a is the temperature of atoms, α_e and α_a is the thermal diffusivity of electrons and ions; E_F is the Fermi energy; K_B is the Boltzmann constant.

For modelling the impact of the femtosecond laser radiation on the skin we used the following boundary conditions:

$$T_e|_{t=0} = T_e|_{t=0} = T_0 \quad (2.1)$$

$$\left.\frac{\partial T_e}{\partial z}\right|_{z=0} = \left.\frac{\partial T_e}{\partial z}\right|_{z=L} = \left.\frac{\partial T_a}{\partial z}\right|_{z=0} = \left.\frac{\partial T_a}{\partial z}\right|_{z=L} = 0 \quad (2.2)$$

The mechanism of effects on eyes

3.1. Features of the effects of radiation on the eye

Eye - is a complex self-regulating optical system, which is the scope of an average of about 24 mm in diameter (Figure 2). [14].

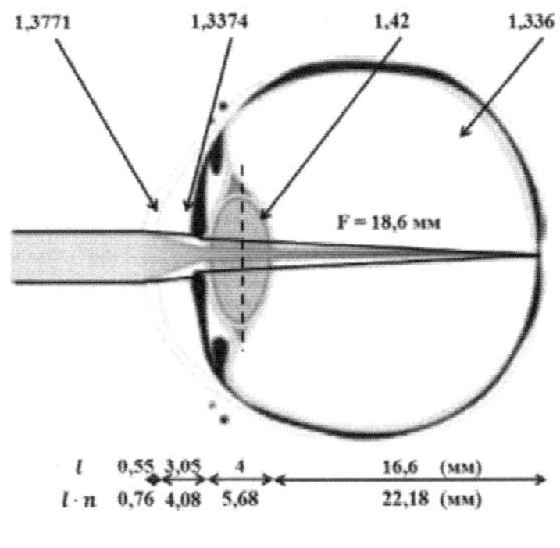

Figure 2. Schematic representation of the eye. In the upper part of the figure shows the refractive indices in the lower part - and the linear dimensions of the optical path-sky in mm [14].

Vitreous humor is mainly composed of water (99%), the balance being acetylcholine, protein, glucose, thiamine, pyridoxine, ascorbic acid, and proteolytic enzymes, little effect on the non-linearity of the medium. Thus, the major contribution to make the optical properties of water molecules.

On the optical path of the dispersion experiences femtosecond pulse broadening, which can be described by the formula [15]:

$$n(\omega) = N_0 + ac\omega^2 - bc\omega^{-2}$$

where N_0, a, b - the dispersion characteristics of the medium.

The main danger for retina if laser radiation is heat exposure, which can lead to thermal and photochemical damage. However, when using pulsed radiation, femtosecond excited by electron subsystem that transfers the energy of the atomic subsystem, increases the time to reach a maximum temperature of [16]. At the moment, considering several theoretical models of optical damage mechanisms and biological objects, which are divided into mechanisms, associated with absorption, which include thermal explosion, and the internal mechanism - impact ionization and photoionization. At 800 nm contribution to damage modify the main impact- and photoionization in the range of durations of 7 to 300 fs [17]. Photoionization is the dominant mechanism in a pulse length of less than 50 fs, and impact ionization becomes predominant in a pulse length of 50 fs more. [18]

1.2 Mathematical methods of calculation

The dynamics of a two-dimensional spectral density of g TE-polarized radiation in a homogeneous isotropic dielectric medium with a fast-response cubic nonlinearity can be described by the equation [19]:

$$\frac{\partial^2 g}{\partial z^2} + (k(\omega)^2 - k_x^2)g =$$
$$-\frac{\omega^2 \varepsilon_{nl}}{c^2}\frac{1}{(2\pi)^4}\int_{-\infty}^{\infty}\int_{-\infty}^{\infty}\int_{-\infty}^{\infty}\int_{-\infty}^{\infty} g(\omega-\omega', k_x - k_x', z)\, g(\omega'-\omega'', k_x' - k_x'', z) \quad (1)$$
$$\times\, g(\omega'', k_x'', z) d\omega' dk_x' d\omega'' dk_x''$$

where

$$g(\omega, k_x, z) = \int_{-\infty}^{\infty}\int_{-\infty}^{\infty} E(t, x, z) e^{-i(\omega t + k_x x)} dt dx$$

w and k_x - temporal and spatial frequency of the radiation, k(ω) - wave number, n (ω) - the frequency dependence of the refractive index of the medium, ε_{nl} - coefficient of its nonlinear dielectric constant, c - speed of light in vacuum, t - time, x, z - Cartesian spatial coordinates, E (t, x, z) - the electric field radiation polarized perpendicular to the x_z plane.
The spectral density of the radiation entering the medium (at z = 0) was set in the form:

$$g(\omega, k_x, 0) = ig_0 e^{\frac{\Delta x^2 k_x^2}{8}} \left(e^{\frac{\Delta t^2(\omega-\omega_0)^2}{8}} - e^{-\frac{\Delta t^2(\omega-\omega_0)^2}{8}} \right)$$

where $g_0 = \frac{\pi \Delta t \Delta x}{4} \sqrt{\frac{8\pi}{3N_0}} I_0$ - maximum spectral density, ω_0 - centre frequency, Δx - the transverse momentum of the size of the inlet to the environment, Δt - its initial duration, I_0 - radiation intensity.
Another boundary condition, which determines the direction of wave propagation, was reviewed in the linearized form:

$$\frac{\partial g}{\partial z}|_{z=0} = -ig(\omega, k_x, 0)\sqrt{k(\omega)^2 - k_x^2}$$

To solve, equation (1) has used numerical scheme Crank-Nicholson adaptive step axis Z. Convolution in the right-hand side of equation (1) is calculated using the fast Fourier transform algorithm. The calculation has been implemented in Fortran using OpenMP parallelization tools.

Results

4.1 For skin
By means of Monte Carlo numerical simulation of propagation of radiation through the skin tissues, we obtained the intensity distributions over the tissue depth. The calculation approximated the solution of the radiation transfer equation by modelling all possible photon trajectories passing through the skin tissue model. The steady-state form of the radiation transfer equation in a homogeneous medium can be written as the integral equation [20, 21]:

$$L(r,s) = \int_{R-0}^{\infty} \exp(-\mu_{tr}R)\, \mu_s \int_{4\pi} p(s,s')L(r-Rs,s')d\Omega' \, \mathrm{dR} + \int_{R-0}^{\infty} \exp(-\mu_{tr}R)Q(r-Rs,s)dR \qquad (4)$$

where L is the luminance at the point r in the direction s, considered for both the light source and for the radiation scattered towards the direction s; R is the path length, Q is the light emitted in the direction s by the source at the point r – Rs. The parameter $\exp(\mu_{tr}R)$ determines the radiation transfer from the point r – Rs to the point r. Thus, the first term in the equation characterises the light scattered at the point r – Rs, from any direction s' to the direction s, reaching the point r, and the second term describes the light emitted from the point r – Rs in the direction s.

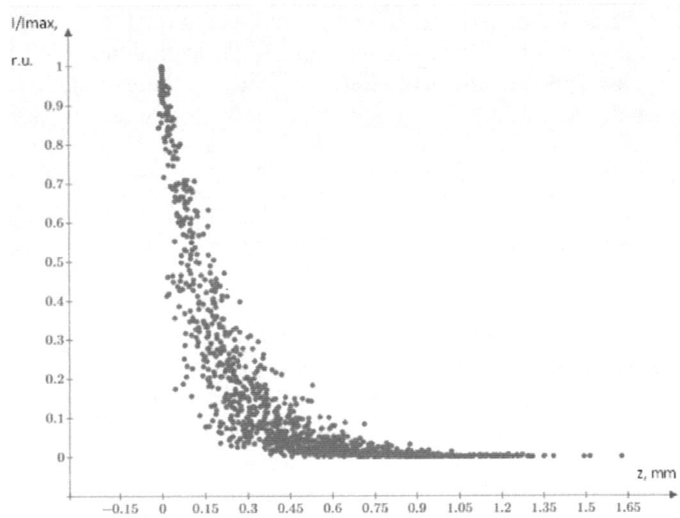

Figure 3. The intensity of the radiation on the depth of penetration (at λ = 800 nm).

From fig. 3 investigation of interaction femtosecond laser pulses with skin and eyes mathematical model can see that at the depth of 0.3 mm the radiation intensity is below 30% of the initial level. It is known that the probability of multiphoton processes is directly proportional to the second, the third and higher powers of the laser intensity. Thus, the main nonlinear processes will take place in the upper skin layer, the epidermis.

The dynamics of the electron system can be described by the energy distribution. The state of the electron subsystem immediately after the impact, part of electrons transfers their energy to the atom subsystem as a result of relaxation more than 3 ps, when most part of the electrons has transferred their energy to the atom subsystem, the whole system tends to thermal equilibrium (fig. 4).

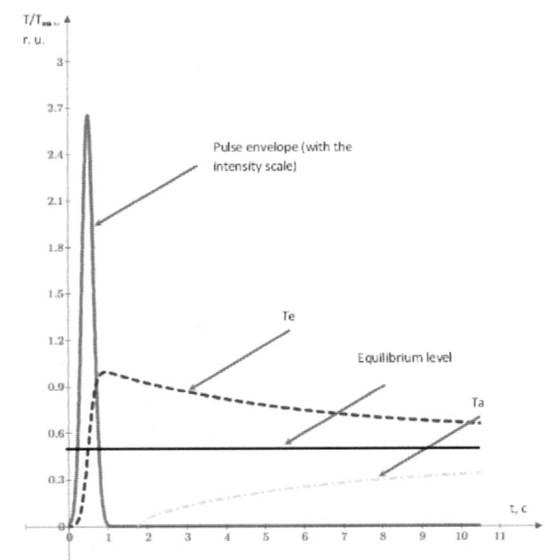

Figure 4. Thermodynamics of laser pulse, the dynamics of the electron gas temperature T_e, the dynamics of the crystal lattice temperature T_a.

The absorption of light quanta by no equilibrium electrons of the dielectric maximizes their kinetic energy. This leads to the increase of electron temperature during the light pulse, whereas the lattice remains practically cold. Therefore, one can group the process of the action of the femtosecond laser radiation on the skin into three stages. At the first stage during the femtosecond pulse the multiphoton excitation of water molecules occurs. The energy of ionisation in this process amounts to 6.5 eV, so that to ionise one water molecule nearly 5 photons (800 nm) are needed, which is a cause for the reduction of the quantum efficiency ($\eta \leq 20\%$). Simultaneously the process of impact ionisation takes place. As a result, before the end of the laser pulse action Nmax $\approx \eta \cdot E/\hbar w$ (E being the pulse energy) electrons will be excited to the upper ionised states of the water molecules with the energy above Ui = 6.5 eV. The second stage occurs after the action of the femtosecond pulse and lasts until the complete transfer of the energy from the electrons to the phonon subsystem. The electron gas is cooled and the collagen molecules are heated (tens of picoseconds). The third stage is the dissipation of heat over the bulk sample (a few microseconds).

We presented a mathematical model of the propagation of femtosecond radiation through the cutaneous covering. Using the Mote Carlo numerical simulation, we determined the dependence of the radiation intensity up-on the penetration depth. By the help of the analytical solution of the system of equations describing the energy balance between the electron subsystem and the atomic one, we considered the mechanism of the effect of femtosecond radiation on the skin and estimated the temporal processes that occur in the course of the femtosecond radiation acting on the skin.

4.2 For eye

The work shows that at wavelengths that are in the area of transparency of the eye media pulse, duration in contact with the retina cannot be less than 100 fs. Thus, the main mechanism of damage to the retina using FLS in this wavelength range will photo ionize.

Calculations show self-focusing in the vitreous occurs when power exceeds 10^{12} W/cm^2. If this power was recorded a significant increase in amplitude of the field on the retina. this near power density retina occurs when input power density in the optical media of the eye of about 10^6 W / cm^2.

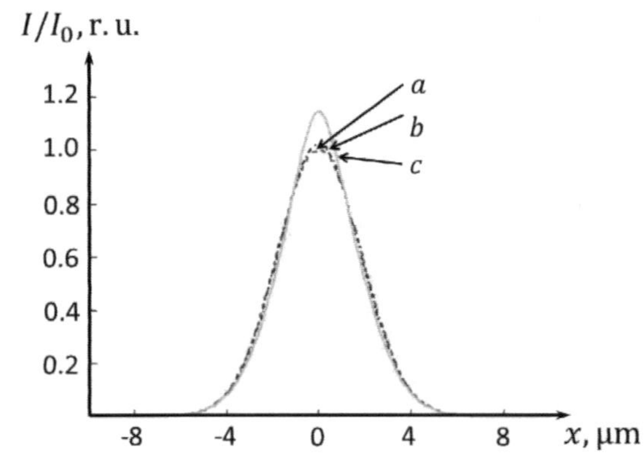

Figure 5. The graph of the spatial distribution of radiation intensity in the area of the retina to the initial duration of 10 fs (input into the eye) with an average wavelength of 800 nm.

Increasing the intensity when the input power density of the radiation 10^{12} W/cm^2 of 3%, and at a power density of radiation $5 \cdot 10^{12}$ W/cm^2 - about 15% of the input power. Note that the process of self-focusing is snowballing, so at higher power, the beam collapse does not give numerical simulation.

Acknowledgments

We acknowledge a financial support from the grant of the Russian foundation for basic research (the agreement No 16-52-52049\16 from 29.01.2016).

Journal of Physics: Conference Series **737** (2016) 012047 doi:10.1088/1742-6596/737/1/012047

References

[1] Akhmanov S A, Vysloukh V A, Chirkin A S 1988 *Optics of femtosecond laser pulses* (Moscow: Nauka) 310 p
[2] Rulliere C 2005 *Femtosecond laser pulses* (Springer Ed. Rulliere C) 438 p
[3] Yilbas B S and A M Arif 2005 *Proceedings of the Institution of Mechanical Engineers Part C: Journal of Mechanical Engineering Science* **219** 73-81
[4] Louis H et al. 2012 *Journal of biophotonics* **5(2)** 194-199
[5] GOST 50723-94: Laser Safety 1995 (General safety requirements for time-processing and operation of laser products, Moscow: Standards Publishing House) 34 p
[6] Sriramoju V Alfano R R 2015 *Journal of biomedical optics* **20(10)** 108001-108001
[7] Pushkareva A E 2008 *Methods of Mathematical Modeling in Biotissue Optics* (SPb: ITMO) 43 p
[8] Ingemar F Larsson M and Strömberg T 2008 *Journal of Biomedical Optics* **13(1)** 014015-014015
[9] Stampfli P and Bennemann K H 1990 *Phys Rev* **42(11)** 7163
[10] Stampfli P and Bennemann K H 1992 Phys. Rev. **46(17)** 10686
[11] Kropman M F and Bakker H J 2001 *Science* **291** 2118
[12] Kropman M F Nienhuys H K and Bakker H J 2002 *Phys. Rev. Lett.* **88(7)** 077601
[13] Lipp V P et al. 2014 *Journal of Optical Technology* **81(5)** 254-255
[14] Artal P 2014 *Advances in Optics and Photonics* **6(3)** 340–367.
[15] Shpolyanskiy Y A Belov D L Bakhtin M A and Kozlov S A 2003 *Applied Physics B* **77(2)** 349–355
[16] Veiko V P Libenson M N Chervyakov G N and Yakovlev E B 2008 *Interaction of Laser Radiation with Matter* (Fizmatlit Moscow)
[17] Manenkov A A 2014 *Optical Engineering* **53(1)** 010901–010901
[18] Ezerskaya A A Ivanov D V Kozlov S A Kivshar Y S 2012 *Journal of Infrared Millimeter and Terahertz Waves* **33(9)** 926–942
[19] Paltsev Y and Levina A 1996 *Proc. SPIE* **2728** 137–147
[20] Meglinski I Doronin A V 2012 *Monte Carlo modeling for the needs of biophotonics and biomedical optics* (Taylor & Francis).
[21] Jacques S, Wang L, Welch A J, Martin J, Van Gemert C 1996 *Monte Carlo modeling of light transport in tissue Optical-thermal response of laser-irradiated tissue* (Plenum Press New York) 73-100

The study of laser induced fluorescence of tooth hard tissues with aluminum phthalocyanine nanoparticles

D S Farrakhova[1], J O Kuznetsova[1] and V B Loschenov[1,2]

[1]National Research Nuclear University MEPhI (Moscow Engineering Physics Institute), 31 Kashirskoe shosse, 115409 Moscow, Russia
[2] A.M. Prokhorov General Physics Institute, Russian Academy of Science, Vavilov str.38, bld.5, 119991 Moscow, Russia

E-mail: farrakhova.dina@mail.ru

Abstract. This work is about the possibility of fluorescence diagnosis application with the use of aluminum phthalocyanine nanoparticles (nAlPc) in order to detect enamel microdamage. For the investigation, five human teeth samples of various age groups were removed for various reasons. The autofluorescence spectrums of these samples hard tissues and fluorescence spectrums of nAlPc mixed with enamel powder were obtained during the experiment. The research shows that sample pathogenic microflora causes nAlPc fluorescence. This fact will allow detecting enamel microdamage in future studies.

1. Introduction

The development of laser diagnostic and treatment technologies for different diseases discover new perspectives for doctors and gives hopes for recovery of many patients. Year after year, the number of procedures using the laser techniques in medicine, including dentistry, increases.

Laser radiation allows for faster healing process and decreases the risks of infections. It is a painless, non-invasive procedure that only influences the affected tooth areas [1]. Another advantage of using laser in medicine is that such kind of treatment does not cause any allergic reaction. In dentistry the laser is used for enamel whitening, normal and carious tooth tissues removing, diagnosing and treating of enamel, dentine and root canals, treating and preventing abnormal germination of tooth and curing mucous membrane of oral cavity [1,2,3]. Up to now the diagnostics of early caries and other oral cavity diseases, which appear due to genetic predisposition and infection processes, continue to be a relevant problem [2,3]. Laser methods provide lesser detection and treatment times for an early caries stage unlike the traditional tooth filling methods.

One of the most important goals of dentistry is the detection of enamel hidden demineralization areas, teeth deposit, calcified debris, stages of early carious lesion and quantification of lesion size, depth and tooth mineralization stages [3]. Carious is a complicated pathological process, which causes serious destruction of hard tooth tissue structure. Untimely detection and treatment of carious lesion result in significant enamel and dentine destruction and subsequent teeth removal. Assembly of pathogenic microflora in microdamaged areas causes mechanics influence and demineralization of tooth that later turns into carious lesions. Traditional diagnosis methods, such as visible inspection and X-ray examination, cannot detect early stage microdamage. In studies [4] and [5] the authors have carried out investigations of normal tooth enamel fluorescence intensity in comparison to microdamaged enamel fluorescence intensity. Fluorescence intensity of the microdamaged enamel was almost the same as normal enamel fluorescence intensity.

nAlPc are used for enamel microdamage localization detection. They have a good chemical homogenity, light absorption in 675-680 nm spectral range, moreover, nAlPc are harmless and may show a repair effect [6]. Pathogenic microflora accumulates in microdamaged enamel. nAlPc interact with pathogenic microflora and acquire fluorescence properties. Among other things, the nAlPc colloid solution does not have fluorescence properties. The investigations of nAlPc interaction with enamel powder are necessary for further enamel microdamage detection studies with the use of nAlPc.

2. Materials and methods

For this investigation, five samples of human teeth of various age groups were removed for various reasons.

The measurement of tooth hard tissues autofluorescence spectrums were taken at 405 nm wavelength laser excitation. One of these teeth was dissected longitudinally on two parts for enamel, dentine and cement autofluorescence spectrums measurement.

Also the fluorescence spectrum of nAlPc (FSUE «SRC NIOPIK», Russia), with enamel powder, was obtained. Water and saliva were added to the enamel powder with AlPc nanoparticles to simulate oral cavity.

The Spectroscopy system LESA-01-BIOSPEC was assembled and adjusted for this investigation (Figure 1). Fluorescence spectrums were taken at excitation by 405 nm semiconductor and 632.8 nm He-Ne lasers. Laser radiation was delivered to biological tissues using optical fiber.

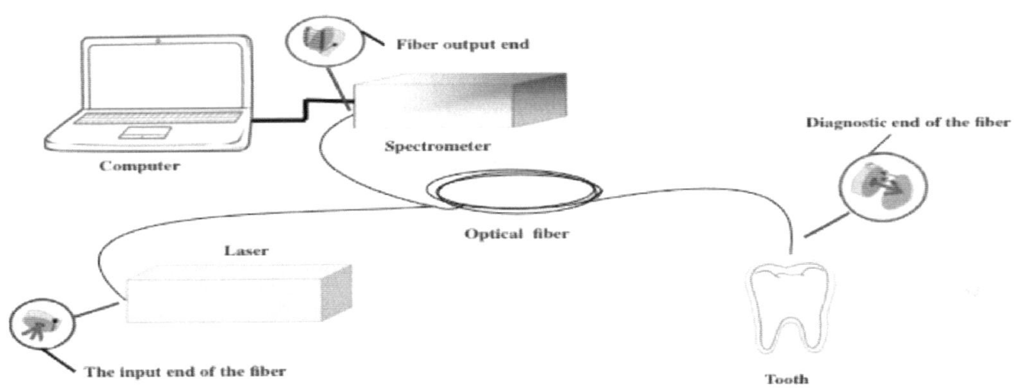

Figure 1. Measurement setup.

During the biological tissues irradiation, a part of it is absorbed by the tissues and the other part is reflected into the receiving fiber. The signal was detected by the spectroscopy system, which was made for fluorescence signal recording and analysis. Than the information goes to computer, which shows it on the display in spectral form by «Uno Momento» software.

3. Discussion

Using the spectroscopy system, fluorescence spectra of normal and carious lesion were investigated.

Hard tooth tissues autofluorescence spectrums are presented in figure 2(a). It can be seen that the cement fluorescence intensity is higher than the enamel and dentine fluorescence intensity because of different optical properties and hard tissues structure.

Enamel autofluorescence spectra of five different samples were obtained at the same excitation wavelength (Figure 2(b)). Enamel autofluorescence peaks were registered at same wavelength 565 nm. As it is known from earlier studies, enamel fluorescence intensity depends on the stage of mineralization and on the patient's age [1].The higher the level of tryptophan in enamel, the higher the fluorescence intensity signal [3].

The fluorescence spectra of nAlPc with enamel powder, water and saliva are presented on figure 2(c).

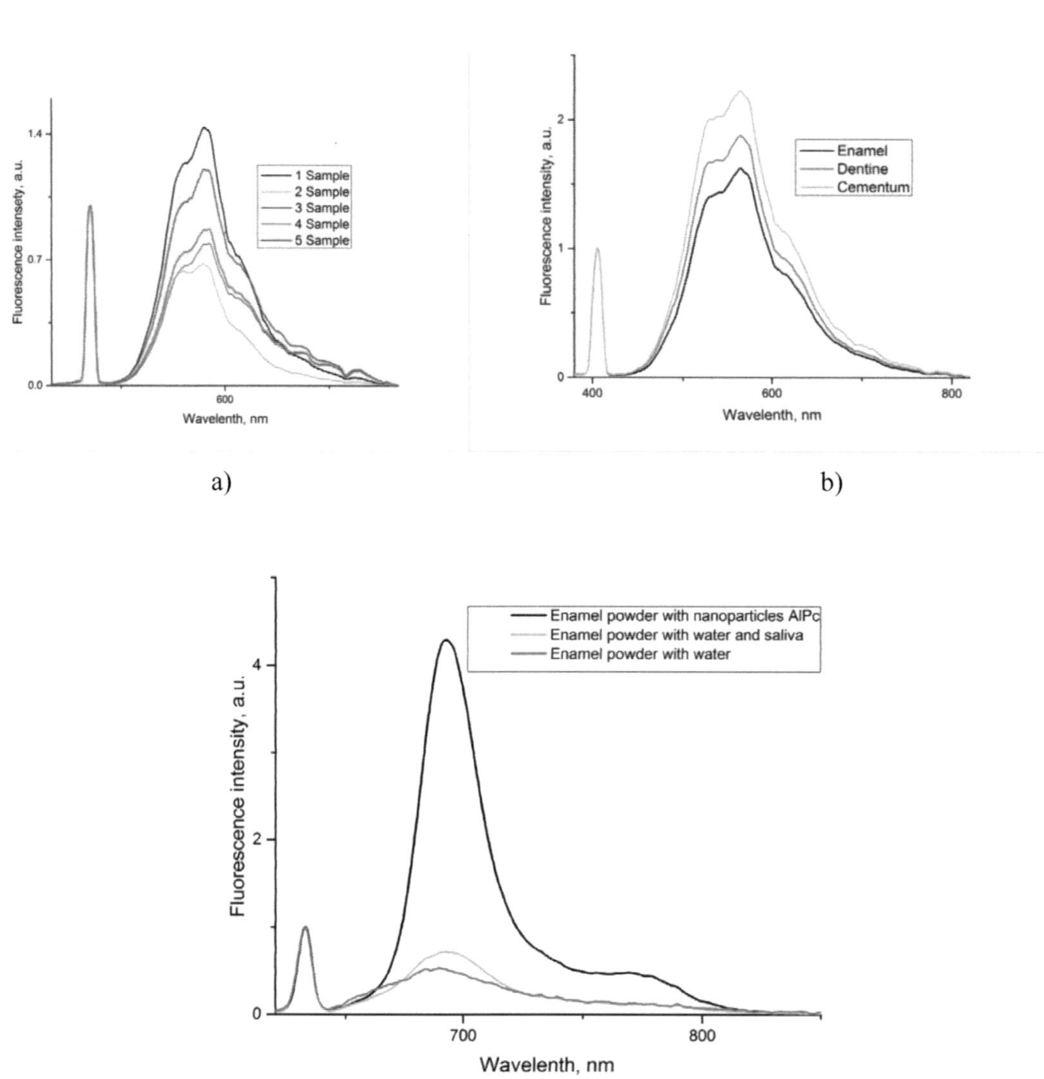

a) b)

c)

Figure 2. a) The autofluorescence spectra of five enamel samples , b) the autofluorescence spectra of enamel, dentine and cement, c) the fluorescence spectra of AlPc nanoparticles mixed with enamel powder, water and saliva.

Thus, microdamaged enamel contains pathogenic microflora. When nAlPc interact with pathogenic microflora, they start to acquire fluorescent properties. It is interesting to note that the nAlPc colloid solution does not fluorescence at 632.8 nm excitation wavelength.

The emergence of fluorescent properties can be explained by external influence. AlPc molecules separate from the nanoparticle and become free, surface molecules transform from para to ortho-position, displaying fluorescence [4,7].

4. Conclusion

During the investigation, the spectroscopy research of hard tooth tissues autofluorescence and fluorescence of nAlPc mixed with enamel powder, were conducted. Normal enamel autofluorescence spectrums of various patients varied due to difference of calcium, magnesium, phosphorus content in tooth enamel. Different enamel, dentine and cement fluorescence intensity was caused because of various mineralization stages and different optical properties. Also, the interaction of nAlPc with enamel powder was studied. This results allows detecting and identifying the location of microdamage on the enamel surface using nAlPc as a marker in the future.

References

[1] Pretty I A 2006 *Journal of Dentistry* **34** pp 727-739
[2] Thomas S 2009 *Spectroscopic Investigation of Tooth Caries and Demineralization* (India, Thiruvananthapuram) Chapter 1 pp 16-20
[3] Sinyaeva M L, Mamedov A A, Vasilchenko S Y, Volkova A I, Loschenov V B 2004 *Laser Physics* **14** pp 1132-1140
[4] Vasilchenko S Yu, Volkova A I, Ryabova A V, Loschenov V B, Konov V I, Mamedov A A, Kuzmin S G and Lukyanets E A 2010 *Journal of Biophotonics* **4** pp 336-346
[5] Sinyaeva M L, Vasilchenko S Y, Volkova A I, Korovin S B, Mamedov A A, Kuzmin S G, Lukyanets E A, Loschenov V B 2007 *Nanotechologies in Russia* **11-12** pp 58-63
[6] Steiner R, Breymayer J, Rück A, Loshchenov V and Ryabova A 2015 *Proceedings of SPIE* **9308** 93080R
[7] Bystrov F G, Makarov V I, Pominova D V, Ryabova A V and Loschenov V B 2016 *Biomedical photonics* **5** pp 3-8

Journal of Physics: Conference Series **737** (2016) 012049 doi:10.1088/1742-6596/737/1/012049

Application of nanophotosensitizers (aluminum phthalocyanine nanoparticles) for early diagnosis and prevention of inflammatory diseases

J O Kuznetsova[1], V I Makarov[2]

[1]National Research Nuclear University MEPhI (Moscow Engineering Physics Institute), 31Kashirskoe shosse, 115409 Moscow, Russia
[2]Prokhorov General Physics Institute, Russian Academy of Science, Vavilov str.38, 119991 Moscow, Russia

E-mail: YOKuznetsova@mephi.ru

Abstract. This paper deals with a possibility of new types of photosensitizers application – Aluminum Phthalocyanine nanoparticles (nAlPc) in clinical practice for diagnosis, prevention and therapy of inflammatory diseases in dentistry and traumatology. It was detected that the aluminum phthalocyanine (AlPc) fluoresces in the nanoparticle form in the presence of pathologic microflora or inflammation process. It will make possible to detect the local accumulation of pathological microflora on the enamel surface and also for diagnostics and treatment of inflammatory diseases. Experimental studies of interaction of NP-AlPc with tooth enamel and with biological joint tissue at arthrosis are presented.

1. Introduction

Nanophotosensitizers application opens new perspectives in the drug delivery system. They can be used for fluorescence diagnostics and treatment by photodynamic therapy (PDT). Aluminum phthalocyanine nanoparticles (nAlPc) are suitable for clinical use because they have a good transportation in aqueous media and penetration into the tissue. Also they are convenient tool for fluorescence diagnostics because nAlPc does not fluoresce in the nanoparticle form but in the monomeric form it does. The fluorescence occurring when AlPc molecules arranged vertically are attached to the surface of the nanoparticles during the dissolution process or the interaction with pathological tissue, pathologic microflora and macrophages [1], [2].

The application possibility of nAlPc in clinic practice are the following: in dentistry for detection of a local accumulation of pathological microflora on the enamel surface and in microcracks, in traumatology for fluorescence diagnostics and the photodynamic therapy of osteoarthrities. These applications are presented in this work.

Early diagnosis of tooth-enamel microcracks is of great importance in modern dentistry for caries prevention. Diagnostic methods being used in clinical dentistry (visual inspection, probing, vital enamel staining) can detect enamel damages of a rather large size. For diagnostics it is possible to use porphyrins autofluorescence when microflora is abundant in the microcracks. Otherwise the microflora autofluorescence is relatively weak. The problem can be solved by using nAlPc as exogenous fluorophore.

For clinical application nAlPc in orthopedics area the model of photoactivity appearance in the affected joint tissue is proposed. Photoactivity occurs in the body after the application of nAlPc only in diseased biological tissues whose microflora and immune system response differ from healthy biological tissue.

2. Materials and methods

The water colloids of AlPc nanoparticles in a concentration 50 μg/ml was received by ultrasonic dispergation large-dispersed AlPc crystals for 50 min using a Bandelin SONOPLUS HD2070 ultrasonic homogenizer with a KE76 attachment (20 kHz, 165 μm amplitude). The hydrodynamic radius of the obtained nanoparticles measured with the use of a dynamic light-scattering spectrometer, Photocor Complex (Photocor Instruments Inc., USA) was 25–100 nm. For fluorescent measurements a fiber-optic spectrometer LESA-01-BIOSPEC was used. He–Ne laser with the wavelength of 632.8 nm and the output power of 5 mW at the fiber end was used as an irradiation source for fluorescence excitation. A fiber-optic probe with one illuminating and six receiving fibers (each of 200 μm diameter) was attached to the spectrometer. Reference [3] contains a detail information about experimental setup.

The sample containing enamel powder and colloidal solution of AlPc nanoparticle was prepared to determine the nature of the interaction. Animal model of osteoarthritis was using for observing AlPc fluorescence enhancement in affected tissue. During experiment 1 ml of AlPc nanoparticles colloid in concentration 50 μg/ml was injected into joint capsule of affected and healthy joint. Before nanoparticles injection the autofluorescence of healthy and pathological joints was detected. The fluorescence spectra of nAlPc from both joints were measured 48 hours after nAlPc injection.

3. Results and discussion

3.1. The nAlPc fluorescence in the presence of enamel powder

Figure 1 presents the fluorescence of colloidal solution of AlPc nanoparticle mixed with enamel powder. As seen the AlPc fluorescence enhancement is present in the enamel powder. This fact confirms the theory of «nAlP activation» when on the enamel surface and in the microcracks there are pathological microflora.

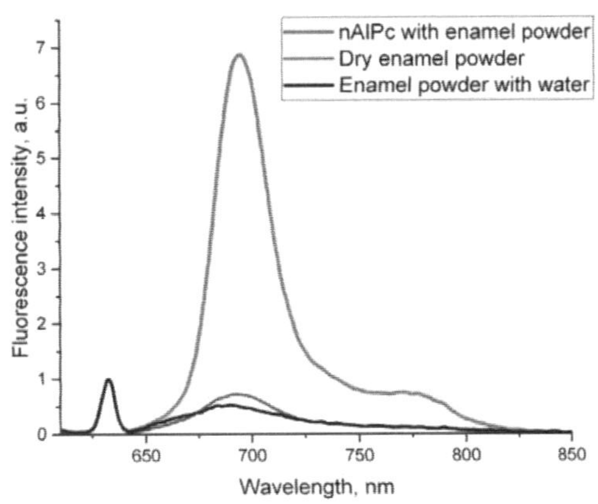

Figure 1. The nAlPc fluorescence in the presence of enamel powder

3.2. The nAlPc fluorescence in pathological joint

Figure 2 demonstrates that 48 hours after nAlPc injection the fluorescent intensity is 5-7 times higher in pathological joint than in healthy joint. It indicates that the nAlPc fluorescence enhancement occurs only in tissue with inflammatory processes.

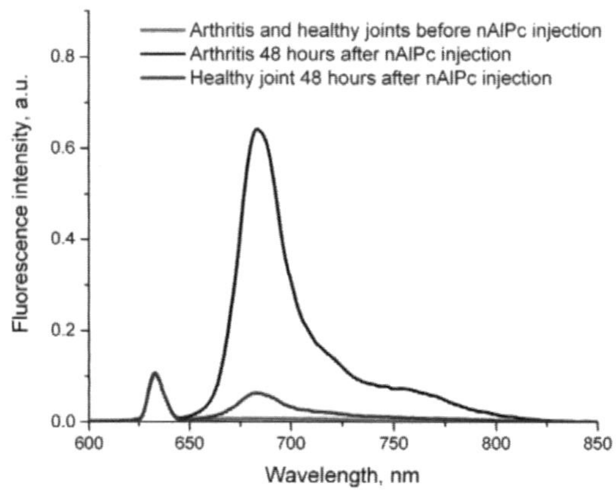

Figure 2. The nAlPc fluorescence in healthy and pathological joint

4. Conclusions

During the experiment it was revealed that nAlPc fluorescence enhancement occurs in the presence of enamel powder with pathological microflora. The nAlPc can be used as exogenous fluorophore for detection of a local accumulation of pathological microflora on the enamel surface and in microcracks in dentistry in the future .

In diseased biological tissue of joint the increasing of nAlPhc fluorescence intensity was obtained. In traumatology nAlPc fluorescence enhancement in pathological tissue can be used for fluorescence diagnostics and the photodynamic therapy of arthrosis.

Acknowledgments

This work was supported by a grants RFBR # 16-32-00930-mol_a. The research was carried out in Laser Biospectroscopy Laboratory at the A. M. Prokhorov General Physics Institute RAS together with the Laser Micro and Nanotechnology Department NRNU MEPhI. We thank Professor Victor B. Loschenov for his scientific support.

References

[1] Breymayer J, Ruck A, Ryabova A V, Loschenov V B and Steiner R W 2014 *Photodiagnosis Photodyn. Ther.* **11 (3)** pp 380–390

[2] Vasilchenko S Y, Volkova A I, Ryabova A V, Loschenov V B, Konov V I, Mamedov A, Kuzmin S G and Lukyanets E A 2010 *J. Biophotonics* **3 (5–6)** pp 336–46

[3] Loschenov V, Konov V and Prokhorov A. 2000 *Laser Physics* **10 (6)** pp 1188–1207

Detailed spectral analysis of decellularized skin implants

E V Timchenko[1], P E Timchenko[1], L T Volova[2], D A Dolgushkin[2], P Y Shalkovsky[1], S V Pershutkina[1]

[1] Samara National Research University, 443086 Russia, Samara, Moscow highway, 34
[2] Institute of Experimental Medicine and Biotechnology (IEMB) 443079 Russia, Samara, ul. Gagarin, 20

E-mail: laser-optics.timchenko@mail.ru

Abstract. The resutls of detailed analysis of donor skin implants using Raman spectroscopy method are presented. Fourier-deconvolution method was used to separate overlapping spectrum lines and to improve its informativeness. Based on the processed spectra were introduced coefficients that represent changes in relative concentration of implant components, which determines the quality of implants. It was established that Raman spectroscopy method can be used in assessment of skin implants.

1. Introduction

Restoration of human skin tissue defects is one of the most important problems of the modern world. [1]

Biomaterials play a crucial role in implantology: maintain the shape and structure of tissues, provide mechanical stability and recovery of integrity of their functions [2]. Use of donor biomatrixes is limited by their immunogenic properties caused by epidermal cells and endothelial cells in tissue. During its implantation into an organism the process can be associated with the occurrence of undesirable effects such as disruption of homeostasis, possibility of implant rejection, its poor integration. Therefore, during the manufacturing of skin implants, tissues treated by decellularization method. This process can be followed by decrease of content of key components of biomatrix: glycosaminoglycans (GAGs) and proteins compared to native samples, which can lead to deterioration of integration during its implantation. At the same time the loss of glycosaminoglycans leads to the loss of tissue elasticity [3], which can also adversely affect the implant engraftment.

Thus, the actual problem of implantology is assessment of biomatrix treatment.

At the present level of development of science and technology optical methods can be applied for the assessment of human skin implants due to its simplicity, efficiency and noninvasiveness. Raman spectroscopy (RS) [6,7,8] finds wide application for quality control of materials in tissue engineering [9,10,11]. Authors of the work [8] used Raman spectroscopy method to study the chemical composition of human skin. The analysis found that the absence of peaks in the spectrum at specific wavenumbers confirms the absence of certain components in human skin. In works [11,12] was used Raman spectroscopy method for assessing the content of key skin components - amide type I and glycosaminoglycans (GAGs) and their corresponding wavenumbers.

At the same time implants, made from human skin, are multicomponent biomatrix, including in its composition a number of components, whose spectra may overlap each other, which, in turn, makes the joint spectrum less informative without special treatment. Thus, to highlight certain spectral lines it is necessary to apply methods of spectral decomposition of profile to highlight each individual component, which is part of skin [4].

One of these methods is method of Fourier-deconvolution. It provides a complete separation of components [4], when precisely known a testing function. However, in practice to improve the resolution of spectrum it can be used without knowledge of form of lines, using various testing functions. The advantage of this method - the consistency of selected components and their parameters when selecting various testing functions that indicates the reliability of the results obtained. As test functions are often occur Gaussian or Lorentz functions or their convolutions.

Detailed analysis can also be carried out using the method of higher derivatives. For this method does not required data about shape of elementary band like in Fourier-deconvolution [5].

Purpose of the work is to study the skin implants using method of Raman spectroscopy with the use of spectral decomposition of profile method.

2. Materials of research

As the objects of study were used allogenic treated skin and control sample (untreated human skin). Chemical treatment of biological tissues was carried out using the decellularization technology from Heinrich-Heine University (Germany) [13].

Control of the qualitative composition of skin surface of implant was carried out using an experimental stand, including a high-resolution digital spectrometer Shamrock sr-303i with built-in cooling chamber DV420A-OE, fiber-optic probe RPB-785 for Raman spectroscopy, combined with laser module LuxxMaster LML-785.0RB-04, which is described in detail in [14]. The stand has the following characteristics: scanning range of wavelengths from 190 nm to 1200 nm; the exposure time from 0.04 to 600s; camera resolution of 1024 * 255 pixels, 50-450 mW laser power with a wavelength of 785 nm.

3. Results and discussion

Spectra were recorded in the range of 500-2500 cm^{-1}. Processing of spectra was carried out in the Mathematica'8 software environment and was smoothing the noises with median filter, polynomial approximation of the fluorescent component in spectrum and subtracting it and obtaining the processed RS spectrum.

Using deconvolution of Lorentz-Gauss functions in MagicPlot Pro software environment were carried out spectral decomposition of profile into spectral lines (fig. 1). In view of the plenty of lines in spectrum and to improve stability of solution, treatment was carried out separately in spectral areas 300-746, 746-1356, 1356-1789 и 1789-2199 cm^{-1}.

Average value of the coefficient of determination of the resulting spectrum, generated by the spectral profiles from the original Raman spectrum for 300-2199 cm^{-1} areas, and average standard deviation of analysis are given in table 1.

Table 1 – Data about deconvolution for different RS areas

RS area (cm^{-1})	Average value of the coefficient of determination (R^2)	The average standard deviation of analysis (σ)
300-746	0.9998	0.5365
746-1356	0.9998	0.2249
1356-1789	0.9987	0.6121
1789-2199	0.9994	0.2658

The degree of treatment of skin implants is determined by the absence of immunogenic cells, such as epithelium, RNA, and preservation of glycosaminoglycan and amide groups in their composition, in this regard, the most interesting for the study is 746-1356 cm^{-1} area, which includes the wave lines we are interested in.

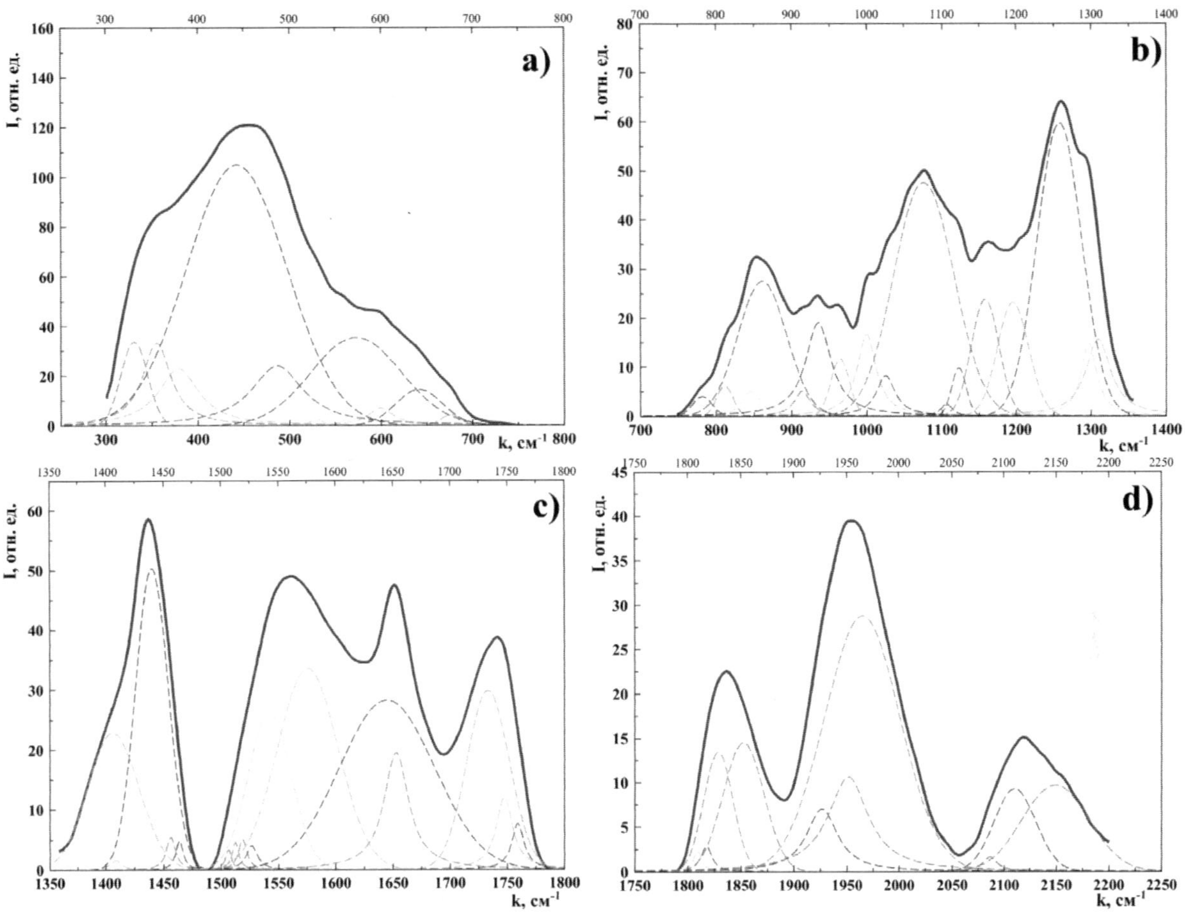

Figure 1. The characteristic spectral profiles for areas:
(a) 300-746 cm^{-1}, (b) 746-1356 cm^{-1}, (c) 1356-1789 cm^{-1}, (d) 1789-2199 cm^{-1}

When studying the spectra the greatest interest have intensities at the wavenumbers 1062 cm^{-1}, 1645 cm^{-1}, 1260 cm^{-1}, 814 cm^{-1}, 1202 cm^{-1}, corresponding to components that are important for the quality of implant: glycosaminoglycans, amide type I, amide type III, asymmetric C-O-S bond oscillations of glycosaminoglycans, RNA and tyrosine, hydroxyproline [14,15,16].

The regenerative properties of implants are determined by their structure and composition. Thus, saved glycosaminoglycans, proline and hydroxyproline affect its structure and elasticity, and the residual RNA and cells are capable of causing the immunologic response of recipient and rejection of implants. The proposed coefficients allow evaluating these parameters. Relatively permanent component in skin implants is amide II [11,12] corresponding to the wavenumber 1553 cm^{-1}, so it was used as denominator (I_{1553}) in the introduced optical coefficients (k):

$$k = \frac{I_i}{I_{1553}},$$

where I_i – intensity values at wavenumbers for analyzed components.

Figure 2 shows the two-dimensional diagrams of the introduced optical coefficients k. Clearly seen the difference between the control and treated samples

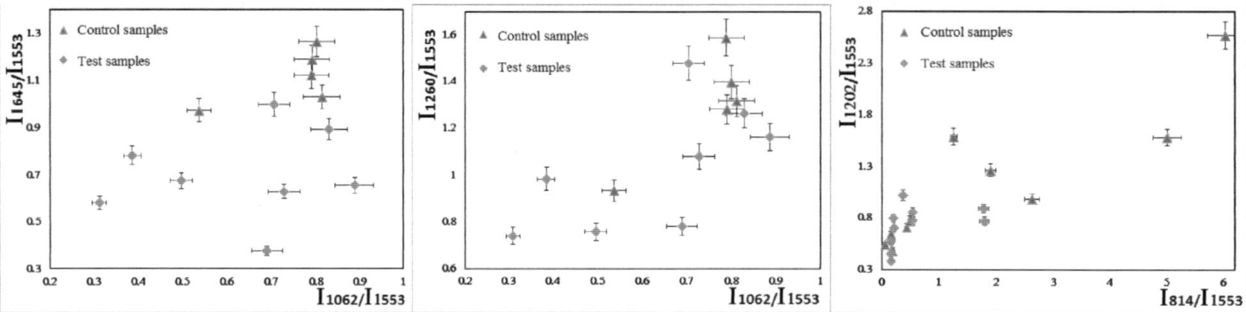

Figure 2. Two-dimensional diagrams of the introduced optical coefficients

The value of line intensities of amide I and amide III is greater in control sample compared to the sample treated. It is seen that the decellularization process of biomatrixes can reduce the content level of immunogenic cells, such as RNA and DNA, which, in turn, is represented on the change of the optical coefficient I_{814}/I_{1553}, while maintaining the immutability of the quantitative content of glycosaminoglycans (GAGs), as can be seen in Figure 2 from relation $I_{1062}/I_{1553.}$
Thus, when using a detailed analysis of the Raman spectra for the assessment of skin implants it is shown that during their processing components, adversely affecting their engraftment, are removed, at the same time retained the necessary level of glycosaminoglycans, proline and hydroxyproline.

4. Conclusion

Deconvolution of spectra using the Gauss-Lorentz functions allows carrying out component structural analysis of Raman spectra. When studying the spectra the greatest interest have intensities at the wavenumbers 1062 cm^{-1}, 1645 cm^{-1}, 1260 cm^{-1}, 814 cm^{-1}, 1202 cm^{-1}, corresponding to components that are important for the quality of implant: glycosaminoglycans, amide type I, amide type III, asymmetric C-O-S bond vibrations of glycosaminoglycans, RNA and tyrosine, hydroxyproline.
Raman spectroscopy can be used to assess skin implants.

This work was supported by the Ministry of Education and Science of the Russian Federation.

References

[1] Gerasimova L.I. 2000 *Scientific and practical journal Combustiology*, **2** p 36-42.
[2] Davis H.E. Leach J.K. 2008 *Topics in Multifunctional Biomaterials and Devices*, **10** p 1–26.
[3] Kurapeev D.I. Lavreshin A.V. S.V. Anisimov 2012 *Cellular Implantology and tissue engineering*, **1** p 34-39.
[4] Guo Y. 2006 High and broad raman gain in the boson peak region 133.
[5] Kurchatov I. S. 2010 Computer analysis of the vibrational spectra of water-alcohol solutions, 42.
[6] Caspers P. J. Lucassen G. W. Puppels G. J. 2001 *J. of Investigative Dermatology*, **116** 434.
[7] Matousek P. Draper E. R. Goodship A. E. Clark I. P. Ronayne K. L. Parker A. W. 2006 Central Laser Facility Annual Report, **4** 133.
[8] Ali S. M., Bonnier F., Tfayli A., Lambkin H., Flynn K. 2013 *J. of Biomedical Optics*, **8** 1.
[9] Assmann, A. Delfs C. Munakata H. 2013 *Biomaterials*, **34** p 6015-6026.
[10] Khmaladze A. Ganguly A. Morris M. D. 2013 *Tissue Engineering Part*, **19** 299.
[11] Zhao J.Y. Chai J.K. Song H.F. Sun T.J. Li D.J. Liu L.Y. Gao Q.W. Liang L.M. Zhonghua Yi XueZaZhi 2011 **91** 1276 .
[12] Kunstar A.A. Leferink A.M Okagbare P.I. 2013 *J. of the Royal Society Interface*, **10** 1 .
[13] Timchenko E.V. Timchenko P.E. Frolova L.T. Milyakova M.N. Maksimenko N.A. Taskina L.A. 2015 *J. Optical Society of America*, **3** 30.
[14] Timchenko E. V., Timchenko P. E., Volova L. T 2014 *J. Quantum Electronics*, **44** 696.
[15] Tfaili S. Gobinet C. Josse G. Angiboust J.-F. Manfait M. Piot O. 2012 *J. Analyst*, **137** 3673.
[16] Zanyar M. Shazza R. Ihtesham U.R. 2007 *J. Applied Spectroscopy Reviews*, **42** 493.

Investigation of the aluminum phthalocyanine nanoparticles colloidal solutions *pH*-dependent photoluminescence kinetics in pico- and nanosecond time range

F G Bystrov[1], **V I Makarov**[2], **V B Loschenov**[1,2]

[1]National Research Nuclear University MEPhI (Moscow Engineering Physics Institute), Kashirskoe shosse 31, 115409 Moscow, Russia
[2]Prokhorov General Physics Institute, Russian Academy of Sciences, Vavilov Str. 38, 119991 Moscow, Russia

E-mail: augustbystrov@gmail.com

Abstract. In this study, the fluorescence intensity decay of aluminum phthalocyanine nanoparticles colloidal solutions at different *pH* was investigated. Hamamatsu Streak-Camera (C10627 – 13 *Hamamatsu Photonics*) with picosecond temporal resolution (15 *ps*) was used to carry out the measurements. For excitation we used *Hamamatsu Picosecond Light Pulser PLP* – 10 with 637 *nm* wavelength and 65 *ps* pulse duration. The changes in fluorescence decay kinetics were found during the experiment. The number of fluorescence lifetime components and duration of lifetimes depends on *pH*. At *pH* 2 the presence of two fluorescence lifetimes was recorded: the first one was 5 *ns*, which corresponded to the molecular form in solution, and 1.5 *ns*, which corresponded to bound state of phthalocyanine molecules. This work is a preparatory step towards a model which describes the interaction of aluminum phthalocyanine nanoparticles with environment in biological tissue.

1. Introduction

Aluminum phthalocyanine nanoparticles (nan-AlPc) are promising type of nanophotosensitizers because of their unique properties. Fluorescence of nan-AlPc injected in biological tissue appears only in pathologically changed areas. It has been shown[1] that nan-AlPc can be used to assess the risks of skin autografts rejection. It was found that application of nan-AlPc colloidal solution on autografts leads to the increase of fluorescence intensity in the case of inflammatory rejection. Furthermore, fluorescence of nan-AlPc is susceptible to photobleaching effect[2] that is absent for molecular form of aluminum phthalocyanine. Therefore, the appearance of nan-AlPc fluorescence cannot be described as dissolution process.

2. Experimental details

This work is dedicated to the study of *pH* dependence of the nan-AlPc colloidal solutions photoluminescence kinetics. For measurements we used a registration system based on Hamamatsu streak camera (Figure 1) with picosecond temporal resolution (15 *ps*), conjugated with the fiber-optic spectrometer and picosecond laser pumping (637 *nm*, 65 *ps*)[3]. For investigations, five nan-AlPc colloids with *pH* 11, 10, 7, 3, 2 were prepared

Journal of Physics: Conference Series **737** (2016) 012051 doi:10.1088/1742-6596/737/1/012051

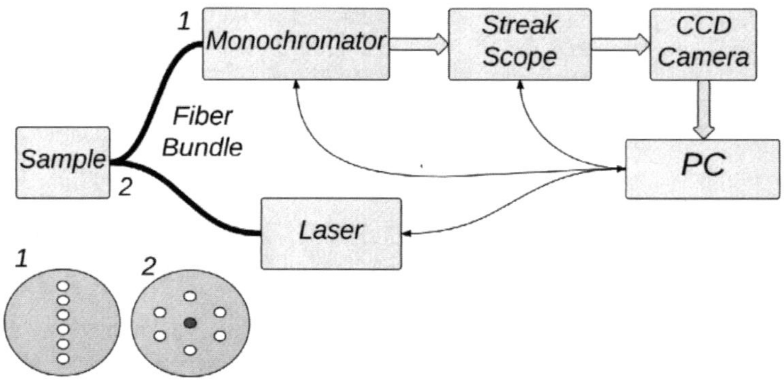

Figure 1. Experimental setup, 1 – end of the optical bundle connected to the monochromator; 2 – distal end of the bundle (1 fiber with laser light and 6 collecting fibers).

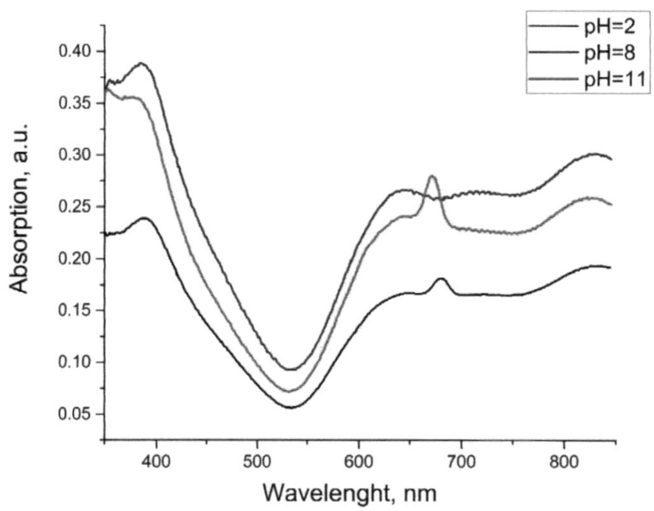

Figure 2. Nan-AlPc colloid absorption spectrum at different *pH*.

3. Results and discussion

Exploring the streak image of the nan-AlPc colloidal solution with *pH* 11 (Figure 3) we registered the appearance of fluorescence with two lifetimes: 3 *ns* and 0.5 *ns*. At *pH* 10 algorithm presents only one fluorescence lifetime 5 *ns*, which corresponds to the lifetime of AlPc molecular form in solution. In the case of acidic *pH* the fluorescence intensity decreased by almost one order of magnitude compared to alkaline *pH*. At *pH* 2 (Figure 4) noticeable disappearance of component at 740 – 770 *nm*, which was observed at *pH* 11. The calculation of the fluorescence lifetime for each case gives the following results: at *pH* 3 only one lifetime - 5 *ns*, which is similar to the fluorescence lifetime of the AlPc molecular form in solution. A more interesting result was found at *pH* 2. In addition to the 5 *ns* component the appearance of one more lifetime of 1,5 ns has been established.

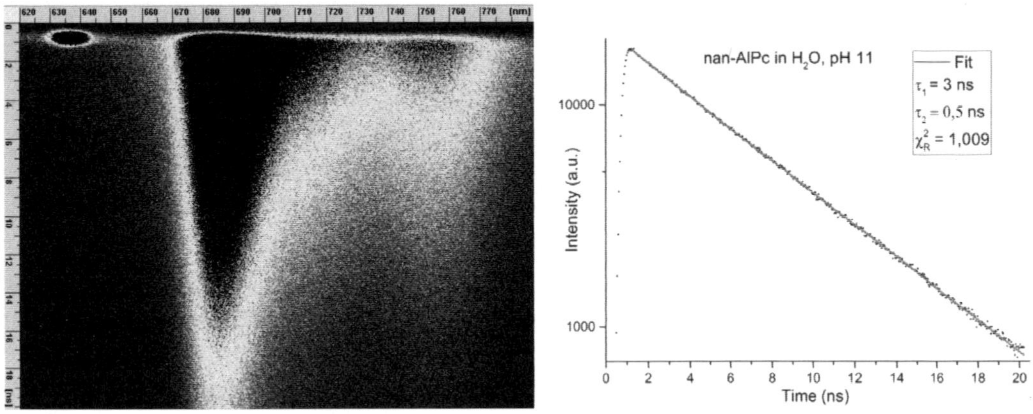

Figure 3. Fluorescence intensity decay of nan-AlPc colloidal solution at *pH* 11.

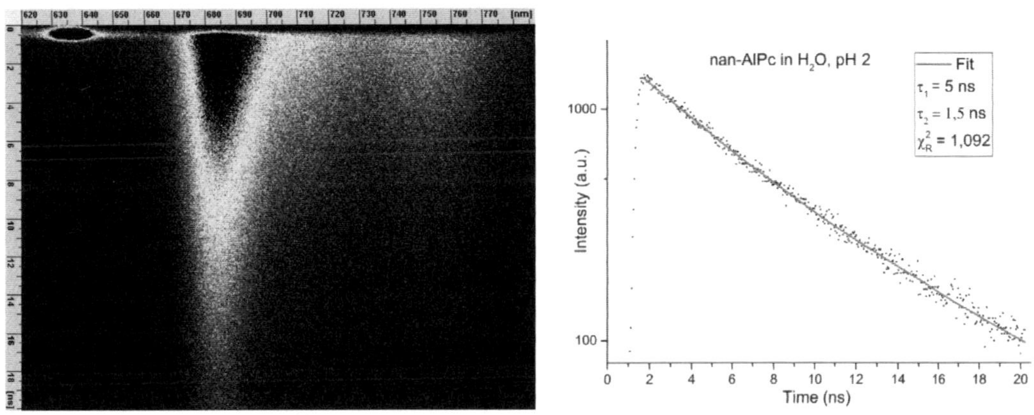

Figure 4. Fluorescence intensity decay of nan-AlPc colloidal solution at *pH* 2.

Based on the fact that fluorescence of nan-AlPc is susceptible to the photobleaching effect, that is absent for molecular form of aluminum phthalocyanine, it can be assumed that the appearance of nan-AlPc fluorescence is not only due to the normal dissolution, but also due to the transitions of surface molecules of nanoparticle to the different states. The appearance of several fluorescence lifetime components each of which corresponds to particular state of the molecule on the surface of the nanoparctile might be the evidence of these transitions.

References

[1] Ryabova A V, Loschenov V B, Konov V I et al 2010 *J. Biophoton.* **3 (56)** pp 336-346
[2] Makarov V I, Borodkin A V, Loshchenov M V et al 2013 *J. Appl. Phys.* **87** 965
[3] Bystrov F.G., Makarov V.I., Pominova D.V., Ryabova A.V., Loschenov V.B. 2016 *Biomedical Photonics* **5 (4)** pp 38

Journal of Physics: Conference Series **737** (2016) 012052 doi:10.1088/1742-6596/737/1/012052

Nanodiamonds + bacteriochlorin as an infrared photosensitizer for deep-lying tumor diagnostics and therapy

A S Sharova[1], YU S Maklygina [2], G V Lisichkin[3], P G Mingalev[3] and V B Loschenov[1, 2]

[1]National Research Nuclear University MEPhI (Moscow Engineering Physics Institute), 31 Kashirskoe shosse, 115409 Moscow, Russia
[2] A.M. Prokhorov General Physics Institute, Russian Academy of Science, Vavilov str.38, bld.5, 119991 Moscow, Russia
[3] Lomonosov Moscow State University (MSU), Chemical faculty, 119991 Moscow, Russia

e-mail: join93@rambler.ru

Abstract. The spectroscopic properties of potentially perspective nanostructure: diamond nanoparticles with a surface layer of IR-photosensitizer, bacteriochlorin, were experimentally investigated in this study. Such specific structure of the object encourages enhancement of the drug tropism to the tumor, as well as increasing of photodynamic penetration depth. The size distribution spectra of diamond nanoparticles; diamond nanoparticles, artificially covered with bacteriochlorin molecules layer, in aqueous solution, were obtained during the study. Based on the absorption and fluorescence spectra analysis, the benefits of functional nanostructure as a drug for deep-lying tumor diagnostics and therapy were reviewed.

1. Introduction

Fluorescence diagnostics (FD) and photodynamic therapy (PDT) are the most effective methods of cancer identification and treatment. These methods are based on the laser radiation influence on a photosensitive agent such as photosensitizer. For successful PDT and FD three key components are necessary: photosensitizer, light and oxygen. [1-4]

1.1. FD and PDT mechanisms

FD is a preparatory stage of PDT. The intensity of the fluorescent signal shows an accumulation rate of the photosensitizer in damaged cells that gives information about tumor localization, its possible proliferation and malignization.

PDT involves two stages. At first, photosensitizer, which is capable to accumulate in tumors at higher concentrations, compared with healthy tissues, is injected into patient's body (intravenously or orally). At the second stage, the damaged areas are illuminated with a certain laser light wavelength, corresponding to the photosensitizer absorption peak. Photosensitizer molecules' absorption of photons in the presence of oxygen (triplet form) induces photochemical reaction with highly active oxygen (singlet form) generation leading to the damage of biomolecules' structures and destruction of sensitized cells.

The need of photodynamic therapy and fluorescent diagnostics methods' improvement makes the task of creating of the photosensitizer with improved properties actual.

2. Purposes

2.1. Increasing of photodynamic penetration depth

The important characteristics of a photosensitizer, other than the ability to accumulate in tissues, are its ability to fluoresce and its spectral range of optical absorption. The greatest penetration depth into the biological tissue can be achieved by infrared (IR) spectral range radiation, where the proper absorption of normal tissue is minimal. In this way, the deepest photodynamic treatment effect can be realized by using photosensitizers with the absorption maximum in infrared (IR) and near-IR (740-1050 nm) spectral ranges. The use of such near-IR photosensitizers will allow analyzing of deep bio tissue layers, what is significant for deep-lying tumor diagnostics and therapy [5, 6].

2.2. Enhancement of drug accumulation selectivity in the tumor

An important role in ensuring of PDT/ FD efficiency and selectivity plays the photosensitizer accumulation level control in cells. In this regard, one of the photodynamic therapy' actual problem is to increase the selectivity of drug accumulation in tumor cells relative to normal tissue.
The mechanism of preferential photosensitizer accumulation in the tumor is associated with the angiogenesis specificity in the area of malignant neoplasms.

Organism cells receive oxygen and nutrients with the use of blood vessels, which normally have pores sizes ~ 2 nm. The principle of photosensitizer accumulation selectivity in the pathological tissues is explained by the fact, that drug molecules are not able to overcome the biological barrier and penetrate into the healthy tissues due to relatively small pores sizes existing in vessels. However, in the region of malignant cells biological barriers change. Tumor cells divide so actively that the blood vessels, "feeding" them, forced to stretch. The pores modify and increase in size (~ 100 ÷ 500 nm). The drug penetrates them and accumulates in malignancy. Furthermore, it should be noted that the capillaries pores of such excretory organs as liver, spleen, lungs and kidneys have sizes of ~ 80-150 nm. Thus, the photosensitizer is able to penetrate not only into the injured targeted organs , where it must show a therapeutic effect in combination with light impact, but also in other healthy tissues where its effect may be negative in certain conditions. This fact reduces the selectivity of drug accumulation in biological target.

Development and improvement of directed delivery system of drug to the tumor cells will reduce the concentration of the administered substance, decrease the drug accumulation time and provide drug accumulation only in malignant neoplasms. Due to the purpose of increasing of drug accumulation selectivity and its tropism to the tumor much attention, currently, is paid to the study of directed drugs delivery to damaged and modified cells using a variety of nanostructures. [7-17]

3. Materials and methods

Diamond nanoparticles are perspective materials for drug delivery due to their optical-spectroscopic, structural, electrochemical and surface properties, size variability within wide limits. [18-20]

An aqueous suspension of diamond nanoparticles was synthesized for this study (*Lomonosov Moscow State University (MSU), Chemical faculty). Prescribed by manufacturer* diameter of nanoparticles is ~ 5 ÷ 10 nm. However, the aggregation tendency (property) of such nanoparticles leads to the following experimental size distribution in an aqueous colloid: 12,8 nm; 215,4 nm. The initial concentration of the diamond nanoparticles in the solution - 8 mg / ml.

Meso-tetra (3-pyridyl) bacteriochlorin*(hereinafter "bacteriochlorin"), was used as the near-IR photosensitizer, chemical structure of which is shown in Figure 1a. The concentration of the aqueous bacteriochlorin solution- 10 mg / ml.

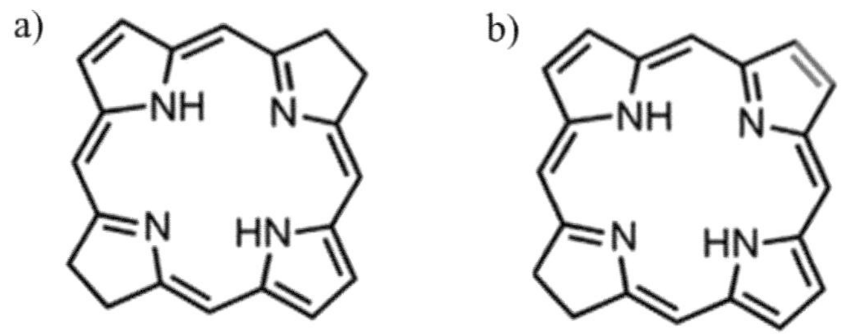

Figure 1. Chemical structure of a) meso-tetra(3-pyridyl)bacteriochlorin* b) chlorin

In order to increase PDT efficiency, for investigation were selected diamond nanoparticles artificially covered with IR -photosensitizer (bacteriochlorin) molecules layer. Such specific structure of the investigated object encourage enhancement of the drug tropism to the tumor, as well as increasing of photodynamic impact depth (Figure 2).

Therefore, the diamond nanoparticles bounded with bacteriochlorin, in aqueous solution, were studied. The initial concentration of diamond nanoparticles in solution – 8 mg/ml, the concentration of bacteriochlorin - 2 mg/ml.

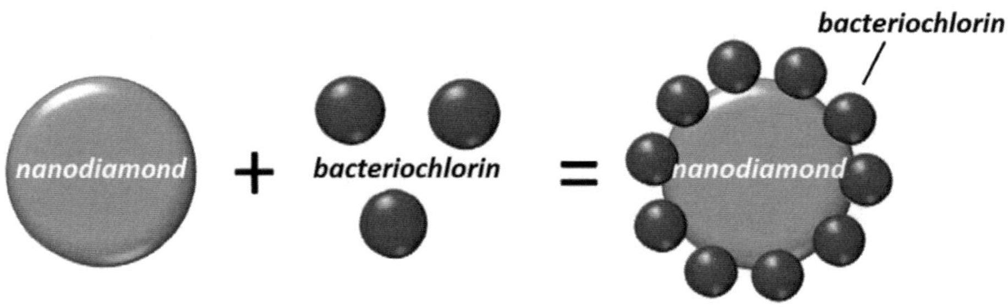

Figure 2. The nanostructure with improved properties (diamond nanoparticles bonded with bacteriochlorin molecules).

4. Results and discussion

4.1. Measuring of nanoparticles sizes

The sizes of original diamond nanoparticles; diamond nanoparticles coated with bacteriochlorin molecules were measured during the study. The measurements were carried out in aqueous solution, with the use of multi-angle dynamic scattering light spectrometer Photocor Complex, which allows getting the size distribution of nanoparticles by analyzing of correlation function of scattered light intensity fluctuations. Obtained slurry was dispersed in an ultrasonic homogenizer Bandelin SONOPLUS HD2070 with KE76 attachment (20 kHz, 165 microns amplitude) before the measurement. Thus, the following experimental size distributions were obtained for various samples:

- The diamond nanoparticles in aqueous solution-12,8 nm; 215,4 nm (Figure 3a).
- The diamond nanoparticles coated with bacteriochlorin, in aqueous solution-12 nm, 71,5 and 495 nm (Figure 3b).

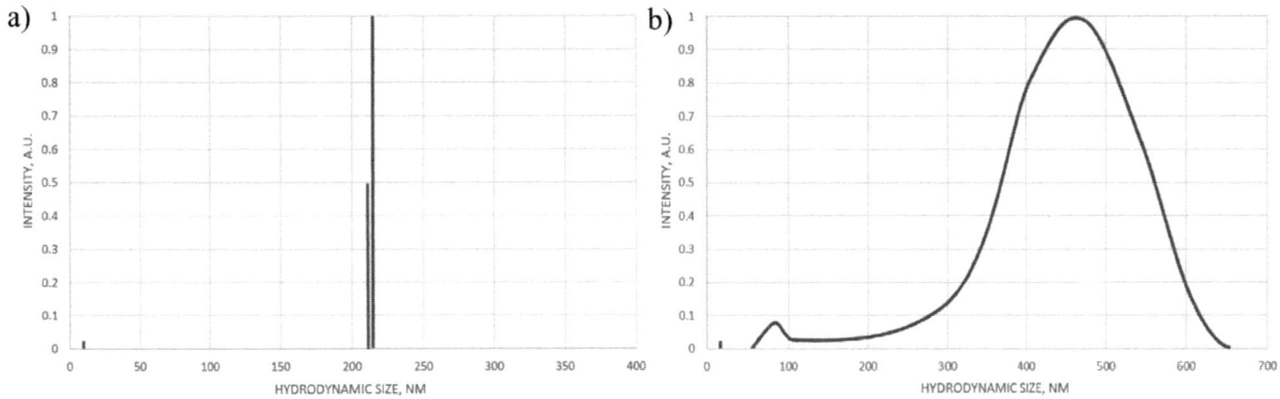

Figure 3. The range of size distribution of a) diamond nanoparticles b) diamond nanoparticles, covered with bacteriochlorin, in aqueous solution. The Spectrum is normalized to the nanoparticles light scattering intensity.

4.2. Determination of absorption and fluorescence spectra

The absorption spectra of the samples were obtained with the use of computerized spectrophotometer HitachiU-3400.

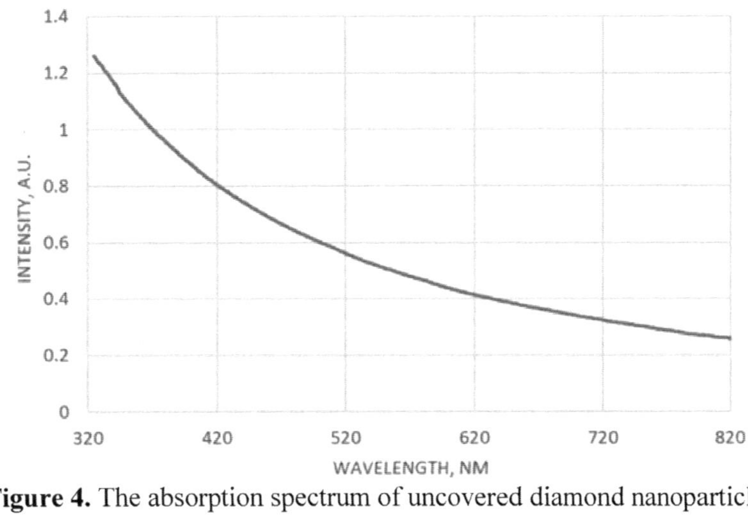

Figure 4. The absorption spectrum of uncovered diamond nanoparticles.

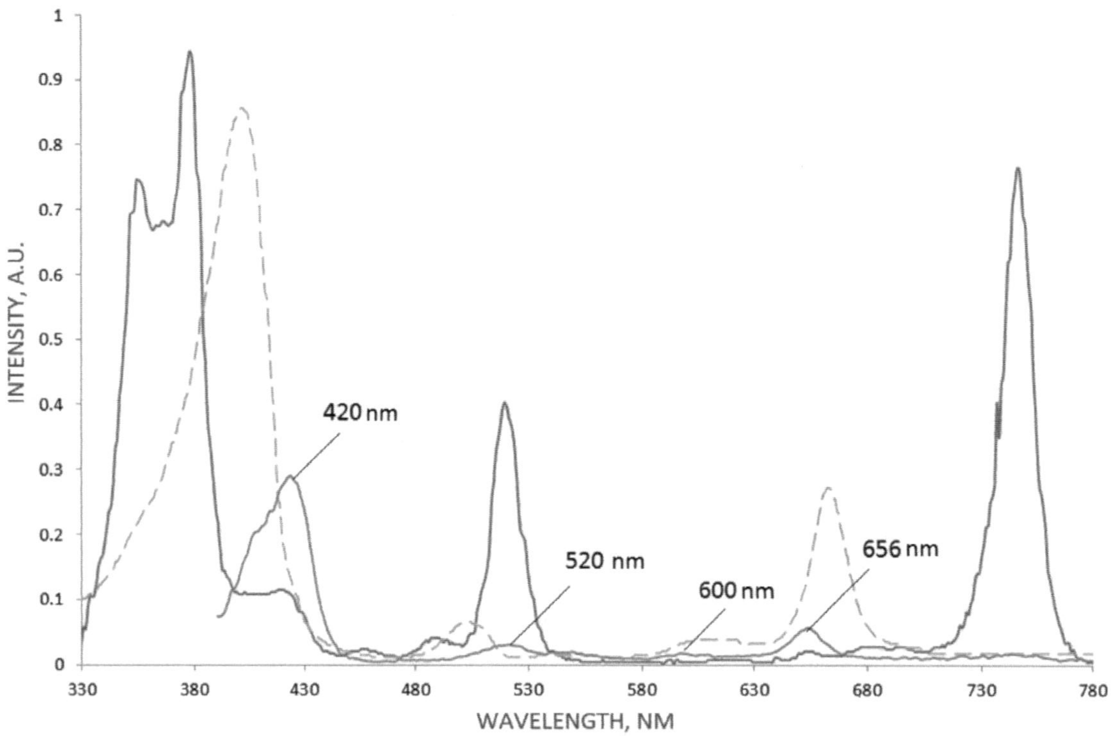

Figure 5. The absorption spectra of bacteriochlorin solution (●); chlorin (---); diamond nanoparticles, covered with bacteriochlorin (●).

The absorption spectrum of diamond nanoparticles changes intensely by coating them with bacteriochlorin. Maximums of absorption spectrum of diamond nanoparticles, covered with bacteriochlorin, occur at 420 nm, 520 nm, 600 nm and 656 nm wavelengths .The most evident peaks correspond to 420 nm, 656 nm wavelengths, where an active absorption of molecular chlorin form is registered (Figure 5). This circumstance can be explained by the fact that bacteriochlorin molecules on the surface of nanodiamond form a kind of antenna that effectively scatters the radiation in the spectral range corresponding to free bacteriochlorin molecules absorption. Evidently, surface bacteriochlorin molecules bond with nanodiamond surface and depending on the surroundings they may "lie" taking para- position according to nanodiamond surface, or "stand up" taking ortho- position, refraining on the nanodiamond top and showing bacteriochlorin and chlorin spectroscopic properties.

By the absorption spectra (Figure 5) analyzing it can be concluded that the most efficient fluorescence excitation of diamond nanoparticles covered with bacteriochlorin, should be realized at λ = 405 nm; 510 nm and 632.8 nm wavelengths.

Using NIR region (740-1400 nm) radiation considers to be the most perspective for fluorescence diagnosis and photodynamic therapy. However, during the study, it was found that the test samples of diamond nanoparticles coated with bacteriochlorin molecules, do not have pronounced IR -absorption peak. Therefore, choosing from three available sources with 400 nm, 510 nm and 632.8 nm wavelengths, in accordance with the absorption spectrum of the colloid test, the most efficient, from the standpoint of fluorescence excitation, would be a laser radiation with 405 nm wavelength. Experimental results confirmed this assumption. Thus, the fluorescence spectrum of colloidal solution of diamond nanoparticles with the surface layer of bacteriochlorin molecules was obtained with the use of fiber-optic spectrum analyzer LESA-01-"Biospec" (Figure 6).

Journal of Physics: Conference Series **737** (2016) 012052 doi:10.1088/1742-6596/737/1/012052

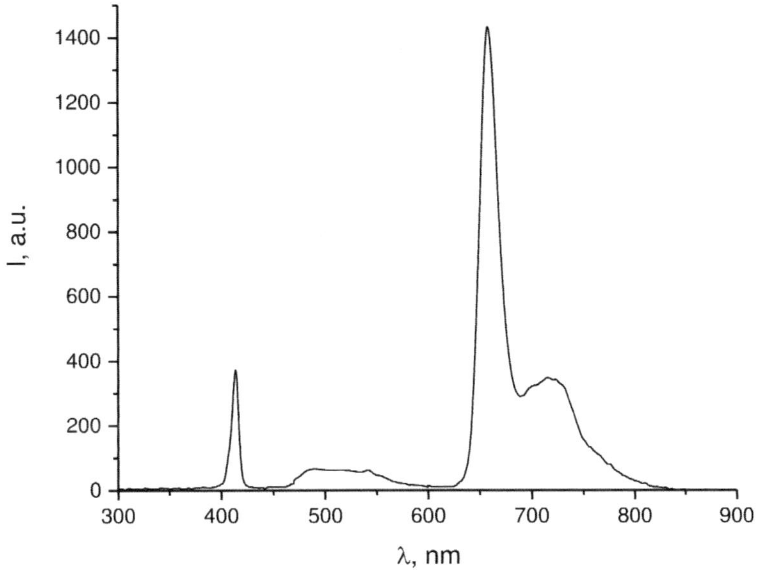

Figure 6. The fluorescent spectrum of diamond nanoparticles, covered with bacteriochlorin, in aqueous solution, with laser excitation at λ_{ex} = 405 nm wavelength.

The fluorescent spectrum of nanostructure created on the basis of diamond nanoparticles with a surface layer of bacteriochlorin molecules, excitated with λ_{ex} = 405 nm laser radiation, has two fluorescence bands at λ_{ex} = 749 nm and λ_{ex} = 662 nm, corresponding to the molecular bacteriochlorin and chlorin forms. Apparently, the surface bacteriochlorin molecules can change their position relative to each other and to the surface of nanodiamond structure, showing the spectroscopic properties of bacteriochlorin (λ_{em} = 758 nm) and chlorin (λ_{em} = 654 nm) solutions. This phenomenon can probably be explained by the surface bacteriochlorin molecules interaction with a complex and heterogeneous surface structure of nanodiamond.

Depending on the localization of molecular bacteriochlorin crystals on the surface of the diamond nanoparticles, surface photosensitizer molecules can "lie", taking the para-position relative to the nanodiamond surface, or "stand up", taking the ortho-position, keeping on the surface and showing the spectroscopic properties of photosensitizer solutions that are demonstrated in Figure 6.

It is seen from Figure 1 that chlorin differs from bacteriochlorin with one more double bond in its structural formula. In this case, probably under the exciting laser radiation influence two close bacteriochlorin molecules on the nanodiamond surface may form joint temporary double bond between them. In this case, one of the molecules may show spectroscopic properties of chlorin. The fluorescence spectrum of the diamond nanoparticles (450-600 nm) has no pronounced peak (Figure 6).

5. Conclusion

During the study, it was found that bacteriochlorin molecules interact with each other and with a complex surface structure of nanodiamond. Depending on the interaction forces, location and surroundings of bacteriochlorin, they may take a different position relative to each other and to the surface of nanodiamond structures, showing spectroscopic properties of chlorin, which does not have an absorption band in IR spectral range. However, it was shown, that the investigated drug has an intense fluorescence signal in the NIR spectral range. In this way, using the drug in the form of specialized nanoparticles covered with IR-photosensitizer (bacteriochlorin) molecules may perspectively enhance the drug accumulation selectivity in biotissues and increase the photodynamic penetration depth.

Acknowledgments
The work was supported by grants of RFBR (Russian Foundation for Basic Research) 14-03-00423.

References

[1] Loschenov V B, Konov V I and Prokhorov A M 2000 Photodynamic *Laser Physics* **10 (6)** pp 1188–1207
[2] Quirk B J, Brandal G, Donlon S, Vera J C, Mang T S, Foy A B, Lew S M, Girotti A W, Jogal S, LaViolette P S, Connelly J M. and Whelan H T 2015 *Photodiagnosis and Photodynamic Therapy* **12 (3)** pp 530-544
[3] Chiaviello A, Postiglione I and Palumbo G 2011 *Cancers* **3** pp 1014-1041
[4] Akimoto J 2016 *Neurol Med Chir* (Tokyo) **56** pp 151–157
[5] Rovers J P, de Jode M L, Rezzoug H, Grahn M F 2000 *Photochem. Photobiol.* **72** pp 358–364.
[6] Maklygina Yu S, Borodkin A V and Loschenov V B 2015 *Biomedical photonics* **3** pp 3–9
[7] Emeje M O, Obidike I C, Akpabio E I and Ofoefule S I 2012 *Recent Advances in Novel Drug Carrier Systems* (InTech, Chapters published) chapter 4
[8] Wilczewska A Z, Niemirowicz K, Markiewicz K H and Car H 2012 *Pharmacological Reports* **64** pp 1020-1037
[9] Bennet D and Kim S 2014 *Application of Nanotechnology in Drug Delivery* (InTech, Chapters published) chapter 4
[10] Swami A, Shi J, Gadde S, Votruba A R, Kolishetti N and Farokhzad O C 2012 *Multifunctional Nanoparticles for Drug Delivery Applications* (Springer Ltd) pp 9-29
[11] Mody V V, Cox A, Shah S, Singh A, Wesley B and Parihar H 2013 *Appl Nanosci.* **4** pp 385–392
[12] Cheng Y 2011 *Case Western Reserve University* pp 1-139
[13] Calixto G M, Bernegossi J, de Freitas L M, Fontana C R and Chorilli M 2016 *Molecules* **21** (342) pp 1-18
[14] Díaz M R and Vivas-Mejia P E 2013 *Pharmaceuticals* **6** pp 1361-1380
[15] Unterweger H, Subatzus D, Tietze R, Janko C, Poettler M, Stiegelschmitt A, Schuster M, Maake C, Boccaccini A R and Alexiou C 2015 *Int J Nanomedicine* **10** pp 6985–6996
[16] Pantiushenko I V, Rudakovskaya P G, Starovoytova A V, Mikhaylovskaya A A, Abakumov M A, Kaplan M A, Tsygankov A A, Majouga A G, Grin M A and Mironov A F 2015 *Biochemistry* **80** pp 752-762
[17] Dou Q Q, Teng C P, Ye E and Loh X J 2015 *International Journal of Nanomedicine***10** pp 419–432
[18] Ho D, Wang C H and Chow E K 2015 *Sci Adv.* 1 (7) pp 1-14
[19] Galli G 2010 Computer-Based Modeling of Novel Carbon Systems and Their Properties: BeyondNanotubes (Springer Ltd) pp 37-56
[20] Mochalin Vadym N, Shenderova Olga, Ho Dean and Gogotsi Yury 2012 *Nature Nanotechnology* **7** pp 11–23

Pulse wave transit time measured by imaging photoplethysmography in upper extremities

M A Volynsky[1,*], O V Mamontov[1,2], I S Sidorov[1] and A A Kamshilin[1]

[1] Computer Photonics and Videomatics Department, ITMO University, 49 Kronverksky Ave., Saint Petersburg, 197101, Russia
[2] Functional Diagnostics Department, Federal Almazov North-West Medical Research Centre, 2 Akkuratova Str., Saint Petersburg, 197341, Russia

*E-mail: maxim.volynsky@gmail.com

Abstract. We describe highly reliable measurement method of the pulse wave transit time (PWTT) to human limbs by using simultaneous recordings of imaging photoplethysmography and electrocardiography. High accuracy of measurements was achieved by access to a larger number of statistically independent data obtained simultaneously in different points. The method is characterized by higher diagnostic reliability because of automatic selection of the regions less affected by environmental noise. The technique was tested in the group of 12 young healthy subjects aged from 21 to 33 years. Even though PWTT in right and left hands was comparable after averaging over the whole group of subjects, significant difference in the time delay of pulse wave between the hands was found in several individuals. The technique can be used for early-stage diagnostics of various vascular diseases.

1. Introduction

Pulse wave velocity (PWV) is an important indicator of arterial stiffness in clinical practice [1]. Moreover, PWV is a marker of cardiovascular risk in hypertensive patients [2] and in diabetics [3]. Pulse wave transition time (PWTT) is a parameter to be measured for assessment of PWV. It can be defined as a time taken for arterial pulse pressure wave to travel from the aortic valve to the peripheral site of the body. Inaccuracy of PWTT measurements affects reliability of medical diagnostics. Development of new, noninvasive, easy for practical use, and cost-efficient methods for measuring this parameter is important for obtaining vital information related to arterial conditions, which can contribute to further improvement of the diagnostic potential of this important parameter. In this paper, we study feasibility of PWTT measurements by imaging photoplethysmography (IPPG) of high spatial resolution [4] simultaneously applied with electrocardiography (ECG). The method possesses higher accuracy of PWTT estimation owing to joint analysis of two-dimensional data obtained with IPPG. In addition, PWTT measured in right and left hands of the same subject was compared.

2. Subjects and methods

2.1. Participants

Twelve apparently healthy volunteers (3 women and 9 men) participated in the experiment. Age of subjects ranged from 21 to 33 years. The study was conducted in accordance with ethical standards presented in the 1964 Declaration of Helsinki. The Ethics Committee of the Federal Almazov

North-West Medical Research Center approved the protocol of this study prior the research. All subjects gave their informed consent in the written form of participation in the experiment.

2.2. Experimental configuration

Spatial distribution of blood pulsations was measured in the area of subject's palm by using custom made IPPG system in reflectance geometry as shown in figure 1. It includes just two elements: illuminator and video camera. The palm was illuminated by light emitting diode (LED) operating at the wavelength of 530 nm. The illuminating light was linearly polarized by means of the film polarizer attached to the illuminator. Another polarizer was attached to the camera lens. The transmission axes of the polarizers were mutually orthogonal to minimize light reflection from the skin. Such polarization filtration allowed us to diminish influence of motion artefacts on the revealed blood pulsations [5]. Light reflected from subject's palm was collected into the digital black-and-white CMOS camera (8-bit model GigE uEye UI-5220SE of the Imaging Development Systems GmbH). The camera was continuously recording a series of video frames with the focused images of the palm at the frame rate of 22 frames per second (fps). The series of images sizing 752 × 480 pixels was saved into a personal computer for digital data processing.

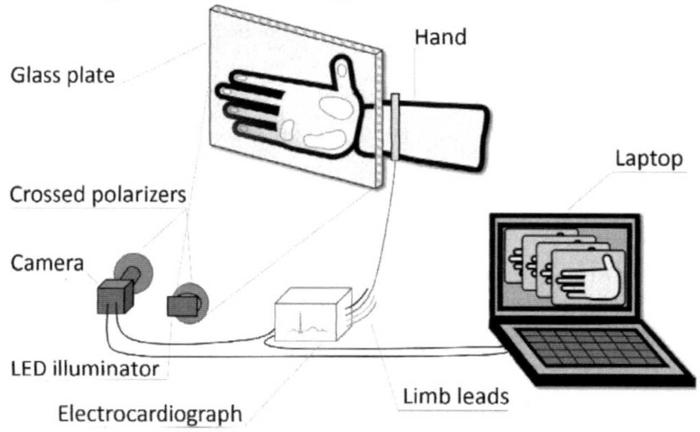

Figure 1. Simplified layout of the experimental setup for simultaneous video and ECG recordings.

Simultaneously with video for each subject we recorded an electrocardiogram (ECG) by means of digital electrocardiograph (model KAP-01-"Kardiotekhnika-EKG" of the Incart Ltd.). ECG was recorded at the data acquisition frequency of 1 kHz by using disposable electrodes attached to the left and right wrists with the reference electrodes on the legs. To synchronize ECG and video recordings, electric pulses corresponding to the beginning of each frame were recorded in one of the free channels of the electrocardiograph. By this way, the difference in the time scale between video and ECG recordings did not exceed 1.0 ms.

2.3. Experimental protocol

Each subject was asked to sit comfortably and lean its palm to the vertically fixed 4-mm thick glass plate (see figure 1). The back side of the palm was adhered to the glass by transparent adhesive tape. Such a procedure was applied to increase the signal-to-noise ratio in IPPG measurements. This was achieved because of (i) diminishing of motion artefacts during video recording, and (ii) significant increase of the IPPG-signal amplitude at the heartbeat frequency due to more efficient modulation of the capillary density in the dermis by pulsatile arterial transmural pressure [6, 7]. Duration of the video recording was 30 s. Measurements were carried out with both left and right hands of each subject.

2.4. Data processing

The whole set of the recorded data was processed offline by custom designed software implemented in the MATLAB® platform. First, we calculated spatial distribution of the amplitude of photoplethysmographic (PPG) signal pulsations over the palm area. In contrast with our previous

IPPG algorithms [4, 6], here we used more straightforward approach for calculations the pulsations amplitude. We considered a frame-to-frame evolution of the pixel values as a PPG-waveform. It was calculated by averaging pixel values in the Region of Interest (ROI) sizing 11 × 11 pixels which corresponds to the area of about 3 × 3 mm^2 at the subject's palm. As any PPG signal [8], it typically consists of the alternating (AC) part at the heartbeat frequency and slowly varying (DC) parts. After detrending and calculating the ratio of AC/DC, one can obtain a waveform which is shown in figure 2 by the red curve. As conventionally accepted in the literature devoted to plethysmography [8], the PPG waveform was inverted to positively correlate with variations of arterial transmural pressure. The ROI was movable over the palm area with the step of a half of the ROI size in both directions.

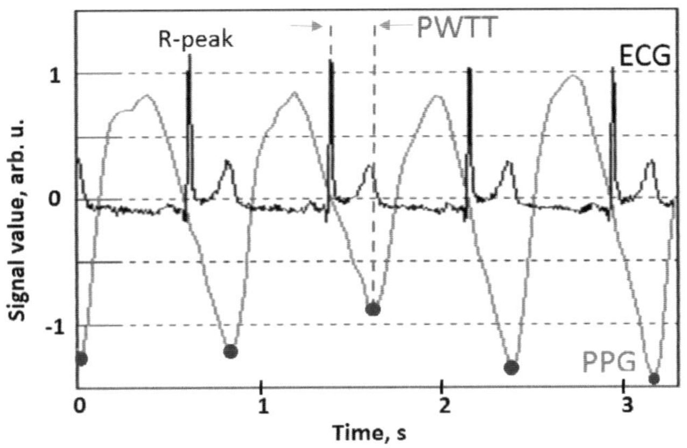

Figure 2. A fragment of simultaneously recorded PPG waveform (red curve) and ECG signal (black curve). Blue circles show end-diastole moments determined from the PPG waveform.

In the next step of the algorithm, we computed a 'goodness metric' for the whole set of ROIs. Two criteria were adapted to select the 'good' ROI in which the PWTT parameter was calculated: (i) the number of the cardiac cycles estimated from PPG and ECG recordings is the same during the measuring window of 30 s, and (ii) cross-correlation between occurrence of the PPG-waveform minima and R-peaks of ECG is significant ($p < 0.05$). To resolve each cardiac cycle from IPPG recordings, positions of the PPG-waveform minima in the time axis were found for every ROI. As an example, blue circles in figure 2 show the end-diastole points determined for the particular PPG waveform. The minimum of the inverted PPG waveform is a physiological equivalent of the R-peak of ECG because the latter corresponds to the electrical systole coinciding with the phase of isovolumic contraction of the left ventricle when the blood pressure is at its minimum (end-diastole). Therefore, one can calculate PWTT as the time delay between the R-peak and respective minimum of the PPG waveform [9, 10] which corresponds to the moment when the pulse wave reaches the selected ROI at the hand. The PWTT was estimated only in the selected 'good' ROIs as the time delay averaged over all cardiac cycles within the measuring windows.

3. Results

3.1. Selection of 'good' ROIs

Spatial distribution of 'good' ROIs in the palm of one of the examined subjects is shown in figure 3. It is seen that all 'good' ROIs are situated in the areas where the skin is contacted with the glass plate. Similar distributions with the most of 'good' ROIs at the palm-glass contact were obtained for all studied subjects. This fact provides an additional confirmation of the recently proposed model of PPG signal formation in which pulsatile arterial pressure deforms the connective-tissue components of the dermis resulting in periodical changes of both the light absorption and scattering [6]. The contact with glass plays a role of the mechanical border for enhancement of the dermis compression thus providing better correspondence to variations of arterial transmural pressure [7]. Only 7% of the total number of

ROIs were selected as 'good' in the particular case shown in figure 3. For other subjects this ratio was varying from 1% to 68% with the mean of 22%. In spite of small percentage of 'good' ROIs in three palms, the absolute number of the selected ROI was high enough (the smallest was 19) for the reliable estimation of the pulse wave transit time because of initially large number of independent ROI in the palm area.

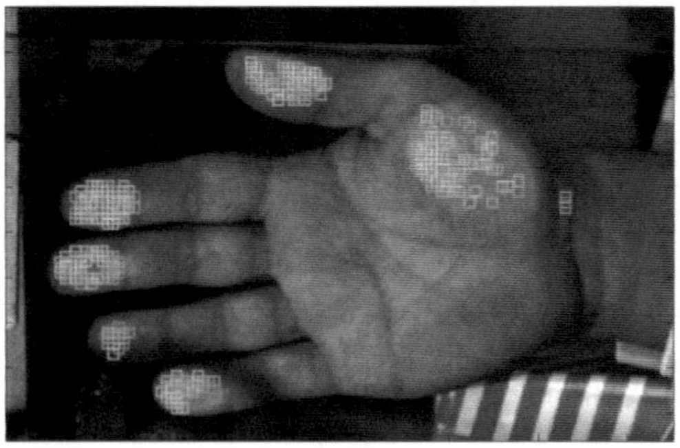

Parameter	Value
Heart rate	68.9 bpm
Mean duration of the cardiac cycle	871 ms
Mean blood pulsation amplitude	1.02 ± 0.04 %
DC level of PPG	103.2 from 256
Total # of ROIs in the palm area	4393
Selected 'good' ROIs	300
PWTT	255 ± 1 ms

Figure 3. Example of data processing: a palm with selected 'good' ROIs and calculated parameters.

3.2. *PWTT measurements*

In each selected ROI, the estimated time delay between the R-peak and PPG-waveform minimum has rather high dispersion: the standard deviation is typically of the order of the time delay, which could be caused by moderate frame rate (22 fps) used for video recording in our experiments. Nevertheless, mean PWTT values (after averaging over all cardiac cycles) possess small standard deviation difference from the magnitude averaged over the all selected 'good' ROIs in the palm. For example, for the palm shown in figure 3, PWTT is 255 ms with the measurement error of 1 ms calculated with the confidence level of 95% for the set of 300 'good' ROIs. The measurement error of PWTT for the palms of other subjects was between 1 and 8 ms.

Figure 4 shows the PWTT parameter for both right and left hand of all subjects participated in the experiment. The PWTT of the right hand averaged over the whole cohort was 244 ± 20 ms, which was comparable with that of the left hand: 240 ± 14 ms. In spite of comparable values in average, the right-left hands difference in PWTT of some individuals was significant (up to 28 ms) and with different sign: the time delay in the right hand was larger in seven subjects of twelve as seen in figure 4. It was found that there is no significant dependence of PWTT on the subjects' age ($r = 0.13$, $p > 0.05$). We hypothesize that it may be related with small dispersion of the volunteers' age in the group under study.

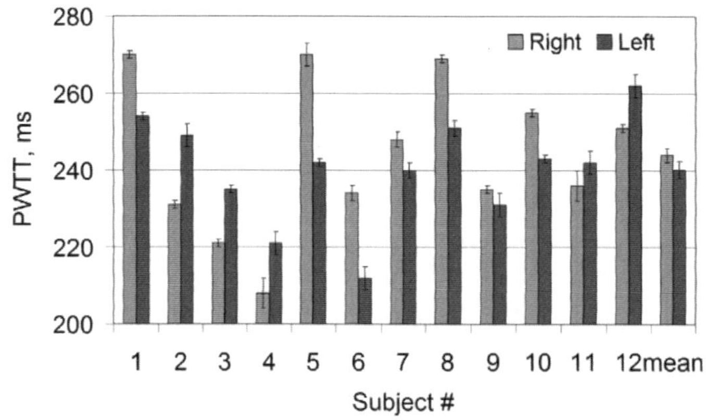

Figure 4. PWTT calculation results. Violet and vinous columns show PWTT of the right and left hands of each subjects, respectively. The last pair of columns represents respective mean values.

4. Discussion

The study revealed that dispersion of the PWTT parameter measured in selected 'good' ROIs for any subject does not exceed 8 ms, which is of the same order of value as synchronization error between video camera and electrocardiograph. In contrast to similar technique of PWTT assessment using classical photoplethysmograph of the contact type (which measures the time delay in a single point, e.g. in the finger) [9, 10], the IPPG system allows us to measure PWTT with higher accuracy due to access to a larger number of statistically independent measurements obtained simultaneously in different points. Moreover, the restrictions imposed on the selection of 'good' ROIs resulted in exclusion of an insecure ROI in which the PPG waveform is spoiled by either motion or physiological type of noise. Therefore, the proposed IPPG method is characterized by higher diagnostic reliability.

Asymmetry and asynchronicity of peripheral blood pulsations in human limbs could be assessed with even higher reliability by simultaneous video recordings using two or more synchronized cameras. The data simultaneously obtained from multiple cameras could reveal the difference in pulse wave propagation with higher accuracy since they are not spoiled by instability of both environmental and internal physiological conditions as it occurs in the case of measuring in different moments. The PWTT difference measured at different hands or legs of the same subject might serve as a good tool for preliminary (screening) diagnostics of various types of problems with the bloodstream such as remodelling, atherosclerosis plaques, stenoses, congenital features, etc.

5. Conclusion

A technique for measuring the pulse wave transit time with increased accuracy has been proposed. It is based on joint processing of the digital data obtained after synchronized recordings of imaging photoplethysmogram and electrocardiogram. Due to calculation PWTT within the large representative set of statistically independent ROIs, the proposed technique is more reliable and of higher diagnostic accuracy compare to the conventional method using ECG and classical single-point PPG. PWTT measured in properly selected 'good' ROIs of any palm are of small dispersion. Even though PWTT in right and left hands averaged over the whole group of subjects was comparable, significant difference in the time delay of pulse wave between the hands was found in some individuals. The technique can be used for early-stage diagnostics of various vascular diseases.

Acknowledgments

The work was financially supported by the Russian Science Foundation (Grant No. 15-15-20012).

References

[1] Laurent S *et al* 2001 *Hypertension* **37** (5) 1236–1241
[2] Blacher J *et al* 1999 *Hypertension* **33** (5) 1111–1117
[3] Cruickshank K *et al* 2002 *Circulation* **106** (16) 2085–2090
[4] Kamshilin A A *et al* 2011 *Biomed. Opt. Express* **2** (4) 996–1006
[5] Sidorov I S *et al* 2016 *Biomed. Opt. Express*. Paper was accepted.
[6] Kamshilin A A *et al* 2015 *Sci. Rep.* **5** 104494
[7] Kamshilin A A *et al* 2015 *Biomed. Opt. Express* **6** (11) 4326-4334
[8] Allen J 2007 *Physiol. Meas.* **28** R1–R39
[9] Loukogeorgakis S *et al* 2002 *Physiol. Meas.* **23** (3) 581–596
[10] Phillips J P and Kyriacou P A 2014 *Proc. 36th Intern. Conf. IEEE EMBC'14 (Chicago)* pp 3809–3812

Journal of Physics: Conference Series **737** (2016) 012054 doi:10.1088/1742-6596/737/1/012054

Compact diffraction phase microscopy for quantitative visualization of cells in biomedical applications

N A Talaikova[1], V P Ryabukho[1,2]

[1] Saratov State University, Astrahanskaya str., 83, Saratov, Russia, 410012
[2] Institute of Precision Mechanics and Control, Russian Academy of Sciences, Rabochaya str., 24, Saratov, Russia, 410028

E-mail: natalyatalaykova@gmail.com

Abstract. We consider a simplified and compact scheme of interference phase microscopy using a diffraction grating and spatial filtering of the diffracted field, i.e., diffraction phase microscopy. The scheme and the parameters of the device with the possibility of using the optical system of a smartphone and its software are analysed. The results of experimental determination of the spatial structure parameters of erythrocytes are presented.

1. Introduction

In the methods of optical phase microscopy the interference image of the studied object is analysed, in which the phase shifts of the light wave, reflected by the object or passed through it, manifest themselves. Such methods include the phase contrast method [1-3], the interference microscopy [4-6] and holographic microscopy [7-10]. In these methods, in one way or another, a coherent reference wave is formed and superposed onto the object image to form an interference image of the object. The computer processing of the interference image allows the reconstruction of the phase profile of the object field and the determination of structure optical characteristics of the object, e.g., morphological parameters of biological objects. The accuracy of such reconstruction and the spatial resolution are increased with the growth of the spatial frequency of the interference fringes of the image. The interference fringes with the highest frequency are obtained in the holographic microscopy with off-axis (oblique) reference beam [8,9]. However, the formation of such reference beam is associated with the complication of the optical scheme of the microscope and decrease of its noise immunity. This drawback is largely overcome in the so-called diffraction phase microscopy, in which a classical light microscope is used, equipped with the additional diffraction module, including a diffraction grating and an optical system for spatial filtering of the wave fields diffracted by the grating [11-15]. The diffraction module in this system is actually an interferometer, in which the diffraction grating plays the role of the amplitude splitter for the initial optical field. At the same time the optical system of the diffraction module by its construction is an imaging system that can operate in the mode of essential magnification of the image, and, therefore, in the microscope mode. In this relation, the diffraction module can be used autonomously without the microscope to form the magnified interference image of the studied object. In this case, the system of diffraction phase microscopy becomes compact, more simple and mobile. Moreover, the use of smartphone optics, computing and data processing software capabilities essentially widens the functional and application potentialities of this phase microscopy system.

During the last decade, the field of application of compact optical microscopes became significantly wider. Such instruments are used for quality control in industry, testing the authenticity of banknotes and books, hand assembling of small-scale elements of complex equipment. The portability of the present instrument provides the convenience of using it in a clinic, laboratory, or plant floor, as well as under the field conditions and for educational purposes.

The aim of the present work was to develop a compact diffraction phase microscope and to study the possibility of using it for phase microscopy without the classical light microscope. In the paper, we consider the scheme of the compact diffraction phase microscope, its possible parameters, the procedures of interference image processing, and the experimental results, obtained with the laboratory prototype of the compact diffraction phase microscope.

2. Scheme design

The diffraction phase microscope is a usual (classical) microscope equipped with additional diffraction module aimed at the formation of interference image and quantitative phase visualisation of the object structure [12,13]. The diffraction module incorporates the diffraction grating placed in the front focal plane of the objective of the telescopic optical system that consists of two objectives (lenses) with coincident focal planes, a spatial filter, placed in the coincident focal planes of the objectives, and a matrix (CMOS or CCD) detector of the interference image. Depending on the used detector, the recording rate can be as large as a few hundreds of frames per second [13,14]. Every frame of the interference image contains all information necessary for subsequent determination of the optical parameters of the object. In the case of using an optical microscope, the module is installed instead of the ocular lens so that the diffraction grating is located in the plane of the object image or in the immediate vicinity of it. The field of the object image incident on the diffraction grating is separated into the diffraction orders. The spatial filter entirely transmits only the first diffraction order, which is used as an object wave, and the wave not scattered by the object in zero diffraction order that plays the role of a reference wave in the back focal plane of the second objective of the telescopic system, where the image of the object it repeatedly formed in the field of the diffraction order, transmitted by the filter. The interference of the reference wave and the field of the object image yields the interference image, which as a result of recording actually yields the hologram of the focused object image. The period of the carrier hologram structure, the interference fringes, equals the period of the diffraction grating image, formed by the telescopic system with the specified magnification.

No essential difference in the operation of the diffraction module will arise, if instead of the object image the object itself will be placed near the grating, and the optical field scattered by the object will be forwarded into the diffraction module. In this case, the diffraction module operates as the primary imaging system, which at sufficiently large magnification can be considered as a microscope. The spatial filtering of the object wave field and the formation of the interference image makes this instrument to be an interference (phase) microscope.

Figure 1 presents a schematic diagram of such use of the diffraction module. The transparent object is illuminated with a plane wave from a laser source. When passing through the object, the wave acquires spatial phase shifts, i.e., the spatial phase modulation, the magnitude of which is dependent on the local thickness of the object and the refractive index of the object and the environment. As a result of this modulation the object field arises, including the scattered and non-scattered components. The diffraction grating decomposes the object field into diffraction orders. To get the interference image, the spatial filter is used that entirely transmits the first diffraction order and the non-scattered component of the object field in the zeroth order of diffraction. In the back focal plane of the second objective the superposition and the interference of non-scattered plane wave and the obliquely incident wave field of the object image occurs.

The light intensity distribution in the interference pattern in the plane of the matrix photodetector can be presented in the following simplified form [16]:

$$I(x,y)=I_0(x,y)+I_R(x,y)+2\sqrt{I_0I_R}\cos\left(\frac{2\pi}{\Lambda}\left[x\sin(\alpha)+y\cos(\alpha)\right]+\varphi(x,y)\right), \tag{1}$$

where the coordinate axes x, y are directed along the rows of photodetector pixels, $I_0(x,y)$, $I_R(x,y)$ are the intensity distributions of the object and reference fields, respectively; $\Lambda = Md$ is the period of interference fringes, M is the image magnification by the optical system, d is the period of the diffraction grating; α is the angle of the grating orientation with respect to the rows of the photodetector pixels; $\varphi(x,y)$ is the phase incursion acquired by the wave passing through the object.

For the plane illuminating wave and sufficiently smooth optical inhomogeneities of the object the function $\varphi(x,y)$ can be expressed as

$$\varphi(x,y)=\frac{2\pi}{\lambda}h(x,y)(n(x,y)-n_0), \tag{2}$$

where λ is the wavelength of light, $h(x,y)$ is the spatial relief of the object surface, $n(x,y)$ is the spatial distribution of the refractive index in the object, n_0 is the refractive index of the environment.

In fact, the resulting system of interference fringes in the object image is a system of carrier fringes of a hologram structure. The digital record of such interference image is a digital hologram of the focused image. The numerical processing of such digital hologram, e.g., using direct and inverse Fourier transforms and spatial filtering, allows the reconstruction of the phase shift spatial distribution of the object field and, therefore, the spatial variations of the optical thickness of the object.

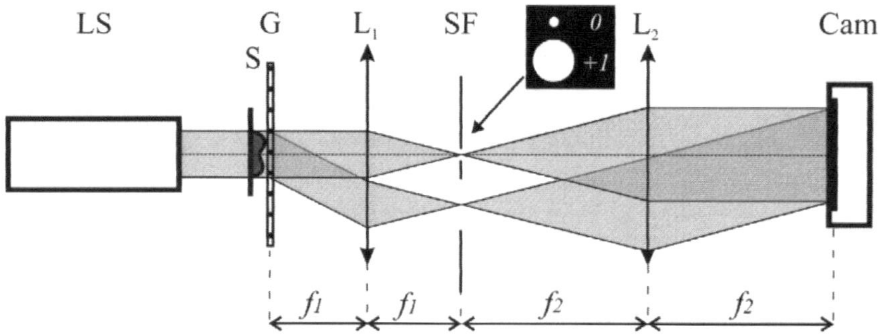

Figure 1. Optical scheme of compact diffraction phase microscope setup. LS, light source; S, sample; G, grating; L_1, L_2, lenses; SF, spatial filter; Cam, CMOS camera, f_1, f_2, focal distances of L_1 and L_2 lenses, respectively.

3. Technical features

In the present work, we considered the possibility of designing a compact diffraction phase microscope resistant to external vibrations, with formation of high-frequency interference fringes in the object image located in the photodetector plane. For implementing these intentions and providing the possibility of further improvement of the compact interference module parameters and/or upgrading the system for 3D visualisation of phase objects and using it in combination with the digital camera of a smartphone, we found the parameters of the interference module that are to satisfy a number of technical requirements.

1. For spatial filtering of wave fields diffracted by the diffraction grating it is necessary that the diffraction orders ±1 hit the aperture of the first objective of the system (L_1 in Fig.1). According to our estimates, this is fulfilled under the condition that the period of the grating $d \geq (3/2)\lambda/NA_1$, where NA_1 is the numerical aperture of the first objective of the interference module. In this case the transverse size Δx_0 of the object resolved by the system, e.g., red blood

cell in the blood smear, should exceed the grating period at least by two times, $\Delta x_0 \geq 2d$. Then the diffraction orders of the object field will not overlap in the back focal plane of the objective, and the efficient spatial filtering of the object light field is possible. In the interference image of the object, formed in the back focal plane of the second lens L_2 of the module (Fig.1), about three interference fringes will appear.

2. For recording a digital hologram it is necessary that the matrix photodetector can resolve the carrier interference fringes of the image with the mean period Λ_H expressed as $\Lambda_H = Md$, where M is the transverse magnification of the system, determined by the ratio of focal lengths of the objectives, $M = f_2/f_1$. This requirement is equivalent to the fulfilment of the Nyquist condition that for the record of a digital hologram can be presented in the form $\Delta x_D \leq \Lambda_H/2$, where Δx_D is the pixels period in the matrix photodetector [17-19]. Taking into account the object image field phase variation that give rise to the variation of the fringes period, this inequality should be written as $\Delta x_D \leq \Lambda_H/3 = Md/3$ [19].
3. The integration of the requirements 1 and 2 leads to the generalized condition of optical spatial filtering of diffracted fields and recording the digital hologram formulated as $\Delta x_0 \geq 2d \geq 6\Delta x_D f_1/f_2$ with $d \geq (3/2)\lambda/NA_1$.

To protect the surface of the diffraction grating one can use a cover glass placed between the object and the diffraction grating. This increases the wear resistance of the diffraction module in the studies of soft objects, e.g., biological slices, or in technical applications, e.g., the control of printed-circuit board soldering. However, both the cover glass and the diffraction grating introduce spherical aberrations, which reduce the resolving power of this optical system.

4. Experimental results

In the present work we used the laboratory prototype of a compact diffraction phase microscope with the overall dimensions 250x100x100 mm and the optical magnification $M = f_1/f_2 = 4$. As the object of study the blood smears on glass slide were used. The visualisation of a blood smear was performed using the first objective with the parameters $NA = 0.5, f_1 = 25$ mm. The object was illuminated by a quasiplane wave from a semiconductor laser module ($\lambda = 0.65$ µm). A cover glass ($d_{cg} \approx 120$µm) was placed between the grating and the object to protect the grating from damage because of the interaction with the object.

Holographic diffraction grating with a period of 10 µm (100 lines per mm) was used. The grating was placed perpendicular to the optical axis so that its grooves form an angle of 45° relative to pixel rows of the matrix.

In the experiments the high-speed digital camera (CMOS, aca2500-14gm, Basler, Germany) with the pixel size $a = 2.2$ µm was used to record interference images. The digital holograms were processes using the software based on double Fourier transform [20,21], written in our laboratory in the LabVIEW programme environment. For recording digital hologram of a focused image the reconstruction procedure for the complex field amplitude includes three sequential numerical procedures, namely, the Fourier transform of the digital hologram that yields the spatial spectrum, the spatial filtering of the hologram spectrum with extraction of the spectrum of the object field complex amplitude, and the inverse Fourier transform for retrieving the complex amplitude of the object image field. As a result of numerical processing the ultimate phase image of the blood smear was presented in the frame, determined by the field of view and having the dimensions 8.25x8.25 mm^2. The visualisation and postprocessing of the recoded interference patterns takes nearly 40 seconds for each frame.

Figure 2d presents the reconstructed image of the human blood smear for the interference pattern, shown in Fig. 2a. Using the above algorithms, the phase shift map $\varphi(x, y)$ is first calculated over the entire plane of the image field of view. Then the latter is used to calculate the distribution of the

Journal of Physics: Conference Series **737** (2016) 012054 doi:10.1088/1742-6596/737/1/012054

sample optical thickness $l(x,y)=h(x,y)n(x,y)$, which is presented as the reconstructed image (Fig. 2d). The mean value of the red blood cell optical thickness in this experiment amounted to $l_m \approx 0.8\,\mu\text{m}$, the spread of values was $\Delta l \approx \pm 0.2\,\mu\text{m}$. Only the separately located erythrocytes were taken into account in the calculation of the optical thickness. The recognition of the appropriate images was executed automatically using the vision algorithms [22].

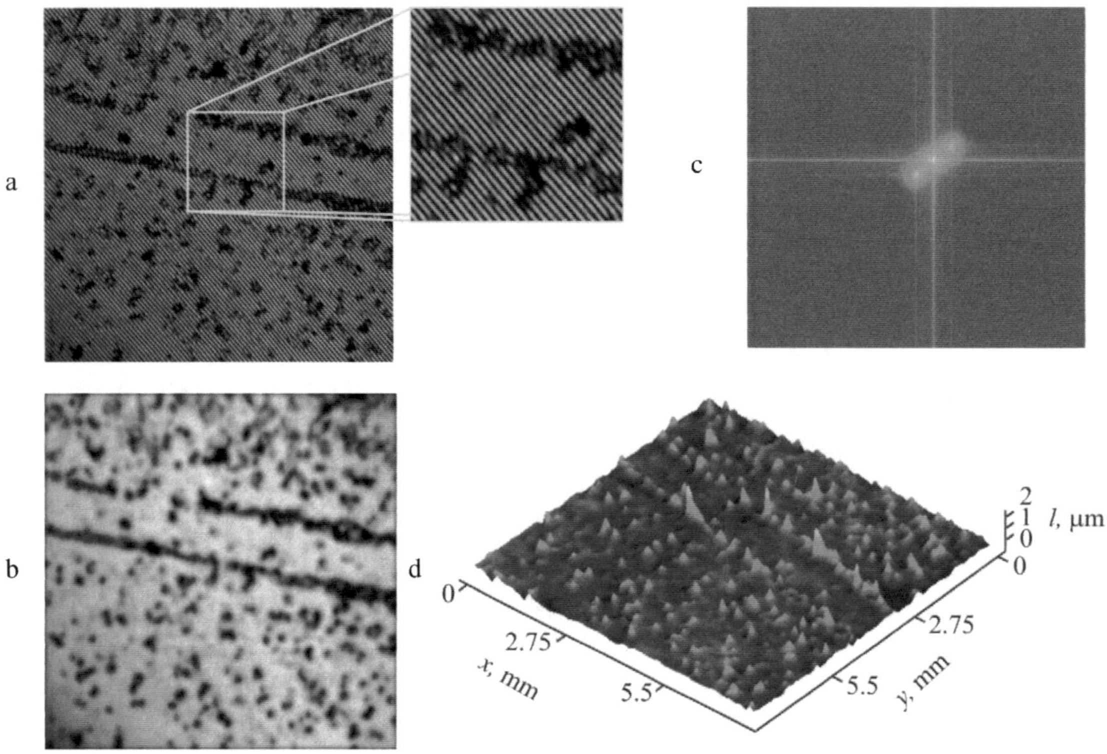

Figure 2. a) Interference pattern recorded using the compact diffraction phase microscope; b) Amplitude image reconstructed from the interference pattern; c) Image of 2D spatial spectrum obtained using the direct Fourier transform of the hologram; d) Reconstructed 3D image of the blood smear in the entire field of view of the microscope.

5. Prospective studies

The positive results of the present study demonstrate the possibility of implementing a compact diffraction phase module as the phase microscope using the optics of a smartphone and its digital camera for subsequent quantitative reconstruction of the object image. Figure 3 presents the possible scheme of such instrument.

For imaging an object in the back focal plane of the smartphone objective in compact interference module an additional lens (L_3 in Fig. 3) must be installed in such a manner that its front focal plane coincides with the plane of the object image in the back focal plane of the lens L_2.

The present instrument can be used as a portable autonomous analyser of blood structure parameters with subsequent transmission of data to clinics and health-care institutions via the mobile network. It can also find application in the technological control of production quality, where the use of large-sized microscopes is hard or impossible. One can expect that the technological progress in the field of processors and photodetectors for the next generations of smartphones will allow the use of already known numerical algorithms for real-time hologram processing in mobile phones.

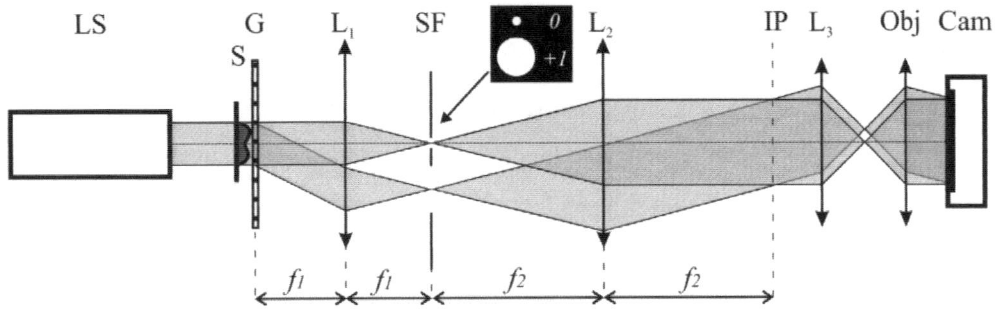

Figure 3. Optical scheme of compact diffraction phase microscope to be used in combination with the optical system and digital camera of smartphone: LS, light source; S, sample; G, grating; L_1, L_2, L_3, lenses; SF, spatial filter; IP, image plane; Obj, smartphone objective; Cam, smartphone camera, f_1, f_2, focal distances of L_1 and L_2 lenses, respectively.

6. Conclusion

In the present study, a compact diffraction phase microscope with the illumination system in transmission of the object was considered. The preliminary results demonstrate the possibility of a new modification of diffraction phase microscope, the compact diffraction phase microscope. The reconstruction of the 3D structure using the compact diffraction phase microscope and the obtained quantitative data on the cell thickness agree with those reported in the literature [23].

Using additional algorithms in the postprocessing of interference images, the microscope can be used to determine the number of leucocytes in blood or to detect the cells with atypical shape, the presence of which may be a symptom of anaemia, cancer, or diseases of the bone marrow. In future we plan to combine the compact diffraction phase microscope with the camera of a cellular phone, which will allow the examination under the conditions, making it impossible to use more perfect methods and instrumentation, e.g., in the areas remote from the laboratory.

Acknowledgement
Research is carried out by a grant Russian Science Foundation (project №161910528)

References

[1] Zernike F 1942 *Physica* **9 (7)** 686–698
[2] Cuche E, Marquet P and Depeursinge C 1999 *Appl. Opt.* **38 (34)** 6994-7001
[3] Born M and Wolf E 1970 *Principles of Optics: Electromagnetic Theory of Propagation, Interference and Diffraction of Light* (Cambridge University Press) p 985
[4] Abdulhalim I 2012 *Ann. Phys. (Berlin)* **524 (12)** 787–804
[5] Gale D M, Pether M I and Dainty J C 1996 *App. Opt.* **35 (1)** 131-148
[6] de Groot P 2015 *Adv. in Opt. and Photonics* **7 (1)** 1–65
[7] Kemper B and von Bally G 2008 *App. Opt.* **47 (4)** A52-A61
[8] Marquet P, Rappaz B, Magistretti P J, Cuche E, Emery Y, Colomb T and Depeursinge C 2005 *Opt. Lett.* **30 (5)** 468-470
[9] Kelly D P, Hennelly B M, McElhinney C and Naughton T J 2008 *Proc. SPIE* **7072** 707215
[10] Kalenkov G S, Kalenkov S G and Shtanko A E 2015 *Quantum Electronics* **45 (4)** 333-338
[11] Lee K R, Kim K, Jung J, Heo J H, Cho S, Lee S, Chang G, Jo Y J, Park H and Park Y K 2013 *Sensors [Internet]* **13 (4)** 4170-4191
[12] Bhaduri B, Pham H, Mir M and Popescu G 2012 *Opt. Lett.* **37 (6)** 1094-1096
[13] Bhaduri B, Edwards C, Pham H, Zhou R, Nguyen T H, Goddard L L and Popescu G 2014 *Adv. in Opt. and Photonics* **6 (1)** 57–119

[14] Rappaz B, Barbul A, Hoffmann A, Boss D, Korenstein R, Depeursinge C, Magistretti P J and Marquet P 2009 *Blood Cells Mol. Dis.* **YBCMD-01288** 1-5
[15] Pham H, Ding H, Sobh N, Do M, Patel S and Popescu G 2011 *Biomed. Opt. Express* **2 (7)** 1781-1793
[16] Kalyanov A L, Talaikova N A and Ryabukho V P 2015 *Proc. SPIE* **9448** 944817
[17] Jacquot M and Sandoz P 2004 *Opt. Eng.* **43 (1)** 214–223
[18] Schnars U, Falldorf C, Watson J and Juptner W 2014 *Digital Holography and Wavefront Sensing* (Berlin & Heidelberg: Springer) p 226
[19] Grebenyuk K A 2014 *J. Phys.: Conf. Ser.* **536** 012005
[20] Takeda M, Ina H and Kobayashi S 1982 *J. Opt. Soc. Am.* **72 (1)** 156-160
[21] Kreis T 1986 *J. Opt. Soc. Am.* A. **3 (6)** 847–855
[22] Gonzalez R C, Woods R E and Eddins S L. 2009 *Digital Image Processing Using MATLAB*, (Gatesmark Publishing) chapter 3,10 p 609
[23] Tuchin V V 2007 *Tussie Optics. Light Scattering Methods and Instruments for Medical Diagnosis* (SPIE Press) chapter 1, 13, 14 p 988

Research of phase changes in process of storage of fine dispersed aqueous suspensions of inorganic substances

R N Chigrin, N V Andreeva and O V Andreeva

ITMO University, Saint Petersburg 197101, Russia

E-mail: rchigreen11@gmail.com

Abstract. In the work were proposed to use the method of digital holographic interferometry on the example of suspensions of fine dispersed silica, developed by different ways, for studying fine dispersed suspensions of inorganic substances in combination with the process of their sedimentation. A comparison of experimental results obtained by this method and by method of dynamic light scattering is presented. The high sensitivity of the interferometric method, particularly during assessing of changes of suspensions with nanosized particles is shown.

1. Introduction

At present time the use of substances in dispersed state is widely used to create new composite materials, to improve the physico-mechanical properties of industrial structural materials, in biomedical technologies [1-3]. The development of nanotechnologies with application of nanodispersed ingredients requires corresponding development of diagnostics of parameters of such small-scale objects.

There is a number of methods of characterization of parameters of fine dispersed inorganic suspensions, such as scanning electron and probe microscopy, dynamic light scattering, x-ray diffractometry. Each of these methods has its methodological peculiarities and during studying of specific objects has both certain advantages and certain disadvantages and conflicts of interest [4-6]. In addition these methods are usually expensive and time-consuming. The development of new methodologies for the study of substances in fine dispersed state, including nanodispersed state, it is highly relevant, because each new method improves the accuracy of measurement results obtained by other methods.

In this work the possibility of characterizing parameters of fine dispersed inorganic suspensions during using the method of digital holographic interferometry for observing the process of sedimentation of suspensions with particles of substance, which size is less than one micron, is considered.

2. Object and methods

2.1. Object of research

The objects of study in this work are aqueous suspensions of fine dispersed quartz, particles of which are obtained by different methods and used for different purposes. The particles of suspension 1 are obtained by method of condensation of silicon dioxide from the gas phase, such particles are used in biomedical technologies and during creation of polymeric composite materials for various purposes with improved physico-mechanical properties. Particles of suspension 2 are obtained from the solid phase by method of grinding in a planetary-centrifugal mills and they are used as industrial additives, which improve the operational properties of building materials.

Initial state of SiO_2 for production of suspension 1 – dry powder; for production of suspension 2 – pasty mixture with water in a weight ratio of 1.3:1.0. Work preparations were 0.1%-0.5% aqueous suspensions of the researched substances.

2.2. Experimental methods

For studying changes of optical characteristics of the sample in the process of radiation exposure and during postexposure period was used experimental stand designed for studying phase transformations of transparent objects by method of digital holographic interferometry (DHI) [7], a block-scheme of which is shown on figure 1a. At this stand the unit of the object (2) placed outside of the main scheme of interferometer (1).

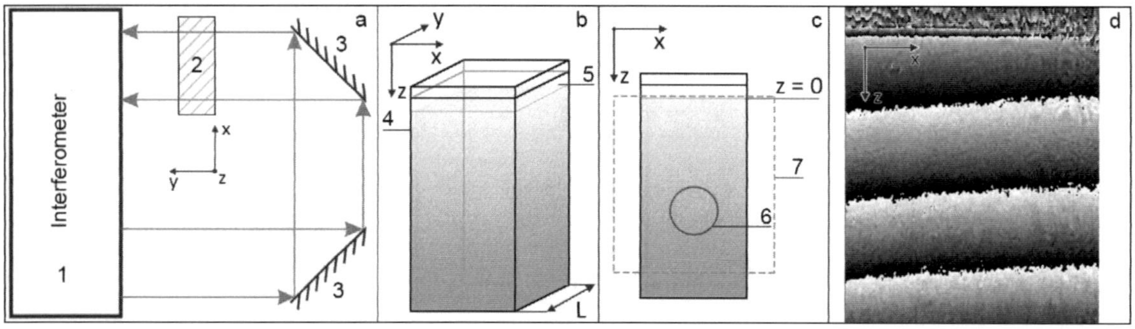

Figure 1. Block-scheme of experimental stand of digital holographic interferometry (a): 1 – block of interferometer, 2 – researched object, 3 – mirrors; the view of cuvette with the researched suspension (b,c): 4 – cuvette, 5 – upper liquid level, 6 – region of monitoring using dynamic light scattering (λ=655 nm), 7 – region of monitoring by method of digital holographic interferometry (λ=532 nm); example of experimental data processing - digital interferogram (d).

The object of study is a preparation of suspension in transparent cuvette with the size of 10x10x40 mm (figure 1b). The cuvette is rigidly mounted in a special cassette and installed in the object beam in accordance with the optical scheme of interferometric experiment. The object beam of interferometer (532 nm) tests the working area of the object with the size at least 25x25 mm and passes through the sample as shown on figure 1c. The stand has a computer terminal to control a registration of sequence of digital holograms and their further processing.

The digital interferogram of the current state of the object represents the spatial distribution of the phase difference of the testing radiation in the plane of the observation of object, corresponding to the change of the sample at time t_i compared to initial (at time t_0) – $\varphi_{i0}(x,z) = \varphi_i(x,z)$ - $\varphi_0(x,z)$. A typical view of the interferogram of the sample state in a certain period of process of sedimentation is shown on figure 1d. Interference pattern in this case takes the form of horizontal interference fringes which characterize the change of optical thickness of the sample, $\Delta(nL)$, by height of the cuvette, z, by changing the concentration and size of particles during their sedimentation.

Interferometric measurements were accompanied by the measurement of the same sample parameters using method of dynamic light scattering. Both samples represented an identical cuvettes filled with the preparation just after its production. The sample area that is tested by laser beam (655 nm) during using method of dynamic scattering is shown on figure 1b. In stable condition of the sample, fixed in the unit Horiba LB-550, were carried out measurements of the distribution of particles of preparation of each suspension by sizes in the initial state (t_0) and at specific moments of time of process of sedimentation (t_i).

3. Experimental results

On figures 2, 3 and in table the results of the research of suspensions 1 and 2 by two methods are shown. The distribution of particles by size (figure 2a and figure 3a) was characterized by the value of the characteristic diameter of the particle, D_{char}, determined at q_{max}, and by half-width of the distribution of ΔD. By method of DHI was determined the dependence of Δn(z) in each experiment. Δn characterized the change of the refractive index towards the upper level of the suspension (z = 0).

Table 1. Estimation of the parameters of suspensions 1 and 2 in the process of sedimentation.

№	Method of dynamic light scattering					Method of digital holographic interferometry	
	t_0		t_i	D_x, nm	ΔD, nm	t_i	Δn, 10^{-5} (z=20 mm)
	D_x, nm	ΔD, nm					
1	220	210	20 days	220	210	72 hours	0.8
2	670	850	20 hours	580	1100	24 hours	16,1

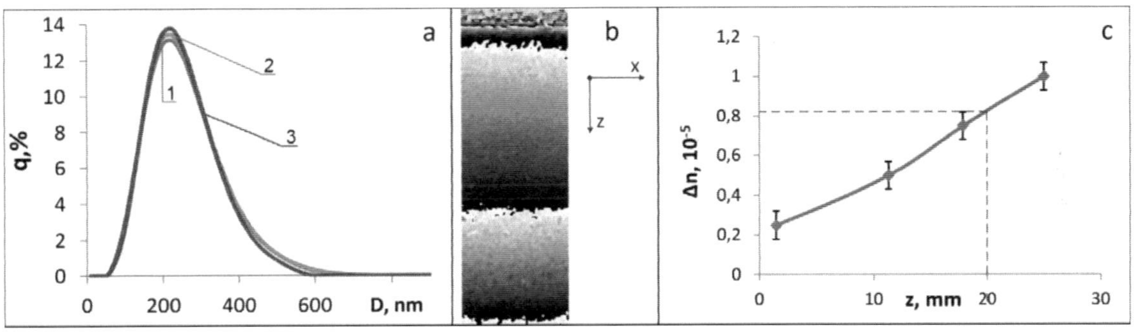

Figure 2. Research of the suspension 1: – distribution of particles by sizes (1 – initial, 2 – after 20 hours, 3 – after 20 days), b – interferogram of the suspension during sedimentation during 72 hours, c - the change of the refractive index of the suspension by height of the liquid column.

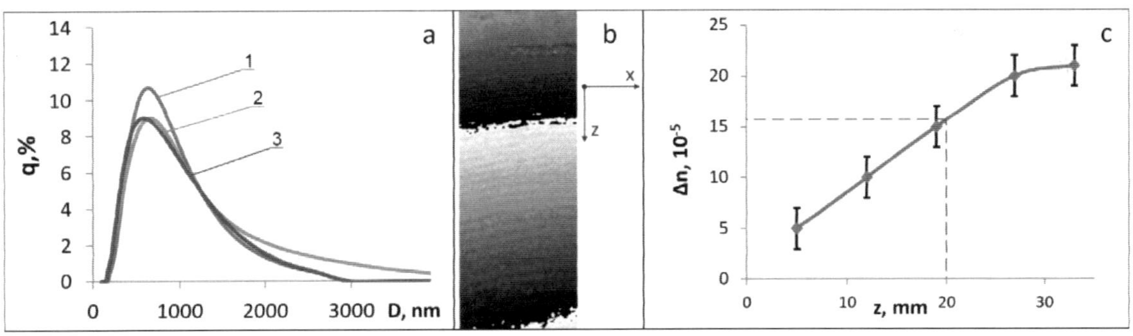

Figure 3. Research of the suspension 2: – distribution of particles by sizes (1 – initial, 2 – after 2 hours, 3 – after 20 hours), b – interferogram of the suspension during sedimentation during 24 hours, c - the change of the refractive index of the suspension by height of the liquid column.

4. Conclusions

The experiments showed that the parameters of the suspension 1 with the SiO_2 particles from 14 nm to 600 nm (D_x=220 nm) determined by the method of dynamic scattering, practically don't change during 20 days of storage. At the same time the method of DHI allows to detect the change Δn(z=20) = 0.8·10-5 after 72 hours of storage of the suspension (remind that Δn=0 at t=0). The parameters of the suspension 2 (D_x=670 nm) demonstrate well diagnosed changes of the parameters by both methods in the first day of storage. Thus, during estimation of changes of suspensions parameters with inorganic nanoparticles with the size of less than 500-600 nm (in the case of substances with density close to

Journal of Physics: Conference Series **737** (2016) 012055 doi:10.1088/1742-6596/737/1/012055

SiO_2) method of DHI allows us to carry out such diagnostics with greater accuracy than the method of dynamic light scattering.

References

[1] Cheow W S, Ng M L, Kho K, Hadinoto K *International Journal of Pharmaceutics* **404 (1–2)** pp 289–300
[2] Kaz'mina O V, Vereshchagin V I, Abiyaka A N 2008 *Glass and ceramics* **9** pp 28-30
[3] Guglielmi M, Kickelbick G, Martucci A 2014 *Advances in Sol-Gel Derived Materials and Technologies* **IX** 227
[4] Morita S, Giessibl F, Sugawara Y, Hosoi H, Mukasa K, Sasahara A, Onishi H 2004 *Springer Handbook of Nanotechnology* **13** pp 141-178
[5] Kirkland A I, Hutchison J L 2007 *Nanocharacterisation* (Cambridge: RSC Publishing) 304 p.
[6] Povstugar V I, Mikhailova S S, Shakov A A 2000 *J. Anal. Chem.* **55** pp 455–467
[7] Manukhin B G, Gusev M E, Kucher D A, Chivilikhin S A, Andreeva O V 2015 *Optics and spectroscopy* **119 (3)** pp 392-397

A testing method for the machine details state by means of the speckle image parameters analysis

A N Malov[1], P V Pavlov[1], A V Neupokoeva[2]
[1]Military Educational and Scientific Center Air Force "Air Force Academy named after prof. N.E. Zhukovsky and Y.A. Gagarin", 394064, Starih Bolshevikov 54-a, Voronezh, Russia
[2]Irkutsk State Medical University, 664003, 3 Iyulya 8, Irkutsk, Russia

E-mail: pavlov.pave@yandex.ru

Abstract. Non destructive testing method, allowing to define a residual resource of power details of mechanical engineering designs under the analysis of registered speckle-image parameters, it is discussed. The "chessboard" algorithm based on calculation of correlation between the given speckle-image and the a chessboard image is considered. Experimental research results of an offered non destructive testing method are presented. It is established, that to increase in quantity of a power detail tests cycles there is an increase in roughness parameters that conducts to reduction of correlation factor between reference and to resultants the image at the given stage of test. Knowing of correlation factor change dynamics, it is possible to define a residual resource of power details while in exploitation.

1. Introduction

Speckle image method is commonly used nowadays. Correlative analysis is usually used for the speckle images processing [1,2,3]. In [4] are defined the effectiveness limits of the correlative algorithms of speckle image processing for the roughness measurement of transparent and non-optically-transparent objects. To increase the sensitivity of speckle image method we brought new algorithm [5]. This algorithm uses the information of whole speckle image. Known methods utilize the information from the only row for wavelet analysis or correlogram along the X or Y axis or diagonal of the image [6]. The presented new method of speckle image processing makes it possible to detect the correlation between speckle image and chessboard structure leading to the possibility of defect parameters history observation.

2. "Chessboard" algorithm

This method makes possible the calculation of the correlation between the spots of speckle image and chessboard structure sqaures. For the simple one-dimensional model:

$$t(p) = \sum_{n=-\infty}^{+\infty} rect\left(\frac{x}{d_{sp}} + nd_{sp}\right), \tag{1}$$

$$t(d) = \sum_{m=-\infty}^{+\infty} a_m rect\left(\frac{x}{d_{ch}} + md_{ch}\right), \tag{2}$$

t(p) – brightness distribution of speckle image pixels, t(d) – brightness distribution of chessboard structure, $a_m = 0$ with even m, d_{sp} – size of speckle spot, d_{ch} – size of chessboard structure square, $rect(x) = \begin{cases} 1 & |x| \le \frac{1}{2}, \\ 0 & rest. \end{cases}$ – rectangular function.

Equations (1) require that speckle image spots are all of the same size (or at least most of them).

To process the indiscrete brightness distribution described by equations (1) and (2) we need to digitalize them, for that we use Dirac comb function [2], as a result, we get the following:

$$t(p) = rect\left(\frac{x}{d_{sp}}\right) * comb\left(\frac{x}{d_{sp}}\right), \quad (3)$$

$$t(d) = rect\left(\frac{x}{d_{ch}}\right) * comb\left(\frac{x}{d_{ch}}\right) \quad (4)$$

* - convolution sign, $comb(x) = \sum_{n=-\infty}^{\infty} \delta(x-n)$ – Dirac comb function.

To find the Fourier spectrum of discrete functions (3) and (4) we use the direct Fourier transform, as a result, we get the following:

$$\hat{F}\{t(p)\} = \mathrm{sinc}(f_x d_{sp}) \mathrm{comb}(f_x d_{sp}), \quad (5)$$

$$\hat{F}\{t(d)\} = \mathrm{sinc}(f_x d_{ch}) \mathrm{comb}(f_x d_{ch}) \quad (6)$$

Getting the Fourier spectrums we multiply them together and use the inverse Fourier transform, which results in function of correlation between speckle image and chessboard structure.

If $d_{ch} = d_{sp}$ then multiplication of spectrums (5) and (6) gives us $\mathrm{sinc}^2(f_x d_{ch})$ components, inverse Fourier transform of which equals $rect^2\left(\frac{x}{d_{ch}}\right)$, which is triangular function $tri(x) = \begin{cases} 1-|x| & |x| \le 1, \\ 0 & \text{in other cases.} \end{cases}$

Switching from one-dimensional case to two-dimensional case, we get the pyramid structure inside the single chessboard square. Chessboard structure with pyramid elements appears in case there are enough speckle spots of the same size (fig. 1).

If $d_{ch} > d_{sp}$ or $d_{ch} < d_{sp}$ then multiplication of spectrums (5) and (6) gives us unbalanced resulting image structure indicating the change of speckle size, and that means the increase or decrease of surface roughness with respect to chessboard square size.

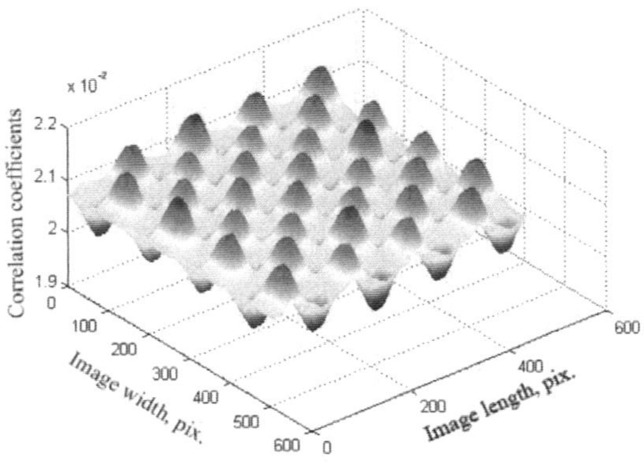

Figure 1. Resulting image of chessboard structure.

3. Numerical simulation

During the research, we conducted the studies of the change process of roughness parameters and micro crack size. We simulated the life cycle of structural elements from initial values of roughness parameters to micro cracks with afterward destruction. To do that we had probed the surface with radiation of 0.65 μm wavelength after 5, 10, 15, 30 and 45 thousands of twist and bend cycles, which gave us speckle images with resolution of 512x512, thereby the surface roughness changed from 0.1 to 1.3 μm.

At the first stages of research before the experiment at Ra=0.1 μm we got the regular pyramid-like structure (fig. 1). We got this picture using correlation between speckle image and chessboard structure with cell size from 5x5 to 100x100 pixels and step of 5 pixels.

The regular structure was taken as reference. In the following analysis steps we analyzed the speckle images with the chessboard chosen after each cycle of the test. After that we calculated the correlation index between the original image and each of the following images, which is shown on Table 1, bottom row along with the chessboard images.

Table 1. Experimental data. Top row – speckle images, middle row – speckle images processed with chessboard method, bottom row – index of correlation between reference image and study sample.

standard	5000 cycles	15000 cycles	30000 cycles	45000 cycles
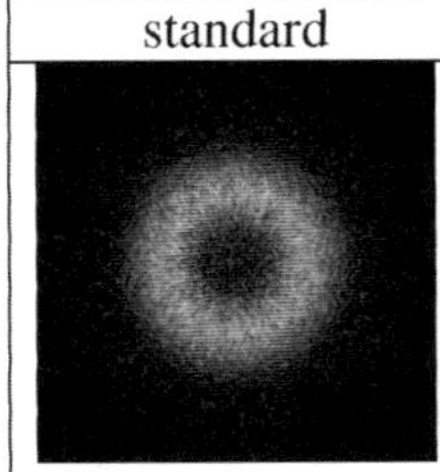	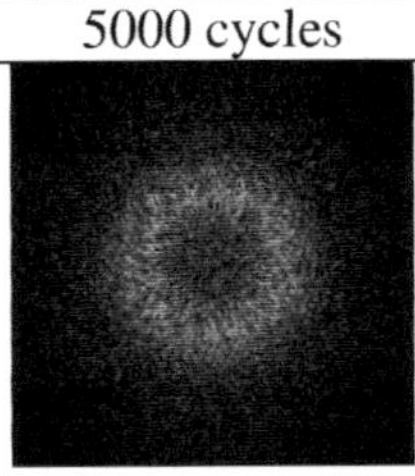	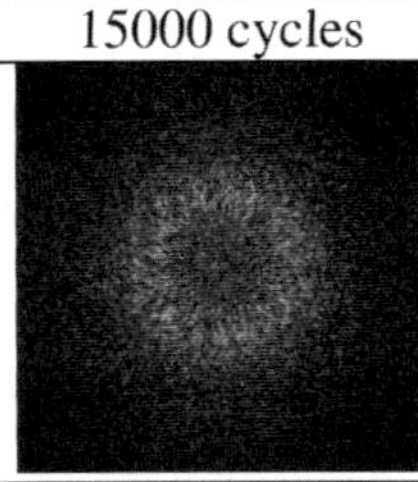	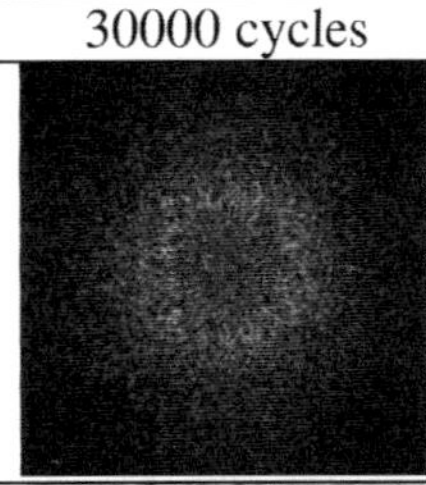	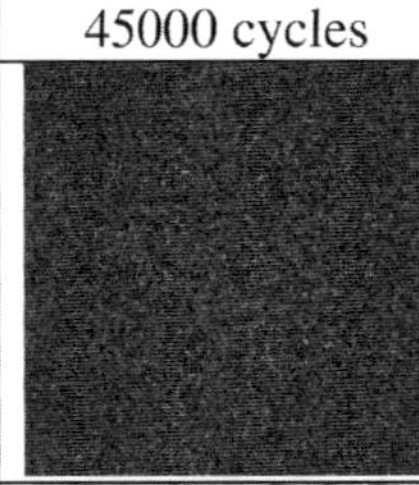
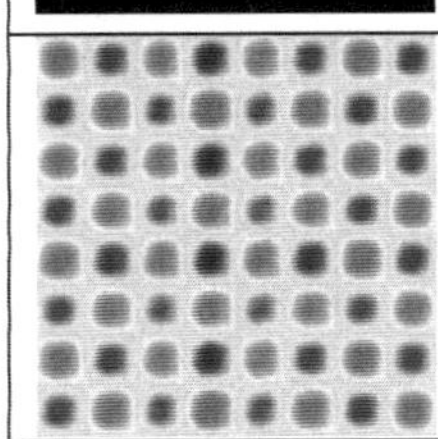	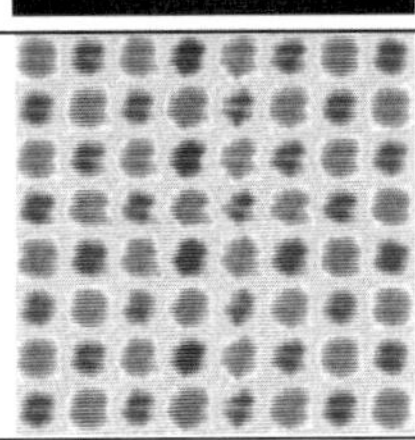	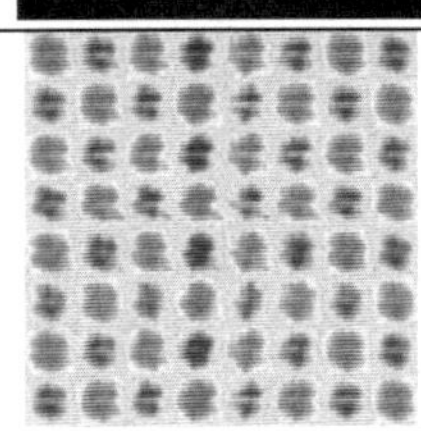	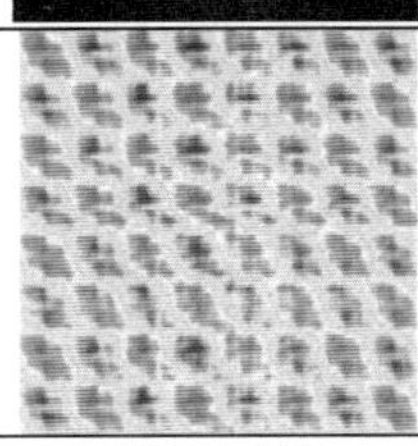	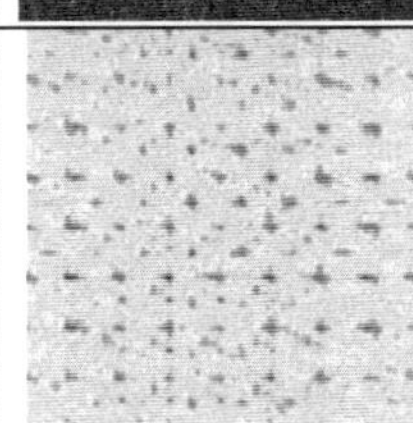
K= 1	K = 0.944	K = 0.87	K = 0.5	K = 0.03

Outcome of the study is that with increase of cycle number roughness Ra also increases, and that leads to the disruption of regular chessboard structure (it turns into the chain). This causes the fall of correlation index from 1 to 0.5. With that micro cracks start appearing on the surface of the test subject resulting in crack after 30000 cycles. Subsequently the complete destruction of regular structure occurs.

Correlation coefficient dependence from of cycles number is presented on fig. 2, its left limit is the initial value of correlation index, its right limit is the test subject destruction.

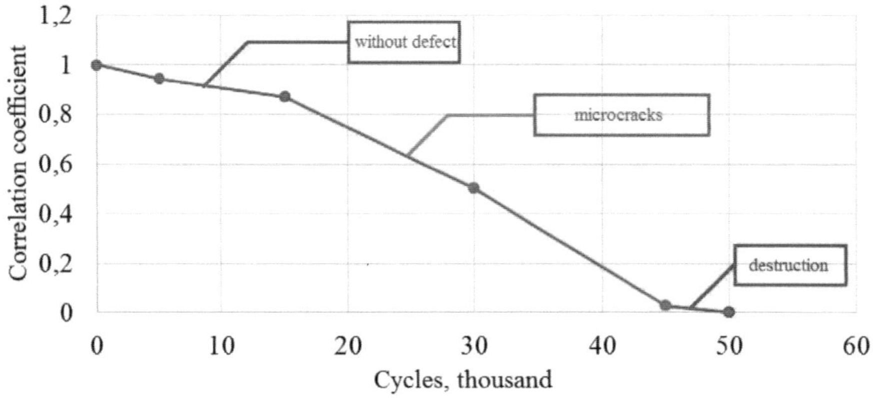

Fig. 2 Experimental data

Summing up, the diagnostic method described above is useful for non-destructive testing and determine to a remaining life of the power details in exploitations.

References

[1] Jones R, Wykes C 1983 *Holographic and Speckle Interferometry. A discussion of the theory, practice and application of the techniques*
[2] Goodman J W 1985 *Statistical Optics*
[3] Pavlov P V, Petrov N V, Malov A N 2011 *Scientific and technical gazette of information technologies, mechanics and optics* **6** 76
[4] Gorjunov A E, Pavlov P V, Petrov N V 2014 *Journal of physics: conference series* **536** 1
[5] Malov A N, Neupokoeva A V 2013 *10th International Conference "Holography EXPO-2013"*
[6] Kayahan E, Gundogdu O, Hacızade F, Nasibov H 2009 *IEEE* **978-1-4244-4210-2**

Investigation of MINACE composite filter capabilities for multicolor images correlation recognition purposes

N N Evtikhiev, E Yu Zlokazov, E K Petrova, R S Starikov, D V Shaulskiy

National Research Nuclear University MEPhI (Moscow Engineering Physics Institute), Kashirskoe shosse, 31, 115409 Moscow, Russia

E-mail: k1121@mail.ru

Abstract. Article represents the results of investigations in the area of distortion invariant images recognition using composite correlation filters. One of the most successive for application is a filter known as MINACE (minimum noise and average correlation energy). The capabilities of MINACE filter synthesis for multicolor image recognition problem are discussed.

1. Introduction

Application of CF (composite filters) allows to increase the degree of target object variance of correlation image recognition method in presence of object image distortions such as plane or spatial rotation, scaling, illumination changes, deformation etc. The main idea of CF synthesis is based on substitution target object single reference image by linear combination of several images with target object under different a priori determined appearance variations. One of the most perspective CF for application is MINACE (minimum noise and average correlation energy) filter [1]. This type of CF demonstrated good results of target object recognition represented by grayscale images with complex backgrounds and spatial noises. Moreover MINACE filter synthesis algorithms are simple by means of computational burden [2-6]. Also the possibility of realization of MINACE filter based correlation image recognition method in optical image correlator attracts a special interest [7-9].

The usage of several color channels in target object images allows to enhance the result of object recognition problem, as multicolor images can provide more information about objects than halftone images in some cases. The present article is dedicated to the problem of multicolor images correlation recognition using MINACE CF.

2. Problem description

The modeling problem conditions assume discrimination of target object images from three types of clutter objects images in conditions of plane rotation. Figure 1 illustrates the examples of images under researches. All the images had resolution of 512x512 pixels. The standard RGB colors model is used in the work. MINACE CFs were synthesized according to the method described in [7-9]. The discrimination capability of filters was estimated using Neyman–Pearson criterion.

Different types of MINACE CFs were synthesized for preliminary tests. All the synthesized CF were trained to invariantly recognize a target object under different plane rotations in the range of 45 degrees. All the preliminary CF had different synthesis conditions such as amount of training images etc.

Figure 1. Examples of test images. Target object (T-true) clutter objects (F-false)

(a) (b) (c)

Figure 2. Target object images in different colors.

(a) (b)

Figure 3. a) Halftone grayscale image of target object; b) red channel image of target object.

3. Preliminary modeling

Different types of MINAE CFs were synthesized for preliminary tests. All the synthesized CF were trained to invariantly recognize a target object under different plane rotations in the range of 45 degrees. All the preliminary CF had different synthesis conditions such as amount of training images etc.

Figure 4 represents the example of MINACE CF discrimination characteristic for red channel images of test objects. In most cases the probability of correct discrimination was estimated to be more than 99.9% according to Neyman–Pearson criterion. Also target object recognition results were obtained for the case of complex scene image of test objects (figure 5a). Figure 5b illustrates the result of correlation function calculation of MINACE CF and the scene from figure 5a. Presented result demonstrates high discrimination stability of MINACE CF in conditions of high similarity of target and clutter objects forms.

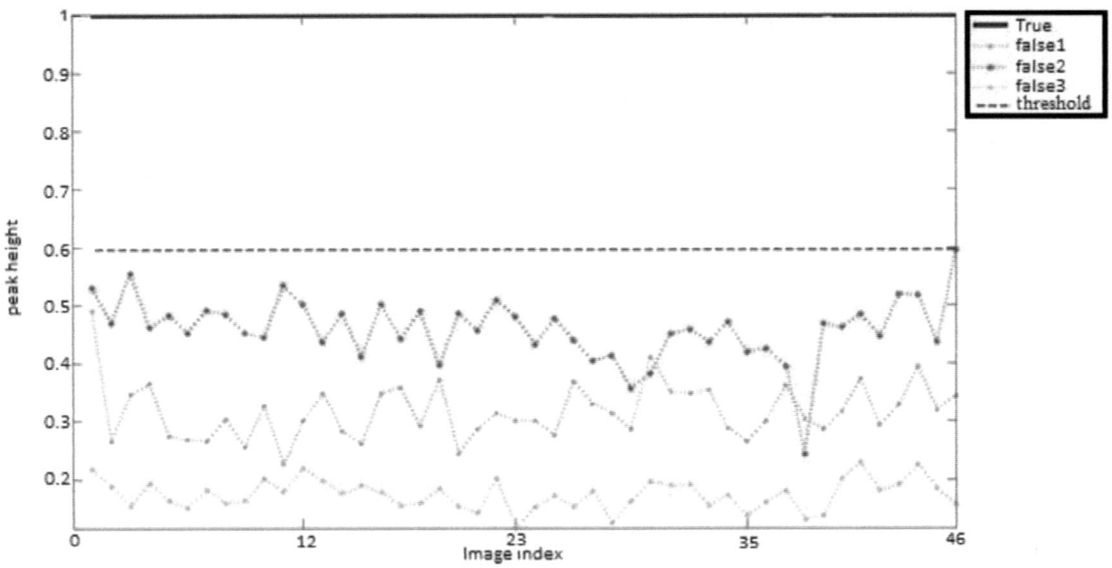

Figure 4. Example of MINACE CF discrimination characteristics – the dependence of correlation peak height from input object rotation angle. Thick solid blue line – target object correlation characteristic, thin lines – clutter objects characteristics, dashed line – threshold value.

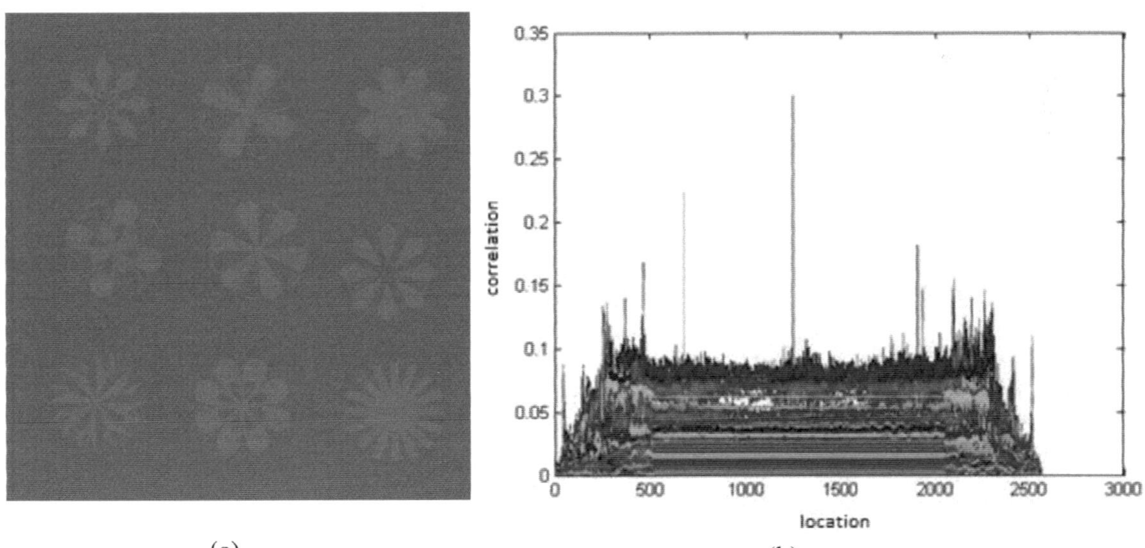

(a) (b)

Figure 5. a) The image of a complex scene contained by target and clutter test images; b) example of calculated correlation function between MINACE CF and scene from figure 5. Target object has the highest correlation peak.

Journal of Physics: Conference Series **737** (2016) 012057 doi:10.1088/1742-6596/737/1/012057

4. Problem description

The discrimination possibility of objects with the same form and different colors was investigated. For simplicity we used an object images with only one non zero color channel, for example the red one or green. An objects were placed on multicolor background with active both red and green channels. See for example figures 2a and 2b. Several variants of a recognition task were discovered:

- both training images and images to be recognized had all three color channels;
- training images had only red channel representation while images to be recognized had three-cannel RGB representation.

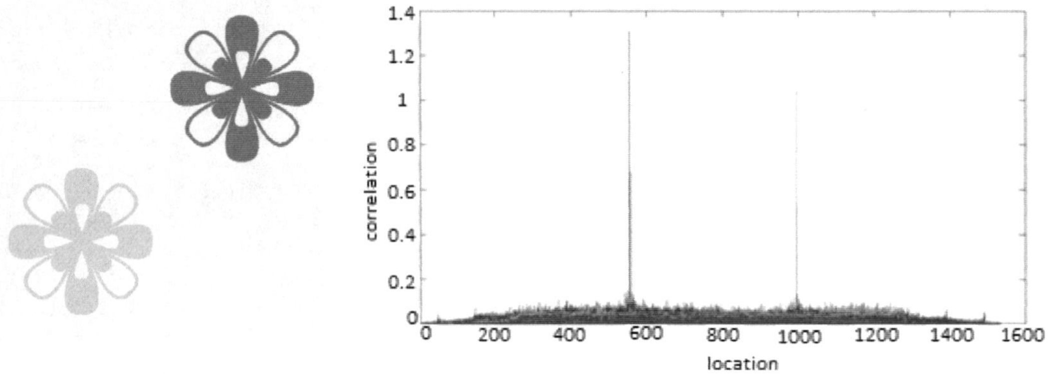

(a) (b)

Figure 6. a) The scene with two target object images in different colors; b) calculated correlation field of this scene and MINACE CF.

A green channel was used for object representation in training images. Images to be recognized had green color to represent object. All the color channels that represent the test object had the maximum sharpness of color. The obtained discrimination characteristics of target object and three clutter objects demonstrated the high obtainable quality of recognition. Also the modeling of recognition of a scene with two images of target object in different colors using CF synthesized on a basis of target object images in all three color channels. The results are illustrated on figures 6–8. These results demonstrate the efficiency of multicolor recognition.

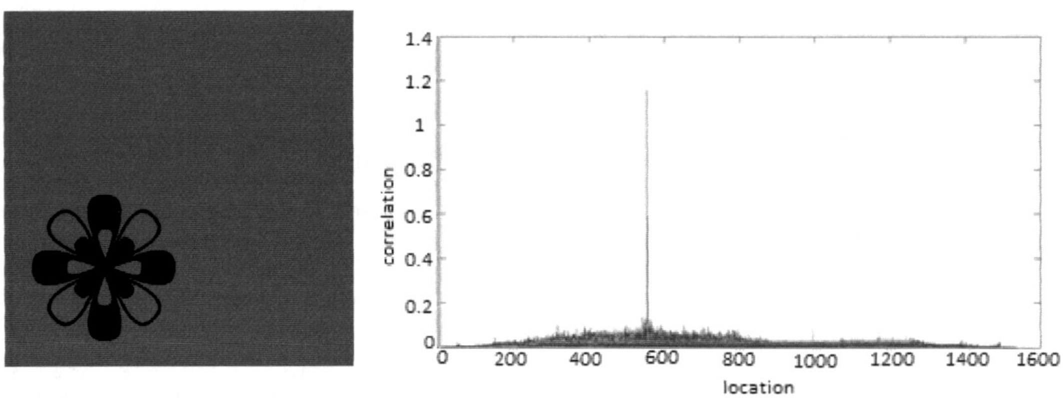

(a) (b)

Figure 7. a) Halftone grayscale image of target object; b) red channel image of target object.

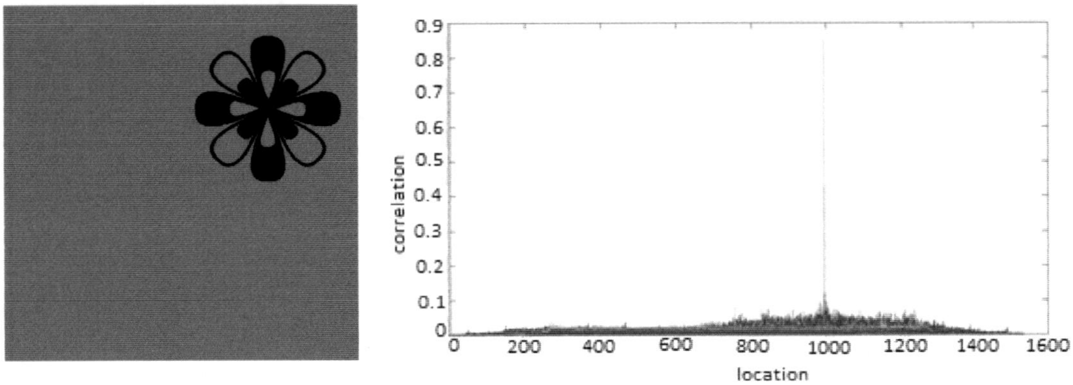

(a) (b)

Figure 8. a) The scene from figure 6a in green color channel; b) correlation field of this scene and filter. A red color object is recognized.

5. Different background and object colors

The possibilities of recognition quality enhancement due to utilization of object and background color differences information is discussed. We discovered the case when green and red channels are used for object representation and blue and green colors are used for background (see figure 2c). The next recognition conditions were studied:

- both training images and images to be recognized had all the RGB channels in representation;
- both training images and images to be recognized had only red channel representation with suppressed green and blue channels;
- training images had only red channel while images to be recognized were represented using all the thee RGB channels.

It was observed that in all cases the probability of correct recognition was more than 99%. The highest recognition quality was observed in the case of one channel selection in training images. Figure 9 represents the example of discrimination characteristic obtained in the case of training set of only red channel images and test images in all three color representation.

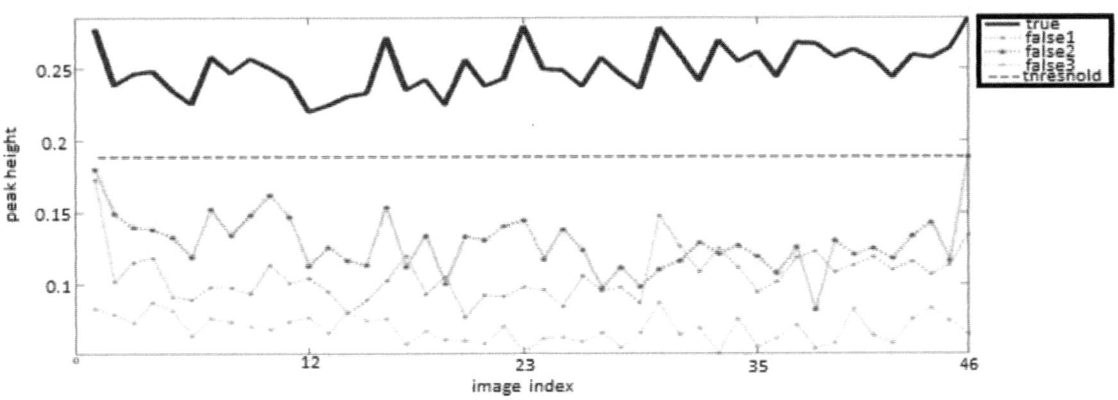

Figure 9. Discrimination characteristic of MINACE CF obtained in modeling the case when training images had only red channel representation while images to be recognized had full RGB colors representation.

6. Conclusion

The represented results show the principal possibilities of multicolor images correlation recognition invariant to target object distortions using MINACE composite filters (CF). A mathematical modeling of these filters was provided. The obtained results demonstrate the increase of flexibility and quality of object recognition using multicolor information of object images.

Further researches assume investigation of more complicated objects recognition and quantitative estimation of possible quality enhancement by varying of color channels information during the filter synthesis

Acknowledgments

Represented researches supported by Russian Scientific Foundation, project №14-19-01751.

References

[1] Ravichandran G, Casasent D 1992 filter *Applied Optics* **31** pp 1823-1833.
[2] Patnaik R, Casasent D 2005 *Proceedings of SPIE* **5816** pp 94-104.
[3] Casasent D, Patnaik R 2006 *Proceedings of SPIE* **6384** 638401.
[4] Casasent D, Patnaik R 2006 *Proceedings of SPIE* **6234** 62340S.
[5] Patnaik R, Casasent D 2007 *Proceedings of SPIE* **6574** 65740H.
[6] Patnaik R, Casasent D 2008 *Proceedings of SPIE* **6967** 696705.
[7] Evtikhiev N, Shaulskiy D, Zlokazov E, Starikov R 2012 *Proceedings of SPIE* **8398** 83980G.
[8] Evtikhiev N, Shaulskiy D, Zlokazov E, Starikov R *Proceedings of SPIE* **8748** 87480O.
[9] Shaulskiy D, Evtikhiev N, Starikov S, Starikov R, Zlokazov E 2014 *Proceedings of SPIE* **9094** 90940K.

Information virtual indicator with combination of diffractive optical elements

Y A Grad, E A Drozdova, L A Nayden, V V Nikolaev, S B Odinokov, A B Solomashenko

Bauman Moscow State Technical University, 2nd Baumanskaya str. 5, 105005 Moscow, Russia

E-mail: art_s87@mail.ru, vnikolaev@hololab.ru

Abstract. A combination of diffractive optical elements for monochrome information virtual indicators is described. To reduce the spectral "blurring" of image in monochrome indicators with OLED-display or LCD-display with LED backlight the possibility of using the volume reflection hologram as a spectral filter is investigated. The theoretical and experimental results show that the volume reflection hologram can be used as part of a monochrome virtual indicator containing OLED-, LCOS- or LCD-display with LED-backlight and relief-phase gratings for output of radiation from substrate to reduce the spectral "blurring" of image.

1. Introduction

In recent years, displays and indicators based on diffractive optical elements (DOEs) are used in the information display systems and visualization systems, where the observed signs-symbolic information is superimposed on the real scene. In such systems the exit pupil of the optical system expands to required size due to the light guide plate with DOEs, which is a significant advantage of such systems [1-3]. Thus, the exit pupil of the virtual sign-symbolic indicator is about 40 mm when the exit pupil of the collimating optical system is relatively small (up to 10 mm) is provided. Due to the weight and size parameters of indicator are reduced.

Now, a number of signs-symbolic information displays and virtual indicators, based on light guide substrates and DOEs, are created. However, these indicators are primarily monochrome [1-4]. The use of the monochrome OLED-displays or LCD with LED-backlight in the indicators leads to spectral "blurring" of displayed image and reduce resolution of indicators. Currently the principles of construction of such devices are developed, as well as problems of radiation output from the light guide substrate by DOE are resolved [3,4]. But the issues of improving the resolution of indicators were considered not fully.

At the same time the combined DOE, combining several different DOEs in one single structure and having a number of advantages compared to single component DOEs, are known [5,6]. Combined DOEs form simultaneously several wavefronts and can provide the spectral selection of radiation. The use of such DOEs as part of the signs-symbolic information systems will improve the resolution of monochrome indicators.

2. Combination of diffractive optical elements in information virtual indicator

To solve these problems the use of combination different types of DOEs, produced on a single substrate and providing the input and output of radiation from the light guide plate and the spectral selection, is suggested. For input and output of radiation the transmissive diffraction grating (DG), which can be relief-phase (including multilevel) or volume, are used [6].

To reduce the spectral "blurring" of image in monochrome indicators with OLED-display or LCD-display with LED backlight the possibility of using the volume reflection hologram as a spectral filter is investigated. It is suggested to use a two-component combination of the DOEs, as shown in Figure 1, including a DG 2 on one surface of the substrate 1 to output radiation from the plate and on the other - a volume reflection hologram 3 with high spectral and angular selectivity. The volume reflection hologram functions as a narrow-band spectral filter. This allows you to select narrow spectrum from propagating radiation and reduce the image "blurring".

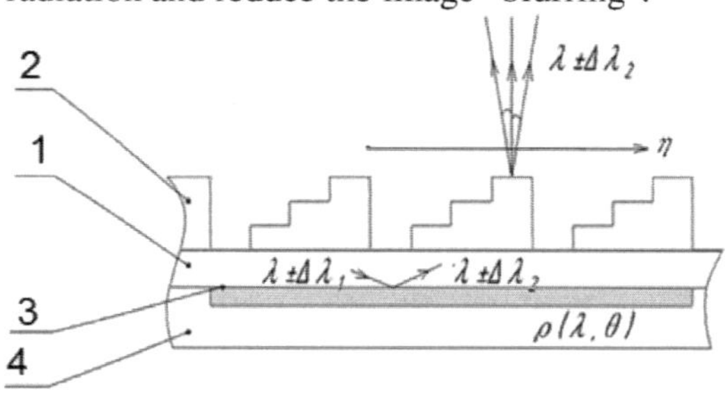

Figure 1. Diagram of combined diffractive optical elements consisting of substrate 1, DG 2, volume hologram 3 and protective glass 4

The holographic virtual indicator scheme consists of LEDs 1, lighting lens 2, cube prism 3, LCOS-display 4, collimating lens 5, light guide plate 5 with a combination of DOEs, i.e. diffraction gratings 6 of variable diffraction efficiency for the output radiation on one side of the substrate and the reflective volume hologram 7 on the opposite side. As a source of the formation of signs-symbolic information in the proposed scheme the LCOS-display that is illuminated by the beam splitter cube via LEDs matrix is used. When using high power LEDs or laser sources, such types of displays and indicators allow to observe the information on the background of bright objects. The disadvantage of this system is its size because of the presence of lighting lenses. OLED-displays, despite its compact size, have low brightness, which prevents their use in displays and indicators to form multicolor and monochrome signs-symbolic information.

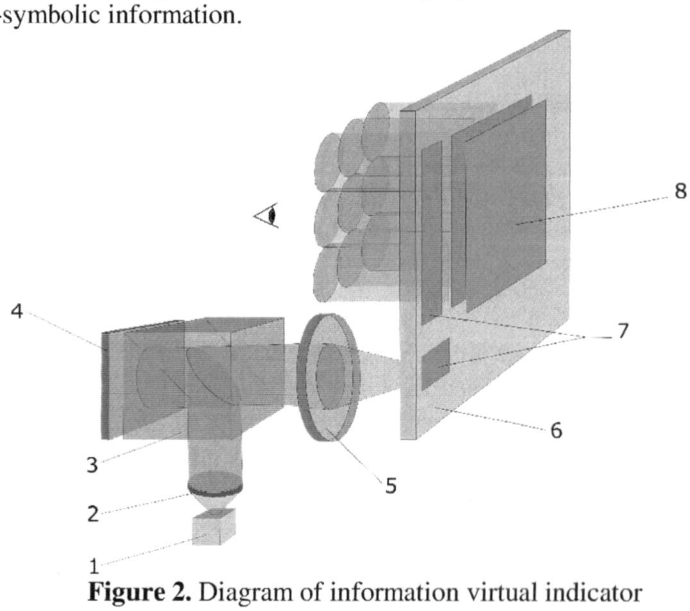

Figure 2. Diagram of information virtual indicator

Diffraction gratings (DGs) in this scheme are multilevel phase relief with triangular profile, i.e., DGs are essentially transmissive echelettes, i.e. DGs gratings having only one working diffraction order. It is necessary to increase the diffraction efficiency and thus reduce the radiation loss at the input and output it from the plate. Such elements can be obtained, for example, on thin transparent films and glued on the surface of light guide plate, which significantly simplifies the process of their manufacture.

For the combined DOEs based on reflection holograms the main parameters are defined: thickness is 20 mm, width of spectrum is 20 nm, the reflection coefficient is about 0.75, period of DG is 500 nm. Figure 3 shows the dependence of the diffraction efficiency and the shrinkage of the photosensitive material thickness for volume reflection holograms.

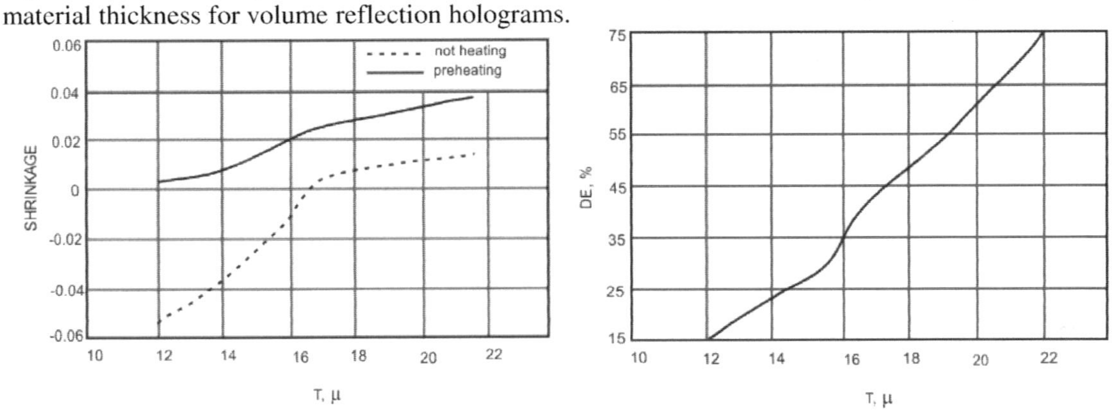

Figure 3. The dependence of the diffraction efficiency and the shrinkage of the photosensitive material thickness

3. The theoretical and experimental results

Figure 4 shows the experimental values of the spectrum of the reflected radiation from the volume hologram, depending on the angle of incidence. When the angle of incidence is about 50 degrees the reflection spectrum is the maximum and it is about 0.75 for a wavelength of 530 nm, and for angles of incidence are 45 and 55 degrees the reflection coefficient is greater than 0.6. The width of the reflection spectrum for the angle of incidence of 50 degrees is 20 nm and for a range of angles from 45 to 55 degrees is 25 nm, which is two times less than the radiation spectrum, for example, of OLED-display or green LED.

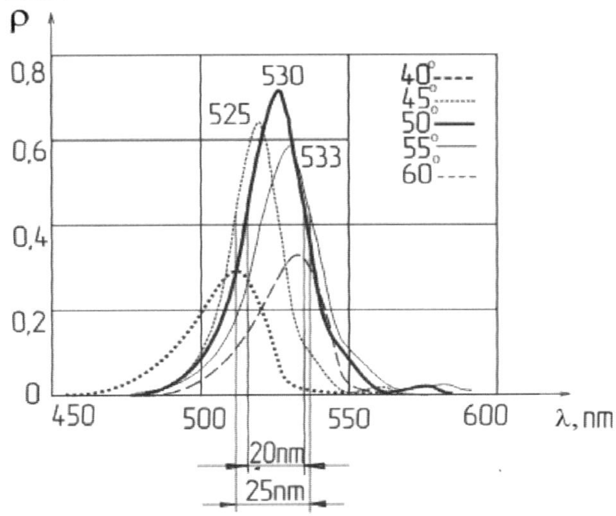

Figure 4. The dependence of the spectrum reflected from the volume hologram for different wavelengths and angles of incidence

Photography of the experimental sample of the light guide substrates with DOEs combination is shown in Figure 5. The developed experimental sample is a parallel plate made of optical glass. A distinctive feature of the construction is that the combined DOE is two flat parallel plates glued together. The first plate contains on one of its surfaces DG, and the second plate - volume hologram. The DOE's production in this case occurs separately and independently at every plates, which simplifies the technological process of manufacturing the substrate.

Figure 5. The experimental sample of the light guide substrate with a combination of DOEs and image of the test-object

The resolution of virtual indicator is controlled visually using the test-object at stand, which shown in Figure 5. The images of the test-object, obtained with combination DOEs, based on volume hologram, and without it are presented in Figure 6. Figure 6 shows images of the test-object, obtained with the combined DOE (left) and without it (right), i.e. with relief-phase DG. When the exit pupil size of about 50 mm and more the blurring and coloring of image are observed. However, when using the combined DOE, this effect is minimized. Thus, the resolution provided by using the experimental sample with a combined DOE is 50 lines/mm, and without it is less than 30 lines/mm.

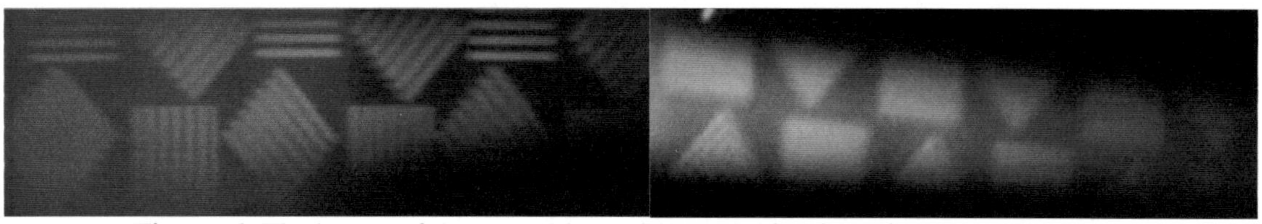

Figure 6. The image of the text-object with a combination of DOEs and without it

4. Conclusion

Thus, the experimental results show that the volume reflection hologram can be used as part of a monochrome virtual indicator containing OLED-, LCOS- or LCD-display with LED-backlight and relief-phase gratings for output of radiation from substrate to reduce the spectral "blurring" of image.

Acknowledgments

This work was supported by a grant from the Ministry of Education and Science of the Russian Federation # 14.574.21.0066 (Project IDRFMEFI57414X0066).

References

[1] Eisen L, Meyklyar M, Golub M, Friesem A, Gurwich I, Weiss V 2006 *Appl. Optics* **45** 4005-4011
[2] Odinokov S, Markin V, Kuznetsov A, Drozdova E, Solomashenko A 2013 *Proc. of SPIE* **8776** 87760Q
[3] Cameron A. 2012 *Proc. of SPIE* **8383** 83830E
[4] Yan Z, Li W, Zhou Y, Kang M, Zheng Z 2011 *Opt. Engineering* **50(9)** 094001
[5] Odinokov S, Kuznetsov A, Kolyuchkin V, Drozdova E, Solomashenko A 2015 *Journal of Physics: Conference Series* **584(1)** 012024
[6] Odinokov S, Zherdev A, Kolyuchkin V, Solomashenko A 2014 *Computer Optics* **38(4)** 704-709

Journal of Physics: Conference Series **737** (2016) 012059 doi:10.1088/1742-6596/737/1/012059

The multi-criteria optimization for the formation of the multiple-valued logic model of a robotic agent

A Yu Bykovsky[1], **A A Sherbakov**[2]
[1] P.N. Lebedev Physical Institute RAS, 53 Leninskij Prosp., Moscow, 119991, Russia
[2] Moscow Institute of Physics and Technology (State University), 9 Institutskiy per., Dolgoprudny, Moscow Region, 141700, Russia

E-mail address: ayubykov@sci.lebedev.ru

Abstract. The k-valued Allen-Givone algebra is the attractive tool for modeling of a robotic agent, but it requires the consensus method of minimization for the simplification of logic expressions. This procedure substitutes some undefined states of the function for the maximal truth value, thus extending the initially given truth table. This further creates the problem of different formal representations for the same initially given function. The multi-criteria optimization is proposed for the deliberate choice of undefined states and model formation.

1. Introduction

The k-valued Allen-Givone algebra (AGA) [1] is the mathematical tool, which is potentially very attractive to be used in optoelectronic data processing and for modeling of multi-agent robotic systems [2-4]. The multiple-valued logic function (MVLF) or a structured set of such functions can be used as the skeleton of the heterogeneous logic model of a robotic agent [3,4], providing exclusively high information capacity both for precise and approximately given parameters of the model. The appropriate structure of MVLFs is firstly to be given as the truth table for the switching function, whose logic expression is further to be written and simplified. This simplification is based on the method of consensus [1] which is to shorten the number of logic product terms and the computing time. But the minimization procedure in AGA use initially undefined states of the MVLF, thus extending its truth table. The problem is especially actual for multi-parametric MVLFs, as nobody will be able to use on practice all possible states for the MVLF with several dozens of input variables and 256 discrete truth values, because the truth table for such a MVLF will have ~10^{70} possible combinations of input variables, given as separate rows in the truth table [2,4]. Thus any real MVLF will contain a large number of initially undefined states, whose choice during minimization can influence greatly on the final set of product terms in the logic model, thus complicating its interpretation and knowledge search.

The aim of this paper is to show that the multi-criteria optimization procedure can be principally used for the choice of undefined states in the consensus minimization procedure of a MVLF, and this procedure can be done autonomously without the direct use of the computer- human expert dialog, which usually takes place in multi-criteria decision making.

2. Basic definitions and procedures of Allen-Givone algebra

The minimization of logic expression is regarded within the model of discrete k-valued Allen-Givone algebra (AGA) [1], where all input variables $x_1, \dots, x_n$ and the output variable $y = F(x_1, \dots, x_n)$ can have k discrete truth values $\{0,1,\dots,k-1\}$. In Boolean logic there are only two possible truth values $\{0,1\}$, but in AGA one can use arbitrary values of k. For the sake of simplicity examples further are given only for k=4, but the procedures to be used will be the same for $k = 256$ and even higher. The full set of non-Boolean operators is the set $\langle 0,1,\dots,k-1, X(a,b), MIN, MAX \rangle$, where $0,1,\dots,k-1$ are the constants, $MIN(x,y) = (*)$, $MAX(x,y) = (+)$ are respectively the choice of minimal and maximal values in the pair of values (x,y), operator $X(a,b)$ is called Literal and is given by

$$X(a,b) = \begin{cases} 0, & if\ b < x < a \\ k-1, & if\ a \leq x \leq b, \end{cases} \quad (1)$$

where $b \geq a$ and $a, b \in L = \{0,1,\dots,k-1\}$.

Arbitrary MVLF in AGA can be given by its truth table [1], like as the function $y = F(x_1, x_2)$ was given for $k = 4$ and for the truth table in figure 1 a), thus the output variable y and input variables x_1, x_2 can have only truth values $\{0,1,2,3\}$. In figure 1 a) the function values y are initially undefined states for $(x_1, x_2) = (0,3)$ and $(x_1, x_2) = (2,3)$ as marked by "-".

x_1	x_2	$y=F(x_1,x_2)$
0	0	1
1	0	1
2	0	3
3	0	1
0	1	1
1	1	1
2	1	3
3	1	1
0	2	1
1	2	1
2	2	3
3	2	2
0	3	-
1	3	2
2	3	-
3	3	2

a)

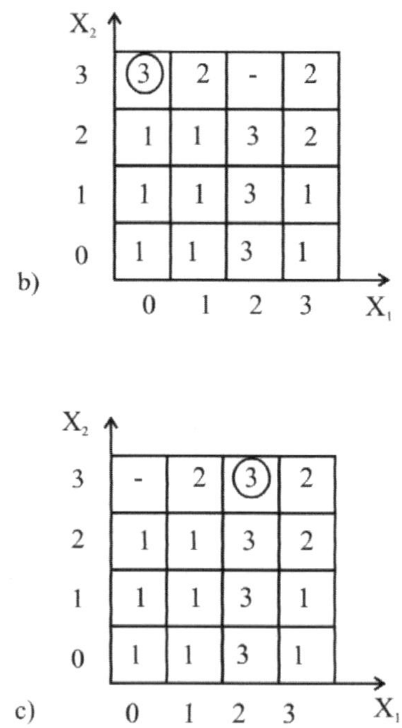

Figure 1. a) The truth table for the MVLF $y=F(x_1, x_2)$, where k=4 and there are initially two undefined function values y marked "- ". Both of them can be taken as the DCS in the process of consensus minimization.
b) The 2D grid representation of the truth table for the new function with the DCS chosen for (x_1, x_2)=(0,3), where "-" was changed for 3 as k-1=3.
c) The grid representation of the truth table for the new function with the DCS chosen for (x_1, x_2)=(2,3), where "-" was also changed for 3.

Any truth table can be transformed into the formal logic expression [1] written via operators MAX, MIN, LITERAL and constants C:

$$F(x_1,\ldots,x_n) = \quad C_1 * X_1(a_{11},b_{11}) * X_2(a_{12},b_{12}) * \ldots * X_n(a_{1n},b_{1n}) + \quad (2)$$
$$\ldots$$
$$C_{k-1} * X_1(a_{k-1,1},b_{k-1,1}) * X_2(a_{k-1,2},b_{k-1,2}) * \ldots * X_n(a_{k-1,n},b_{k-1,n}).$$

The function $y = F(x_1,x_2)$ in figure 1 a) can be composed [1] from Literal expressions, given for every row of the truth table

$$y = 1 * X_1(0,0) * X_2(0,0) + 1 * X_1(1,1) * X_2(0,0) + 3 * X_1(2,2) * X_2(0,0) + 1 * X_1(3,3) * X_2(0,0) + \quad (3)$$
$$1 * X_1(0,0) * X_2(1,1) + 1 * X_1(1,1) * X_2(1,1) + 3 * X_1(2,2) * X_2(1,1) + 1 * X_1(3,3) * X_2(1,1) +$$
$$1 * X_1(0,0) * X_2(2,2) + 1 * X_1(1,1) * X_2(2,2) + 3 * X_1(2,2) * X_2(2,2) + 2 * X_1(3,3) * X_2(2,2) +$$
$$2 * X_1(1,1) * X_2(3,3) + 2 * X_1(3,3) * X_2(3,3),$$

where, e.g. the first row with $(x_1,x_2) = (0,0)$ in the truth table in figure 1 a) respond to $1 * X_1(0,0) * X_2(0,0)$. Literal expressions for "-" rows are not to be written into (3).

The method to obtain the minimized logic expression for MVLF`s truth table is given in [1] and is grounded on definitions 1-4, cited from [1] to disclose the principle of subsuming of product terms and the use of consensus method.

Definition 1. One product term $r_1 * X_1(a_1,b_1) * \ldots * X_n(a_n,b_n)$ **subsumes** another product term $r_2 * X_1(c_1,d_1) * \ldots * X_n(c_n,d_n)$, if and only if both conditions are true:

1) $r_1 \le r_2$, (4)
2) $c_i \le a_i \le b_i \le d_i$ for all X_i, $i = 1,\ldots,n$.

Example 1: in exp.(3) the product term $1 * X_1(0,0) * X_2(0,0)$ subsumes term $1 * X_1(0,1) * X_2(0,0)$ as its parameters a_i, b_i are between c_i and d_i both for x_1 and x_2. Thus, the first product term is included into the second one and is to be deleted from the formal notation.

Definition 2. **Consensus** $j * U_j *^i k \cdot U_k$ in the i-th coordinate for product terms $r * U_1 = r * X_1(a_1,b_1) * \ldots * X_n(a_n,b_n)$ and $s * U_2 = s * X_1(c_1,d_1) * \ldots * X_n(c_n,d_n)$ is given by $j * U_j *^i k * U_k = q * X_1(e_1,f_1) * X_2(e_2,f_2) \ldots * X_n(e_n,f_n)$ if and only if there exist q, e_k, f_k that

$$q = j * k, \quad (5)$$
$$X_k(e_k,f_k) = X_k(a_k,b_k) + X_k(c_k,d_k) \text{ for } k = i,$$
$$X_k(e_k,f_k) = X_k(a_k,b_k) * X_k(c_k,d_k) \text{ for all } k \ne i.$$

Definition 3. The operator **union** of Literals is defined as $X(a,b) = X(c,d) + X(e,f)$, and it exists if $a = MIN(c,e)$, $b = MAX(d,f)$, $e - 1 \le d, c - 1 \le f$. (6)

Definition 4. The **intersection** of Literals is defined as $X(a,b) = X(c,d) * X(e,f)$, which exist if $a = MAX(c,e)$, $b = MIN(d,f)$, $e \le d, c \le f$. (7)

Definitions given above provide the calculations results given in *Example 2* and further in Table 1.

Example 2: Consensus $1 * X_1(0,0) * X_2(0,0) *^1 1 * X_1(1,1) * X_2(0,0) = 1 * X1(0,1) * X2(0,0)$,
Consensus $1 * X_1(0,0) * X_2(0,1) *^2 1 * X_1(1,1) * X_2(0,0)$ does not exist.

Namely undefined states (i.e. "-" values of y) can be used for further minimization [1].

3. The problem of choosing undefined states of the MVLF for its minimization

Using definitions of subsuming (4) and consensus (5) the minimization procedure can be done, which demonstrate the dependence of minimization results from the choice of undefined states. According to [1] the new modified MVLF should be received by substitution of $k - 1$ instead of "-" for one or more undefined states, and adequate product terms are to be added to exp.(3). Authors of [1] named undefined states of y as "don`t care states" (DCSs) because additional values for the modified MVLF does not change the initial truth table. For simplicity the authors further use only DCS abbreviation.

For the example in figure 1 a) DCSs can be chosen both for $(x_1,x_2) = (0,3)$ and (0,3), thus product terms $3 * X_1(0,0) * X_2(3,3)$ or $3 * X_1(2,2) * X_2(3,3)$ should be added to exp. (3) providing function modification. The new MVLF can be simplified quite efficiently. In order to show this more visible, lets firstly re-write truth table in figure 1 a) as two grids b) and c), containing two different variants of DCSs for the function $y = F(x_1,x_2)$. Product terms $3 * X_1(0,0) * X_2(3,3)$ and $3 * X_1(2,2) * X_2(3,3)$ were added respectively into rows N15 of the left and right parts of Table1.

Table 1. Minimization results are given for two possible variants of DCS choice for the MVLF from figure1 a). Left part of the table respond to the figure 1 b) and to the DCS taken at $(x_1, x_2) = (0,3)$. Right part (marked cursive) respond to figure 1 c) and to the DCS taken at $(x_1, x_2) = (2,3)$. Both left and right columns "Consensus $j *^i k$" demonstrate what tags j and k from columns "N" were substituted into consensus expression for the variable i.

N	IMPLICANT	Consens. $j*^ik$	Subsumes	*N*	*IMPLICANT*	*Consens. $j*^ik$*	*Subsumes*
1	~~1*X1(0,0) *X2(0,0)~~		Subsumes 16	1	~~1*X1(0,0) *X2(0,0)~~		Subsumes 16
2	~~1*X1(1,1) *X2(0,0)~~		Subsumes 16	2	~~1*X1(1,1) *X2(0,0)~~		Subsumes 16
3	3~~*X1(2,2) *X2(0,0)~~		Subsumes 18	3	~~3*X1(2,2) *X2(0,0)~~		Subsumes 18
4	~~1*X1(3,3) *X2(0,0)~~		Subsumes 17	4	~~1*X1(3,3) *X2(0,0)~~		Subsumes 17
5	~~1*X1(0,0) *X2(1,1)~~		Subsumes 19	5	~~1*X1(0,0) *X2(1,1)~~		Subsumes 19
6	~~1*X1(1,1) *X2(1,1)~~		Subsumes 19	6	~~1*X1(1,1) *X2(1,1)~~		Subsumes 19
7	~~3*X1(2,2) *X2(1,1)~~		Subsumes 18	7	~~3*X1(2,2) *X2(1,1)~~		Subsumes 18
8	~~1*X1(3,3) *X2(1,1)~~		Subsumes 20	8	~~1*X1(3,3) *X2(1,1)~~		Subsumes 20
9	~~1*X1(0,0) *X2(2,2)~~		Subsumes 21	9	~~1*X1(0,0) *X2(2,2)~~		Subsumes 21
10	~~1*X1(1,1) *X2(2,2)~~		Subsumes 21	10	~~1*X1(1,1) *X2(2,2)~~		Subsumes 21
11	3*X1(2,2) *X2(2,2)		Subsumes 23	11	~~3*X1(2,2) *X2(2,2)~~		Subsumes 23
12	~~2*X1(3,3) *X2(2,2)~~		Subsumes 22	12	~~2*X1(3,3) *X2(2,2)~~		Subsumes 22
13	~~2*X1(1,1) *X2(3,3)~~		Subsumes 24	13	~~2*X1(1,1) *X2(3,3)~~		Subsumes 25
14	~~2*X1(3,3) *X2(3,3)~~		Subsumes 26	14	~~2*X1(3,3) *X2(3,3)~~		Subsumes 27
15	**3*X1(0,0)*X2(3,3)**			15	~~3*X1(2,2)*X2(3,3)~~		Subsumes 23
16	~~1*X1(0,1) *X2(0,0)~~	$1*^{1}2$	Subsumes 28	16	~~1*X1(0,1) *X2(0,0)~~	$1*^{1}2$	Subsumes 28
17	~~1*X1(2,3) *X2(0,0)~~	$3*^{1}4$	Subsumes 28	17	1~~*X1(2,3) *X2(0,0)~~	$3*^{1}4$	Subsumes 28
18	~~3*X1(2,2) *X2(0,1)~~	$3*^{2}7$	Subsumes 23	18	~~3*X1(2,2) *X2(0,1)~~	$3*^{1}7$	Subsumes 31
19	~~1*X1(0,1) *X2(1,1)~~	$5*^{1}6$	Subsumes 30	19	~~1*X1(0,1) *X2(1,1)~~	$5*^{1}6$	Subsumes 29
20	~~1*X1(3,3) *X2(1,2)~~	$8*^{2}12$	Subsumes 25	20	~~1*X1(3,3) *X2(1,2)~~	$8*^{2}12$	Subsumes 26
21	1~~*X1(0,1) *X2(2,2)~~	$9*^{1}10$	Subsumes 30	21	~~1*X1(0,1) *X2(2,2)~~	$9*^{1}10$	Subsumes 32
22	**2*X1(2,3) *X2(2,2)**	$11*^{1}12$		22	~~2*X1(2,3) *X2(2,2)~~	$11*^{1}12$	Subsumes 37
23	**3*X1(2,2) *X2(0,2)**	$11*^{2}18$		23	~~3*X1(2,2) *X2(2,3)~~	$11*^{2}15$	Subsumes 31
24	**2*X1(0,1) *X2(3,3)**	$13*^{1}15$		24	1~~*X1(1,1) *X2(2,3)~~	$13*^{2}21$	Subsumes 33
25	~~1*X1(3,3) *X2(1,3)~~	$14*^{2}20$	Subsumes 29	25	~~2*X1(1,2) *X2(3,3)~~	$13*^{1}23$	Subsumes 35
26	**2*X1(3,3) *X2(2,3)**	$14*^{2}22$		26	1~~*X1(3,3) *X2(1,3)~~	$14*^{2}20$	Subsumes 36
27	~~1*X1(0,0) *X2(2,3)~~	$15*^{2}21$	Subsumes 32	27	~~2*X1(3,3) *X2(2,3)~~	$14*^{2}22$	Subsumes 37
28	1~~*X1(0,3) *X2(0,0)~~	$16*^{1}17$	Subsumes 34	28	**1*X1(0,3) *X2(0,0)**	$16*^{1}17$	
29	**1*X1(3,3) *X2(0,3)**	$17*^{2}25$		29	~~1*X1(0,2) *X2(1,1)~~	$18*^{1}19$	Subsumes 38
30	~~1*X1(0,1) *X2(1,2)~~	$19*^{2}21$	Subsumes 32	30	~~2*X1(2,2) *X2(0,2)~~	$18*^{2}22$	Subsumes 31
31	~~1*X1(2,3) *X2(0,2)~~	$23*^{1}29$	Subsumes 34	31	**3*X1(2,2) *X2(0,3)**	$18*^{2}23$	
32	~~1*X1(0,1) *X2(1,3)~~	$24*^{2}30$	Subsumes 33	32	**1*X1(0,3) *X2(2,2)**	$21*^{1}22$	
33	**1*X1(0,1) *X2(0,3)**	$28*^{2}32$		33	~~1*X1(1,1) *X2(1,3)~~	$24*^{2}29$	Subsumes 39
34	**1*X1(0,3) *X2(0,2)**	$31*^{1}33$		34	~~1*X1(1,3) *X2(3,3)~~	$25*^{1}26$	Subsumes 35
				35	**2*X1(1,3) *X2(3,3)**	$25*^{1}27$	
				36	~~1*X1(3,3) *X2(0,3)~~	$26*^{2}28$	Subsumes 40
				37	**2*X1(2,3) *X2(2,3)**	$27*^{1}31$	
				38	~~1*X1(0,2) *X2(0,1)~~	$28*^{2}29$	Subsumes 41
				39	~~1*X1(1,1) *X2(0,3)~~	$28*^{2}33$	Subsumes 43
				40	1~~*X1(2,3) *X2(0,3)~~	$31*^{2}36$	Subsumes 44
				41	~~1*X1(1,2) *X2(0,3)~~	$31*^{2}40$	Subsumes 44
				42	~~1*X1(1,3) *X2(2,3)~~	$32*^{2}35$	Subsumes 44
				43	**1*X1(0,2) *X2(0,2)**	$32*^{2}38$	
				44	**1*X1(1,3) *X2(0,3)**	$40*^{1}41$	

After this the minimization was done for both variants. During it some product terms with numbers in columns "N" subsumed other product terms with numbers given in the column "Subsumes", that is why they are shown crossed out. This procedure for both variants of DCS has given 8 and 7 final product terms, which are not crossed out and are shown bold. Direct substitution of all possible pairs (x_1, x_2) into resulting expressions demonstrates that both grids in figure 1 b) and c) are equivalent to reproduce the initial truth table in figure 1 a).

Resulting sets of product terms are written in the descending order of logic constants in figure 2 c) and d). They respond to grid variants b) and c) in the figure 1. AGA based consensus minimization [1] has provided double reduction of product terms number, and each of minterms respond to the specific rectangular group of cells in the 2D space (multidimensional in general case!). Sets of minterms in figure 2a) and b) are drastically different, what can be seen from the structure of truth values in rectangular segments separated by ellipses in figure 2 c) and d), respectively. As here $k = 4$, then truth values can be only within the range from 0 up to 3. The number signed on the outer part of each ellipse designates the serial number of the product term in columns of the figure 2 a) and b).

It follows from the data given above that one step minimization for one added DCS has formed the set of segments with equal values of truth values, and these rectangular segments also contain "islands" with greater truth values. Thus the choice of DCS modifies not only the rectangular segment for the truth value $k - 1$, but also can change all segments with $\{1, 2, \ldots, k - 2\}$.

a) y=3∗X1(0,0)∗X2(3,3)+
3∗X1(2,2)∗X2(0,2)+
2∗X1(2,3)∗X(2,2)+
2∗X1(0,1)∗X2(3,3)+
2∗X1(3,3)∗X2(2,3)+
1∗X1(3,3)∗X2(0,3)+
1∗X1(0,1)∗X2(0,3)+
1∗X1(0,3)∗X2(0,2).

b) y=3∗X1(2,2)∗X2(0,3)+
2∗X1(1,3)∗X2(3,3)+
2∗X1(2,3)∗X(2,3)+
1∗X1(0,3)∗X2(0,0)+
1∗X1(0,3)∗X2(2,2)+
1∗X1(0,2)∗X2(0,2)+
1∗X1(1,3)∗X2(0,3).

c)
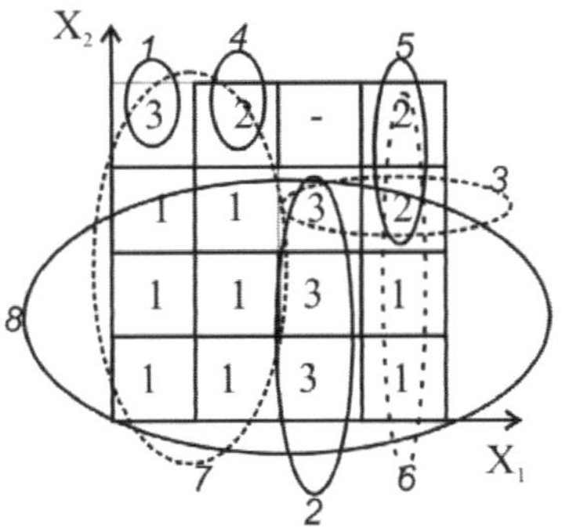

d)
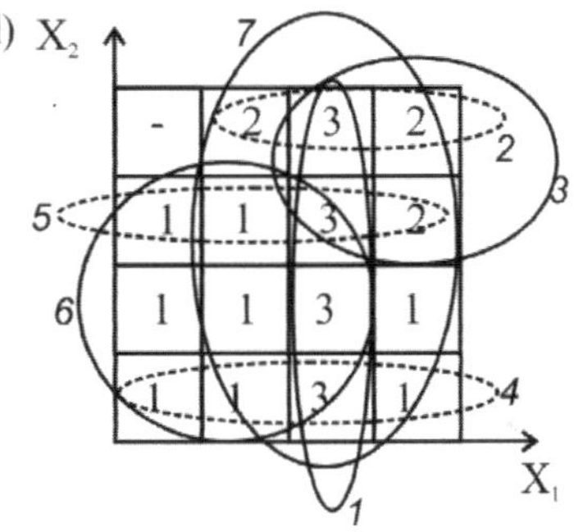

Figure 2. Minimized sets of minterms for DCSs chosen for $(x_1, x_2) = (0, 3)$ and $(x_1, x_2) = (2, 3)$ are given at a) and b), respectively. Grid segments structures produced by them are shown in c) and d). Group of cells produced by specific minterms in cases a) and b) are marked by ellipses in c) and d), where cursive number near the ellipse shows the serial number of this minterm in appropriate columns of minterms in a) and b). 8 ellipses are given for a) and 7 ones are given for b), respectively.

It seems reasonable that the obtained segments structure like given above can be potentially used for processing of multi-grade digital images, but its interpretation is difficult in contrast to traditional digital maps, which usually contain united groups of equal parameters. This stimulates to search for a method of intended choice of DCSs within specific rectangular segments or at their boundaries, which would form deliberately specific product terms to depict some object classes. Such a work can be done as a separate step of processing, providing additional pages of the MVLF digital map [2]. But for a model with large $k = 256$, the preparation of such data will need to process the large set of different segments with various truth values. That is why in general case one should regard the choosing of DCS as the multi-criteria problem.

4. The multi-criteria optimization as the potential method to choose DCSs for the MVLF

It follows from the previous section that the consensus minimization [1] can produce not visual structure of product terms even for very simple MVLFs. As this procedure takes into account DCSs position and truth values of nearby cells and segments of the grid, it will need to compute the large number of formal criteria and to apply multi-criteria choosing of DCSs. Now there are different methods of multi-criteria optimization (MCO) [5], where the most difficult cases refers to Pareto optimization methods. The general idea of Pareto method [5] is to find alternatives of $x: x_1, x_2, \dots, x_n$ which obey the constraints $Q(x) = (q_1(x), \dots, q_{1k}(x)) \rightarrow minimum\ or\ maximum$, $h_j(x) \geq 0$, where $q_j(x)$ are quality criteria and $h_j(x)$ are the constraints for the feasible solutions set D. Pareto set is regarded as the set for which the value of any criteria can be improved only at the cost of reduction of another criteria. The solution x_0 is Pareto optimal if there are no solutions $x \in D$, for which $q_i(x) \geq q_i(x_0)$ and the value at least for one criteria is strictly greater than for x_0.

However there is one substantial limitation for the robotic agent applications of MCO, as there are no universal multi-criteria methods for all tasks and mainly such methods involve direct participation of a human decision maker [5,6], whose expert preferences are used to choose the optimal solution.

That is why robotic agents are to use only "autonomous" methods of decision making, e.g. described in [6, 7] and based on axioms and special rules instead of direct dialog PC-human expert. The recent development of this technology was given in [7], where 5 basic axioms have been proven to provide Pareto optimal solutions. MCO problem was regarded in [7] as the choosing of possible solutions x in the arbitrary set X, where $f(x) = (f_1(x), \dots, f_m(x))$ is the criteria vector, and binary relation $\succ_X$ given for X means preferences of the decision making system, i.e. the fact that solution x_1 is more preferable than solution x_2 is expressed by the notation $x_1 \succ_X x_2$. Solution of the problem is called the set of selected solutions $C(X)$, i.e. the set of best alternatives chosen from X by the decision making system. Also sets of possible vectors $Y = f(X)$ and of chosen vectors $C(Y) = f(C(X))$ were used [7].

Substantial hardship of MCO application for the choice of DCSs is in the fact, that MCO procedures are to be adapted to the specifics of MVLFs and heterogeneous logic model. Earlier proposed heterogeneous logic model [4] was designed to unite together precise (Boolean and multiple valued logic based) and approximate (fuzzy logic based) computing in an robotic agent, as in it all multi-parametric computing processes are to be controlled by means of the special skeleton structure of MVLFs. As AGA uses very specific non-binary operators set, the direct integration of MCO in AGA is impossible, but the architecture of heterogeneous logic modeling [4] provides the mapping of MCO calculations results on the subset $N` = \{0,1,2, \dots, 256\}$ of the natural numbers set $N = \{0,1,2,3 \dots\}$. First of all the task is to provide high enough precision for MVLF images of MCO.

The method to provide necessarily high precision for MVLF images was earlier proposed in [2] for high precision excessive description of space coordinates x, y, z by several interrelated parameters. E.g., the coordinate x can be represented as the triple (x_1, x_2, x_3) with different scales, where x_1 has kilometer scale, x_2 has meter scale and x_3 has millimeter scale. Thus the criteria $f(x) = (f_1(x), \dots, f_m(x))$ can be mapped into AGA model as the extended set of interrelated functions $f(x) = (f_1^1(x), f_1^2(x), f_1^3(x))$, which are defined on the subset of $N` = \{0,1,2, \dots, 256\}$. In AGA [1] new variables always can be correctly added to the model as additional Literal operator X(0, k-1) with parameters $a = 0$ and $b = k - 1$ just written into every product term in the model. For example, the product term $1 * X_1(0,0) * X_2(0,0)$ for $k = 4$ always can be extended as $1 * X_1(0,0) * X_2(0,0) * X_3(0,3)$ to enlarge the number of variables in the model.

Another aspect is to appreciate, if switching functions of AGA are "good" enough for mapping of criteria of MCO. Let`s see possible constraints drawn by 5 basic axioms of autonomous MCO procedure, which are cited from [7] and accompanied by commentaries.

Axiom 1 (Pareto axiom). *For every pair of vectors $y`, y`` \in R^m$, which satisfy inequality $y` \geq y``$, the ratio $y` \succ_Y y``$ is satisfied too.*

Interpretation. R here is the set of real numbers. Due to [7] correct notation $y` \geq y``$ means that $y_i` \geq y_i``$ for all $i = 1,2, \dots, m$, and the preference relation $y` \succ_Y y``$ is true, which provides the correct choice of one of y in the pair. Besides this the axiom impose that the preference is done even for equal vectors, thus reducing the number of vectors to be further analyzed. Also formal expressions for criteria can`t be changed and there are no other criteria to be used for choosing of solutions. MVLF mappings on to $N` = \{0,1,2,\dots,256\}$ respond to Pareto axiom as it will always provide comparable values of MVLF criteria images according to inequality $y` \geq y``$, and can be done by means of comparison of output variables for MVLF, written by its truth table. Also it follows that not only vectors *y* but their MVLF images must be comparable and they can`t contain any DCSs.

Axiom 2. *For every pair of vectors $y`, y`` \in Y$, which satisfy relation $y` \succ_Y y``$, the condition $y`` \notin C(Y)$ is also satisfied.*

Interpretation. If one vector in the pair $y`, y``$ was not chosen, it is to be excluded from the whole set of possible vectors $C(Y)$ and out of all processing procedures. Thus the main task of Pareto set reduction method [7] is to shorten the set $C(Y)$ and to exclude one vector from every analyzed pair during every step of MCO.

Axiom 3. *The relation $\succ_Y$ has irreflexive and transitive extention $\succ$ for the whole space R^m.*

Interpretation. It means [7] that for the set Y there is no difference between $\succ_Y$ and $\succ$, what makes possible to work with the whole set *R* of real numbers. As the set of natural numbers *N* is the subset of *R*, then one can also use preference relation for the whole set $N`$. Also if the first solution is preferred to the second solution, and the second solution is preferred to the third one, than the first one is preferred to the third solution mapped on $N`$ [7]. Thus the process of Pareto set reduction can be regarded simply as a consequence of preference selections for all values of MVLF variables.

Axiom 4. *Each of criteria $f_1, f_2, \dots, f_m$ is consistent with preference relation $\succ$.*

Interpretation. Due to [7] it means that the criteria vector $f(x) = (f_1(x), \dots, f_m(x))$ is given as the formal expressions set, and if the pair of vectors $y`, y``$ has not equal criteria values for one criterion, but for all other criteria values are equal, than the decision maker system always prefer the maximal criterion value. Note that minimum criterion if necessary can be received from maximum criterion by the reverse function. Thus mapping criteria functions can`t have undefined DCS in it, as for each of criteria $f_1, f_2, \dots, f_m$ their images also always to be compared and chosen.

Axiom 5. *Preference relation $\succ$ is invariant relative to the positive linear transformation.*

Interpretation. Due to [7] for every pair of vectors $`, y`` \in R^m$, *for* the relation $y` \succ y``$, for every $\alpha > 0$ and for an arbitrary vector $c \in R^m$ the expression $\alpha y` + c \succ \alpha y`` + c$ is performed. As multiplication and summation are not defined in MVLF, linear transformations can`t be done directly within AGA, but MVLF images should correctly sustain the same preference rule after the mapping procedure. That is why criteria functions are to be mapped on the MVLF switching functions with high enough precision, what needs to use further large k.

Generalizing the requirements of MCO axioms (1-5) [6, 7], one can conclude the following criteria functions limitations actual for AGA models:

- only bijection mapping one-to-one should be used for MVLFs images of criteria functions,
- all criteria values and their MVLF mappings should be comparable and any DCSs are prohibited,
- only single brunch and continuous functions should be selected for criteria mapping,
- criteria functions can`t be changed during minimization, any alternatives are regarded separately.

If MCO axioms (1-5) given above are satisfied, then the method [7] can be applied for further deliberate choosing of DCS, basing on the preference relation $\succ$ and criteria vector $f(x) = (f_1(x), \dots, f_m(x))$. Briefly it can be described as follows.

If there are two groups of criteria $A = (f_1(x), f_2(x), \dots, f_{q-1}(x))$ and $B = (f_q(x), f_{q+1}(x), \dots, f_m(x))$, where:

- the group of criteria $A = (f_1(x), f_2(x), \dots, f_{q-1}(x))$ is preferred to the group of criteria $B = (f_q(x), f_{q+1}(x), \dots, f_m(x))$ for relative increase in criteria parameters $\{w_1^+, w_2^+, \dots, w_{q-1}^+\}$ and relative decrease in criteria parameters $\{w_q^-, w_{q+1}^-, \dots, w_m^-\}$,
- the group of criteria *B* is preferred to the group *A* for relative increase in criteria parameters $\{v_1^+, v_2^+, \dots, v_{q-1}^+\}$ and relative decrease in criteria parameters $\{v_q^-, v_{q+1}^-, \dots, v_m^-\}$,
- all these sets of mutually dependent data are not contradictory,
- for two solution vectors x^1 and x^2 the criteria vector takes values $F(x^1) = (r_1, r_2, \dots, r_m)$ and $F(x^2) = (s_1, s_2, \dots, s_m)$,

then the task can be transformed into the problem to choose the solution from the set $X = (x^1, x^2)$ using two new criteria $g(x) = \{g_1(x), g_2(x)\}$, where values $g_1(x)$ and $g_2(x)$ can be calculated via

$$\begin{aligned}
g_1(x^1) &= min\left\{\frac{r_1}{w_1^+}, \frac{r_2}{w_2^+}, \dots, \frac{r_{q-1}}{w_{q-1}^+}\right\} + min\left\{\frac{r_q}{w_q^-}, \frac{r_{q+1}}{w_{q+1}^-}, \dots, \frac{r_m}{w_m^-}\right\}, \\
g_2(x^1) &= min\left\{\frac{r_1}{v_1^+}, \frac{r_2}{v_2^+}, \dots, \frac{r_{q-1}}{v_{q-1}^+}\right\} + min\left\{\frac{r_q}{v_q^-}, \frac{r_{q+1}}{v_{q+1}^-}, \dots, \frac{r_m}{v_m^-}\right\}, \\
g_1(x^2) &= min\left\{\frac{s_1}{w_1^+}, \frac{s_2}{w_2^+}, \dots, \frac{s_{q-1}}{w_{q-1}^+}\right\} + min\left\{\frac{s_q}{w_q^-}, \frac{s_{q+1}}{w_{q+1}^-}, \dots, \frac{s_m}{w_m^-}\right\}, \\
g_2(x^2) &= min\left\{\frac{s_1}{w_1^+}, \frac{s_2}{w_2^+}, \dots, \frac{s_{q-1}}{w_{q-1}^+}\right\} + min\left\{\frac{s_q}{w_q^-}, \frac{s_{q+1}}{w_{q+1}^-}, \dots, \frac{s_m}{w_m^-}\right\}.
\end{aligned} \quad (8)$$

One solution from x^1, x^2 is to be chosen further [7] according to greater values obtained for criteria $g_1(x)$ and $g_2(x)$. Before the practical testing of the method, however, one should also to analyze the method to retrieve non contradicting data from mutual related data [7] for criteria groups of A and B.

5. Conclusions

K-valued Allen -Givone algebra is attractive for modeling of robotic agents and optoelectronic data processing, but it needs to simplify logic expressions by the consensus minimization method. This procedure includes the substitution of arbitrary taken initially undefined states of the MVLF into the truth table. That finally results in alternative formal representations for the same initially given truth table, what is discomfort for knowledge analysis. In order to control the formation of logic model expressions and to compare quickly them with templates, one should deliberately choose these DCS (initially undefined states), but this problem needs to analyze the complicated structure of product terms and segments, obtained for the large number of logic constants and variables. That is why this problem is to be regarded as the MCO task based on the Pareto set methods.

As any autonomous robotic system can`t be based on the direct dialog with the human expert, traditionally used in MCO tasks, than special autonomous decision making method is to be applied here [7], which bases on the special set of axioms and rules. In order to apply this version of MCO within the heterogeneous logic model [2-4] of a robotic agent, one should use only bijection (one-to-one) and high enough precision mapping of MCO criteria functions on to MVLFs, excluding any DCSs for them. Also only one brunch and continuous functions are possible for the criteria set. For this conditions to be fulfilled the MCO problem can be reduced to comparatively simple expressions and to the comparison of relative increase and decrease for different criteria groups.

The MCO application in AGA minimization is supposed to be used in robotic agents computer vision and pattern recognition systems.

References

[1] Allen C M and Givone D D 1984 *Computer Science and Multiple-Valued Logic: Theory and Applications* (North Holland Amsterdam) pp 262-283
[2] Bykovsky A Yu 2013 *Bull. of Lebedev Physics Institute* **11** pp 9-18
[3] Bykovsky A Yu 2015 *Physics Procedia* **73** pp 257-263
[4] Bykovsky A Yu and Rager B Yu 2014 *XII-th Russian Conference on the Control Problems VSPU-2014* (Moscow) pp 3917-3928
[5] Figueira J, Greco S and Ehrgott M 2005 *Multiple criteria decision analysis: State of the art surveys* (Springer)
[6] Noghin V D 1997 *J.Multi-Criteria Decision Analysis* **6** pp 355-363
[7] Klimova O N 2009 *Naučn. vedom. Belgor. gos. univ. seria Mat. fiz.* **9 (64)** pp152-159

Journal of Physics: Conference Series **737** (2016) 012060 doi:10.1088/1742-6596/737/1/012060

Experimental study of the use of multiband acousto-optic filters for spectral encoding / decoding the optical signals

V V Proklov[1], O A Byshevski-Konopko[1], A L Filatov[1], A V Lugovskoi[1], Yu V Pisarevsky[2]

[1]Institute of Radioengineering and Electronics of Russian Academy of Sciences, Vvedensky sq. 1, Fryazino, 141190 Moscow region, Russia
[2]Shubnikov Institute of Crystallography, Russian Academy of Sciences, Leninskii pr. 59, 119333 Moscow, Russia

E-mail: proklov@inbox.ru

Abstract. A prototype of the acousto-optic (AO) decoder of optical signals is created on the base of the multiband AO filter. The joint work of the decoder with the developed previously AO coder has been verified experimentally. The main qualitative and quantitate characteristics of the spectral coding and decoding by Walsh sequences of the industrial LED radiation in the near infrared range are investigated. It is shown, that in the proposed data transmission system realization Signal-to-Interference Ratio (SIR) is not less than 13 dB.

1. Introduction

At present acousto-optic tunable filters (AOTF) [1], which are based on a highly selective interaction of broadband light with a frequency tunable sound wave, are well known and widely used. The absence of moving elements and a high enough speed of tuning is an important advantage of these filters. Only in recent years [2], it was proposed to use such devices (MAOF - multiband acousto-optic filters) for multi-band (but not tunable) AO filtration, which is based on multi-frequency acoustic wave excitation. It was supposed in [3] that the MAOF can become a key element of a new component base for creating optical communication systems with code division of multiple access channels (O-CDMA). In [4] the first prototype of the AO encoder for O-CDMA, which uses a bulk MAOF, based on TeO_2, has been realized and investigated experimentally.

Figure 1 shows the proposed in [3] the concept of AO decoder, that realizes the parallel decoding of spectrally encoded optical signal. The light before decoding splits into two parallel optical channels with equal power. Further, the signal filtering is carried out simultaneously by two MAOFs. Transmission spectra of the MAOFs, formed by specially synthesized RF electrical signals, correspond to two complementary Walsh sequences W_{i+} and W_{i-}. After the filtration, the radiation intensity in both channels is converted by the photodetectors PD1 and PD2 into electrical currents. The differential amplifier converts the difference between this currents into a voltage U_{out}.

In paper [3] the numerical simulation of the decoder operation was carried out and it was shown, that values of U_{out} are grouped in three well separated from each other ranges, that allowed to determine exactly which Walsh sequence was used for the signal coding.

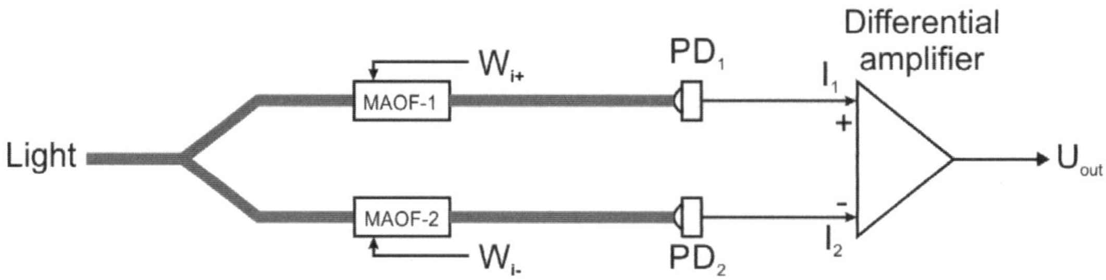

Figure 1. Schematic diagram of AO decoder proposed in [3].

In this paper the joint operation of the proposed earlier AO coder [5] and a new AO decoder prototype, functionally identical to one in [3], is studying to verify the quality of spectral coding/decoding with the use of coding Walsh sequences and a standard LED as a source of incoherent radiation.

2. Experimental setup

In designing the experimental setup, it was taken into account, that a phased verification and testing of proposed in [3] decoder circuit is appropriate and possible. At the first stage the stationary quality of spectral encoding / decoding processes was testing, whereas temporal characteristics were supposed to be studied later. A new circuit of AO decoder, proposed and realized in the experimental setup, which is shown in Figure 2, performs decoding in serial mode. Used in the previously proposed scheme, the parallel operation provides better performance, but it is sensitive to not the full identity of work in parallel channels. Systematic errors associated with the not full identity may be associated with not exact division of light in half, with the incomplete spectral identity of two MAOFs and photodetectors operation etc.

As shown in Figure 2, a new scheme uses for decoding one instead of two MAOFs. Synthesized RF electrical signals, forming multiband optical transmission function according to the selected complementary Walsh sequences in the MAOF, are fed to it sequentially. Fed to encoder MAOF RF electric signal does not change during sequential sending of these two signals on decoder MAOF. After each feeding of the electrical signal to the single encoder MAOF, the detecting of the optical signal occurs on the same photodetector, then the detected values are stored and their difference determines the value of the autocorrelation function of the expected signal. Therefore, the two circuits shown in Figures 1 and 2 have identical functionality when operating in the stationary mode.

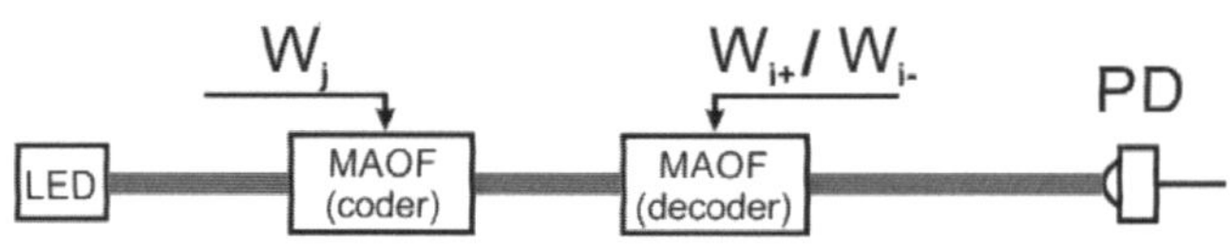

Figure 2. Schematic diagram of the experimental setup.

Figure 3 shows a block diagram of the experimental setup that realizes a schematic diagram shown in Figure 2.

The light from GaAs/GaAlAs LED TSAL6200 with 940 nm central wavelength, 100 mW power and an angle of divergence of $\pm$ 17^0, after passing the collimator with a focal length F = 10 cm forms a beam with the aperture D = 10 cm, which enters the MAOF-1 (Coder) and right after it at a distance of 1 cm enters MAOF-2 (Decoder), both with the input aperture of 8 cm. The radiation, passing through the MAOF-1 and MAOF-2, is focused on the photodiode (PD). A specialized ADC module digitizes the electrical signal from the PD output. The measured data is stored, processed and displayed on a computer.

The wide non-collinear AO geometry, used in MAOF on the basis of TeO_2 single crystal, allowed to realize the following parameters: the spectral range of 900-1600 nm, the range of control frequencies 55

- 100 MHz, a spectral resolution of 2 nm at a wavelength of 1152 nm. Control electrical RF signals for both MAOFs in the range of 70-90 MHz are formed by the two-channel arbitrary waveform generator Tektronix AFG3252 and amplified by two electronic power amplifiers. To increase the signal/noise ratio the possibility of additional low-frequency (LF) modulation of RF signals with subsequent amplification and the LF filtering of the signal by Selective Nanovoltmeter type 237 was realized. Since the LED's working spectral emission band (λ = 900-980 nm) is in the invisible to the eye near-infrared range, we used a standard night vision device for the system adjustment.

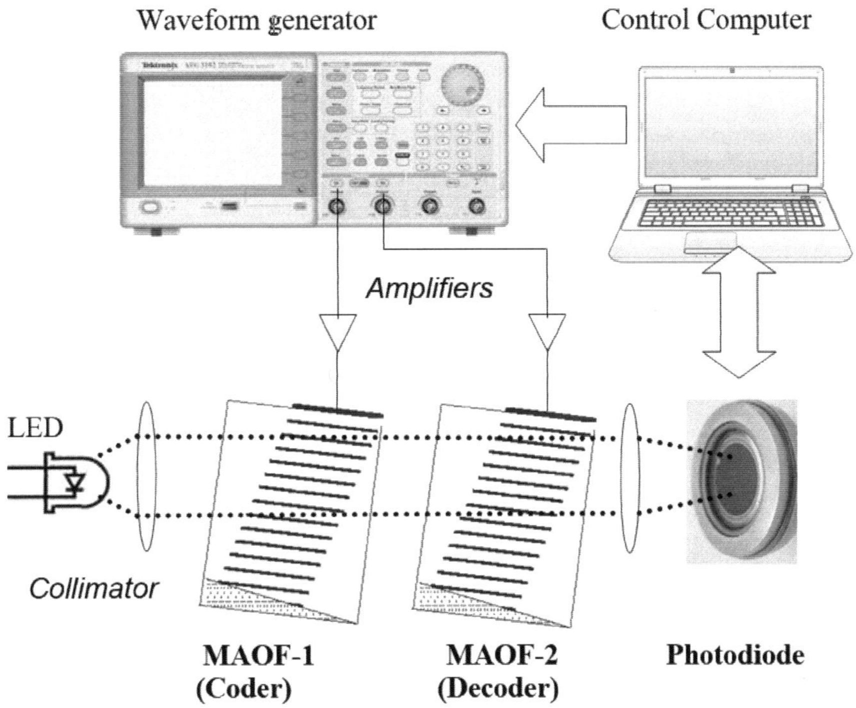

Figure 3. Block diagram of the experimental setup.

3. Experimental results and discussion

At the initial stage of the joint operation of the previously proposed encoder and a new decoder control measurements of the stationary single-band radiation spectrum were carried out at the output of encoder MAOF-1 when the decoder MAOF-2 worked in the mode of slowly tunable AOTF.

Figure 4 shows an experimental graph, obtained by the MAOF-1 in the single frequency mode, with a stationary sound wave at the frequency F = 81.0 MHz. In this case MAOF-2 operated at AOTF regime with a slowly tuning in time sound frequency F from 79.6 to 82.4 MHz. The abscissa represents the frequency of the sound, which was excited in the MAOF-2. The ordinate - optical power in relative units, digitized at the output of photodetector PD. The slight difference of the control frequencies in two MAOFs, that emerged due to technological reasons, was taken into account by software.

It can be seen that the resulting dependence of the intensity of radiation transmitted through both the MAOFs is in accordance with the transmission formula for two AO filters [3]:

$$I(\lambda) = \mathrm{sinc}^4(\Delta\lambda),$$

where $\Delta\lambda = 2\pi\Delta n L(\lambda_F - \lambda)/\lambda_F^2$, Δn - the difference between the refractive indices of the ordinary and extraordinary rays in the MAOF (AOFT) material, L - the AO interaction length, $\lambda f = \lambda_F F = const$.

Figure 5 shows experimental plots obtained when MAOF operated in multi-frequency mode. These plots correspond to the spectra of two Walsh sequences with dimension 8: $W5_+$ (red graph) and $W5_-$

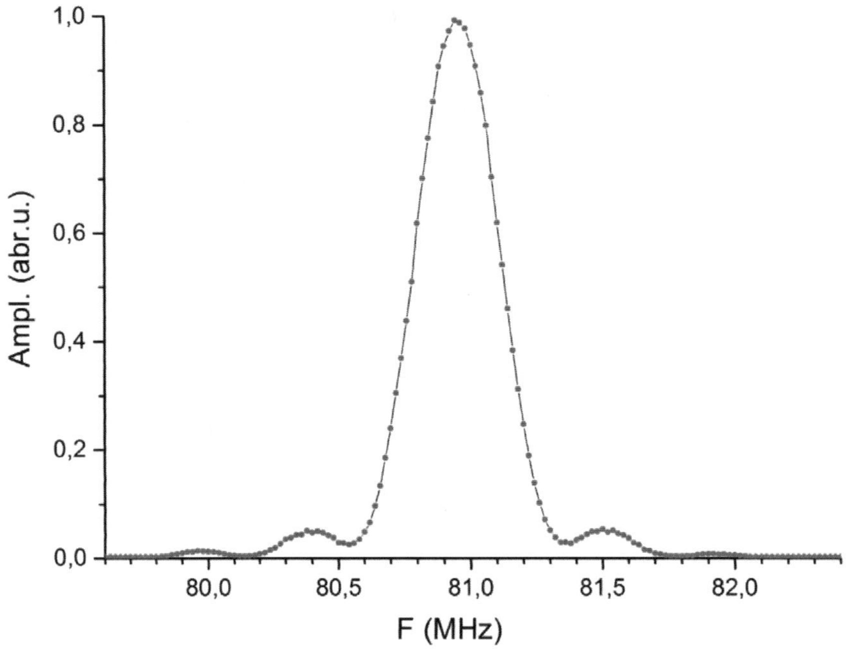

Figure 4. Transmission spectrum of the MAOF-1 at F=81.0 MHz

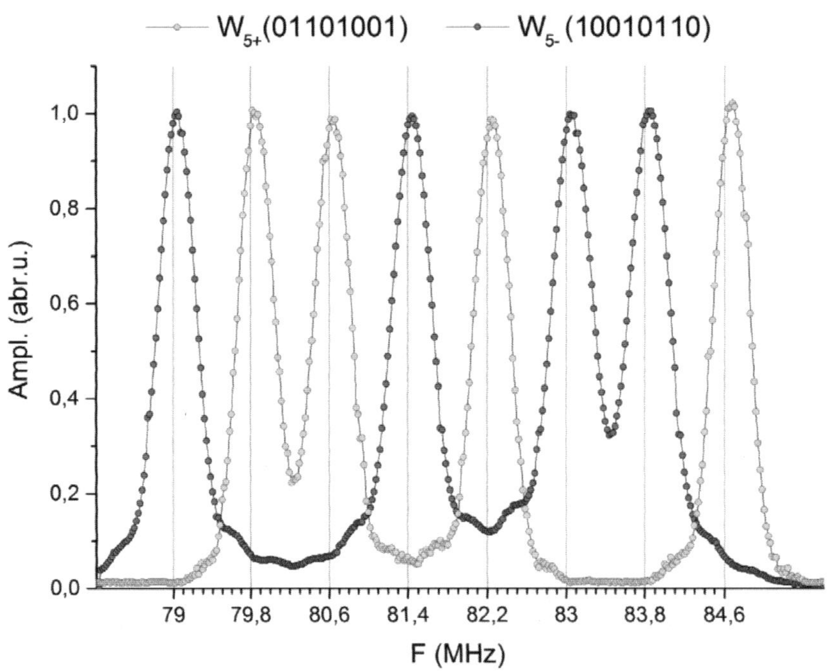

Figure 5. Transmission spectra of MAOF-1 for W_{5+} (red) and W_{5-} (green).

(green line). The abscissa represents frequencies F in the range of 78.2 to 85.4 MHz, that were fed on MAOF-2, operating in AOTF mode. The optical power in arbitrary units, measured at the output of the photodiode PD, is plotted along the ordinate axis. The spectra were formed when the MAOF was fed by the electric signal synthesized by Tektronix AFG3252. The synthesis procedure, having the purpose to align the LED spectral emission nonuniformity, was described in [5]. Each Walsh sequence was put in correspondence with unique set of four of the eight equidistant frequencies between 79 to 84.6 with 0.8 MHz step. This step between the adjacent frequencies was chosen to be two-time Rayleigh criterion, which is enough for confident decoding, according to theoretical calculations [3]. Methods of LED spectral emission alignment, proposed in [5], allowed to reduce the nonuniformity of all spectral lines maximum amplitudes down to 5%.

Table 1 shows the correspondence between the Walsh sequences and frequencies of the control signals fed to MAOF-1 and MAOF-2 for realization any of the 10 chosen Walsh sequences with the dimension 8 $W_{i(+/-)}$ $i = 1..5$. Note that sequences Wi with the same index i, but different colors (red or blue), correspond to the complementary Walsh sequences.

Table 1. Correspondence between the Walsh sequences and frequencies of the control signals.

Walsh	F (MHz)							
	84.6	83.8	83.0	82.2	81.4	80.6	79.8	79.0
W_{1+}	0	1	0	1	0	1	0	1
W_{1-}	1	0	1	0	1	0	1	0
W_{2+}	0	0	1	1	0	0	1	1
W_{2-}	1	1	0	0	1	1	0	0
W_{3+}	0	1	1	0	0	1	1	0
W_{3-}	1	0	0	1	1	0	0	1
W_{4+}	0	1	0	1	1	0	1	0
W_{4-}	1	0	1	0	0	1	0	1
W_{5+}	0	1	1	0	1	0	0	1
W_{5-}	1	0	0	1	0	1	1	0

The main results of the experiment on the decoding of the optical spectra, obtained by passing an incoherent LED light through MAOF-1 (Coder) and then through MAOF-2 (Decoder), followed by the detection at PD, are shown in Table 2.

The table headings indicate labels of ten optical Walsh sequences $W_{i(+/-)}$, which have been realized in MAOF-1 in accordance with Table 1 data. The first column shows the labels of five pairs of complementary Walsh sequences W_{j+}, W_{j-}, that have been sequentially realized in MAOF-2. Besides, the five labels of Δ_j is shown below each pair (W_{j+}, W_{j-}). The data in the Δ_j line correspond to the output results of the differential amplifier shown in Figure 1. The data in the table cells (in arbitrary units) were obtained as a result of the following measurement algorithm:

1. Properly shaped and amplified electrical RF signal was fed to the MAOF-1 and generated the filter's transmission spectrum, equivalent to one of the ten of Walsh sequences, shown in Table 1.
2. The transmission spectra, equivalent to a pair of complementary Walsh sequences Wj_{+}, W_{j-}, was formed sequentially in MAOF-2.
3. In cells ($W_{i(+/-)}$, W_{i+}), ($W_{i(+/-)}$, W_{j-}) the digitized measurement signals from PD are shown. The light from LED passes through the MAOF-1 with the $W_{i(+/-)}$ transmission spectrum and through the MAOF-2 with W_{j+} ,W_{j-} transmission spectra.
4. Line Δ_i shows the result of subtracting values in the cells ($W_{i(+/-)}$, W_{i+}) and ($W_{i(+/-)}$,W_{j-}), which is equivalent to the difference signal digitized after the differential amplifier in the decoder [3] in Figure 1. The cell data in the lines Δ_i are colored by red or blue colors, respectively, when $i = j$, and green, when $i \neq j$.

It can be seen from Table 2, that the data in the rows labelled by "Δ_i", are grouping according to the values in three spaced far apart ranges that completely coincide with the colour bands. The values of coloured in red data lie in the range from 619 to 699, in blue from -688 to -642 and in green from -35 to 30. In each of the 10 columns only one coloured data is significantly different from zero in absolute value, exactly when formed in MAOF-1 sequence $W_{i(+/-)}$ is identical or complementary to the one from pair of sequences (W_{j+}, W_{j-}), that are formed sequentially in MAOF-2.

Table 2. Main experimental results.

MAOF-2 (decoder)	MAOF-1 (coder)									
	W_{1+}	W_{1-}	W_{2+}	W_{2-}	W_{3+}	W_{3-}	W_{4+}	W_{4-}	W_{5+}	W_{5-}
W_{1+}	655	38	337	324	335	334	303	343	363	337
W_{1-}	36	691	325	322	339	326	319	327	335	362
Δ_1	619	-653	12	2	-3	8	-15	16	28	-25
W_{2+}	337	327	713	37	341	344	310	340	367	349
W_{2-}	335	337	33	700	323	326	299	331	338	352
Δ_2	12	-10	680	-663	18	18	11	9	29	-4
W_{3+}	335	337	338	332	694	44	321	339	349	355
W_{3-}	339	331	328	326	42	691	290	338	364	343
Δ_3	-4	6	10	4	652	-647	31	1	-35	12
W_{4+}	303	342	336	328	346	325	648	47	349	358
W_4	319	333	343	326	336	355	43	689	354	345
Δ_4	-16	9	-7	2	10	-30	605	-642	-5	13
W_{5+}	363	327	353	323	329	358	299	343	740	43
W_{5-}	335	342	342	336	352	328	310	330	41	731
Δ_5	28	-15	11	-13	-23	30	-11	13	699	-688

The experimental results in Table 2 are fully corresponding to mathematical modeling, carried out in [3]. It can be seen that the data in the red/ blue band are concentrated in a narrow range of sufficiently large positive/negative values. So, one can see, that the signal power and, consequently, the ratio of the transmitted signal power to single-signal interference power $SIR = 10\,Log\left(699/35\right) \cong 13\ dB$.

4. Conclusions

A new prototype of multiband AO decoder, based on MAOF, is proposed and investigated experimentally. This prototype is free of systematic errors caused by the asymmetry of the two parallel optical channels in a previously introduced prototype.

A model of the transmission line (with two based on TeO_2 controlled MAOFs and industrial LED as key elements) was created for researching the optical signal spectral encoding/decoding.

It has been demonstrated, that the SIR value, calculated on the base of measured data, was about 13 dB in the transmission line realization.

Acknowledgements

This work was supported by RFBR grants - 14-07-00163-a and 16-02-00124-a.

References

[1] Pozhar V E and Pustovoit V I 2015 *Bulletin of the Russian Academy of Sciences Physics* **79** p 1221
[2] Yin H, Richardson D J 2007 *Optical Code Division Multiple Access Communication Networks: Theory and Applications* (Tsinghua University Press, Beijing and Springer-Verlag GmbH Berlin Heidelberg)
[3] Proklov V V, Byshevski-Konopko O A, Grigorievski V I 2013 *Quantum Electronics* **43** p 542
[4] Ziemann O, Iversen K 1995 *Proc. SPIE* **2614** p 142
[5] Proklov V V, Byshevski-Konopko O A, Filatov A L 2015 *Technical Physics Letters* **41** p 987

Optical encryption of series of images using a set of encryption keys using scheme operating with spatially-incoherent illumination based on two LC SLMs

A P Bondareva[1], P A Cheremkhin[1], N N Evtikhiev[1], V V Krasnov[1], D Yu Molodtsov[1] and S S Nalegaev[2]

[1]National Research Nuclear University MEPhI (Moscow Engineering Physics Institute), Kashirskoe shosse 31, 115409 Moscow, Russia
[2]ITMO University, Kronverkskiy pr. 49, 197101 St. Petersburg, Russia

E-mail: vitaly.krasnov@mail.ru

Abstract. Optical encryption and numerical decryption of series of test images using a set of different encryption keys is carried out using scheme operating with spatially-incoherent illumination based on two LC SLMs. Results of experiments on images optical encryption and numerical decryption are presented. Satisfactory average decryption error over 49 encrypted images equal to 0.20±0.05 is achieved.

1. Introduction

Methods of optical encryption have been attracting researchers attention for many years [1–12]. Most of them operate not only with light intensity, but also with its phase. Original method from whom many varieties were created is double random phase encoding technique [1–6]. Its best feature is transformation of spectrum of image to be encrypted into white spectrum with random phase mask, also, encryption key, which is point spread function (PSF) of second random phase mask, has also white spectrum, consequently encrypted image also has white spectrum. This ensures best security with given number of elements and phase levels a random phase mask. Unfortunately, there are also serious disadvantages in form of holographic setup to register not only light intensity distribution, but also its phase and speckle noise coming with coherent illumination. These factors results in bad decryption quality in optical implementations. Elimination of these disadvantages is possible via usage of incoherent illumination instead of coherent one. In this case, phase registration is not required and speckle noise is gone.

Optical encryption with spatially-incoherent illumination technique was presented in [7]. Its basic principle consists of placing diffractive optical element (DOE) into optical imaging system. DOE acts as an encryption element, its PSF as an encryption key. Light passing through DOE forms convolution of object image and DOE PSF in photosensor plane. This convolution is encrypted image. This principle was used for the first time in [13] not for encryption purpose though, but to improve optical-digital system characteristics. Later, similar principle was used for optical-digital correlators [14,15] in [16].

Optical encryption in spatially incoherent light with two liquid crystal (LC) spatial light modulators (SLM) was experimentally implemented in [12] and later with better quality in [17]. However, despite the fact that satisfactory quality of decryption was achieved, this claim was based on only few encrypted images. It must be supported with sufficient number of experiments. Therefore, purpose of this paper is optical encryption and numerical decryption of series of test images using a set of different encryption keys using scheme operating with spatially-incoherent illumination based on two LC SLMs.

The rest of the paper organized as follows. In Section 2 experimental setup of optical image encryption with digital information input and dynamic encryption key based on two LC SLMs is described. In Section 3 results of optical encryption and numerical decryption of series of test images using a set of different encryption keys are presented. Main results are given in Conclusion.

2. Experimental setup of optical image encryption with digital information input and dynamic encryption key based on two LC SLMs

Optical encryption with spatially incoherent illumination is based on optical convolution. Mathematical description is given in detail in [18].

Experimental setup of optical image encryption with spatially incoherent illumination is presented in Fig. 1.

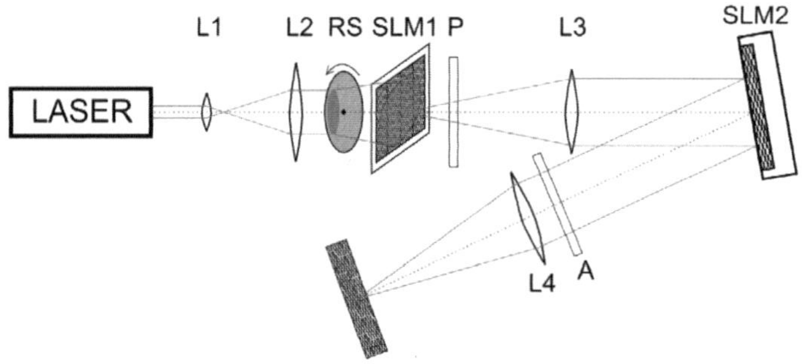

Figure 1. Scheme of optical image with spatially incoherent illumination

Spatial coherence is destroyed by rotating scatterer (ground glass) RS. SLM1 serves for display of image to be encrypted. SLM1 and SLM2 are located in focal planes of lens L3. SLM2 is used for display of encryption DOE. SLM2 and photosensor allocated in focal planes of lens L4. Optical convolution of SLM1 image and PSF of DOE displayed on SLM2 forms on photosensor. Specifications of experimental setup are given in [18].

3. Experiments on optical image encryption

Set of 7 test images used in experiment on optical encryption is shown in left column of Fig. 2. They have 128×128 elements and 256 halftones.

Direct search with random trajectory (DSRT) method [19–22] was used for encryption keys generation as described in [20]. Seven encryption keys were used in experiments. They had 64×64 elements and 2 gray levels. Their normalized average energy (NAE), i.e. ratio of matrix mean value to its maximum value, equaled to 0.005.

After encryption keys generation, kinoforms with PSF containing these keys were synthesized using Gerchberg-Saxton method [23–25]. Due to intensive zero diffraction order, which occurs due to SLM temporal phase fluctuations [26–28], useful first diffraction order was placed in the corner of restoration field so that image contained within zero diffraction order was located outside camera shaft. Kinoforms had 1080×1080 elements and 256 phase levels. One of the encryption keys and corresponding kinoform are shown in Fig. 3.

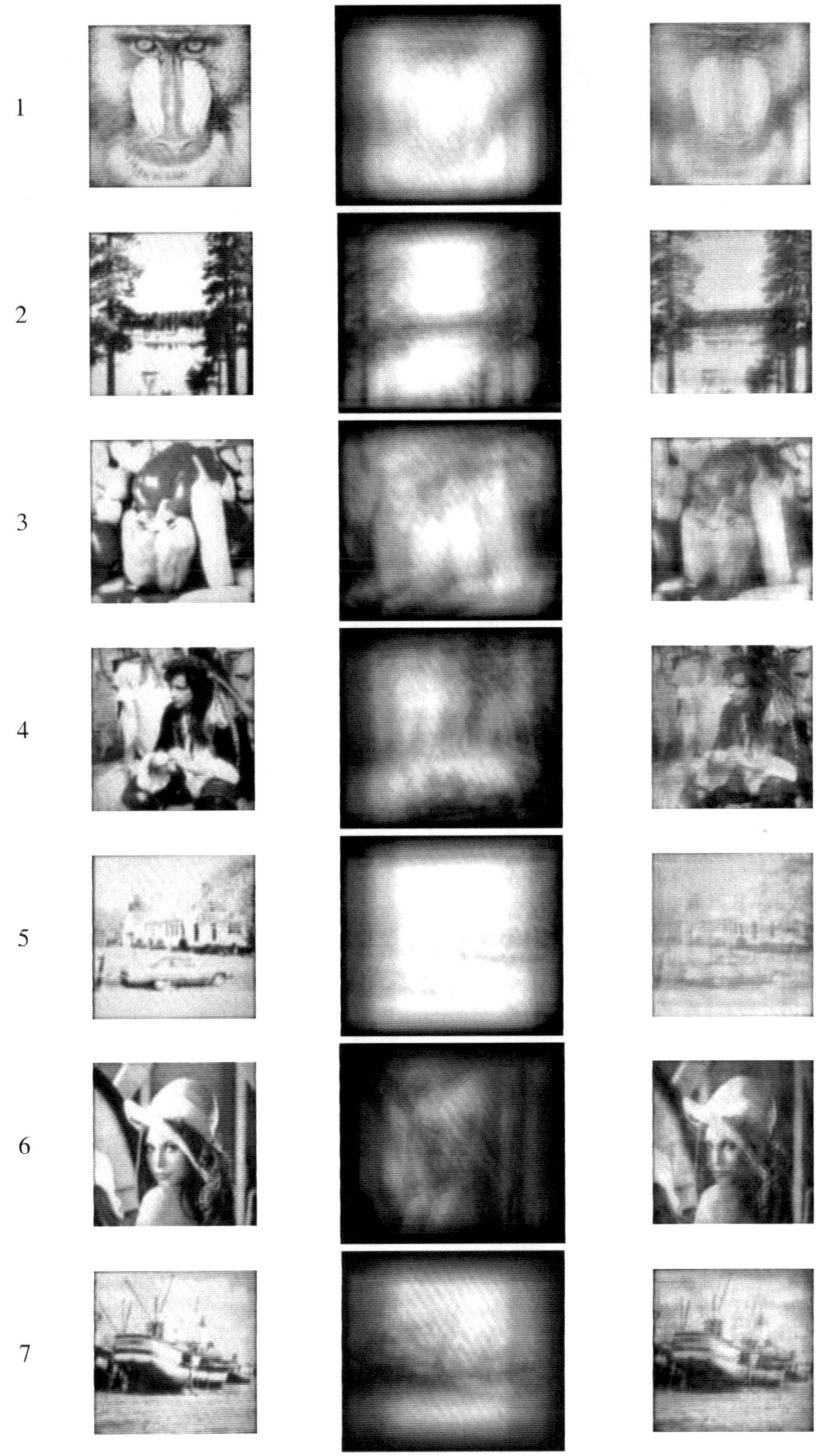

Figure 2. Original images (left), encrypted images (middle) and decrypted images (right)

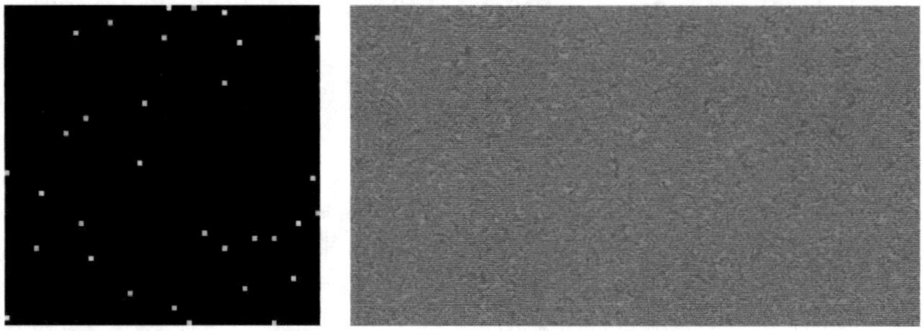

a b

Figure 3. One of the encryption keys (a) and corresponding kinoform (b)

Total of 49 encrypted images was obtained because each original image was encrypted with seven different keys. Examples of encrypted images (one for each original image) are shown in middle column of Fig. 2.

Inverse filter with Tikhonov regularization [29] implemented in MATLAB programming environment was used for digital decryption. In capacity of smoothing function the maximum value of PSF power spectrum was used.

Decrypted images are contaminated with intense interference fringes (see Fig. 4a) which are probably caused by double reflection in protective glass plate mounted right in front of camera's photosensor. To eliminate interference fringes, Fourier spectrum filtration was used (see Fig. 4c and Fig. 4d). Filtered decrypted image is shown in Fig. 4b. It can be seen that fringes are gone from filtered image. This filtration was used for all decrypted images. Examples of decrypted images (one for each original image) are presented in right column of Fig. 2.

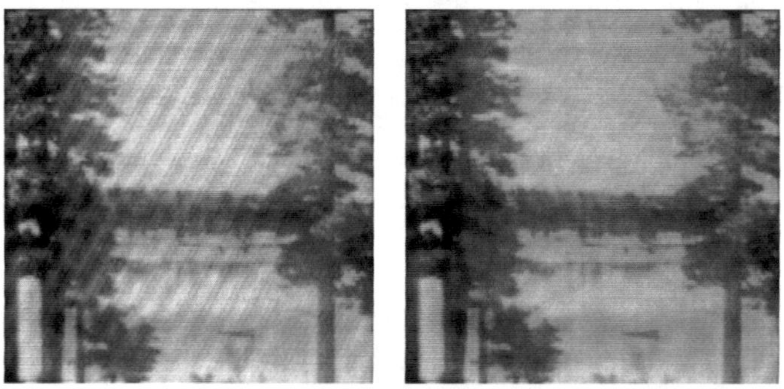

a b

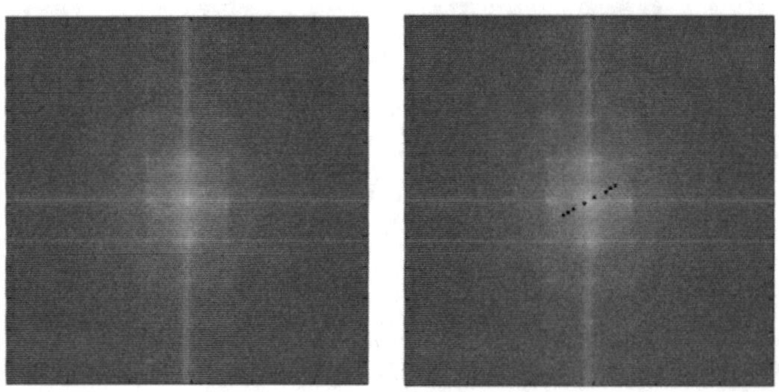

c d

Figure 4. Decrypted image contaminated with interference fringes before (a) and after (b) filtration, and corresponding Fourier spectra (c) and (d)

NSTD [30] between decrypted images and original ones were calculated to assess quality of decrypted images.

NSTD values of all 49 decrypted images are presented in Fig. 5. It can be seen that NSTD values substantially differ between different images and less so for one image but different encryption keys. Averaged NSTD values for 7 images are given in Table 1.

Table 1. Averaged NSTD values for 7 images

Image number	1	2	3	4	5	6	7
Average NSTD	0.154 ±0.007	0.212 ±0.005	0.234 ±0,012	0.271 ±0.008	0.155 ±0.015	0.231 ±0.012	0.145 ±0.007

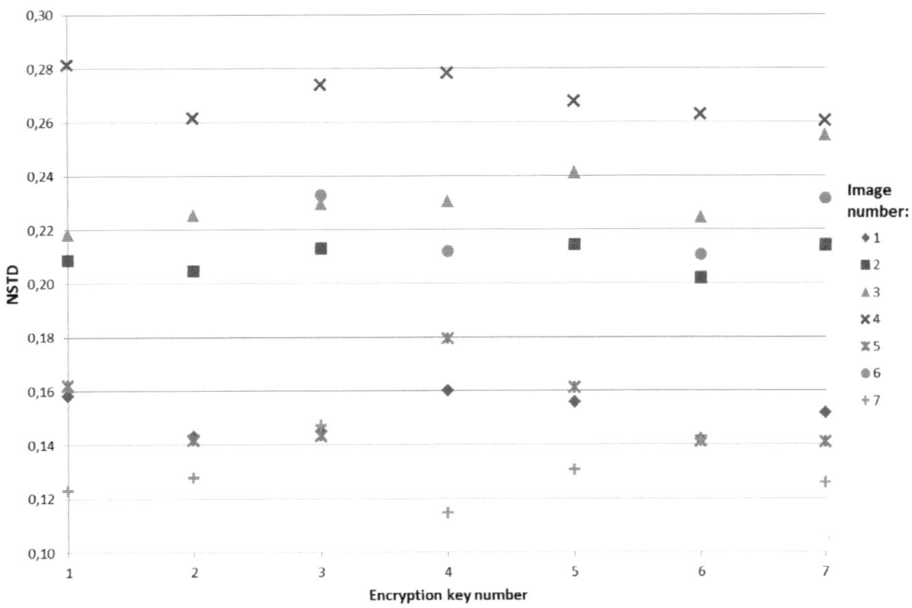

Figure 5. NSTD values of all 49 decrypted images

Conducted experiments demonstrate successful optical encryption and numerical decryption of series of test images using a set of different encryption keys. Decrypted images are contaminated with interference fringes. Fourier spectra filtration was used to eliminate them. NSTD values of 49 decrypted images lie in range 0.14÷0.27. Average NSTD value equals to 0.20±0.05.

4. Conclusion

Optical encryption and numerical decryption of series of test images using a set of different encryption keys is carried out using scheme operating with spatially-incoherent illumination based on two LC SLMs.

Seven test images with 128×128 elements and 256 gray-levels and seven encryption keys with 64×64 elements and 2 gray levels were used in experiments. Keys normalized average energy, i.e. ratio of matrix mean value to its maximum value, was to 0.005.

Results of experiments demonstrate successful optical encryption and numerical decryption of series of test images using a set of different encryption keys. Decrypted images are contaminated with interference fringes. Fourier spectra filtration was used to eliminate them. NSTD values of 49 decrypted images lie in range 0.14÷0.27. Average NSTD value equals to 0.20±0.05.

Acknowledgements
This work was supported by grant 14-19-01751 from the Russian Science Foundation (RSF).

References

[1] Refregier P, and Javidi B 1995 *Opt. Lett.* **20** p 767.
[2] Javidi B, Sergent Arnaud, Guanshen Z, and Guibert L 1997 *Opt. Eng.* **36** p 992.
[3] Unnikrishnan G, Joseph J, and Singh K 2000 *Opt. Lett.* **25** p 887.
[4] Javidi B, Towghi N, Maghzi N, and Verrall S C 2000 *Appl. Opt.* **39** p 4117.
[5] Hennelly B M, and Sheridan J T 2004 *Opt. Eng.* **43** p 2239.
[6] Rajput S K, and Nishchal N K 2014 *Appl. Opt.* **53** p 418.
[7] Evtikhiev N N, Starikov S N, Cheryomkhin P A, Krasnov V V., and Rodin V G 2012 *Proc. SPIE.* **8429** p 84291P.
[8] Li J, Li J, Shen L, Pan Y, and Li R 2014 *Opt. Express.* **22** p 4849.
[9] Yang B, Liu Z, Wang B, Zhang Y, and Liu S 2011 *Opt. Express.* **19** p 2634.
[10] Chen W, Chen X, and Sheppard C J R 2010 *Opt. Lett.* **35** p 3817.
[11] Qin Y, and Gong Q 2014 *Opt. Commun.* **310** p 69.
[12] Bondareva A P, Cheremkhin P A, Evtikhiev N N, Krasnov V V, Rodin V G, and Starikov S N 2014 *Proc. SPIE.* **9249** p 924915.
[13] Cathey W T, and Dowski E R 2002 *Appl. Opt.* **41** p 6080.
[14] Evtikhiev, N. N.; Ivanov, P. A.; Lyapin, A. S.; Reyzin, B. M.; Shevchuk, A. V.; Sirotkin, S. I.; Starikov, R. S.; Zaharcev A V. 2005 *Proc. SPIE.* **5851** p 242.
[15] Evtikhiev Nikolay N.; Shaulskiy Dmitriy V. ; Zlokazov Evgeny Yu.; Starikov Rostislav S. 2012 *Proc. SPIE.* **8398** p 83980G.
[16] Starikov S N, Balan N N, Konnik M V, Rodin V G, Solyakin I V, and Shapkarina E a 2007 *Proc. SPIE.* **6574** p 65740J.
[17] Bondareva A P, Cheremkhin P A, Evtikhiev N N, Krasnov V V, and Starikov S N 2015 *Phys. Procedia.* **73** p 320.
[18] Krasnov V V., Starikov S N, Starikov R S, and Cheremkhin P A 2016 *Russ. Phys. J.* **58** p 1394.
[19] Evtikhiev N N, Krasnov V V., and Starikov S N 2013 *J. Opt. Technol.* **80** p 294.
[20] Cheremkhin P A, Evtikhiev N N, Krasnov V V., Rodin V G, and Starikov S N 2014 *Proc. SPIE.* **9131** p 913125.
[21] Bondareva A P, Cheremkhin P A, Evtikhiev N N, Krasnov V V., Rodin V G, and Starikov S N 2014 *Proc. SPIE.* **9216** p 92161J.
[22] Cheremkhin P A, Evtikhiev N N, Krasnov V V., Rodin V G, and Starikov S N 2015 *Proc. SPIE.* **9386** p 93860R.
[23] Gerchberg R W, and Saxton W O 1972 *Optik (Stuttg).* **35** p 237.
[24] Fienup J R 1982 *Appl. Opt.* **21** p 2758.
[25] Cheremkhin P A, Evtikhiev N N, Krasnov V V., Porshneva L A, Rodin V G, and Starikov S N 2014 *Proc. SPIE.* **9131** p 913124.
[26] Bondareva A P, Cheremkhin P A, Evtikhiev N N, Krasnov V V., Starikov R S, and Starikov S N 2014 *J. Phys. Conf. Ser.* **536** p 012011.
[27] Cheremkhin P A, Evtikhiev N N, Krasnov V V., Rodin V G, and Starikov S N 2014 *Proc. SPIE.* **9006** p 900615.
[28] Cheremkhin P A, Evtikhiev N N, Krasnov V V., Rodin V G, and Starikov S N 2015 *Proc. SPIE.* **9508** p 95080U.
[29] Tykhonoff A N, and Arsenin V Y 1977 Solution of Ill-posed Problems, Washington: Winston & Sons, Washington, .
[30] Fienup J R 1997 *Appl. Opt.* **36** p 8352.

A study of the particularities of an authentication system with a method of an asymmetric holographic encryption based on the DRPE to protect the passwords of the technical devices

S S Nalegaev[1] and V V Krasnov[2]

[1]ITMO University, Kronverkskiy pr. 49, 197101 St. Petersburg, Russia
[2]National Research Nuclear University MEPhI (Moscow Engineering Physics Institute), Kashirskoe shosse 31, 115409 Moscow, Russia

E-mail: s.nalegaev@gmail.com

Abstract. The present work is devoted to research the peculiarities of an authentication system of the technical devices with the use of an optical asymmetric holographic encryption on the basis of the Double Random Phase Encoding (DRPE). The series of the numerical experiments was performed to implement the encryption and the decryption of the initial image. The angular spectrum approach was used as a tool for the simulation of the propagation of the light in the free space forward and backward along the whole optical scheme of the DRPE. In the present work the particularities of the authentication system based on an asymmetric approach of the holographic encryption method DRPE were studied.

1. Introduction

The identification and the authentication are a foundation of all the information protection systems, since all the information protection mechanisms are designed to work with the named subjects and objects of the automated system (AS) [1]. The identifiers (IDs) or the usernames is often used as a way of naming of the corresponding subjects and objects of the automated system. In a general sense, the authentication is not intended to establish a correspondence between the entered IDs and subjects of AS which are matched with them, but to authenticate the submitted data [2]. And this is it essential difference from the identification. The authentication systems are often used as an intermediate element that is used to implement an access control to anything between the authorized and the unauthorized subjects. In this case both the users and the processes [1] or the technical devices can act as the subjects of AS. And the information and the other resources of the system [1] including the ability to manage anything can act as the objects of AS.

Such the authentication system can be used, for example, within the concept of the home automation, where it is required to ensure a secure communication of the general system of the monitoring and control with the various technical elements and devices: the temperature and humidity control sensors of the indoor and outdoor air; the detectors of the fire alarm, the water leakage or the short circuit; the sensors used to control the breaking and the entering or the presence of a people in the rooms; for the processing signals from the domestic commercial devices (the electric kettles, the refrigerators, the heating system) or to control such the devices. In all these cases it is necessary to ensure the secure data exchange between this system and all the listed devices, taking into account the fact that the communication can take place over an insecure communication channel. To accomplish this, it is necessary to bring to the correspondence a set of the unique IDs and the passwords with each of these devices, as well as with the general monitoring and control system to perform the authentication process and to be able to set the interaction privileges among themselves and the general monitoring and control system.

All of these problems can be solved by an authentication system based on the use of some asymmetric optical encryption technique, for example, the asymmetric holographic image encryption based on the method of the Double Random Phase Encoding (DRPE) [3,4]. In the method DRPE, an amplitude object with the spatial distribution of the transmittance usually serves as the source image, and the encryption process is based on the diffraction of the wavefield of the optical radiation transmitted through the object, as it travels along the optical axis [3-5]. In contrast to the methods of the optical encryption in spatially incoherent light (for instance see [6-8]) in the case of DRPE the hologram of the wavefield recorded at the output of the optical system serves as an encrypted image of the object [3-5]. Such the optical encoding system can be performed within an experimental scheme or using a numerical simulation of the light propagation process through the whole optical system, for example, like in these works [9-11].

Theoretically, the information encoded by the method DRPE can be decrypted by the use of the simple brute force attack. In this case the choice of the proper criterion or criteria may play an important role for the estimation of the set of the key parameters of the optical scheme which have been used to obtain the encrypted image [5,12]. On the other hand, because of the multidimensionality of the encryption and decryption process [13], the numerical methods of the breaking DRPE through the direct brute force attack can not be used in the present time due to a high computational complexity of this task [3-5]. But there are several much more efficient approaches of the breaking DRPE based on the iterative techniques [14-21]. Within the frame of this work, we investigated the particularities of an optical authentication system based on the use of an asymmetric method of the optical holographic encryption. Such authentication systems can be used in practice to protect the passwords of the various commercially-rated technical devices within the concept of the home automation.

2. Numerical scheme and results

For the studied asymmetric holographic encryption method the process of the encryption is based on the optical transformations of the initial wavefield at the output plane of the object O with flat spatial distribution of the phase during its propagation along the optical system as it shown in the figure 1. To perform the encryption, the information for which the encryption is required is converted into a two-dimensional image which is used as a two-dimensional spatial distribution of the amplitude of the field in the output plane of the object. This initial amplitude distribution is shown on the figure 2 (*a*).

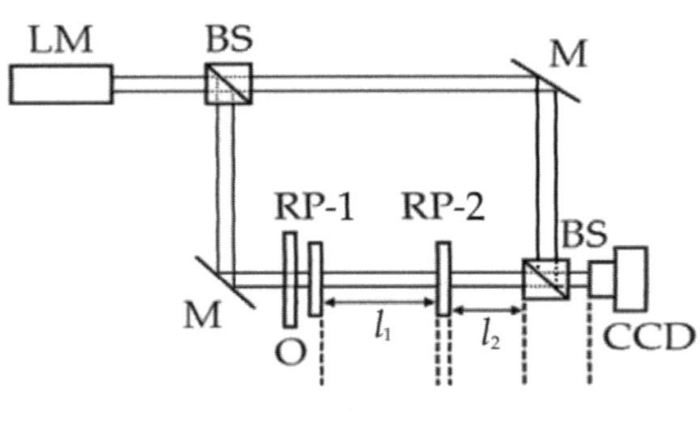

Figure 1. The scheme used in the simulation.

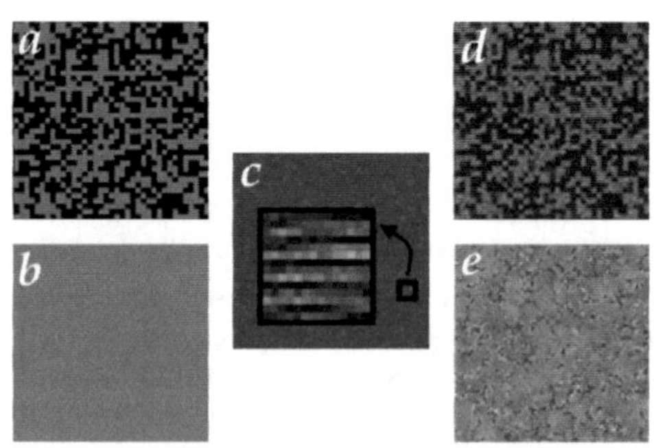

Figure 2. The obtained results.

Due to the asymmetry of the encryption method there are two keys one of which (privacy key) includes (without any protection) all the parameters of the optical scheme that are required for the recovery of the encrypted information. For example, these are such parameters as (see the figure 1): propagation distances in free space l_1 and l_2; spatial phase distribution in the output plane of the second random phase mask RP-2; wavelength λ. It is interesting to note that there is a possibility to develop many variations of the holographic asymmetric encryption scheme based on the classical scheme DRPE but with some additional key parameters of the optical set-up. For example, it is possible to assume that the properties of the DKDP and the KDP crystals investigated in these works [22, 23] also could be used as the additional parameters. And many of such the variations of this method were developed at the present time already. The private key with all of such the parameters is not intended for the storage or the transmission. The second key (public key) is intended to send it over an insecure communication channel together with the encrypted images. This public key is represented by the digital hologram obtained by the method DRPE from the private key using a set of the key parameters that differs from those used to protect the target images. The pictures (*b*) and (*e*) on the figure 2 represents the spatial phase distributions of the wavefield in the object plane: (*b*) is for initial phase distribution and (*e*) is for reconstructed one by the decryption of the encrypted hologram (*c*). The same is for the pictures (*a*) and (*d*), which represents the spatial amplitude distributions of the wavefield in the object plane. The hologram (*c*) (with an enlarged fragment) represents the result of the holographic encryption used to the the initial wavefield with amplitude (*a*) and phase (*b*), which is measured in the CCD plane. The figure 1 also includes the following optical elements: LM - laser; BS – beam splitter; M – mirror.

To perform the encryption and the decryption process it is necessary to take into account the diffraction of the light while it propagates through the free space with the distances l_1 and l_2. And the angular spectrum method [24,25] was used to calculate the results of such the numerical forward (encryption) and backward (decryption) propagation. The reconstruction of the full information about the wavefield in the plane of the CCD (the amplitude and phase distributions) was performed from the hologram (*c*) according to the method described in [26]. As can be seen from the figure 2, the initial wavefield has a flat spatial phase distribution for the current results of the numerical experiment.

3. Conclusion

The particularities of the authentication system based on an asymmetric approach of the holographic encryption method DRPE were investigated. The encrypting and the decrypting of the images were numerically demonstrated. To take into account the diffraction effects of the light the angular spectrum method was used to calculate the results of the numerical propagation of the wavefield in free space. The use of the some asymmetric variations of the known holographic encryption methods DRPE for the developing new authentication systems is of interest because of the several points of view. For example, a multidimensionality of the process of the encryption and the decryption leads to a high degree of the information security. And the use of the asymmetric algorithms reduce the requirements to the data channel allowing the use of the common unsecured communication channels.

Acknowledgments
S S Nalegaev acknowledge a financial support from the grant of the Russian foundation for basic research (the agreement No 16-52-52049\16 from 29.01.2016).

References

[1] Gladkih A A and Dementiev V E 2009 *The basic principles of information security of computing networks* (Ulyanovsk: Ulyanovsk state technical university) 156 p
[2] Merkle R C 1979 Secrecy, authentication, and public key systems (Stanford: Stanford university) 182 p
[3] Refregier P and Javidi B 1995 *Opt Lett* **20** 767-769
[4] Chen W, Javidi B and Chen X 2014 *Adv Opt Photon* **6** 120-155

[5] Nalegaev S S and Petrov N V 2015 *Physics Procedia* **73** 281-286
[6] Krasnov V V, Starikov S N, Starikov R S and Cheremkhin P A 2016 *Russian Physics Journal* **58** 1394-1401
[7] Evtikhiev N N, Starikov S N, Cheryomkhin P A, Krasnov V V and Rodin V G 2012 *Proceedings of SPIE* **8429** 84291P
[8] Bondareva A P, Cheremkhin P A, Evtikhiev N N, Krasnov V V and Starikov S N 2015 *Physics Procedia* **73** 320-327
[9] Nalegaev S S and Petrov N V 2015 *Russian Journal of Physical Chemistry* B **9** 555-557
[10] Nalegaev S S, Petrov N V and Bespalov V G 2014 *Journal of Physics: Conference Series* **536** 012025
[11] Nalegaev S S, Petrov N V and Bespalov V G 2014 *Fringe 2013: 7th Int. Workshop on Advanced Optical Imaging and Metrology* ed W Osten (Berlin, Heidelberg: Springer-Verlag) 321-324
[12] Nalegaev S S and Petrov N V *Russian Physics Journal* **58** 128-131
[13] Barrera J F, Mira A and Torroba R 2013 *Opt Express* **21** 5373
[14] Carnicer A, Montes-Usategui M, Arcos S and Juvells I 2005 *Opt Lett* **30** 1644-1646
[15] Peng X, Wei H Z and Zhang P 2006 *Opt Lett* **31** 1044-1046
[16] Peng X, Tang H Q and Tian J D 2007 *Acta Physica Sinica* **56** 2629-2636
[17] Peng X, Zhang P, Wei H Z and Yu B 2006 *Opt Lett* **31** 3261-3263
[18] Frauel Y, Castro A, Naughton T J and Javidi B 2007 *Opt Express* **15** 10253-10265
[19] Qin W, He W, Meng X and Peng X 2009 *Proc of SPIE* **7512** 751203
[20] Qin W and Peng X 2009 *J Opt* A: *Pure Appl Opt* **11**
[21] Nalegaev S S, Petrov N V and Bespalov V G 2012 *Scientific and Technical Journal of Information Technologies, Mechanics and Optics* **82** 30-35
[22] Nalegaev S S, Putilin S E and Bespalov V G 2012 *Scientific and Technical Journal of Information Technologies, Mechanics and Optics* **81** 29-35
[23] Nalegaev S S, Putilin S E and Bespalov V G 2013 *Proc SPIE* **8699** 869914
[24] Goodman J W 2005 *Introduction to Fourier Optics* (Englewood: Roberts & Company) 491 p
[25] Poon T C and Banerjee P P 2001 *Contemporary Optical Image Processing with MATLAB* (Elsevier Science & Technology) 262 p
[26] Liebling M, Blu T and Unser M 2004 *J Opt Soc Am* **A21** 367–77

Journal of Physics: Conference Series **737** (2016) 012063 doi:10.1088/1742-6596/737/1/012063

Application of input amplitude masks in scheme of optical image encryption with spatially-incoherent illumination

A V Shifrina, N N Evtikhiev, V V Krasnov

National Research Nuclear University MEPhI (Moscow Engineering Physics Institute), Kashirskoe shosse 31, 115409 Moscow, Russia

E-mail: vitaly.krasnov@mail.ru

Abstract. Optical encryption with spatially incoherent illumination does not have drawbacks of coherent encryption techniques. In this case however, one of the factors affecting decrypted image quality is original image spectrum. In most cases, majority of image energy is concentrated in area of low spatial frequencies. Therefore, only this area in spectrum of encrypted image contains information about original image, while other areas contain only noise. Additional amplitude encoding of input scene can be used for increase of the size of the area of spatial frequencies containing useful information. Numerical simulation demonstrates reduction of decryption error up to 2.7 times.

1. Introduction

Majority of optical encryption techniques operate not only with light intensity, but also with its phase, thus requiring coherent illumination. Original method from whom many varieties were created is double random phase encoding technique [1–6]. Its best feature is transformation of spectrum of image to be encrypted into white spectrum with random phase mask, also, encryption key, which is point spread function (PSF) of second random phase mask, has also white spectrum, consequently encrypted image also has white spectrum. This ensures best security with given number of elements and phase levels a random phase mask. Unfortunately, there are also serious disadvantages in form of holographic setup to register not only light intensity distribution, but also its phase and speckle noise coming with coherent illumination. These factors results in bad decryption quality in optical implementations.

Elimination of these disadvantages is possible via usage of incoherent illumination instead of coherent one. In this case, phase registration is not required and speckle noise is gone. Basic principle of this approach is illustrated in Figure 1. Object is illuminated with spatially-incoherent monochromatic light. In the absence of diffractive optical element (DOE), lens forms object image on photosensor. DOE acts as an encryption element, its PSF is an encryption key. Light passing through DOE forms convolution of object image and DOE PSF in photosensor plane. This convolution is encrypted image.

This technique is simpler in experimental implementation and does not suffer from speckle, but it has its own weaknesses. Major one is it operates only with positive real numbers and not complex ones, which is the case with coherent techniques. This leads to two drawbacks. First, mean value of image to be encrypted is always above zero which leads to intensive zero spatial frequency peak in image spectrum. Second, in DRPE based techniques first random mask transforms spectrum of any image into white one, here we have no control over image spectrum. Because encryption is based on convolution Fourier spectrum amplitude distribution of encryption key should overlap Fourier spectrum amplitude distribution of image to be encrypted, in other case some spatial frequencies will be lost among with information they carried. Moreover, correct decryption of every spatial frequency is possible only if key spectrum do not have values below noise level. Additionally, heights of zero frequency peaks is defined by average energies of image and key. Because encrypted image contains noise, ratio of its average spectrum energy to noise average energy defines signal to noise ratio of decrypted image. In other words if image contains more white than others, then after encryption-decryption it will contain more noise compared to others. Also, ratio of amplitude at zero frequency to average spectrum amplitude of encryption key impacts decrypted images quality [7].

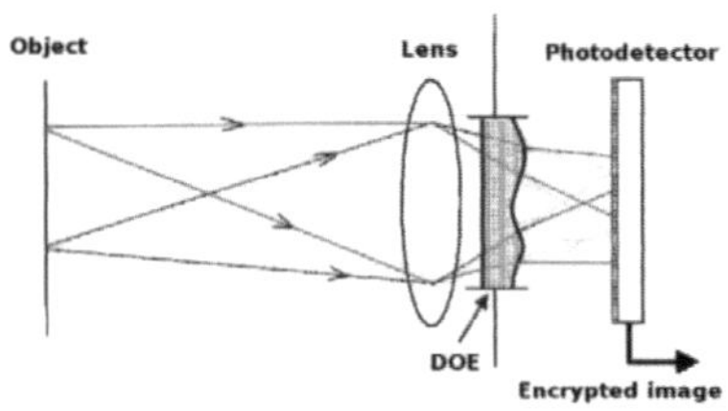

Figure 1. Example of basic scheme of optical image encryption using monochromatic spatially incoherent illumination

Phase masks cannot be used in case of spatially incoherent illumination, but amplitude ones can. Additional amplitude encoding of input scene can be used for increase of the size of the area of spatial frequencies containing useful information. This allows to increase signal-to-noise ratio in encrypted image and, therefore, leads to better quality of decrypted images.

Therefore, purpose of this paper is application of additional encoding of input scene and analysis of its effects on decryption quality and encryption strength of encrypted image for optical encryption with spatially-incoherent illumination technique.

Rest of the paper organized as follows. In Section 2 description of applied methods of input scene encoding is given. In Section 3 results of numerical simulation of optical encryption with spatially incoherent illumination and input scene additional encoding are presented. Main results are given in Conclusion.

2. Methods of Additional encoding of input scene

Input scene encoding is accomplished by overlaying image to be encrypted with amplitude mask.

Two different types of amplitude masks were used. First one is rectangular grating mask (RGM). This mask increases area of spatial frequencies containing useful information by duplicating original image spectrum in high spatial frequencies corresponding to RGM frequency. Test image before and after application of RGM is shown in figure 2.

(a) (b) (c)

Figure 2. Test image before (a) and after (b) application of RGM, magnified fragment of encoded test image (c)

Second one is random mask (RM). Unlike RGM, RM has white spectrum (excluding zero spatial frequency). Two encryption-decryption procedures were applied with two inverse RM, and then resulting encrypted images were subtracted. This is equivalent to application of random binary phase mask. Resulting encrypted image has basically white spectrum. RM mask can be used for increased security. Example of random mask fragment is shown in figure 3a. Fragment of test image before and after application of RM is shown in figure 3b and figure 3c. Encoded images for both AM are shown in figure 4.

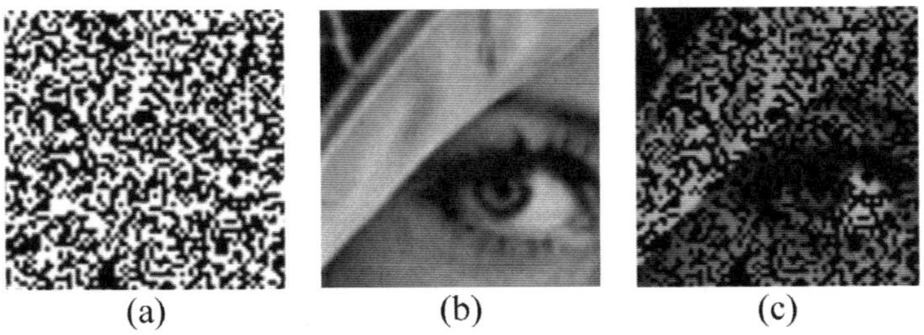

Figure 3. Random mask (a), fragments of test image before (b) and after (c) application of random mask

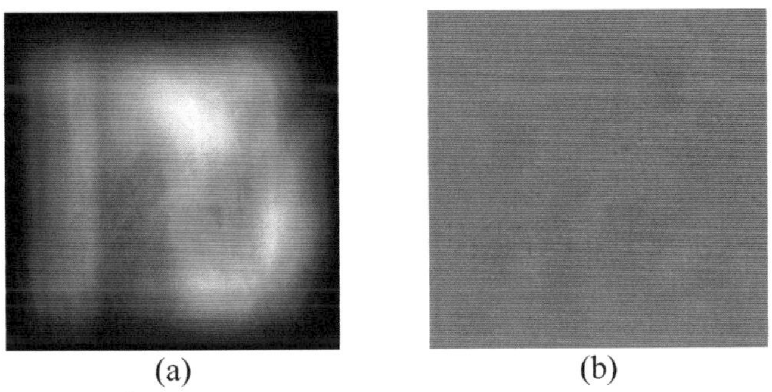

Figure 4. Images encoded using RGM (a) and RM (b)

Encryption strength for RGM (figure 4a) is much lower then for RM (figure 4b). Some fragments of original image can be recognized in first case. It should be noted that improvement provided by AM requires registration of 4 frames in case of RGM and 2 frames in cases of RM (instead of 1 frame originally).

3. Results of numerical simulation of optical encryption with input scene additional amplitude encoding

The set of 160 random encryption keys with RZA values varying in range 1.6÷191.5 were used in experiments. They had 128×128 elements each with normalized average energy (NAE) ranging from 0.0001 to 0.5. Ten test images (for example see figure 2a) had 512×512 elements. Noises of scientific camera Megaplus ES II 11000 with high maximum linear signal-to-noise ratio (SNR) equal to 141 were simulated.

Numerical simulation is described as follows. First, input scene is encoded with AM, then convolution with encryption key is calculated. Then, simulated camera noises are added. Measured cameras noise characteristics were taken from [8,9]. Then, decryption process takes place. Inverse filter with Tikhonov regularization [10] was used for decryption. After decryption, AM effects are compensated and normalized standard deviations (NSTD) [11] between decrypted and original image is calculated.

Results of encryption-decryption procedure without additional input AM are shown in figure 5. It is clear that encryption keys with low RZA values provide high SNR in the decrypted image.

Results of application of additional input AM are shown in figure 6. Figure 6a corresponds to RGM, figure 6b corresponds to RM and figure 6c corresponds to joint application of RGM and RM in order to increase SNR and encryption strength at the same time.

Application of RGM provides greatest NSTD decrease up to 2.3 times at cost of 4 frames instead of one. Application of RM provides lower NSTD decrease up to 1.6 times at cost of 2 frames instead. Joined application of RGM and RM reduce decrease to 1.9 times at cost of 8 frames.

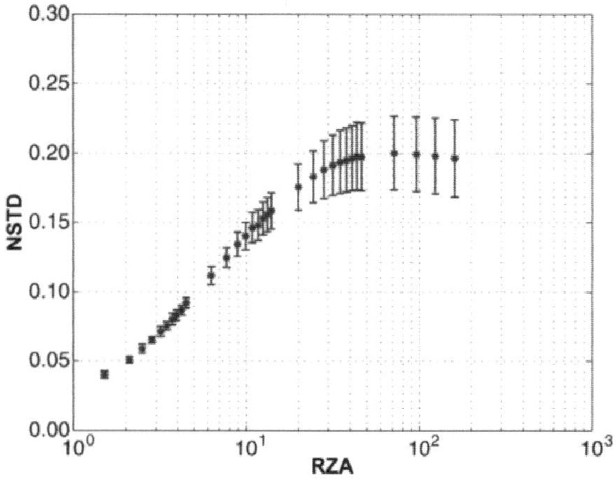

Figure 5. Results of encryption-decryption procedure without additional input amplitude masks

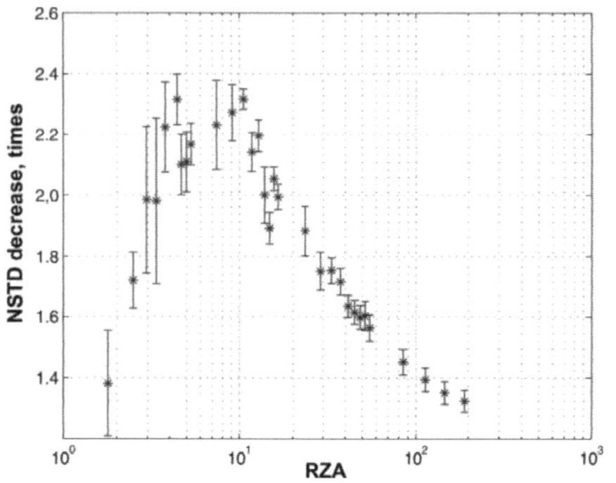

(a)

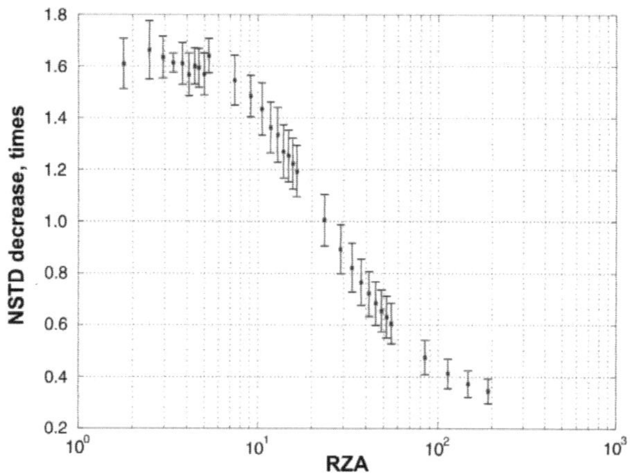

(b)

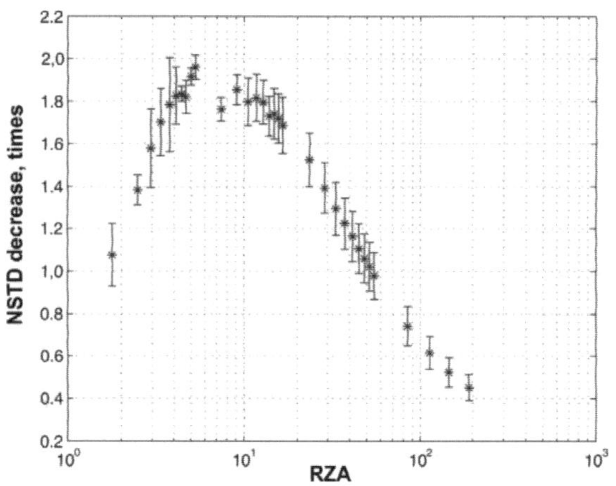

(c)

Figure 6. Results of application of additional input amplitude masks. Rectangular grating mask (a), random mask (b) and joint application of rectangular grating mask and random mask (c)

While keys with low RZA values provide highest NSTD decrease for each AM, encryption strength of such keys is too weak. But joint application of RGM and RM allows encryption with low-RZA (in range 3.0÷20.0) keys: decrease in encryption strength compensates by usage of RM and at the same time it keeps high NSTD decrease.

4. Conclusion

Input scene additional amplitude encoding provides significant reduction of decryption error. Rectangular grating mask provides greatest decryption error decrease up to 2.3 times at cost of 4 frames instead of one. With additional application of random mask decryption error decreases up to 1.9 times at cost of 8 frames, but encoded image has higher encryption strength than image encrypted with only rectangular grating mask or no additional mask at all.

Acknowledgments
This work was partially supported by grant 14-19-01751 from the Russian Science Foundation (RSF).

5. References

[1] Refregier P, and B Javidi 1995 *Opt. Lett.* **20** p 767.
[2] Javidi B, Sergent Arnaud, Z Guanshen, and L Guibert 1997 *Opt. Eng.* **36** p 992.
[3] Unnikrishnan G, J Joseph, and K Singh 2000 *Opt. Lett.* **25** p 887.
[4] Javidi B, N Towghi, N Maghzi, and S C Verrall 2000 *Appl. Opt.* **39** p 4117.
[5] Hennelly B M, and J T Sheridan 2004 *Opt. Eng.* **43** p 2239.
[6] Rajput S K, and N K Nishchal 2014 *Appl. Opt.* **53** p 418.
[7] Cheremkhin P A, N N Evtikhiev, V V. Krasnov, V G Rodin, and S N Starikov 2014 *Proc. SPIE.* **9131** p 913125.
[8] Evtikhiev N N, S N Starikov, P A Cheryomkhin, and V V. Krasnov 2012 *Proc. SPIE.* **8301** p 830113.
[9] Cheremkhin P A, N N Evtikhiev, V V. Krasnov, V G Rodin, and S N Starikov 2014 *Opt. Eng.* **53** p 102107.
[10] Tykhonoff A N, and V Y Arsenin 1977 Solution of Ill-posed Problems, Washington: Winston & Sons, Washington, .
[11] Fienup J R 1997 *Appl. Opt.* **36** p 8352.

Journal of Physics: Conference Series **737** (2016) 012064 doi:10.1088/1742-6596/737/1/012064

Wave front sensor based on holographic optical elements

M S Kovalev, G K Krasin, P I Malinina and S B Odinokov and H R Sagatelyan

Bauman Moscow State Technical University, 2nd Baumanskaya str. 5, 105005 Moscow, Russia

E-mail: m.s.kovalev@gmail.com

Abstract. A wavefront sensor (WFS) based on holographic optical elements, namely computer generated Fourier holograms is proposed as a perspective alternative to the Shack-Hartmann sensor. A possibility of single and multimode sensor and the dependence of their characteristics were investigated.

1. Introduction

Creating reference wavefronts (WF) with an error of less than 0.01λ is an actual task for many fields of optics [1, 2]. Typically, WF distortion occurs during the passage of radiation and optical components and is a consequence of the refractive index distribution inhomogeneity and the presence of streaks. The technology of obtaining elements with high requirements on the deviation values of the refractive index of the calculation and the absence of streaks is very time-consuming, long and expensive. Therefore, the development of optical systems with high beam quality using different ways to measure and correct aberrations of the wave function. This correction is usually performed by the adaptive optics [3, 4], which gives the possibility of dynamic correction.

Wavefront control by Hartmann method and its kind - by Shack-Hartmann method - is currently one of the most simple, yet accurate and effective way. The difference between the methods is to use a variety of masks, mounted in front of the radiation detector: in Hartmann sensors mask may be a raster small holes in an opaque screen [5] or multi-element amplitude mask with apodized apertures [6], and in Shack-Hartmann sensors masks are performed as matrices of microlenses of square or circular aperture [7].

Shack-Hartmann sensor application is effective when wavefront distortions are static or slowly change - for example, in ophthalmology, in a number of applications in laser technology, etc. However Shack-Hartmann sensor has limitations, firstly, when the distortion changes at high speed (for example, in the case of atmospheric turbulence, which varies at a rate above a few KHz structure), the method encounters serious problems even when using modern high-speed computer systems [8]. Secondly, it is not possible to register a phase change, when the spatial scale is less than the distance between the holes or matrices of microlenses.

Perspective alternative is using computer-generated holograms (CGH) in the problem of WFS. In this case, CGH can replace more expensive and complex in terms of technology of obtaining and control, lens and raster components. In [9] a method of measuring the WF based on the principle of the mode holographic sensor provided by the author shows a linear response to the amplitude of the wave function contained in the Zernike directly measures the numerical value of Zernike modes, which can reduce the processing time. Another version of WFS has been proposed in [10] based on the holographic filtering the input wavefront. However, such devices have been demonstrated only in laboratory performance, their practical application was restrained by imperfection of the technology used

The present work is devoted to the development of principles of construction of specialized WFS based on Fourier CGH. The physical implementation of such holograms is based on using liquid crystal spatial light modulators (SLM) and is described in detail in [11, 12].

2. Analysis of WF phase distortion

There are many approaches to the description of the distorted wavefront [13,14], the most common expansion in Zernike polynomials orthogonal basis. Every polynomial is a smooth function is uniquely described by the expression, responsible for its own type of aberrations. Then the representation of the input radiation in the form of a linear combination of Zernike modes is:

$$W(x,y)=\sum_n a_n \cdot Z_n(x,y).$$

Distorted wavefront:

$$B(x,y)=B_{plane}(x,y)\cdot \exp\left(i\cdot\frac{2\pi}{\lambda}\cdot W\right)=B_{plane}(x,y)\cdot \exp\left(i\cdot\frac{2\pi}{\lambda}\cdot a_n\cdot Z_n\right).$$

The orthogonality of Zernike polynomials gives them great advantages in the analysis of aberrations comparing with the power basis. The main advantage is that each factor contributes to a number of aberrations given type and order in the overall wave aberration from a position of balance of all types of aberrations. This means that certain types of aberration represented by Zernike polynomial expansion, affect the image quality quite independently from each other.

25-30 first polynomials are quite enough for a complete description of the WF.

Table 1. An example of the WF in the representation of Zernike polynomials. ρ – normalized pupil size и $\Theta=tan^{-1}(y/x)$ – angle coordinate.

	0,1λ	0,25λ	0,5λ	1,0λ
Defocus $Z_3=2\cdot\rho^2-1$	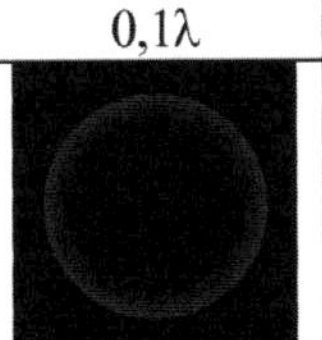	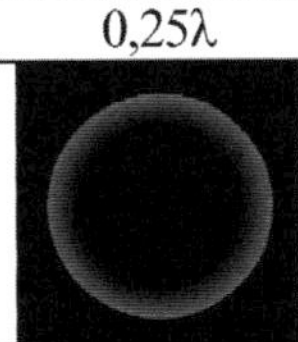	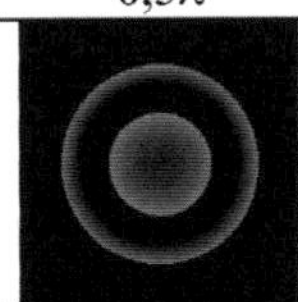	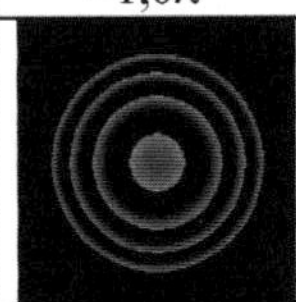
Coma $Z_6=\left(2\cdot\rho^3-2\cdot\rho\right)\cdot\cos 2\theta$		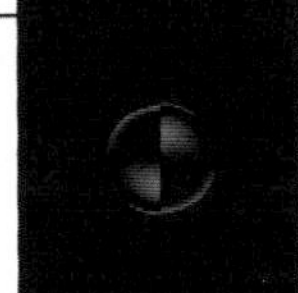	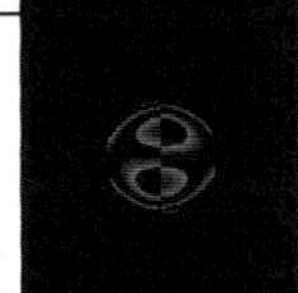	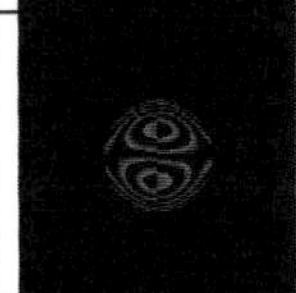
Spherical $Z_8=6\cdot\rho^4-6\cdot\rho^2+1$	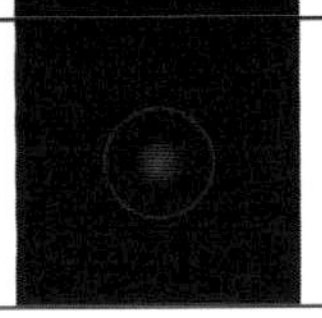	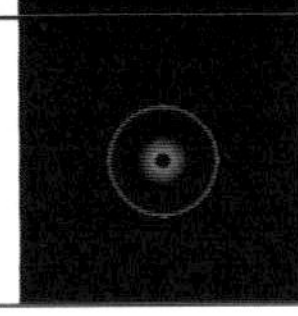	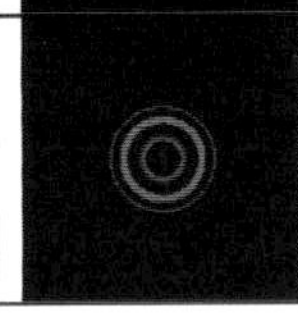	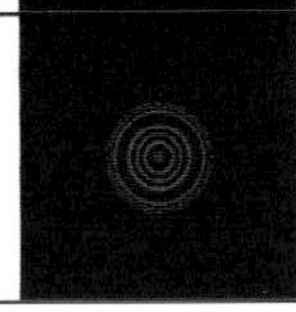

3. The principle of constructing a computer-synthesized Fourier hologram

As known, there are two types of computer-generated holograms - Fourier and Fresnel hologram. In both cases, the synthesis is the calculation of one or more fast Fourier transform operations. The disadvantage of Fresnel holograms is the presence of spatial noise due to zero order, which could have a critical impact on the accuracy of measurements of aberrations, the size of which is not more than four pixels.

Journal of Physics: Conference Series **737** (2016) 012064 doi:10.1088/1742-6596/737/1/012064

Therefore, the author considers the principles of the sensors only based on computer-synthesized Fourier holograms, implemented in accordance with the diagram shown in Figure 1a. In the synthesis method of such holograms uses zero order suppression.

In this case the synthesized hologram is an intensity distribution of the two waves - the reference (distorted – $B(x,y)$) and the object ($O(x,y)$):

$$I(x,y)=|O(x,y)+B(x,y)|^2=|B(x,y)|^2+|O(x,y)|^2+O(x,y)\cdot B^*(x,y)+O^*(x,y)\cdot B(x,y).$$

The reference wave we assume WF, object wave is a radiation of a point source, the position of the source is given by encoding. Thus, illuminated hologram with distorted WF restores image of the object - the point. In his position can judge the presence of aberrations in the sensor.

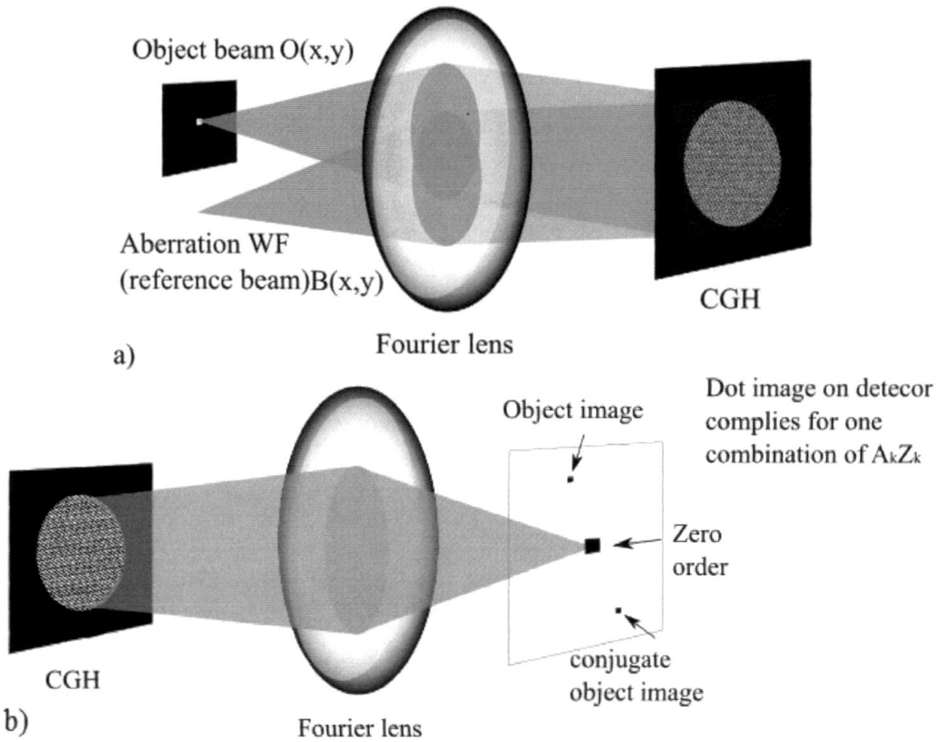

Fig. 1. Principles of recording (a) and reconstruction (b) CGH for WFS

4. Mathematical modeling of the sensor

Simulation of the sensor was produced in Mathlab, using an FFT algorithm. Fig. 2-4 shows the synthesized binary hologram, which demonstrates the ability to create CGH containing various combinations of aberrations. Type of the hologram image (binary or 8bit grayscale) determine by the characteristics of the LCD SLM used in the projection recording scheme. The quality of the recorded hologram to determine the parameters of the optical components of the circuit, but not the calculated structure.

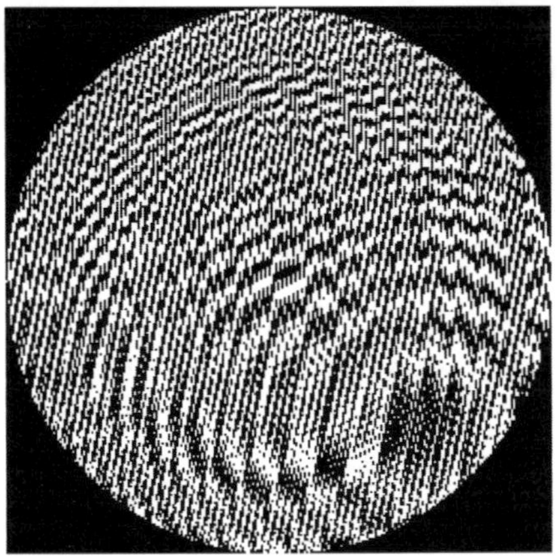

Fig. 2. Binary CGH for defocus with amplitudes from -2λ to 2λ with step 0,5λ

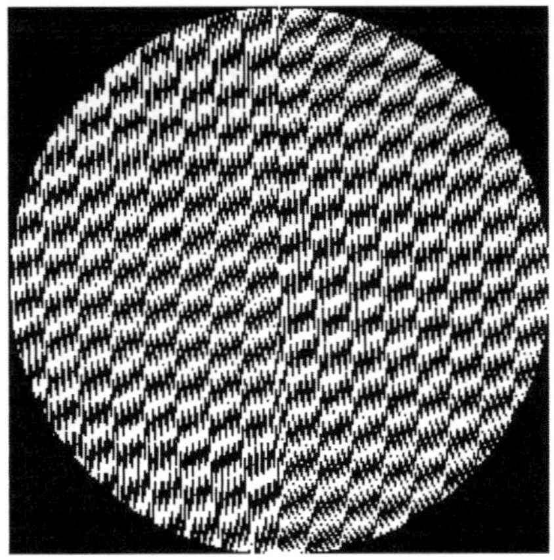

Fig. 3. Binary CGH for coma with amplitudes from 0.1λ to 1.0λ with step 0.1λ

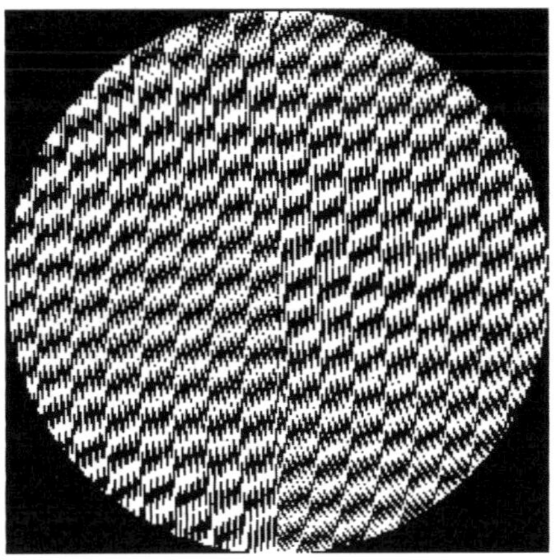

Fig. 4. Binary CGH for defocus and coma with amplitudes from 0.1λ до 1.0λ wih step 0.1λ

The total hologram at coordinates (100,200) has been encoded defocus aberration, the wave aberration with a value of 2λ, and at the point with coordinates (400, 200) the same type of aberration with a value of 1λ.

Fig. 5,6 shows the results of research. The ordinate shows normalized intensity, the abscissa shows coordinate on the light receiver. When additional illumination hologram different distorted waves diffraction patterns were obtained, the position of different points.

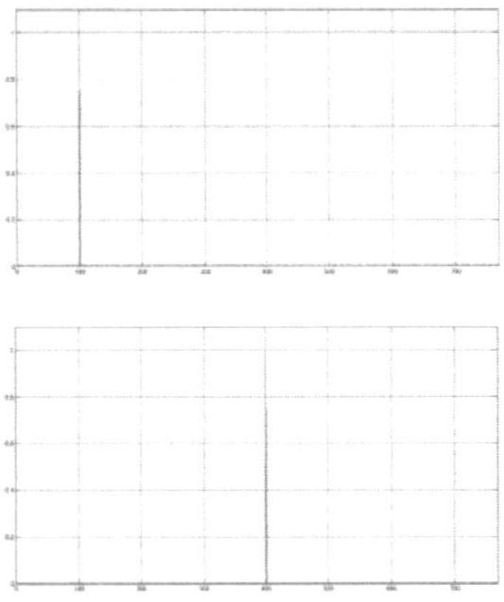

Fig. 5. Reconstruction of the hologram on computer (perfect case)

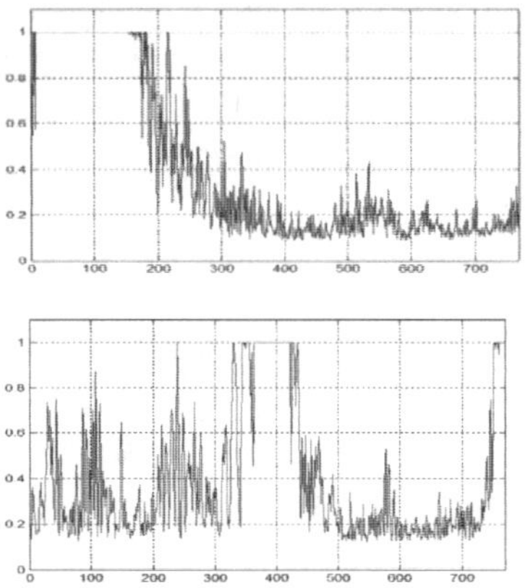

Fig. 6. Reconstruction of the hologram using distorted wave (real case)

Centre position of the reconstructed spot corresponds to encoded. Blur and noticeable distortions arising from the use of the arithmetic addition of a hologram, further work will focus on the study of methods for creating multiplex and composite holograms.

Acknowledgments

The work was completed in BMSTU with financial support from Russian Ministry of Education as a part of the grant agreement № 14.577.21.0115 (ID RFMEFI57714X0115).

References

[1] Golub M A, Jivopiscev E S, Karpeev A M, Sisakyan I I, Soifer V A 1980 *Doklady AN SSSR* **253** 1104–1108.

[2] Nikitin A. Sheldakova J, Kudryashov A, Borsoni G, Denisov D, Karasik V, Sakharov A 2016 *Photonic Instrumentation Engineering III*

[3] Beckers J M 1993 *Annual Review of Astronomy and Astrophysics* **31 (1)** 13–62

[4] Booth M J 2007 *Philosophical Transactions of the Royal Society A: Mathematical, Physical and Engineering Sciences* **365** 2829–2843

[5] Malacara D 1989 *Optical shop testing* (Wiley) 258-261

[6] Poleshchuk A G, Sedukhin A G, Trunov V I, Maksimov V G 2014 *Computer optics* **38 (4)** 695-703

[7] Platt B and Shack R 2001 *Journal of Refractive Surgery* **17**

[8] Venedictov V Y 2016 *Photonics* **1** 132-141

[9] Neil M A A, Booth M J and Wilson T 2000 *Journal of the Optical Society of America A* **17 (6)** 1098-1107

[10] Andersen G, Ghebremichael F, Gurley K 2007 *Proc. of SPIE* **6488** 64880I

[11] Betin A Y, Dontchenko S S, Kovalev M S, Odinokov S B, Solomashenko A B, Zlokazov E Y 2015 *Digital Holography and Three-Dimensional Imaging* 410

[12] Betin A Y, Bobrinev V I, Verenikina N M, Dontchenko S S, Evtikhiev N N, Odinokov S B, Zlokazov E Y, Starikov S N, Starikov R S 2015 *Quantum Electronics* **45 (8)** 771–775

[13] Born M Wolf E 1970 *Principles of optics*

[14] Hopkins H H 1949 *Wave theory of aberrations*

A new phase unwrapping method

Igor Shevkunov

St. Petersburg State University, 79, Universitetskaya nab., St. Petersburg, 199034, Russian Federation

E-mail: shevkunov_igor@mail.ru

Abstract. This work is devoted to an elimination of the phase discontinuities problem of the reconstructed object wavefront. Novel robust phase unwrapping method is proposed. It is based on the phase shift implementation of the reconstructed wavefront. The unwrapped phase field of the amplitude-phase object is presented. Comparison with another known methods is shown and discussed. Proposed method resistant to noise and object phase field inhomogeneities. Its advantages are robustness and simplicity.

1. Introduction

Holography is the universal method for the complete reconstruction of the object wavefront [1]. Nowadays digital holography successfully replaces the analog one and becomes more popular due to the rapid development of digital technologies. Hologram recording is performed with the use of CCD or CMOS matrices, and holograms subsequent processing is realized on computers [2]. The advantage of using digital technology is significant acceleration of the hologram recording and processing owing to the absence of the chemical preparation of the recording media [3, 4], and numerical reconstruction of the object wavefront greatly simplifies holograms post-processing. Due to the universality of the holography scientific interest to this method is not quenched, but rather attracting more attention of researchers [5, 6, 7, 8, 9].

The main goal of holographic research methods is an absolute phase calculation of object wavefronts. This task is especially difficult because of the specificity of the holographic technique for the phase reconstruction: it always comes down to calculation of inverse trigonometric functions. As a result, it gives the phase values, that differ from the absolute through the periodicity of these functions: $\Theta_{abs} = \Theta_i + 2\pi * n$, Θ_{abs} – is the absolute phase of the object wavefront, Θ_i – wrapped phase, n – integer number. Thus, after reconstruction the absolute phase is divided into separated regions with phase values in the region $(-\pi,\ \pi]$ (these values depend on a period of trigonometric function used in holographic reconstruction), and researchers' task is to unwrap these phase regions in a single continuous field. Unwrapping is complex problem due to the presence of a noise in registration systems and phase inhomogeneities of the object.

Phase unwrapping problem was set long ago, and many methods were suggested to solve it. The main task of the phase unwrapping is a choice of the place where the phase of the field should be shifted. In real experiment conditions such choice is often complicated by a phase noise which can lead to erroneous phase unwrapping shifts. Since phase noise often have a higher frequency than the desired signal, initial filtering of the wrapped phase field is the easiest and most intuitive way to simplify unwrapping, for example it could be done by the Fourier filtering [11], or least square method [12]. But the filtering could result in a loss of the useful high-frequency signal and smoothing of the whole wavefront. Also there is popular method that based on the finding of local phase errors by a specific scanning of the phase wavefront [13], but this method has difficulties when spatial size of errors is much larger than the size of recording system pixel. An alternative to the search of local errors method is an approach based on the energy minimization algorithm with initial assumption that the phase difference between adjacent pixels is less than the value of π radians [14], in fact this assumption is not correct in many real experiment cases. There is also a hardware method for obtaining of the absolute phase by utilization of two or more radiation light wavelengths with close values [15]. Increasing of the wavelengths number gives possibility to create a synthetic wavelength with bigger value, and therefore it is possible to enlarge continuous phase regions of the reconstructed wavefront. Often in this method the unwrapping procedure is not required or reduced to the simplest one. It should be noted, however, that the holograms registration with different light wavelengths greatly complicates this technique both in a hardware and software. Despite the significant number of proposed methods for phase unwrapping problem solving, there is no unequivocal reconstruction method of the absolute phase from holograms of objects with phase inhomogeneities.

Especially it is hard to unwrap the phase of the amplitude-phase object wavefront, that reconstructed with high resolution, because amplitude field creates phase inhomogeneities in areas of complete absorption of the light radiation. The higher resolution the bigger number of such inhomogeneities. Such high resolution wavefront of the amplitude–phase object (the wing of a fly) is shown in figure 1. It was reconstructed by phase–shifting digital holography [16].

$$\Theta = \arctan\left[\frac{I_{3\pi/2} - I_{\pi/2}}{I_0 - I_\pi}\right] \tag{1}$$

The phase of the object wavefront was calculated by equation 1, where Θ - object wavefront phase, I_p - holograms with shifted reference beam by 0, $\pi/2$, π and $3\pi/2$ radians. Arctangent has a period of π radians, but with proper consideration of numerator and denominator signs, the phase has the period of 2π. Actually this procedure could be considered as a first step in the unwrapping, because it decreases the number of wrapped lines twice.

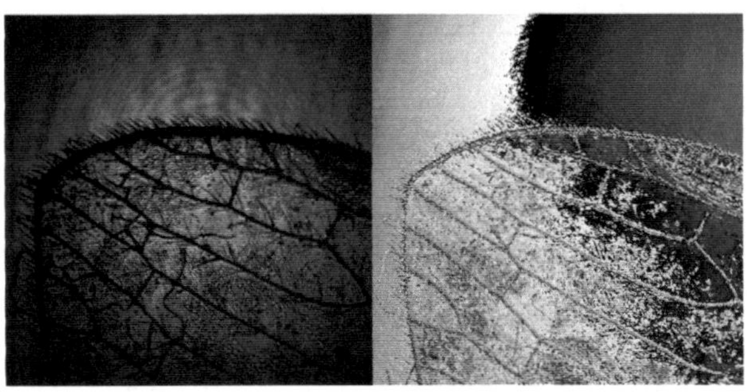

Figure 1. Amplitude (left) and phase (right) of the object wavefront

2. Method description

This paper presents a phase unwrapping method by a Phase Shift of the Reconstructed Field (PSRF) that shows reliable unwrapping results for amplitude–phase objects. Since wrapped phase field is defined by a periodic function, it is possible to add a constant phase to whole wrapped phase field, thus realizing a shift of the entire field without disturbing its general order. It is obvious that addition of a half period of the function will result in phase shift of the field, and ruptured area of the field will be replaced by the continuous area. Steps of the PSRF method application to phase field of the amplitude–phase object are shown in figure 2. Fig. 2(a) represents the initial phase reconstruction of the object wavefront: $E = A\exp^{i\Theta}$ (where E – object wavefront, A – amplitude, Θ – phase), and Fig. 2(b) shows the same wavefront's phase field, but with an addition of the phase constant π to its phase: $E = A\exp^{i\Theta+\pi}$.

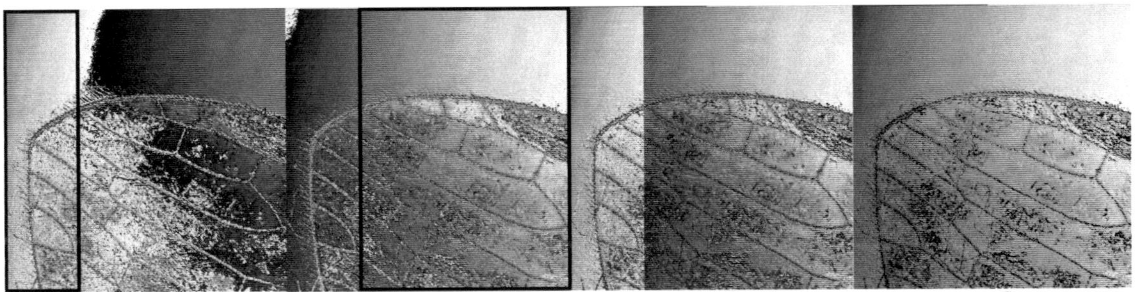

Figure 2. Method PSRF. Wrapped phase field (a), phase field shifted by π rad (b), phase field after stitching of continuous areas (c), resulting unwrapped phase field (d). Black squares mark out regions of the undisturbed phase.

As it shown in Fig. 2(b) corrupted region moved to another part of the field. Thus, the addition of the phase shift to whole field leads to a realization of the phase unwrapping by stitching only undisturbed areas of the phase field (areas marked by rectangles in Fig. 2(a) and (b)). The stitching result is presented in Fig.2(c). Final unwrapped phase field is presented in Fig.2(d), where the phase additive constant is added only to the shifted phase area. The π radians shift is a half period of the wrapped function, therefore, this shift will always lead to a replacement of the ruptured region by continuous, that makes it possible to assemble a complete picture of the absolute wavefront phase by a successive stitching of shifted continuous fragments.

3. Results and discussion

The main problem of all unwrapping methods is phase discontinuities presence in the reconstructed wavefront. The cause of this discontinuities is in high noise levels of the hologram and/or characteristics of investigated objects. To evaluate the effectiveness of the PSRF method, results of the phase unwrapping by modern methods: Goldstein [13] and PUMA [14], are presented in Fig. 3. Detailed analysis of absolute phases, obtained by these methods, shows that the large number of non-smooth phase regions leads to erroneous unwrapping results. Processing by Goldstein in this case leads to unsatisfactory results due to false detection of discontinuities in the spatial phase distribution. Method PUMA, based on energy minimization, shows better result than the Goldstein method. However, it provides the distorted unwrapped phase due to assumption the difference between adjacent pixels smaller than π radians. This assumption leads to a smoothing of existing phase inhomogeneities and unjustified unwrapped phase jump on the border of the object. The proposed PSRF method do not form phantom phase shifts, and unwrapping is reliably performed even in the case of the phase inhomogeneities presence, through the implementation of a global phase shift, without a consideration of the individual field points.

For the qualitative comparison in Fig.4 longitudinal cross-sections are presented: initial phase field distribution (*1*) and unwrapped phase fields by methods: (*2*) – Goldstein, (*3*) – PUMA, (*4*) - PSRF. Fig.4(a) presents cross-sections from the part of the undisturbed phase, that corresponds to dashed line from fig. 3(a), and fig.4(b) presents cross-sections from the

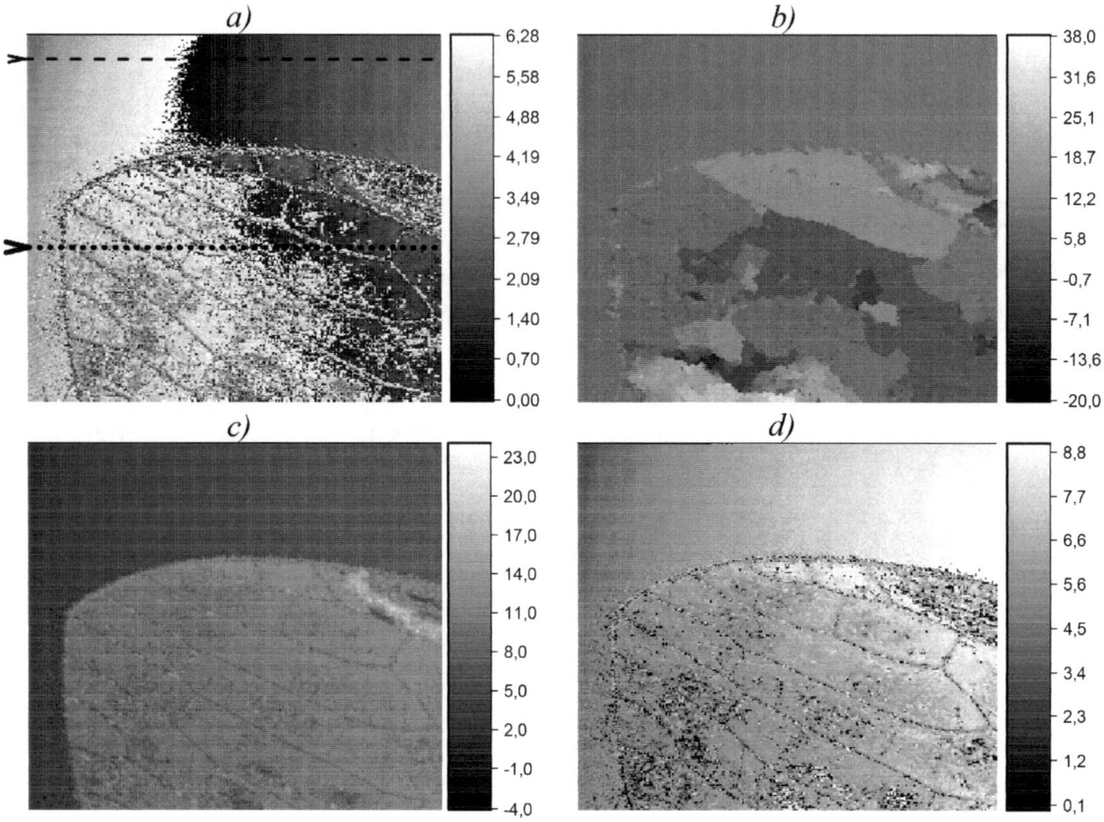

Figure 3. Reconstructed object phases – wrapped (a) and unwrapped by: Goldstein(b), PUMA (c), PSRF (d)

object phase field, dotted line from fig. 3(a). As it can be seen from the cross sections shown in Fig.4(a), in undisturbed conditions all methods work reliably and accurately, unwrapping results are performed equally. But Goldstein's and PUMA's results from the area with many phase inhomogeneities, related to characteristics of the object, are erroneous (Fig.4(b)). Goldstein shows good results only on small areas of the phase field, but the whole phase field is unwrapped wrong. Due to small height of the object the phase shift on the "object–undisturbed wavefront" line should not be bigger than π (see fig.4) , but PUMA recognized it as about 8π (fig.4(b)). This error is explained by that the border of the object is not transparent, and it creates inhomogeneities on the phase field, but PUMA considers this border as place of the phase shift and wrongly unwraps it. The proposed PSRF method is free of such shortcomings and gives legitimate result. Thus, it is shown that the PSRF method can be used to eliminate phase discontinuities as for undisturbed (Fig.4(a)), so for noised (Fig.4(b)) wavefronts. Furthermore PSRF could be realized in real-time unwrapping owing to relatively simple calculations.

4. Conclusion

I have described the new phase unwrapping method based on the use of the phase shift of the reconstructed wrapped field. This method demonstrates unwrapping robustness to a noise and features of the object. Due to the fact that phase information more often lost in high amplitude areas of the radiation absorption, phase anomalies take place in these areas and make difficulties

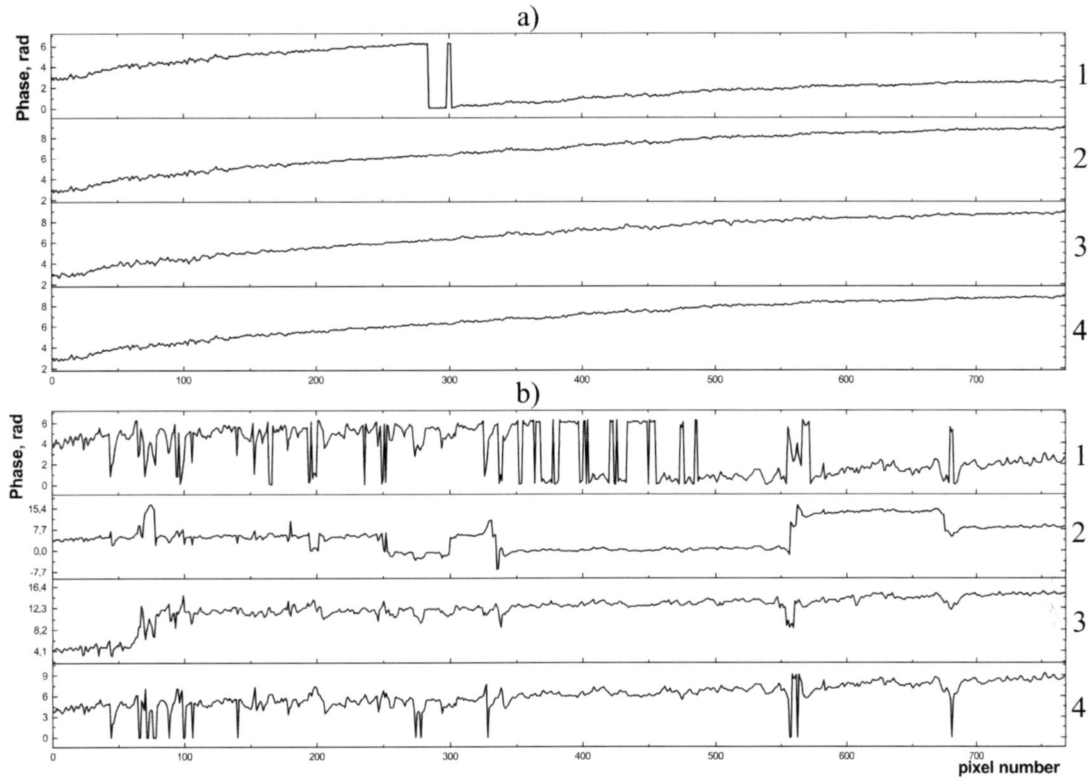

Figure 4. Longitudional cross-sections of phase fields: (*a*) – 50-th row pixels (dashed line from fig. 3, *a*), (*b*) – 440-th row (dotted line from fig. 3, *a*). Here *1* – cross-section of wrapped phase, and cross-sections of the unwrapped phase by methods: *2* – Goldstein, *3* – PUMA, *4* – PSRF.

to existing unwrapping methods. Based on the demonstrated results, proposed PSRF method is the most promising in amplitude-phase object cases.

References

[1] D. Gabor, Proceedings of the Royal Society of London A: Mathematical, Physical and Engineering Sciences **197**, 454 (1949).
[2] J.W. Goodman, R.W. Lawrence, APL. **11**, 77 (1967).
[3] V.P. Smaev, Y.A. Vavilova, A.D. Galpern, Opt.-Mekhan. Prom.**3**, 44 (1990).
[4] V.P. Smaev, A.D. Gal'Pern, Yu.A. Vavilova, A.A. Paramonov, "Some Features of Production of Monolayers for Multilayer Photographic Materials" Opt. and Spectr. **76**, 738 (1994).
[5] A.V. Belashov, N.V. Petrov, I.V. Semenova, Opt. Express, **22**, 28363 (2014).
[6] V. Katkovnik, I.A. Shevkunov, , N.V. Petrov, K. Egiazarian, Opt. Lett. **40(10)**, 2417 (2015).
[7] S. Hakim, M. Yamaguchi, and F. Kimura, Information Optics (WIO), 2015 14th Workshop on. IEEE, 1 (2015).
[8] D. Merrill, R. An, J. Turek, D.D. Nolte, Appl. Opt. **54**, A89 (2015).
[9] P.A. Cheremkhin, I.A. Shevkunov, N.V. Petrov, Physics Procedia **73**, 301 (2015).
[10] W.J. Hwang, H.Y. Chen, C.J. Cheng, Appl. Opt. **54(1)**, A67 (2015)
[11] M. Kujawinska, J. Wojciak, Opt. Lasers Eng. **14**, 325 (1991).
[12] M.D. Pritt, J.S. Shipman, Geoscience and Remote Sensing, IEEE Transactions on, **32**, 706 (1994).
[13] R. Goldstein, H. Zebreker, C. Werner, Radio science, **23**, 713 (1988).
[14] J.M. Bioucas-Dias,G. Valadao, Image Processing, IEEE Transactions on 2007, **16**, 698 (2007).
[15] Y. Li, W. Xiao, F. Pan,Appl. Opt.**53**,979 (2014).
[16] I. Yamaguchi, T. Zhang, Opt.Lett. **22**, 1268 (1997).

Accurate measurement of spatial noise portraits of photosensors of digital cameras

P A Cheremkhin, N N Evtikhiev, V V Krasnov, M N Kulakov, R S Starikov

National Research Nuclear University MEPhI (Moscow Engineering Physics Institute), Kashirskoe shosse 31, 115409 Moscow, Russia

E-mail: PACheremkhin@mephi.ru

Abstract. Method of measurement of accurate portraits of light and dark spatial noise of photosensors is described. The method consists of four steps: creation of spatially homogeneous illumination; shooting light and dark frames; digital processing and filtering. Unlike standard technique, this method uses iterative creation of spatially homogeneous illumination by display, compensation of photosensor dark spatial noise portrait and improved procedure of elimination of dark temporal noise. Portraits of light and dark spatial noise of photosensors of a scientific digital camera were found. Characteristics of the measured portraits were compared with values of photo response and dark signal non-uniformities of camera's photosensor.

1. Introduction

Basic element of any digital photo- and videocamera is the photosensor [1-3]. Photosensor consist of pixels and the number of pixels in modern digital cameras is equal to tens of millions. Due to imperfections of production of photosensors, photosensitivities of separate pixels are slightly different. In result any photosensor has inherent light spatial noise portrait (LSNP) [4] and dark spatial noise portrait (DSNP). LSNP represents the array of relative differences of photo responses of separate pixels. DSNP is array of dark spatial non-uniformities of pixels.

The knowledge of LSNP and DSNP potentially allows to increase the signal-to-noise ratio of the images [5-7]. To suppress temporal and spatial noises, flat-field correction [8-10] or dark correction [8] techniques are used. We propose modification of flat-field correction technique for image signal-to-noise ratio increasing. Unlike the sky flats or twilight flats that commonly used in flat-field correction technique [9-10] we proposed to use compensation of LSNP and DSNP of camera's photosensor for spatial noise suppression [5-7]. As the proposed modification is aimed to reduce spatial noise, it is effective after application of others methods of SNR increasing that suppress temporal noise.

Other variant of application of the LSNP and DSNP is identification of digital photocameras based on a number of images [4] and identification of digital videocameras based on available video [11]. Besides identification of cameras, the LSNP is suitable for detection of existence and nature of editing contents of a picture, an assessment of age of pictures, and etc. [12-15].

In this paper we measured the LSNP of photosensor of scientific camera using the proposed modification.

2. Description of the modification of method of measurement of camera's photosensor LSNP

Modification of method of measurement of camera's photosensor LSNP is proposed in [5-7]. It is based on technique described in [4]. Main differences between proposed modification and presented in [4] technique are:

- iterative creation of spatially homogeneous illumination using display,
- using of compensation of photosensor dark spatial noise portrait,
- improved procedure of elimination of dark temporal noise.

Our modification consists of 4 steps:

1. Creation of a spatially homogeneous illumination of camera's photosensor (Homogeneous distribution displayed on monitor is used for sensor illumination. On results of shooting, spatially homogeneity across sensor is analyzed, map of illumination is created and new distribution on monitor is displayed. These actives are repeated iteratively.);
2. Shooting two series of frames: with opened lid (light frames under created illumination) of and with closed lid (dark frames);
3. Digital processing of registered frames for elimination of:
   3a) dark and light temporal noise (by frames averaging)
   3b) dark spatial light (by elimination averaged dark frame from light one)
4. Digital filtering of obtained array for elimination of residual non-uniformity of illumination (by dividing on median filtered array and unity subtracting).

3. Experimental estimation of spatial noise portraits of scientific digital camera

For experimental estimation of photosensor spatial noise portraits, scientific camera MegaPlus II ES11000 was used. Its main technical characteristics are given in Table 1 [16].

Table 1. Technical characteristics of camera MegaPlus II ES11000.

Characteristic	Value
Sensor type	CCD
Quantity of pixels	4008 × 2672 (10.7 MP)
Pixel size	9μm× 9μm
Output bit depth of ADC	Up to 12 bits
Spectral	Monochromatic
Minimum integration time	140μs
Maximum integration time	> 5 s

Noise and radiometric characteristics of this camera measured in [17-19] were used. They are shown in Table 2, where DN is digital numbers of signal, e-/DN is the ratio of quantity of emitted electrons in one pixel to one digital number of signal.

Table 2. Noise and radiometric characteristics of camera MegaPlus II ES11000.

Characteristic	Value
Dark temporal noise	2.3 ± 0.4 DN
Camera gain	11.88 ± 0.11 e-/DN
Dark spatial non-uniformity	0.56 ± 0.04 DN
Photo response non-uniformity	0.0053±0.0008
Black level offset (BLO)	3.4 ± 0.6 DN
Maximum linear signal with BLO subtracted	3916 ± 30 DN
Dark temporal noise	2.3 ± 0.4 DN

Using the described procedure LSNP of photosensor of camera MegaPlus II ES11000 was obtained. 400 light and 400 dark frames were used. Figure 1 shows a fragment of light spatial noise portrait of MegaPlus II ES11000 photosensor. Figure 2 shows a fragment of dark spatial noise portrait of MegaPlus II ES11000 photosensor. Sizes of the fragments are 512×512 pixels.

The histogram of brightness of pixels both in DSNP and in LSNP approximately corresponds to Gaussian distribution. Mean square deviation of values of LSNP of a photosensor of the MegaPlus II ES11000 camera is equal to 0.0051. This value is average photo response non-uniformity of pixels of a photosensor of this camera. It corresponds to earlier measured value [17-19]. Mean square deviation of values of LSNP of photosensor is equal to 2.3 digital signal levels (or 26 e^- using camera gain value). This value corresponds to average dark signal non-uniformity of pixels of the camera.

Figure 1. A fragment of light spatial noise portrait of camera MegaPlus II ES11000 photosensor. Size is 512×512 pixels.

Figure 2. A fragment of dark spatial noise portrait of camera MegaPlus II ES11000 photosensor. Size is 512×512 pixels.

4. Conclusion

Thus, experiments on measurement of portraits of light and dark spatial noises of photosensor of the scientific digital camera were performed. Characteristics of the measured portraits correspond to photo response and dark signal non-uniformities of the photosensor. The obtained portraits allow to increase the signal-to-noise ratio of the images registered by this camera.

Acknowledgments

This work was partially supported by grant 14-19-01751 from the Russian Science Foundation (RSF).

References

[1] Janesick J 2001 *Scientific Charge-Coupled Devices* (Bellingham-Washington: SPIE Press)
[2] Nakamura J 2006 *Image sensors and signal processing for digital still cameras* (Boca Raton, FL: CRC Press)
[3] Baker R J 2011 *CMOS: Circuit Design, Layout, and Simulation: Third Edition* (John Wiley and Sons)
[4] Fridrich J 2009 *IEEE Signal Processing Magazine* **26** 26-37
[5] Evtikhiev N N, Starikov S N, Cheryomkhin P A, Krasnov V V and Rodin V G 2014 *Proceedings of SPIE* **9025** 90250X
[6] Cheremkhin P A, Evtikhiev N N, Krasnov V V, Rodin V G and Starikov S N 2015 *Proceedings of SPIE* **9406** 94060O
[7] Starikov S N, Cheremkhin P A, Krasnov V V, Kurbatova E A and Starikov R S 2015 Physics Procedia **73** 264–268
[8] Widenhorn R, Dunlap J C and Bodegom E 2010 *IEEE Transactions on Electron Devices* **57** 581-587
[9] Schröder S E, Mottola S, Matz K-D and Roatsch T 2014 *Icarus* **234** 99-108
[10] Rasmussen A 2014 *Journal of Instrumentation* **9** C04027
[11] Hyun D-K, Ryu S-J, Lee M-J, Lee J-H, Lee H-Y and Lee H-K 2012 *Proceedings of SPIE* **8303** 0E1-0E8
[12] Fridrich J and Goljan M 2011 *Proceedings of SPIE* **7880** 788006
[13] Sugiyama T, Araie M, Riva C E, Schmetterer L and Orgul S 2010 *Acta Ophthalmologica* **88** 723-729
[14] Evtikhiev N N, Shaulskiy D V, Zlokazov E Y and Starikov R S 2012 *Proceedings of SPIE* **8398** 83980G
[15] Kodovský J, Fridrich J, Holub V 2012 *IEEE Transactions on Information Forensics and Security* **7** 432-444.
[16] Princeton Instruments, http://www.princetoninstruments.com/Uploads/Princeton/Documents/Brochures/MEGAPLUS_camera_brochure_Princeton_Instruments_B0.pdf
[17] Evtikhiev N N, Starikov S N, Cheryomkhin P A and Krasnov V V 2012 *Proceedings of SPIE* **8301** 830113
[18] Cheremkhin P A, Evtikhiev N N, Krasnov V V., Rodin V G. and Starikov S N 2014 *Optical Engineering* **53** 102107
[19] Cheremkhin P A, Evtikhiev N N, Krasnov V V., Rodin V G, Starikov R S and Starikov S N 2015 *Proceedings of SPIE* **9648** 96480R

Study of objects visualization and image inversion by the phase-contrast method with linear and nonlinear filters

E L Bubis[1], **V V Lozhkarev**[1], **A N Stepanov**[1], **A I Smirnov**[1], **I V Kuzmin**[1], **O A Malshakova**[1], **S A Gusev**[2], **E V Skorokhodov**[2]

[1]The Institute of Applied Physics of the RAS, Nizhny Novgorod, 603950, Russia
[2]The Institute for Physics of Microstructures of the RAS, Nizhny Novgorod region, Kstovsky district, 603087, Russia

E-mail: bel@appl.sci-nnov.ru

Abstract. The adaptive phase-contrast method with nonlinear (photothermal) and linear Zernike filters was investigated. Liquid and polymer media partially absorbing radiation served as photothermal Zernike filters. Efficient visualization and inversion of images of small-scale model objects were demonstrated experimentally. Growth-sector boundary in a nonlinear crystal was visualized.

1. Introduction

The phase-contrast method is widely used for visualizing small-scale weak (phase $\varphi<1$) objects [1-3]. To transform the phase modulation induced by the studied object into the illuminating beam to the amplitude modulation, a phase screen (Zernike filter) is placed in the focal plane of the lens (Fourier plane). The Zernike filter introduces a selective shift $\theta_Z=\pm\pi/2$ (for objects with phase $\varphi<<1$) between the zero and higher spatial frequencies participating in imaging the object. The phase-contrast method proposed by Frits Zernike in 1934 was initially used mainly in microscopy. Later its application was extended to plasma physics (measurement of weak fluctuations), acoustic waves, gas flows, and so on.

This work concerns investigation of the phase-contrast method using nonlinear Zernike filters [2,4-25], when spatial frequency phase shift occurs in a cubic nonlinear medium (nonlinear Zernike filter) situated in the Fourier plane where the spatial harmonics are separated. As compared to the schemes with linear Zernike cells, the nonlinear schemes are more flexible adaptive systems where a needed phase shift is attained by selecting appropriate light intensity in the nonlinear medium. Phase objects are visualized in phase-contrast schemes with Zernike filters using the following types of nonlinearity: Kerr [4-7], bacteriorhodopsin [8,9], photorefractive [10], orientational liquid crystal [11-16], thermocapillary [17], supersaturation nonlinearity [18], and others (see, e.g., [19,20]). In the present work photothermal filters [9,21-25] based on the thermal mechanism of nonlinearity were used in the phase-contrast scheme. This mechanism is the most low-threshold one for cw and quasi-cw laser generation in simple available media.

2. Experimental study

The optical schemes used in experiments are shown in fig. 1 (solid line – direct light, dash line – diffracted wave on the object). A green laser module SLM-417 with wavelength λ=530 nm was used as a radiation source.

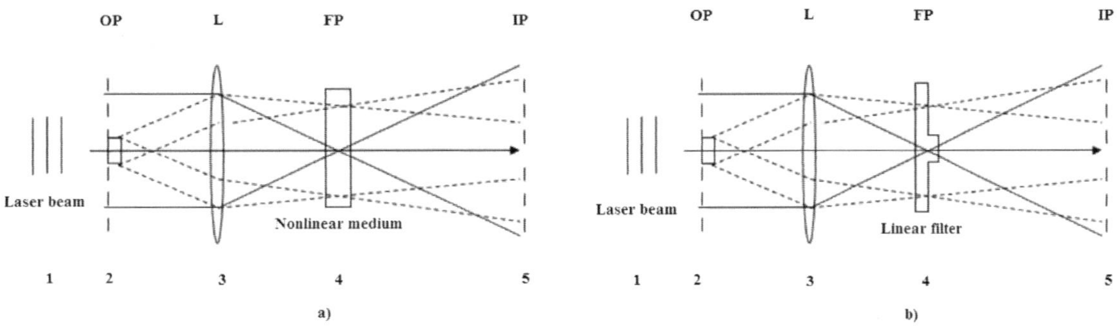

Figure 1. Optical schemes used in experiment.

The source power did not exceed $P \leq 10$ mW. Object 2 was located in the object plane OP. The object was exposed to laser radiation focused by lens 3 on a cell with absorbing liquid (5,a) or on a linear filter with a 265 nm high step (5,b) located in the Fourier plane FP. The obtained image was recorded by a CMOS-camera situated in the image plane IP.

The results of phase object visualization are presented in fig. 2. It follows from this figure that the visualization is efficient, which is important for constructing schemes for measuring weak medium absorption [18]. It is also seen that acceptable image quality persists at the initial stage of thermal self-action (fig. 2 b,c,d) and further deteriorates due to a strong thermal lens.

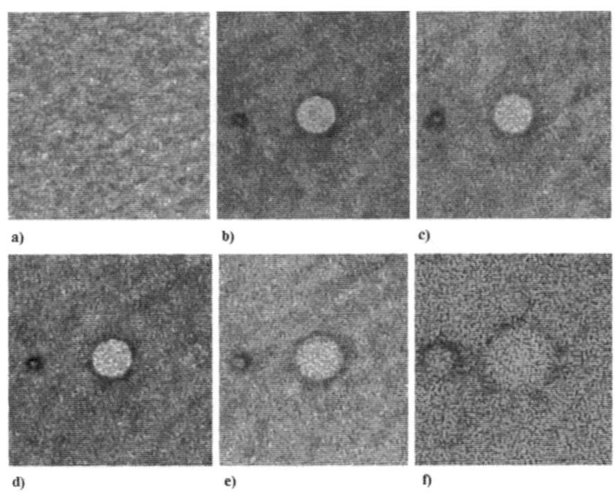

Figure 2. Visualization of transparent object with phase shift $\lambda/2$ using a liquid cell (alcohol doped with iodine) as a function of power of laser radiation. Circle diameter d = 150 μm.

Visualization of a phase object (grating) by means of a linear filter with a step 265 nm high in the Fourier-plane of the lens is depicted in fig. 3. Structures for measurements were formed by the method of electron-beam lithography from thin PMMA 950 films deposited on fused silica or glass substrates. The structure was exposed on the SUPRA 50VP (Carl Zeiss) electron microscope equipped with a system for electron lithography ELPHY PLUS. At the finishing stage, the required relief depth (PMMA film thickness) was obtained by means of dry etching in oxygen plasma.

The relief depth was controlled either by an atomic force microscope or by the optical interference microscope Talysurf CCI 2000. As follows from fig. 3, the image obtained with a photothermal filter is free from false structures caused, in particular, by boundary effects arising during alignment of a Zernike phase disk with a diameter of 150 μm.

a) b) c)

Figure 3. Visualization of a phase object by means of linear and a photothermal filters: a) without filters, b) linear filter (Zernike disk) placed in the Fourier-plane, c) photothermal filter also placed in the Fourier-plane.

Direct and inverted microimages of the abbreviated names of the institutions with which the authors of the paper are affiliated are presented in fig. 5. The height of the letters is 30 μm. The images were obtained on a 2 mm thick polymer plate used as phototermal filter.

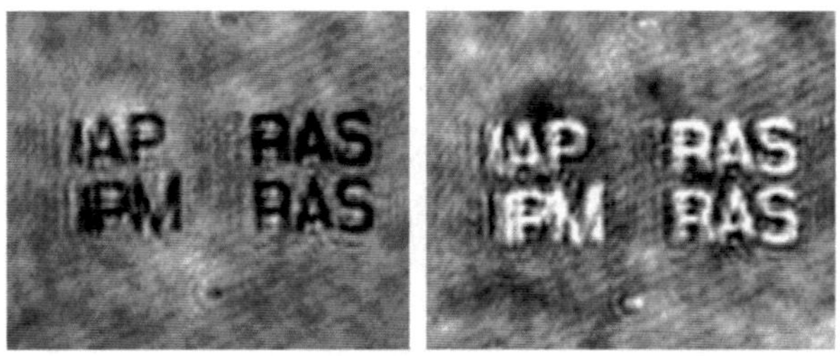

Figure 5. Direct and inverted microimages of the name of the institutions. Letter size 30 μm.

The visualized image of the growth-sector boundary in a nonlinear KDP crystal is shown in fig. 6. Such a boundary was earlier recorded by the schlieren method and by the fluorescence technique in the works [26, 27].

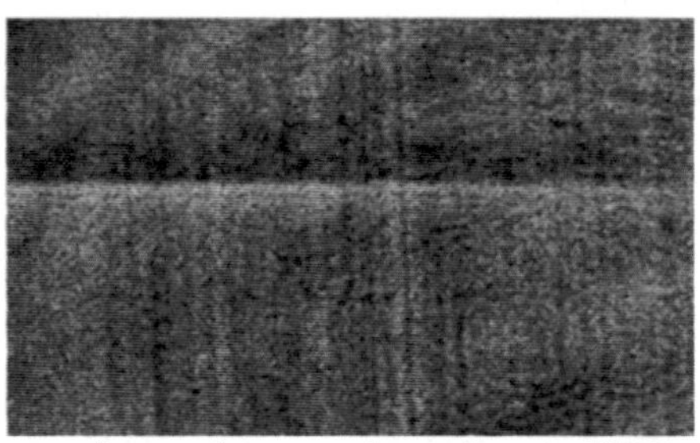

Figure 6. Image of growth-sector boundary in KDP crystal.

3. Conclusion

To conclude, we have registered good-quality images of transparent objects and structures and also inverted images of nontransparent objects using a simple phase-contrast technique. Model objects as well as an intersectorial boundary in a nonlinear KDP crystal have been visualized.

The structures were prepared and studied using the equipment of the Multiple-access Center "The Physics and Technology of Micro- and Nanostructures". The authors appreciate the support obtained from the Russian Foundation for Basic Research (grant No. 15-02-08212).

References

[1] Fransson M 1952 *Phase-contrast and interference microscopes* (Paris: Revue d'Optique)

[2] Vorontsov M A., Koryabin M A, Shmalgausen V I 1988 *Controlled optical systems* (M.: Nauka) [in Russian]

[3] Davydenko V I, Ivanov A A, Weissen G 1999 *Experimental methods of plasma diagnostics. The Lecture. Part 1* (Novosibirsk, NSU) [in Russian]

[4] Chernega N V, Brekhovskikh G L, Kudryavtseva A D, Kirsanov B P, Sokolovsky A I 1989 *Quantum.electronics* **16** 2530

[5] Goy A, Psaltis D 2011 *Physical Review A* **83**, 031802(R)

[6] Zhengyan L., Chih-Hao P., Yen-Yu C. 2013 *Optics Letters* **38 (23)** 5157

[7] Bubis E L, Korytin A I, Stepanov A N, Malkov Y A, Arsenev A A, Yashunin D A 2015 *Proceedings of the XIII International conference "OMFI -2015"* (Moscow) p 152

[8] Iturbe Castillo M D, Sanchez-de-la-Llave D, Ramos Garcıa R, Olivos-Pérez L, González L A, Rodríguez-Ortiz M. 2001 *Optics Engineering* **40 (11)** 2367.

[9] Treviño-Palacios C G, Iturbe-Castillo M D, Sánchez-de-la-Llave D, Ramos-García R, Olivos-Pérez L I 2003 *Applied Optics* **42** 5091

[10] Liu J, Xu J, Zhang G, Liu S 1995 *Applied Optics* **34** pp. 4972-4975

[11] Vorontsov M A, Justh E W, Beresnev L A. 2001 *J. Opt. Soc. Am. A* **18** 1289

[12] Bubis E L, Budagovsky I A, Zolot'ko A S, etc. 2016 *Pribory i tekhnika experimenta* **4** 4

[13] Porras-Aguilar R, Ramirez-San-Juan J C, Baldovino-Pantaleon O, May-Arrioja D, Arroyo Carrasco M L, Iturbe Castillo M D, Sánchez-de-La-Llave D, Ramos-Garcia R 2009 *Optics Express* **17 (5)** 3417

[14] Yu M, Vorontsov M A 2005 *Proceedings of SPIE*, **5998**

[15] Yelleswarapu C S, Kothapalli S, Aranda F J and Rao D V G L N 2006 *Appl. Phys. Lett.* **89**

[16] Yelleswarapu C S, Kothapalli S, Rao D V G L N 2008 *Optics Communications* **281** 1876

[17] Bubis E L 2008 *Technical Physics Letters* **34 (6)** 510.

[18] Pavlov A, Pavlov A, Golubev M. 2012 *16th Int Symp on Application of Laser Techniques to Fluid Mechanic* (Lisbon, Portugal)

[19] Kurian P A and Vijayan C 2009 *Applied Optics* **48 (28)** 5259

[20] Komorowska K, Miniewicz A, Parka J, Kajzar F. 2002 *J.of Appl.Phys.* **92** 5635

[21] Bubis E L 2011 *Quantum electronics* **41 (6)** 568.

[22] Bubis E L, Matveev A Z 2012 *Quantum electronics* **42 (4)** 361

[23] Bubis E L, Gusev S A, Loshkarev V V, Kozhevatov I E, Martynov V O, Silin D E, Stepanov A N 2014 *Instruments and Experimental Techniques* **57 (5)** 620

[24] Bubis E L, Gusev S A, Kozhevatov I E, Martynov V O, Mamaev A Yu, Sergeev A S 2012 *Prikladnaya fizika* **5**, 16

[25] Bubis E L, Gusev S A, Kozhevatov I E 2013 *Uspekhi prikladnoi fiziki* **1(4)** 408

[26] Pommie`s M, Damiani D, Le Borgne X 2007 *Optics Communications* **275** 372

[27] Chang J, Zhao Y, Hu G, Wang Y, et al 2015 *Chinese Optics Letters* **13 (8)** 081601

Journal of Physics: Conference Series **737** (2016) 012068 doi:10.1088/1742-6596/737/1/012068

Precise baseline correction using statistical criteria in the measurement of characteristics of laser beams

I S Matsak[1], V V Kapranov[1], V Y Tugaenko[1], E M Kudryavtsev[2]

[1]S.P. Korolev Rocket and Space Corporation Energia, 4a Lenin street, Korolev, Moscow area, 141070, Russia
[2]National Research Nuclear University MEPhI (Moscow Engineering Physics Institute), Kashirskoe shosse 31, 115409 Moscow, Russia

E-mail: ISMatsak@mephi.ru

Abstract. Laser beam measurement using transmissive diffusion screen is the most appropriate method for wide aperture laser beams. Precise measurement requires minimizing the instrumental and methodical errors. Structure of instrumental errors and its reducing were considered in works [1,2]. In this work the method of precise baseline correction is discussed, which allow to reduce methodical error and to avoid clangers caused by highlights and uneven illumination.

1. Introduction

Laser beaming employed for wireless power transmission application requires precise optical system with high accuracy determined characteristics. The optical system can be characterized by measurement of the laser beam along its propagation. Direct laser beam measurement allows to make a correct qualification of system.

Vast majority of laser beam measurement techniques are well described in Refs. [3–8]. Recently an extension for camera-based methods for wide-aperture laser beams measurement was reported [1,2,9]. Precision of verification methods plays an important role in qualification of precise optical system. Thus an analysis of measurement error and an approach for its compensation is necessary. A number of works in this area is available. Several important results related to instrumental and methodical errors were reported in Refs. [1,4,10]. In particular, baseline offset is a factor with a great impact on the measurement error [11]. In this work we report an algorithm for ranging the data which allows to distinguish a data for high repeatability of measurement.

2. Measurement setup for wide-aperture laser beams

Wide-aperture laser beam measurement technique requires experimental setup based on camera and diffusing screen [1,9]. The scheme of the setup is shown on figure 1.

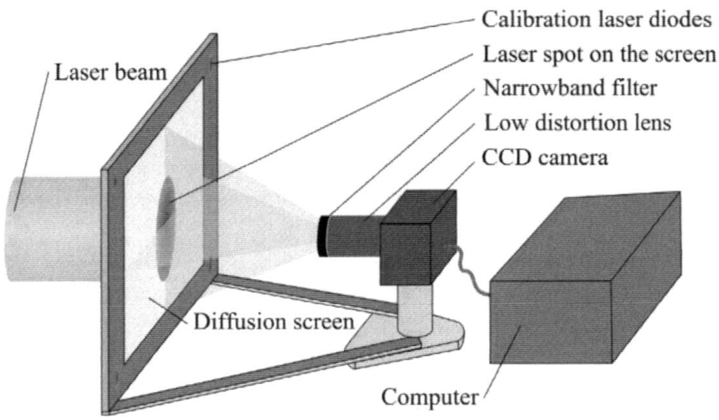

Figure 1. Experimental setup for wide-aperture laser beam measurement

3. Standard methods of baseline correction

In beam measurement baseline offset is mainly caused by background light and noise of CCD sensor. Detailed description of noises and its calibration are given in Refs. [12,13]. Standard methods of baseline offset calculation are described in Refs. [7,8]. Namely, these methods are: noise map of dark frame (1), mean of dark frame (2), approximation method (mean of unlighted pixels) (3), mean plus dispersion (4), statistical method (5), histogram method (6) or corner area mean (7). Any of these methods can be employed depending on purpose specifics. Each method is discussed in more details below.

Baseline offset as noise map of dark frame ($\overline{E_B^M}$) can be calculates using

$$\overline{E_B^M}(i,j)=\frac{1}{n_B}\sum_{k=1}^{n_B}E_B(i,j)_k\,, \tag{1}$$

where n_B is a number of dark frames, $E_B(i,j)$ - is a dark frame.

Equation for mean of dark frame $\overline{E_B^D}$ is given by:

$$\overline{E_B^D}=\frac{1}{m^2\cdot n_B}\sum_{k=1}^{n_B}\sum_{i,j}^{m}E_B(i,j)_k\,, \tag{2}$$

where m - is resolution of squared frame.

Background calculated in accordance to approximation method $\overline{E_B^A}$ (mean of unlighted pixels) is given by

$$\overline{E_B^A}=\frac{1}{\sum_{i,j}\Omega(i,j)}\sum_{i,j}^{m}E(i,j)\cdot\Omega(i,j), \tag{3}$$

where $E(i,j)$ is a frame with a beam, $\Omega(i,j)$ is a map of unlighted area of frame

$$\Omega(i,j)=\begin{cases}1 \text{ if } (i,j)\in\Omega\\ 0 \text{ if } (i,j)\notin\Omega\end{cases}.$$

Implementation of this method requires correctness verification using

$$\frac{\left|\overline{E_B^A}-\overline{E_B^{D\Omega}}\right|\sqrt{N}}{\sigma_B^{D\Omega}}<t\,,$$

where $t = 2...4$, $\overline{E_B^{D\Omega}}$ is a mean of dark frame in corresponding to unlighted area. The mean of dark frame is given by

$$\overline{E_B^{D\Omega}} = \frac{1}{\sum_{i,j}\Omega(i,j)}\sum_{i,j}^{m} E_B(i,j)\Omega(i,j),$$

$$N = \sum_{i,j}\Omega(i,j).$$

Method of mean plus dispersion can be implemented using

$$\overline{E_B^{MD}} = \overline{E_B^D} + t \cdot \sigma_B, \tag{4}$$

where σ_B is a dark frame dispersion given by

$$\sigma_\phi = \left[\frac{1}{m^2 \cdot n_B}\sum_{k=1}^{n_B}\sum_{i,j}^{m}\left(E_B(i,j)_k - \overline{E_B^D}\right)^2\right]^{1/2}.$$

Background calculation using statistical method is given by

$$\overline{E_B^H} = \frac{1}{\sum_{i,j} S(i,j)}\sum_{i,j}^{m} E(i,j)S(i,j), \tag{5}$$

where $S(i,j)$ is defined as

$$S(i,j) = \theta\left[\overline{E_B^D} - E(i,j)\right],$$

where $\theta(x)$ is a Heaviside function, which defines unlighted area.

Method of histogram is given by formula (6).

$$E_f = \frac{1}{E_T}\sum_{0}^{E_T} E_l \cdot I(E_l) + t \cdot \sigma_{E_l}, \tag{6}$$

where $I(E_l)$ is a histogram of intensity distribution of frame with a beam, E_l is a boundary of a column of histogram, E_T is boundary of clusters of noise and signal, σ_{E_l} is a dispersion of noise distribution on the histogram, which is defined as

$$\sigma_{E_l} = \frac{1}{E_T}\sum_{0}^{E_T} I(E_l) \cdot \left(\overline{E_l^T} - E_l\right)^2.$$

In many cases it is convenient to define baseline offset at specified area on the frame with the beam. Usually square areas located in corners of frame can be used. Calculation is performs similarly to equation (3):

$$\overline{E_B^{\Omega}} = \frac{1}{\sum_{i,j}\Omega(i,j)}\sum_{i,j}^{m} E(i,j) \cdot \Omega(i,j) \tag{7}$$

where $E(i,j)$ is a frame with a beam, $\Omega(i,j)$ is a specified area (square areas in corners of the frame),

$$\Omega(i,j) = \begin{cases} 1 \text{ if } (i,j) \in \Omega \\ 0 \text{ if } (i,j) \notin \Omega \end{cases}$$

4. Baseline correction using ranging experimental data with statistical criteria

For the precision baseline correction it is necessary to redefine integration area by cutting a region symmetrically with regard to the beam as shown on figure 2. The size of region required depends on beam profile and varies typically from 2 to 10 times the diameter of a beam. For Gaussian beam profile it is recommended to check that the area is at least 5 times greater than beam.

Algorithm of baseline correction includes 8 steps:

1) evaluation of the background
2) determination of a beam center and diameter
3) cutting the region which is symmetric with regard to the beam
4) verification of relative beam size and optional second cutting
5) baseline correction using mean of dark frame
6) noise measurement in corners of frame
7) statistical comparison of noise values
8) second baseline correction

In order to cut a part of the area symmetric relative to the beam it is necessary to evaluate and correct baseline using dark frame mean or corner area mean of background. Next, beam center and diameter should be evaluated. For calculation of the diameter second moment or percentage of power criteria can be used. Double of minimum distance from center of the beam to the edge of frame is required as a square side, which is cut from the frame as shown on fig 2.

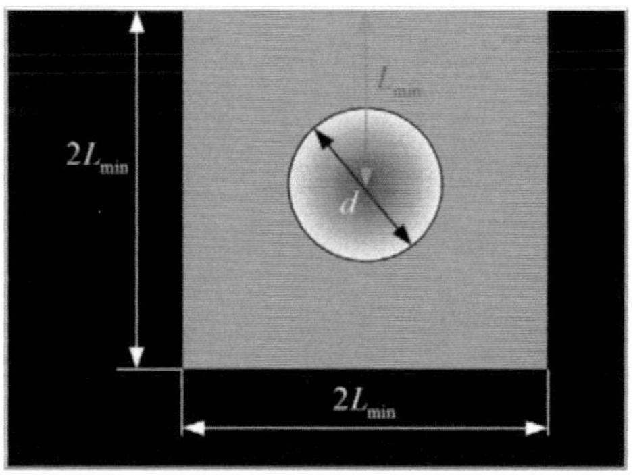

Figure 2. A scheme of cutting of the frame area part symmetric relative to beam

For achieving adequate results of beam parameters calculation it is important to check relative beam diameter (diameter to size of frame ratio).

In order to avoid errors caused by size of integration a following ratio should be verified:

$$d_{отн} = \frac{d_\sigma^O}{2L_{\min}} < \gamma , \quad (8)$$

where $\gamma = 0.1...0.5$ is the limit value of relative diameter, $d_\sigma^O = 2\sqrt{2}\cdot\left(\sigma_x^2 + \sigma_y^2\right)^{0.5}$ is beam diameter according to second moment criteria. If the inequality (8) is not satisfied it is necessary to reject the frame from consideration. In case of high frequency of appearance of reject decision it is better to consider the change in experiment condition.

After cutting the frame it is proposed to correct baseline offset by means of implementation of the method of mean of dark frames (Eq. 2) according to

$$E_{i,j}^O = E_{i,j} - \overline{E_B^D} , \quad (9)$$

where $E_{i,j}$ is experimental power density distribution, $\overline{E_B^D}$ is calculated baseline offset, $E_{i,j}^O$ is distribution with corrected baseline.

Result of baseline correction need to be verified by measuring of residual background in corner areas, as shown on figure 3. The first step includes background evaluation $\overline{E_{B\,res}}$ using equation (7). Then acquired results should be statistically compared with background of dark frames. Comparing is performs using equations (10) - (13):

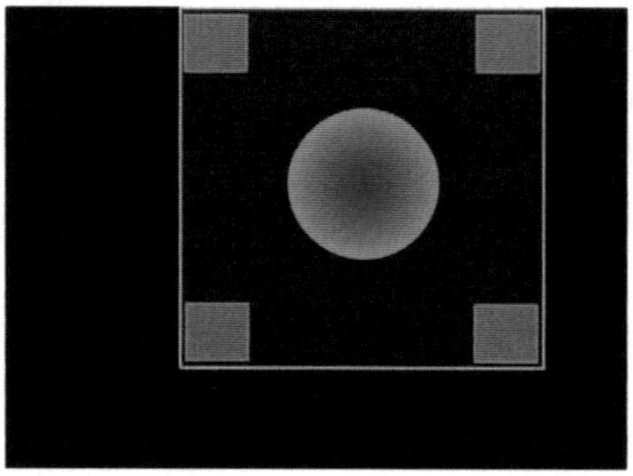

Figure 3. Areas of residual background checking

$$\left|\overline{E_{B\,res}}\right| < t\frac{\sigma_B^D}{\sqrt{N}}, \tag{10}$$

$$\chi_{p_1} < \frac{\sigma_{Bres}}{\sigma_B^D}\cdot\sqrt{(N-1)} < \chi_{p_2}, \tag{11}$$

where dark frame dispersion σ_B^D is defined as

$$\sigma_B^D = \left[m^{-2}\cdot n_B^{-1}\sum_{k=1}^{n_B}\sum_{i,j}^{m}\left(E_B(i,j)_k - \overline{E_B^D}\right)^2\right]^{0.5} \tag{12}$$

and residual dispersion σ_{Bres} is defined as

$$\sigma_{Bres} = \left[\frac{1}{4h^2}\sum_{\Omega}\left(E_{i,j}^O - \overline{E_{B\,res}}\right)^2\right]^{0.5}, \tag{13}$$

Here $N = 4h^2$ is the area of chosen corner region s Ω, h is side of corner square, t is quantile of Student's distribution, χ_{p_1} and χ_{p_2} are quantiles of χ^2-distribution, with $p = p_2 - p_1 = 1-\alpha = 0.9$, $p_2 = 1-\alpha/2 = 0.95$, $p_1 = \alpha/2 = 0.05$.

To obtain precise results of laser beam measurement it is important to satisfy the inequalities 10 and 11. If inequalities are not satisfied it is necessary to decide whether to reject experimental data or to make second baseline correction using $\overline{E_{B\,res}}$ as new level of baseline. After second baseline correction it is necessary to use equation (10)-(13) again.

In order to ensure the background values of all 4 corners, their equality and absence of occasional uneven illumination or highlights the residual background $\overline{E_{B\,res}^q}$ of each corner should be compared with each other using equation (14):

$$\left|\overline{E_{B\,res}^q} - \overline{E_{B\,res}^w}\right| < t\frac{\sigma_B^D}{h}, \tag{14}$$

where $q, w = 1...4$ is number of corner, $q \neq w$.

If inequality (14) is not satisfied it indicates presence of occasional uneven illumination or highlights or some other factor. It is necessary to provide better experimental conditions if inequality (14) is not satisfied frequently.

5. Example of using criteria for blink detection

Data with high dispersion were obtained during verification of beam forming system. Checking the data processing algorithm doesn't identify any errors. Then statistical comparison of values of residual background in the corners was performed. Due to performed procedure the highlights in frames with beam was detected as shown on figure 4a and 4b. Amplitude of highlight is 100 times less then maximum value of signal. Thus to show highlight the image transformation using (15) was performed.

$$f(E)=\begin{cases} E_{i,j} \; if \; E_{i,j}>1 \\ 255 \; if \; E_{i,j}<1 \end{cases} \tag{15}$$

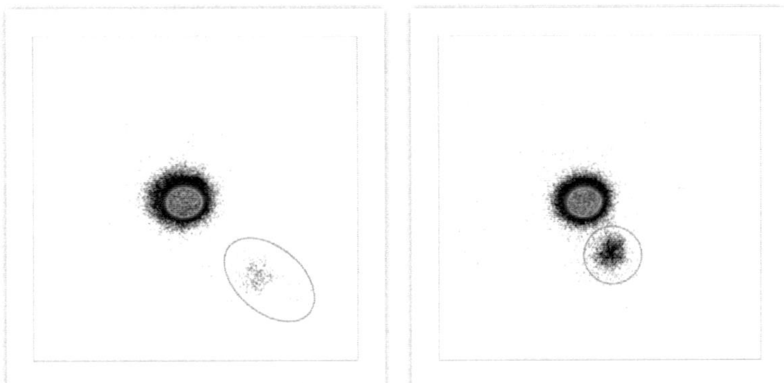

(a) (b)

Figure 4. An example of highlights detected on frames with a beam.

After detection of highlight the lens in measurement setup was changed to lens with a cover appropriate for used wavelength of laser beam.

Statistical comparison can detect such small effects, which may cause error of baseline correction, shift of calculated center of the beam and error of diameter measurement.

6. Conclusion

An algorithm for precise baseline correction with selection of experimental data proposed in this work allows to improve the repeatability and traceability of laser beam measurement. The algorithm is highly useful for verification of beam forming systems. Small dispersion of beam diameter measurement can provide appropriate conditions for adjustment of optical system especially in case of system based on off-axis mirror and extrafocal shifts of laser light source.

The algorithm was tested with a setup for wide aperture laser beam measurement. The setup was used for off-axis beam forming system verification and adjustment in laboratory condition.

Acknowledgements

The study was supported by the Russian Foundation for Basic Research, project 15-08-07484 "A comprehensive study of methods of forming extremely narrow beams of monochromatic radiation from space to Earth for the wireless power transmission and studying the atmosphere".

Journal of Physics: Conference Series **737** (2016) 012068 doi:10.1088/1742-6596/737/1/012068

References

[1] Matsak I S 2015 *Proc. SPIE* **9526** 95260K
[2] Matsak I S, Kudryavtsev E M and Tugaenko V Y 2015 *Comput. Res. Model.* **7** 113–124
[3] Kirkhamand K D and Roundy C B 2014 *Laser Beam Shaping: Theory and Techniques* ed F M Dickey (CRC Press) pp 463–524
[4] Roundy C B 1998 *Proc. SPIE* **3405** 1045–1055
[5] ISO 11145:2006 *Optics and photonics -- Lasers and laser-related equipment -- Vocabulary and symbols*
[6] ISO 11146-1:2005 *Lasers and laser-related equipment -- Test methods for laser beam widths, divergence angles and beam propagation ratios -- Part 1: Stigmatic and simple astigmatic beams*
[7] ISO/TR 11146-3:2004 *Lasers and laser-related equipment -- Test methods for laser beam widths, divergence angles and beam propagation ratios -- Part 3: Intrinsic and geometrical laser beam classification, propagation and details of test methods*
[8] ISO 13694:2000 *Optics and optical instruments -- Lasers and laser-related equipment -- Test methods for laser beam power (energy) density distribution*
[9] Matsak I S, Tugaenko V Y and Sergeev E S 2014 *Meas. Tech.* **57** 651–657
[10] Sheldakova Y V., Cherezova T Y and Kudryashov A V. 2002 *Proc. SPIE* **4900** 367–372
[11] Roundy C B 1997 *Proc. SPIE* **3110** 860
[12] Konnik M V., Manykin E A and Starikov S N 2010 *Quantum Electron* **40** 314–320
[13] EMVA Standard 1288. *Standard for Characterization of Image Sensors and Cameras, European Machine Vision Association* Release 3.0, 2010

Journal of Physics: Conference Series **737** (2016) 012069 doi:10.1088/1742-6596/737/1/012069

Multimodal combinational holographic and fluorescence fluctuation microscopy to obtain spatial super-resolution

V V Dudenkova[1], **Yu N Zakharov**[1,2]
[1]Radiophysical department, Lobachevsky State University of Nizhni Novgorod, 23 Gagarin avenue, Nizhny Novgorod, Russia
[2]Center for Advanced Biomedical Imaging and Photonics. HMS/BIDMC, Harvard University, 330 Brookline Ave, Boston, MA 02215, US

E-mail: orannge@mail.ru

Abstract. Ways of combination of holographic and super-resolution fluorescent techniques in the same optical scheme are described. The key parameters influencing achievement of maximum possible resolution are considered. The possibility to choose different fluorescence technic for different types of fluorophores without any optical scheme changes is presented. As a result in case of visualization of the samples, which transparent in optical band, three-dimensional super resolution is received that significantly expands possibilities of the noninvasive analysis of biological samples.

1.Introduction

Fluorescence microscopy allows three-dimensional investigation of living cell cultures and tissues. But the resolution in such sort of systems is limited by optical Abbe's diffraction limit. Other techniques, like atomic force microscopy, electron microscopy and scanning tunnelling microscopy achieve molecular-level resolution, but are not fit for the live cells imaging. Last decades, the optical diffraction limit has been overcome by using a several new methods, pioneered by stimulated emission depletion (STED) [1], structured illumination microscopy (SIM) [2,3], image interference microscopy[4] and ground-state depletion[5]. photo-activated localization microscopy (PALM)[6] and (fPALM) [7], stochastic optical reconstruction microscopy (STORM) [8], and variants thereof [9,10] also improved stochastic techniques using photoswitchable probes. PALM and STORM techniques reach nanometre resolution, but with slow registration speed (minutes to hours). STED has got video frame rate but the method is extremely exacting of dyes and labelling procedures [11]. Recently, superresolution microscopy at 11 Hz has been exposed by using SIM, achieving a 2-fold increased longitudinal resolution[12]. All this approaches are able achieving three-dimensional super-resolution, but with the serious technical modifications to the microscope optical scheme[13-15].

All of these methods require special sample preparation techniques making these methods more elaborate and complicated than traditional fluorescence wide-field microscopy. One of the another superresolution techniques is bleaching/blinking assisted localization microscopy (BaLM). It relies on the intrinsic bleaching and blinking behaviours characteristic of commonly used fluorescent probes [16]. The main idea of this approach is that all fluorophores blink: at some instant of time they are exited and begin to shine and early or late they are bleached. And they do it independently of one another. This fact is the base of BaLM image series acquisition and data analysis (gSHRImP [17], 3-B [18]). This technique is suitable for all commonly used synthetic fluorescent dyes and genetically expressed fluorescent proteins. But in the result of reconstruction researcher have only the most probable location of fluorophore with artificial brightness. Other methods based on fact, that the signal from object space is convolved with the point spread function (PSF) of the imaging apparatus. Digital image recording leads to a projection of the convolved signal onto a discrete set of points in space (pixels) and intensities. Thus the microscopy images are a vector of real data of finite length. Regularization techniques have been applied by Facciolo et al [19]. The power of the new incomplete Dykstra's algorithm (ICD) is demonstrated by using it as preprocessor for the super-resolution optical fluctuation imaging (SOFI) method by Dertinger et al [20]. The negative side of this approach is, that the number of frames for SOFI applications has to be large enough to resolve the statistical properties of the fluorescence signal [21,22] and crucial for the feasibility of applying it as preprocessor. On the other hand, SOFI is not exacting to level of signal to noise and is suitable for endogen fluorophores.

Though BaLM and SOFI need a sufficient amount of images to resolve the statistical behaviour of the temporal fluctuations of the observed signal, its strength is that any of them can be combined with a variety of microscopy techniques without any modifying the experimental setup , but using proper methodics and next data processing. That's why, holographic scheme is suitable for use of these methods too. Application of holographic techniques for creation of the three-dimensional structure of biological samples is free of these limitations so it is the perspective direction in study of biological objects with achievement of ultrahigh three-dimensional and necessary time resolution.

We show optical scheme, which consist of off-axis digital holography and super resolution fluorescent setup. Such combination allows to research live biological objects dynamic changes. As a result of application of the holographic recording method and further reconstructing of phase and amplitude information of object, there was an opportunity to carry out visualization of the dynamic changes in live biological samples. The developmental results on the example of neuronal plasticity research were obtained. Application of holographic approach allows receiving ultrahigh sensitivity to longitudinal changes in optical length. But, transversal resolution is the same as in diffraction limited microscopy. For resolve of this restriction it is possible to combine holographic method with one of approaches to improve transversal resolution, which meets certain requirements. Depend of biological object we able choose data analysis method to correct reconstruction of sequence of fluorescent images.

2.Methods to be combined

2.1 Holography

Holography makes it possible to reconstruct variations of the light intensity and the phase increment during the passage through the transparent or semi-transparent object. Holographic reconstruction allows obtaining information on the three-dimensional structure of an object using only one hologram in the transverse plane. Holography is divided into digital and analog ones by the recording media and reconstruction approach. The main drawback of digital holography as compared with the analog one is a lower information capacity [23,24]. The main advantage of digital holography is high rate of registration and numerical processing of the results[23,24]. So, the digital recording type was chosen for combined scheme. In this case it is possible to record holographic video (a set of sequential holograms) with minimum possible delays between frames and then reconstruct them for the phase dynamics visualization. The recording rate is determined by the using digital matrix. The available time resolution is inversely - proportional to a chosen active area of these digital detectors.

2.2 Bleaching/blinking assisted localization microscopy

BaLM fluorescence analysis consists in excitation of fluorescence and registration and analysis of fluorophore blinking. Then consecutive frames are subtracted the each other. As a result of this operation, signals only from blinking or bleaching molecules stay on the difference images . So, there are images of separate molecules in the form of their point spread function (PSF). All difference elements that are less, than the estimated PSF size, aren't considered. It makes possible to determine the location of individual fluorophore molecules with the diffraction limit. Processing consists in dilation of the difference image in order to eliminate intensity dips [25]. For the pixels which do not undergo dilation, i.e., retain their initial intensity, the mean intensity within a circle of several pixels should be calculated. The obtained mean value is compared with a certain threshold value determined by the noise of the initial data. If the mean value exceeds a specified threshold, then this diffraction spot is approximated by a two-dimensional Gaussian distribution using the Levenberg–Marquardt algorithm [26]. The coordinates of the centre of this function are assumed to be the desired position of the molecule. Using the Gaussian filter, the coordinates of the found centre are converted into a new image with resolution, that is already beyond the diffraction limit[27]. The filter width corresponding to the resolution of the new image is mainly determined by noise. This algorithm appears to be resistant to additive noise with an amplitude of up to 40% of the maximum signal amplitude. Under these conditions, pixelization (efficient resolution) of the new image is performed with density exceeding the initial one by 10 times.

2.3 Super-resolution optical fluctuation imaging

A different approach is SOFI (super-resolution optical fluctuation imaging) [20,28], which also improve resolution due to the temporal behaviour of the imaged object signal. SOFI does not require controlled or synchronized photoactivation, and relies on the independent stochastic fluctuations of the emitters. It works without the need for sophisticated electronics or special acquisition setup. The main aspect of SOFI is that the fluorescent label has to exhibit at least two different emission states. Different emitters have to switch between states over and over again and independently from each other in a stochastic manner. A SOFI image pixel value (of the order n) is received from the n-th order cumulant of the original pixel time series. Using standard imaging applications the signal in a pixel is represented by the superposition of the fluorescence coming from different, closely emitters. The n-th order cumulant is quantity linked to the n-th-order correlation function. It's filtering the signal on the base of its fluctuations and only highly correlated fluctuations are left over.

3. Combination of the holographic and superresolution fluorescent microscopy in the same optical setup

We show optical scheme consist of off-axis digital holography and super resolution fluorescent setup on Figure 1. This combination allows to research live biological objects dynamic changes. Optical scheme consist of several basic elements: 1- a laser, 2 and 3- a collimator lenses, 4, 4'- beam splitters, 5 and 5'- mirrors, 6- Petri dish with cells, 7 – objective, 8- beam splitter cube, 9 – filter, 10 and 10'- digital array matrixes for holographic and fluorescent channels.

Journal of Physics: Conference Series **737** (2016) 012069 doi:10.1088/1742-6596/737/1/012069

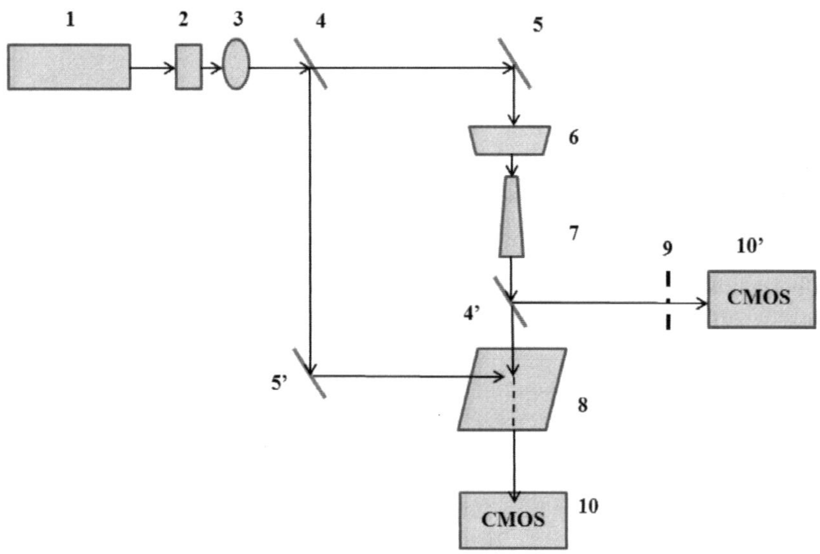

Figure 1. Optical setup for combination of the holographic and superresolution fluorescent microscopy.

In this case, reference beam, which is coherent with the object beam, is introduced at the optimum convergence angle from the outside[24]. For combination we have to upgrade standard off-axis holography scheme to divide object beam on two parts by beam splitter 4'. First of resulting wave is an object beam and interfering with reference beam on one digital detector 10. It should be mentioned, that the intensity of object and reference beam should not differ significantly. "VEC-545" digital cameras based on the complementary metal–oxide–semiconductor structure (CMOS camera) with a matrix diagonal of 0.4 inch, 2592×1944 pixels, and a 2.2μm pixel dimension were used. Second beam after filtering excitation wavelength is registering by another digital detector 10' - Flea-3 8,8 megapixel CMOS camera with 1.55 um pixel size. The coherent laser beam was used both for recording of holograms and for fluorescence excitation. He–Ne laser 1 with a wavelength of 633 nm was used, which could be replaced by a laser operated at any other wavelength, depending on the tasks assigned. As for a holography, illumination power influences only to exposure time. But for the super resolution fluorescent microscopy is extremely important to select the source of light type and parameters. And it's necessary to select optimum excitation wavelength, depending on an absorption spectrum characteristic of a specific fluorophore.

4 Features of using

4.1 Excitation parameters in case of Bleaching/blinking assisted localization microscopy

The most important for BaLm method to estimate the excitation light power correctly[29]. Record of a frame series with blinking is necessary for the localization of separate fluorophore molecules accurately, i.e. it is necessary to detect acts of separate molecules light emission on all field of view in order to define their centres. Therefore parameters of a fluorophore, exciting light, and also registering digital matrix shall be factored into in calculation of optimum excitation power. First of all it is necessary to select correct excitation wavelength λ. It has to be guided by excitation spectrum of visualizing fluorophore. Fluorescence from the molecules which are in "the diffraction volume" V_d, depending of λ and numerical aperture of a lens *NA*, will get to a diffraction spot from which all information registers. Knowing concentration *n* of a fluorophore in a sample, it is possible to calculate quantity of molecules of *N* in volume of excitation. From all photons of excitation which propagate through the volume only part of them will be absorbed by molecules of a fluorophore. It is proportional to the fluorophore absorption section σ. Also it is necessary, that only one molecule of a fluorophore would be excited in a diffraction volume. Power of exciting light will be equal to sum energy of photons radiated during time *t* between two subsequent frames of record with *f* frequency, divided on this time. It is able to calculate the number of photons, flying from a source through the focal diffraction volume during time *t*, respectively. Also we should keep in mind, that another fluorophores shouldn't be excited around a distance equal to PSF diameter near the PSF of blinked one. Then the optimal excitation power is

$$P = \frac{3fhcSNA^4}{16\theta_F \sigma n \lambda^4 \pi^2}, \tag{1}$$

where θ_F is the quantum yield of a fluorophore. So, optimal excitation wavelength power depends of several different parameters. Concentration of *n* is controlled by dye loading. Parameters θ_F, λ and σ depend on the selected fluorophore. Value of *NA* depends on the selected lens. And value of a beam area of *S* is defined by the chosen collimator. Digital camera influence on frame rate frequency *f*. Thus, for receiving cross ultrahigh resolution, possible to calculate the necessary laser line power for correct operation of the BaLM fluorescent analysis. Using a formula (1) it's able to select optimum excitation light power in each experiment.

4.2 Reconstruction

For holography, the image of object is reconstructed numerically by double Fourier transformation algorithm with filtering in the frequency area [24]. As a result of application of the holographic recording method and further reconstructing of phase and amplitude information of object, there is an opportunity to carry out visualization of the dynamic changes in live biological samples. For fluorescent channel is able to use BaLM or SOFI method. It depends of fluorophores (endogenous or exogenous), necessary speed of reconstruction, period of object changes and type of reconstructed data.

5. Discussion

In contrast to STORM (PALM) method SOFI approach allows to develop overlapping fluorophores and not requires single photon sensitivity. So, unlike above we used not scientific CMOS camera, but choose with the smallest pixel size one. The most important aspect of BaLM is that the fluorophore molecule images can overlap due to diffraction blur, but lighting up and extinguish events during the inter-frame time gap should be located at a distance exceeding the diffraction limit. The number of blinking molecules is determined by the excitation intensity, absorption cross section, the quantum efficiency and the amount of marker per volume unit. SOFI is devoid of this disadvantages, but in case of fluorophores with difference brightness able to lost the faintest signal during increasing cumulant order. So, we are able to choose different approach for different situation without any changes in optical setup, excitation or recording parameters. Therefore, by combining these advantages with the holographic microscopy method, one can achieve very high sensitivity to optical path difference along Z-axis with superresolution in XY plane due to fluorescence fluctuation analysis algorithms. Moreover since different approaches are used in the combination they not only improve resolution but reveal different sample properties making possible to get complementary information, e.g. structural and functional imaging simultaneously with high sensitivity and space-temporal resolution.

Acknowledgments
This work has been financially supported by Russian Foundation for Basic Research (grant No. 15-42-02658).

References

1. Hell S W and Wichmann J. 1994 *Optics Letters* **19** 780-782
2. Gustafsson M G L 2000 *Journal of Microscopy* **198** 82-87
3. Heintzmann R, Jovin T M and Cremer C 2002 *Journal of the Optical Society of America A* **19** 1599-1609
4. Gustafsson M G L, Agard D A and Sedat J W 1999 *Journal of Microscopy* **195** 10-16
5. Hell S W and Kroug M 1995 *Applied Physics B* **60** 495-497
6. Betzig E et al 2006 *Science* **313** 1642-1645
7. Hess S T, Girirajan T P K and Mason M D 2006 *Biophysical Journal* **91** 4258-4272
8. Rust M J, Bates M and Zhuang X 2006 *Nature Methods* **3** 793-796
9. Folling J et al 2008 *Nature Methods* **5** 943-945
10. Heilemann M. et al 2008 *Angewandte Chemie International Edition* **47** 6172-6176.
11. Westphal V. et al 2008 *Science* **320** 246-249
12. Kner P, Chhun B B, Griffis E R, Winoto L and Gustafsson M G L 2009 *Nature Methods* **6** 339-342
13. Huang B, Jones S A, Brandenburg B and Zhuang X 2008 *Nature Methods* **5** 1047-1052
14. Juette M F et al 2008 *Nature Methods* **5** 527-529
15. Shtengel G et a. 2009 *Proceedings of the National Academy of Sciences* **106** 3125-3130
16. Burnette D T, Sengupta P, Dai Y, Lippincott-Schwartz J and Kachar B 2011 *Proceedings of the National Academy of Sciences* **108** 21081-21086
17. Simonson P D, Rothenberg E and Selvin P R 2011 *Nano Letters* **11** 5090-5096
18. Cox S et al 2012 *Nature Methods* **9** 195-200
19. Facciolo G, Almansa A, Aujol J and Caselles V 2009 *Multiscale Modeling & Simulation* **7** 1574-1608
20. Dertinger T, Colyer R, Iyer G, Weiss S. and Enderlein J 2009 *Proceedings of the National Academy of Sciences* **106** 22287-22292
21. Frick K, Marnitz P and Munk A 2012 *Electronic Journal of Statistics* **6** 231-268
22. Frick K, Marnitz P and Munk A 2013 *Journal of Mathematical Imaging and Vision* **46** 370-387
23. Dudenkova V V, Murav'eva M S, Rybnikov A I and Zakharov Y N 2015 *Radiophysics and Quantum Electronics* **57** 551-558
24. Rybnikov A I, Dudenkova V V, Murav'eva M S and Zakharov Y N 2013 *Journal of Optical Technology* **80** 457-462
25. Serra J. 1983 *Image Analysis and Mathematical Morphology* (Academic Press, Inc.)
26. Mao H and Zhao D 2009 *Optics Express* **17** 4540-4552
27. Samoilin E A 2007 *Journal of Communications Technology and Electronics* **52** 773-778
28. Dertinger T, Colyer R, Vogel R, Enderlein J and Weiss S 2010 *Optics Express* **18** 18875-18885
29. Dudenkova V. and Zakharov Y. 2015 *Proceedings of SPIE* **9529** 95290V-95290V-5

The influence of diffraction gratings relief noise on the intensity distribution in diffraction orders

V V Kolyuchkin, S B Odinokov

Bauman Moscow State Technical University, 2nd Baumanskaya str. 5, 105005 Moscow, Russia

E-mail: vasilykolyuchkin@gmail.com

Abstract. The method of security hologram quality inspection, in which the relief phase parameters are determined by registering results of the intensity distribution in diffraction orders, is known. The profile of relief as a harmonious distribution is represented. Deviation of the real relief profile from the ideal profile, another words phase relief noise, influence on the accuracy of this method. In the paper, on the assumption of homogeneity of the phase relief noise, the mathematical expressions for evaluating the influence of the phase relief noise on the intensity distribution in the diffraction orders are represented. Parameters of the correlation functions approximation describing the phase relief noise are determined. The dependence of the intensity values from the standard deviation of the phase relief noise is represented.

1. Introduction

The quality inspection method of security hologram based on indirect measurements of diffraction grating parameters is described in the paper [1]. The principle of the method is to estimate the parameters of the diffraction grating by measuring the intensity distribution values in the diffraction orders, the orientation of the plane and the angular size of the diffraction orders. In the paper [1] we accepted the profile has a perfect sinusoidal shape. But the real profile is always different from the ideal, and these differences are random. Therefore, there is the important task of estimating the influence of diffraction grating profile random distortion on the intensity distribution in the diffraction orders. The aim of the paper is to develop the mathematical expressions that describe the intensity distribution in diffraction orders with considering the phase relief noise that characterizes deviation of the real relief profile from the ideal profile. This is necessary to analyze the phase relief noise influence.

2. The intensity distribution in diffraction orders with considering the phase relief noise

We accepted, the ideal relief profile of diffraction grating is describes by the following expression:

$$Z(\mathrm{x}) = \mathrm{A}\left[1-\cos\left(\frac{2\pi x}{T}\right)\right] = 2A\sin^2\left(\frac{\pi x}{T}\right), \quad (1)$$

where A – amplitude of the sinusoidal profile, T – period of the sinusoidal profile.

As shown in [1, 2], the dependence of the phase shift from the coordinate x of wavefront is determined as

$$\Delta_1(x) = r_0(x)\cdot\left[\sin(\alpha+\varphi)+\sin(\beta+\varphi)\right], \tag{2}$$

where α and β – the angle of the incidence plane and the diffraction angle respectively,

$$r_0(x) = \sqrt{x^2 + \left[2A\sin^2\left(\frac{\pi x}{T}\right)\right]^2}. \tag{3}$$

In the particular case when $\alpha = 0$

$$\operatorname{tg}\varphi = \frac{2A\sin^2\left(\frac{\pi x}{T}\right)}{x}. \tag{4}$$

We assumed that the ideal sinusoidal profile of the diffraction grating is distorted by the noise $\xi(x)$ which is a uniform random field, described by the Gauss's law $w_\xi(\xi)$ with a zero expectation value $\langle\xi(x)\rangle = 0$ and a variance $\sigma_\xi^2 = \langle\xi^2(x)\rangle$.

To calculate the wave amplitude in diffraction orders we need to find an expression for the phase shift $\Delta\xi_1(x)$ with considering the random field $\xi(x)$ of phase relief noise. The phase shift from analogy with (2) can be determined by the expression

$$\Delta\xi_1(x) = r'_0(x)\left[\sin(\varphi')+\sin(\beta+\varphi')\right], \tag{5}$$

where

$$r'_0(x) = \sqrt{x^2 + \left[2A\sin^2\left(\frac{\pi x}{T}\right)+\xi(x)\right]^2}; \tag{6}$$

$$\operatorname{tg}\varphi' = \frac{2A\sin^2\left(\frac{\pi x}{T}\right)+\xi(x)}{x}. \tag{7}$$

We assumed the standard deviation $\sigma_\xi << A$. Then the expression (5) can be expanded in a Maclaurin series by ξ:

$$\Delta\xi_1(x,\xi) = \Delta\xi_1(x,0) + \left.\frac{\partial\Delta_1(x,\xi)}{\partial\xi}\right|_{\xi=0}\cdot\xi + \left.\frac{\partial^2\Delta_1(x,\xi)}{\partial\xi^2}\right|_{\xi=0}\cdot\xi^2 + \ldots \tag{8}$$

Using only two first terms of the series, we get an approximate expression for calculation of the phase shift spatial distribution when the noise of sinusoidal relief profile is appeared.

$$\Delta\xi_1(x) = \Delta_1(x) + \Delta_{1\xi}(x), \tag{9}$$

where

$$\Delta_1(x) = \sqrt{x^2 + \left[2A\sin^2\left(\frac{\pi x}{T}\right)\right]^2}\cdot\left[\sin(\varphi)+\sin(\beta+\varphi)\right], \tag{10}$$

$$\Delta_{1\xi}(x) = \left.\frac{\partial\Delta_1(x,\xi)}{\partial\xi}\right|_{\xi=0}\cdot\xi(x), \tag{11}$$

Next, we get an expression for the intensity distribution in diffraction orders with the influence of aforementioned phase relief noise $\xi(x)$ described by the Gauss's law with a zero expectation value and a variance σ_ξ^2.

As observed in [2] the complex amplitude value for diffraction angle β and number of periods N in diffraction grating is determined as

$$U = U_0 \cdot U_1 \cdot U_2 \tag{12}$$

where U_0 – amplitude of plane wave;

$$U_2 = \sum_{j=0}^{N} e^{ikj\Delta_0} = \frac{1 - e^{ikN\Delta_0}}{1 - e^{ik\Delta_0}}; \tag{13}$$

$$U_1 = \int_0^T e^{ik\Delta_{1\xi}(x)} dx; \tag{14}$$

$$\Delta_0 = T \cdot (\sin\alpha + \sin\beta). \tag{15}$$

Respectively, the intensity value is determined as

$$I = I_0 \cdot I_1 \cdot I_2, \tag{16}$$

where $I_0 = |U_0|^2$;

$$I_2 = |U_2|^2 = \frac{\sin^2\left(\pi \frac{N\Delta_0}{\lambda}\right)}{\sin^2\left(\frac{\pi\Delta_0}{\lambda}\right)}. \tag{17}$$

Since the component U_1 of the wave amplitude is a random variable, then the intensity is determined by averaging over the ensemble of realizations.

$$I_1 = \langle U_1 \cdot U_1^* \rangle = \left\langle \int_0^T \exp\left\{ik\left[\Delta_1(x_1) + \Delta_{1\xi}(x_1)\right]\right\} dx_1 \cdot \int_0^T \exp\left\{-ik\left[\Delta_1(x_2) + \Delta_{1\xi}(x_2)\right]\right\} dx_2 \right\rangle. \tag{18}$$

Using the property of the linearity for expectation operation, equation (18) can be written as

$$I_1 = \langle U_1 \cdot U_1^* \rangle = \int_0^T \int_0^T \exp\left\{ik\left[\Delta_1(x_1) - \Delta_1(x_2)\right]\right\} \left\langle \exp\left\{ik\left[\Delta_{1\xi}(x_1) - \Delta_{1\xi}(x_2)\right]\right\} \right\rangle dx_1 dx_2 \tag{19}$$

After calculating mathematical expectation, part of the expression (19), we obtain the expression for the intensity of a particular diffraction order.

$$I_1 = \int_0^T \int_0^T \exp\left\{ik\left[\Delta_1(x_1) - \Delta_1(x_2)\right]\right\} \exp\left\{-\frac{2\pi^2\sigma_\Delta^2(x_1, x_2)}{\lambda^2}\right\} dx_1 dx_2. \tag{20}$$

The analysis of the expression (20) shows that the random relief noise of the grating reduces diffraction efficiency. To quantify the influence of the phase relief noise we should know the variance σ_ξ^2 and the normalized correlation function $\rho_\xi(x_1 - x_2)$ of the noise random field. These characteristics are determined using processed measuring results of the profile realizations.

3. Quantitative estimation of the phase relief noise

To quantify the influence of the random relief noise using (20) intensity for the first-order diffraction was calculated with the following initial data:

grating period $T = 1$ *μm*;

wavelength длина волны $\lambda = 0{,}55$ *μm*;

incident angle $\alpha = 0$;

first-order diffraction angle $\beta \approx 33°$;

amplitude of the phase relief $A = 0{,}1$ *μm*.

Root-mean-square (RMS) deviation σ_ξ of the phase relief noise is varied during the research. Normalized correlation function is approximated by the following function:

$$\rho(x_1 - x_2) = \exp\left(-\frac{|x_1 - x_2|}{k_P}\right)\cos\left[2\pi\frac{(x_1 - x_2)}{\gamma T}\right], \qquad (21)$$

where k_P and γ — approximation parameters.

Figure 1 shows the section of correlation function for the case $k_P = 0{,}2$ and $\gamma = 0{,}5$.

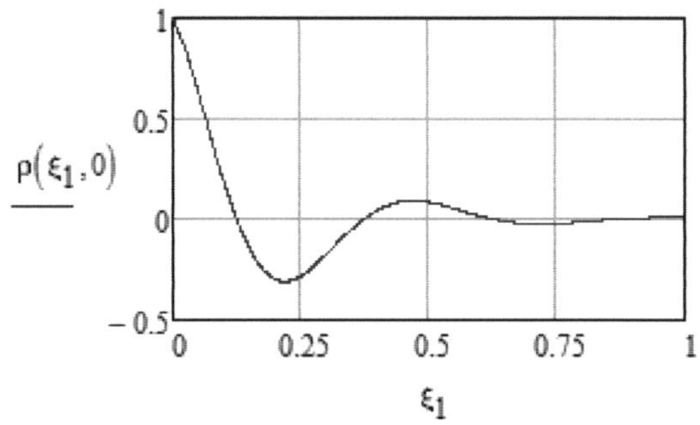

Figure 1. Normalized correlation function of phase relief noise

The phase relief noise influence is estimated by intensity percentage change δ_I in the first diffraction order according to following expression:

$$\delta_I = \frac{|I_{\xi 1} - I_1|}{I_1} \cdot 100\%. \qquad (22)$$

Figure 2 presents functional relation intensity percentage change versus RMS deviation of the phase relief noise in the setting of:

- $k_P = 0{,}2$ и $\gamma = 0{,}5$ (solid line, $\delta 1_I$);
- $k_P = 0{,}1$, $\gamma = 0{,}5$ (dotted line, $\delta 2_I$);
- $k_P = 0{,}2$, $\gamma = 0{,}25$ (cross-hatching line, $\delta 3_I$).

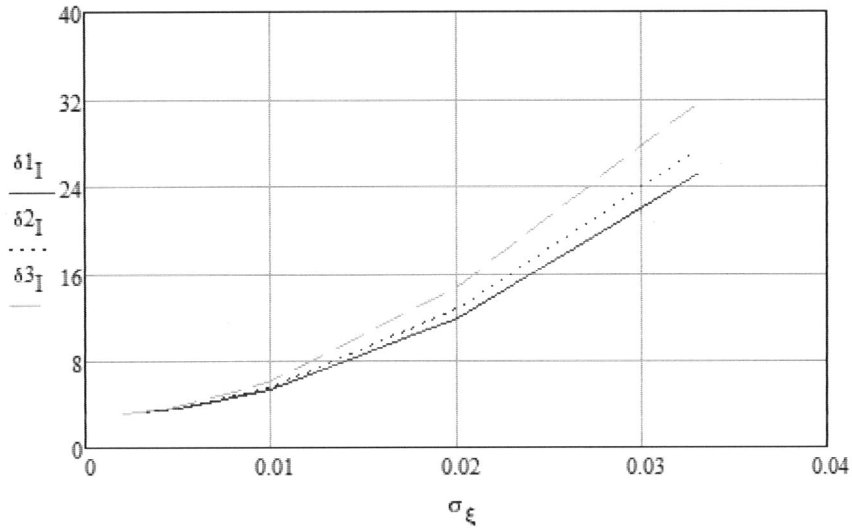

Figure 2. Functional relation intensity percentage change versus RMS deviation of the phase relief noise

Obtained results led us to conclusion that RMS deviation of the phase relief noise affects to intensity value as well as correlation function signature.

It is worth mentioning that during research the method error of (20) is quantified for first-order intensity calculations in the presence of random relief noise. For RMS deviation approaching to zero ($\sigma_{\xi} \to 0$) the percentage error of (20) doesn't exceed 3%.

4. Conclusions

The phase relief grating noise of security holograms has a significant impact to intensity distribution in diffraction orders. Therefore it is necessary to consider the phase relief noise influence inspecting quality of security holograms.

Acknowledgments

The research was carried out at BMSTU. This work was supported by a grant of the main part of the state order "Organization of scientific research" and project # 3.1426.2014K from the Ministry of Education and Science of the Russian Federation.

References

[1] Kolyuchkin V V, Zlokazov E Y, Odinokov S B, Talalaev V Y, Tsyganov I K 2015 *Computer Optics* **39 (4)** 515-520

[2] Khorokhorov A M, Klishyuk A M, Tsiganov I K, Odinokov S B, Shirankov A F 2005 *Herald of the Bauman Moscow State Technical University. Insrtument Engineering* **4 (61)**

Methods of compression of digital holograms, based on 1-level wavelet transform

E A Kurbatova, P A Cheremkhin, N N Evtikhiev

National Research Nuclear University MEPhI (Moscow Engineering Physics Institute), Kashirskoe shosse 31, 115409 Moscow, Russia

E-mail: CheremhinPavel@mail.ru

Abstract. To reduce the size of memory required for storing information about 3D-scenes and to decrease the rate of hologram transmission, digital hologram compression can be used. Compression of digital holograms by wavelet transforms is among most powerful methods. In the paper the most popular wavelet transforms are considered and applied to the digital hologram compression. Obtained values of reconstruction quality and hologram's diffraction efficiencies are compared.

1. Introduction

The photography allows to register amplitude (intensity) of the waves. In result information about objects can be retrieved in 2D-form [1-2].In the same time the holography allows to obtain information about object in 3D-form [3-7]. As a result now digital and computer holography are popular techniques in different fields of scientific research: an interferometry [8-9], microscopy [10-11], optical and optoelectronic information processing [3, 12], reconstruction static and dynamic 2D- and 3D-scenes [13-16], and etc.

Characteristics of digital photo- and videocameras are improving. Currently size of images may be more than ten megabytes [17-18]. For transfer of holographic video with a standard frequency of movies (24 Hz) it is necessary to have the channel with capacity of 2-4 Gbit/sec. This size is significantly more, than in standard communication channels. Size of ten minutes of holographic video will be more than 1 terabyte. Thus large archival memory sizes are required to store holographic video.

To compress digital and computer-synthesized holograms, various types of compression methods [19-24], including methods based on different transforms (cosine, wavelet, etc.) are used. In case of compression of standard photoimages, this group of methods allows to reach high compression ratios with the minimum losses of quality [25-27]. Compression on the basis of wavelet transform consists in decomposition of the initial photoimage on arrays of the approximating coefficients and the related coefficients of specification. These coefficients allow to consider dependences between values of a signal (intensity) in the next pixels.

In this paper the most popular methods of wavelet compression of digital holograms are considered and applied.

2. Application of wavelet transforms for digital holograms compression

The described approach of images compression is focused on digital processing with gradient of brightness since it is possible to describe gradient of brightness by small number of coefficients. Thus, in case of compression of standard photos, high compression ratios (up to 200-300 times) with the minimum losses of quality or even without quality loss are achieved.

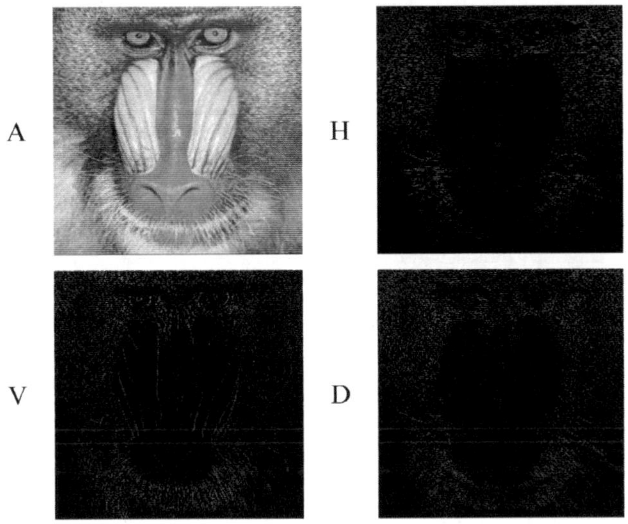

Figure 1. Wavelet transform of standard grayscale image: approximating (A), horizontal (H), vertical (V) and diagonal (D) coefficients of decomposition

However holograms are actually interference patterns with a large number of small details with different brightness. Such images are described by a large number of the approximating coefficients. Thus, the methods based on wavelet transforms provide smaller compression ratios for holograms, than that one for standard photos. However, despite focus of a method on compression of images with gradient of brightness, it is possible to choose the wavelet transform considering features of a certain type of images as in case of digital holograms.

Wavelet transform coefficients of decomposition of standard grayscale image are shown in Fig. 1 and coefficients of decomposition of digital hologram are shown in Fig. 2.

There are different types of wavelet transforms [25-27]:

- orthogonal,
- semi-orthogonal,
- biorthogonal,
- symmetric,
- asymmetric,
- and etc.

Also wavelet transforms differ with smoothness degree, range of definition, computer resource intensity. In practice it is necessary to choose the wavelet transforms having algorithms of fast calculation. Usually the best practical results are achieved with using orthogonal symmetric or asymmetric wavelets. These properties only Haar's wavelet possess, but Gabor, Daubechies, Meyer's wavelets, biorthogonal and others are used also.

Journal of Physics: Conference Series **737** (2016) 012071 doi:10.1088/1742-6596/737/1/012071

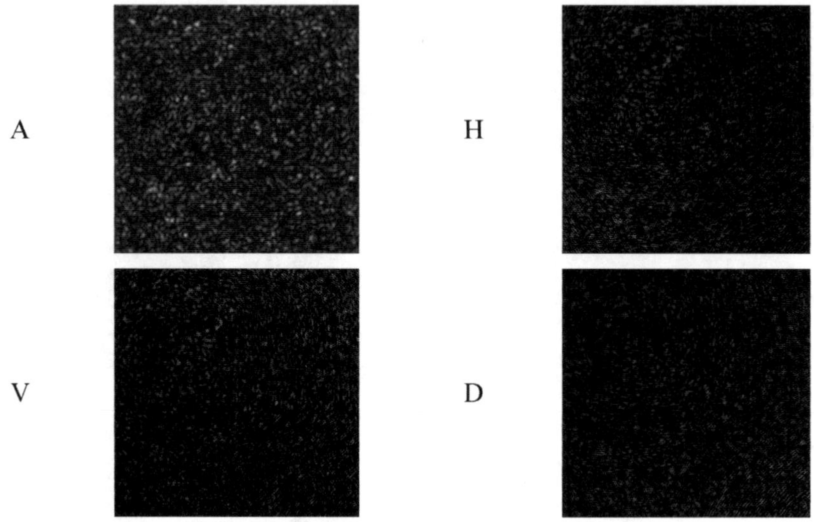

Figure 2. Wavelet transform of digital hologram: approximating (A), horizontal (H), vertical (V) and diagonal (D) coefficients of decomposition

3. Parameters used in numerical experiments on digital holograms compression

Seven methods of compression of holograms based on various wavelet transforms have been realized in program environment. They are:

- Haar,
- Daubechies,
- Symlet,
- Coiflet,
- Biorthogonal,
- Reverse biorthogonal,
- Meyer.

For numerical experiments parameters which can be used in case of the experimental recording of off-axis digital Fresnel holograms were selected:

- wavelength of the laser illumination is 633 nm,
- pixel size is 8 μm × 8 μm,
- size of the hologram is 1024 × 1024pixels,
- distance from object to the digital camera's photosensor (hologram) plane is 0,94 m.

Standard grayscale test images with size of 128×128 pixels was used as the object image.

As quantitative criterion of reconstructed images quality, the normalized standard deviation (NSTD) [28] between reconstructed and initial images was used:

$$NSTD = \sqrt{1 - \frac{\left(\sum_{\xi,\eta=1}^{N,N} E[\xi,\eta]F[\xi,\eta]\right)^2}{\left(\sum_{\xi,\eta=1}^{N,N} E^2[\xi,\eta]\right)\left(\sum_{i,j=1}^{N,N} F^2[\xi,\eta]\right)}} \tag{1}$$

where $E[\xi,\eta]$ and $F[\xi,\eta]$ are matrices with intensities values of the initial and the reconstructed images respectively,

$[\xi,\eta]$ are discrete coordinates,

$N\times N$ is quantity of pixels of images.

NSTD value shows degree of visual resemblance to original image, including similarity in reproduction of halftones.

Diffraction efficiency (DE) of modeled digital holograms was calculated also. DE can be found as product of two values:

- DE_H – relative part of power of illumination that was passed through the hologram,
- DE_O – relative part of power of informative diffraction order under reconstruction.

DE was found as follows [14]:

$$DE = DE_H \cdot DE_O = \left(\frac{\langle H[u,v] \rangle}{max\{H[u,v]\}} \right)^2 \cdot \left(\frac{\sum_{\xi,\eta=1}^{N,N} F_O[\xi,\eta]}{\sum_{\xi,\eta=1}^{N,N} F[\xi,\eta]} \right) \quad (2)$$

where

- $H[u,v]$ is matrix with intensities values of the hologram pixels,
- $[u,v]$ are discrete coordinates in the hologram plane,
- $max\{*\}$ – maximum value of intensity of pixels of the hologram,
- $<*>$– average value ofthe intensities of the object beam in hologram plane,
- $N \times N$ is quantity of pixels of reconstructed image,
- $F_O[\xi,\eta]$ is matrix with intensities values of the reconstructed object image (informative diffraction order)only,
- $F[\xi,\eta]$ is matrix with intensities values of the full reconstructed image.

4. Comparison of quality of reconstructed image from compressed digital holograms

Quality of reconstructed images from compressed digital holograms by seven wavelet transform was estimated. Obtained dependences of the NSTD values on quantity of gradations of wavelet-coefficients are shown in Fig. 3. Digital holograms were compressed by 1-level wavelet transform.

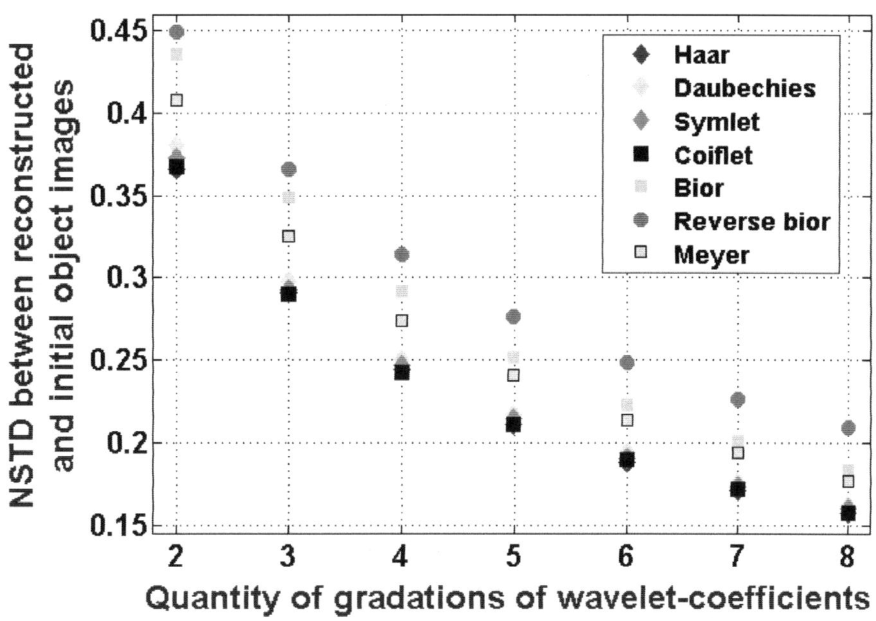

Figure 3. Obtained dependences of quality reconstructed from the compressed holograms images on quantity of gradations of wavelet-coefficients.

Journal of Physics: Conference Series **737** (2016) 012071 doi:10.1088/1742-6596/737/1/012071

As can be seen, the best quality of reconstruction is achieved with using coiflet, symlet and Haar's wavelet transforms. The worst quality of restoration is obtained with the reverse biorthogonal wavelet and biorthogonal wavelet. Reconstruction of images from compressed with different wavelets holograms in case of 8 gradations of wavelet-coefficients are shown in Fig. 4.

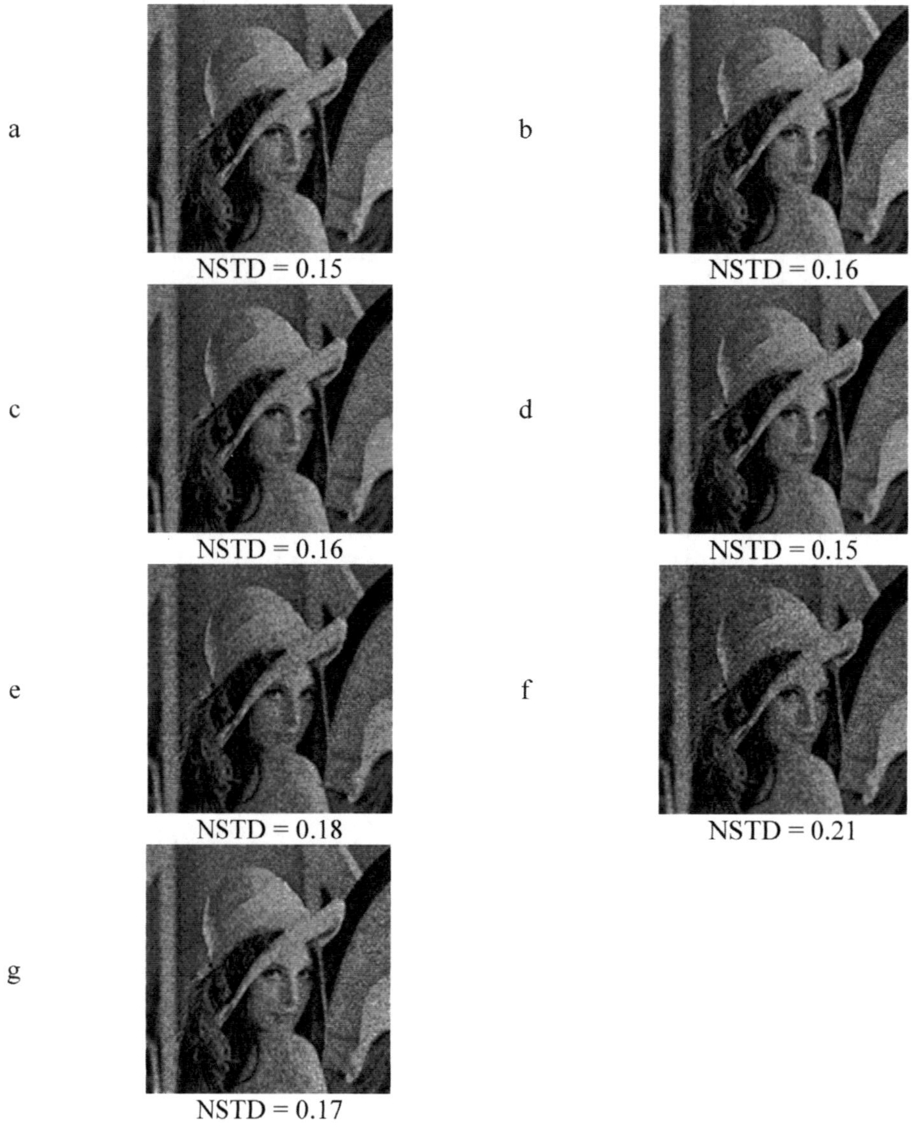

Figure 4. Reconstructed images of the halftone object for different wavelet transform: Haar (a), Daubechies (b), symlet (c), coiflet (d), biorthogonal (e), reverse biorthogonal (f),Meyer (g) in case of 8 gradations of wavelet-coefficients

5. Comparison of diffraction efficiency of compressed digital holograms

In addition to comparison reconstruction quality, diffraction efficiencies were estimated also. Obtained dependences of diffraction efficiency of the compressed holograms on quantity of gradation of wavelet-coefficients are shown in Fig. 5. Digital holograms were compressed by 1-level wavelet transform.

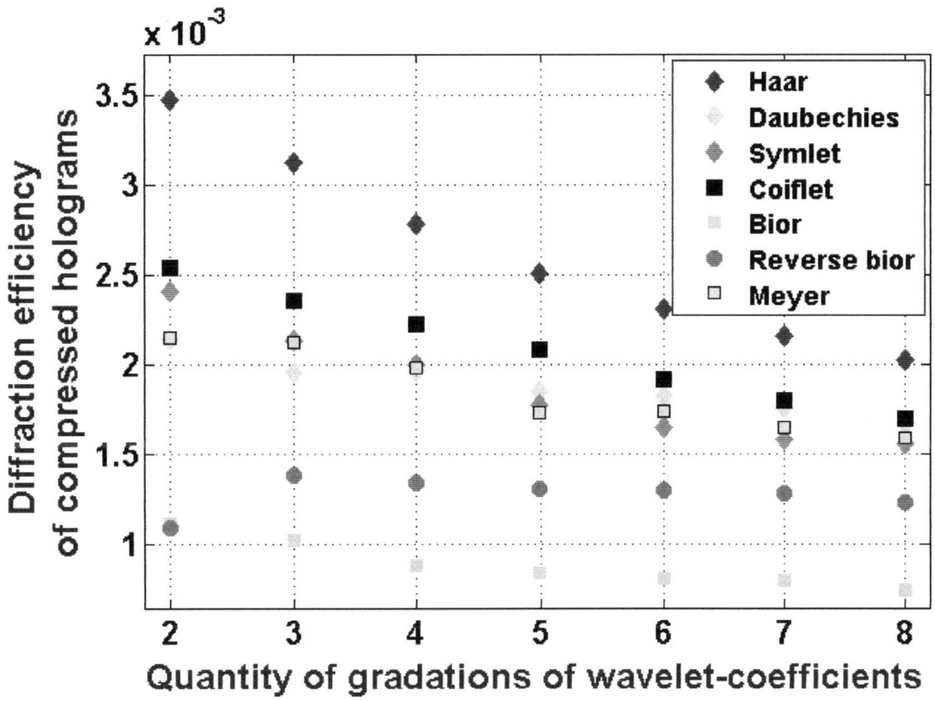

Figure 5. Obtained dependences of diffraction efficiency of compressed holograms on quantity of gradation of wavelet-coefficients.

The highest values of diffraction efficiency are achieved using Haar wavelet transform. The lowest values of diffraction efficiency are obtained when biorthogonol and reverse biorthogonol wavelet transforms wavelet are used.

To further compress digital holograms several techniques can be used. Possible method of such compression is use of multiple levels of wavelet transform. Also zerofilling on a threshold (thresholding) can be used. In this case all values of wavelet-coefficients less certain value of a threshold are nullified. Increase of a threshold will raise compression ratio, but, at the same time, increase quality losses. Reduction of the threshold allows to reduce quality losses of compression, but reduces its efficiency. Except thresholding of coefficients of wavelet-decomposition it's quantization is possible.

6. Conclusion

In this paper the most popular wavelet transforms were applied for digital holograms compression. Quality of the reconstructed images and diffraction efficiency of compressed holograms were obtained. Comparing these parameters, it is possible to determine the most optimal type of a wavelet transform for compression of digital holograms.

In case of the compression based on 1-level wavelet transform, the most optimal methods (according to the best quality of reconstruction and the highest diffraction efficiency) are Haar's wavelet transform and coiflet. However using other wavelet transforms with more levels of compression can provide other results than in case of 1-level compression.

Acknowledgments

This work was partially supported by the Russian Foundation for Basic Research (RFBR) №16-07-00389.

References

[1] Ebner M 2007 *Color Constancy*, 390 p.
[2] Reinhard E, Ward G, Pattanaik S and Debevec P 2006 *High Dynamic Range Imaging,* 502 p
[3] Picart P and Li J 2012 *Digital Holography* (Wiley, Hoboken-London)
[4] Juptner W and Schnars U 2005 *Digital Holography: Digital Hologram Recording, Numerical Reconstruction, and Related Techniques* (Berlin-Heidelberg: Springer-Verlag)
[5] Schnars U, Falldorf C, Watson J and Jüptner W 2015 *Digital Holography and Wavefront Sensing* (Berlin-Heidelberg: Springer-Verlag)
[6] Evtikhiev N N, Starikov S N, Sirotkin S A, Starikov R S and Zlokazov E Yu 2008 *Quantum Electronics* **38** 191-193
[7] Gusev M E and Alekseenko I V 2015 *Radiophysics and Quantum Electronics* **57** 543-550
[8] Monroy-Ramirez F and Garcia-Sucerquia J 2014 *Optik* **125** 2113
[9] Hÿtch M, Houdellier F, Hüe F and Snoeck E 2008 *Nature* **453** 1086-1089
[10] Molder A, Sebesta M, Gustafsson M, Gisselson L, Wingren A G and Alm K 2008 *Journal of Microscopy***232** 240–247
[11] Bianco V, Paturzo M, Finizio A, Calabuig A and Javidi B 2014 *IEEE Journal of Selected Topics Quantum Electron* **20** 6636048
[12] Kim M K 2012 *Optics Letters* **37**2 694-2696.
[13] Locatelli M, Pugliese E, Paturzo M, Bianco V, Finizio A, Pelagotti A, Poggi P, Miccio L, Meucci R and Ferraro P 2013 *Optics Express***21** 5379-5390
[14] Evtikhiev N N, Starikov S N, Cheremkhin P A and Kurbatova E A 2015 *Radiophysics and Quantum Electronics* 57 635-649
[15] Rivenson Y, Stern A and Javidi B 2013 *Applied Optics* **52** A423-A432.
[16] Cheremkhin P A, Evtikhiev N N, Krasnov V V, Porshneva L A, Rodin V G and Starikov S N 2014 *Proceedings of SPIE* **9249** 92491B
[17] Brady D J, Gehm M E, Stack R A, Marks D L, Kittle D S, Golish D R, Vera E M and Feller S D 2012 *Nature* **486** 386-389.
[18] Colomina I and Molina P2014 *ISPRS Journal of Photogrammetry and Remote Sensing* **92** 79-97
[19] Naughton T J, Frauel Y, Javidi B and Tajahuerce E 2002 *Applied Optics* **41** 4124-4132
[20] Naughton T J, Javidi B 2004 *Optical Engineering* **43** 2233-2238
[21] Ren Z, Su P and Ma J 2013 *Optical Review* **20** 469-473
[22] Xing Y, Kaaniche M, Pesquet-Popescu B and Dufaux F 2015 *Applied Optics***54**A98-A109
[23] Kurbatova E A, Cheremkhin P A, Evtikhiev N N, Krasnov V V and Starikov S N 2015 *Physics Procedia* **73** 328–332
[24] Zhang H, Zhou W, Leber D, Hu Z, Yang X, Tsang P WM and Poon T-C 2015 *Journal of Micro/ Nanolithography, MEMS, and MOEMS* **14** 41304
[25] Mallat S G 1989 *IEEE Transactions on Pattern Analysis and Machine Intelligence* 11 674-693
[26] Daubechies I 1990 *IEEE Transactions on Information Theory* **36** 961-1005
[27] Mallat S A 2009 *Wavelet Tour of Signal Processing*, 832 p.
[28] Fienup J R 1997 *Appl. Opt.* **36** 8352-8357

Journal of Physics: Conference Series **737** (2016) 012072 doi:10.1088/1742-6596/737/1/012072

The internal defects detection in crystals by digital holographic methods

V V Dyomin, I G Polovcev, D V Kamenev

Tomsk State University, Tomsk, Russia

E-mail: dyomin@tsu.ru, polovcev_i@mail.ru, kamenev87@mail.ru

Abstract. The internal defects detection method is suggested for crystals intended for the use in the IR part of spectrum. The method is tested on samples of the $ZnGeP_2$ monocrystals, the experimental results are shown.

1. Introduction

Nowadays the production of new materials is rapidly developing, and these materials are widely used in different areas of science and technology. To provide the preset materials useful properties and to ensure the reliability of their further use, the well-honed and certified procedures of their manufacturing and quality control are required. In the case of growing crystals for infrared range of wavelength (IRC) the procedure of quality control is required at different stages of their growth and processing. Worth noting that such procedure is not always implemented in practice.

One of the most important indicators of the quality of optical materials is presence or absence the surface and internal defects. The first type of defects includes scratches, roughness, chips, surface non-flatness and tends to occur due to imperfections in the surface treatment of the sample and / or improper working. The presence of such defects can be quite easily founded using microscopic and interference methods [1,2]. They can be eliminated by additional processing of the sample surface (if they are not internal irregularities, such as cracks and striae that appeared on the surface). The second type of defects is internal defects that include internal cracks and inhomogeneity of chemical composition, and physical inhomogeneity of the crystal structure. Their appearance is primarily dealing with the material manufacturing technology and defects can not be technically removed from the final sample. The presence of internal defects in the crystal is the main factor in terms of its use availability. For example, if a crystal is used as a laser active medium the high energy pumping may lead to its degradation

Conventionally, the detection of defects in the sample is performed after full manufacturing cycle. The quality control is usually performed by its testing on destruction. For example, a powerful laser radiation is injected in optical crystals, and defective samples are degraded. Thus there is a chance when the sample with defects passed such test. In this situation the sizes of defects after tests may be increased, and they can cause unexpected degradation during the sample further operation. Therefore, a reliable quality control of manufactured samples is an actual task.

At cutting blanks stage the quality visualization of existing defects (if any) is quite important. The knowing the spatial location of the defects and their sizes can allow performing the cutting procedure when the resulting samples have no defects. Such approach leads to increasing the efficiency of sample preparation and allows improving the technological process. Examples of a single crystal $ZnGeP_2$ before and a blank after the cutting stage are shown in Fig. 1

In this paper, we propose to detect contactless "defective" spaces in blank on the crystal cutting stage before testing on radiation strength. This method conducts without destruction (degradation) of blank, and will significantly improve the efficiency of crystal quality control. The use of the results of this method can lead to reduction of required resources (financial and time) due to optimization of single crystal cutting procedure.

The aim of this work is to develop methods, hardware and software for the detection, measurement and identification of internal defects in crystals intended for the use in the infrared wavelength range (IRC).

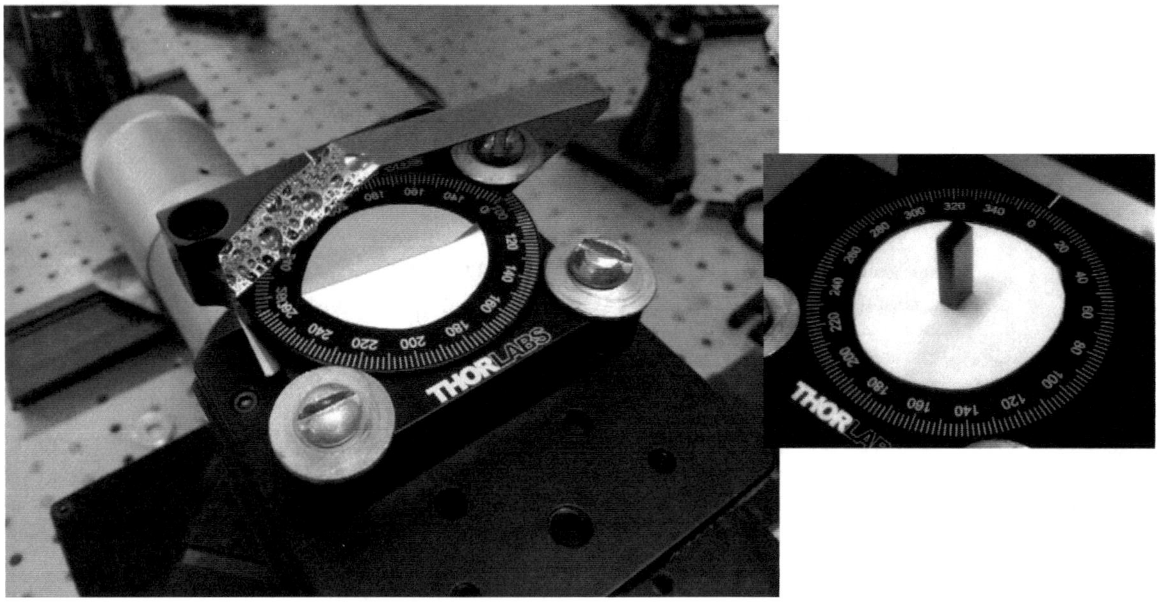

a b

Figure 1. Example of $ZnGeP_2$ single crystal before (a) and blank after (b) the blank cutting stage

2. Techniques description

To detect internal defects in crystals a method based on in-line digital holograms registration is suggested. Thus the wavelength of the coherent illumination has to be selected in the area of sample transparency.

The procedure of further hologram reconstruction and data processing allow evaluating the size, shape and 3D location of all internal defects in field of vision. The method compares favorably with analogs as it is non-destructive and allows exploring the entire volume of the sample recorded on the hologram with high resolution.

Technique of detection of internal defects in crystal is worked on $ZnGeP_2$ single crystals. These single crystals have research and applied interest [3,4], because they are most effective nonlinear optical materials for the near and mid-infrared region (wavelengths of 0.7-12.0 μm). They are used for CO_2 and CO lasers harmonics generation, optical parametric oscillator (covering the wavelength range of 3-10 microns) and for coherent THz radiation sources creation. The main advantages of the single crystal are high threshold of optical breakdown, good thermal conductivity, mechanical strength, and resistance to high humidity and aggressive media.

To register digital hologram of internal volume of the sample it is necessary to choose the light source and receiver correctly. On the one hand, the sample has to be transparent to the laser source radiation; on the other hand, the receiver must be able to register this radiation.

Fig. 2 shows the spectral transmission of single crystals ZnGeP2 [5] and the typical spectral sensitivity of the CCD camera (for example, Videoscan 4021) [6]. The diagrams show that the use of the YAG: Nd laser, generating 1053nm wavelength may provide the registration of digital holograms of the sample. At the given wavelength single crystal transmits radiation, although it has a significant absorption. The curve in Figure 2b does not allow saying definitely that the CCD receiver is able to register this radiation. So this fact further will be confirmed experimentally.

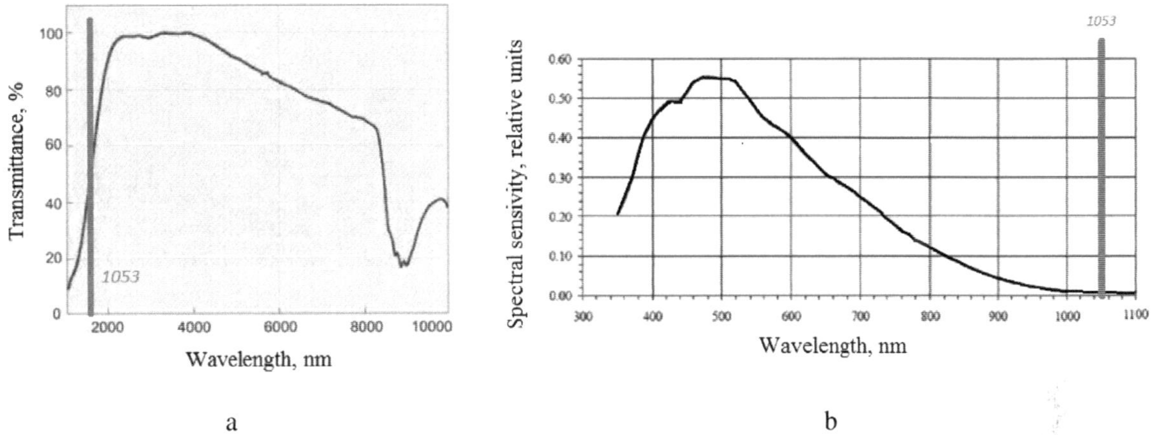

Figure 2. The spectral transmittance of the single crystal sample $ZnGeP_2$ (a) [5] and the curve of Videoscan 4021 CCD-camera spectral sensitivity [6]

The information can be extracted from digital hologram about size, shape 3D position of all defects located in volume of the sample and registered on hologram The procedure of holographic data processing and information extracting is described in detail in [7,8], and is not the subject of investigation in this paper.

3. Results and discussion

To detect internal defects in IRC the in-line scheme of digital holograms registration is implemented. YAG: Nd DTL-399QT diode-pumped laser that generates at three wavelengths (1053, 527 and 351 nm) simultaneously is used. To receive the laser radiation CCD camera Videoscan 4021 is used. The optical system adjustment is carried out on 527 nm wavelength, and after adjustment this radiation is eliminated by a filter. As objects of holographing in this experiment a single crystal $ZnGeP_2$ and opaque bolt behind it are used (Fig. 3a). An example of a registered digital hologram is shown in Fig. 3b It is seen that the radiation passed through the sample is recorded CCD receiver. This fact confirms that the used laser source and CCD camera provide recording digital holograms of the internal volume of single crystal.

A number of digital holograms of optical single crystals $ZnGeP_2$ are recorded in a laboratory experiment. To increase the image contrast the background was subtracted, that was previously recorded in the absence of testing crystal (this procedure is illustrated in Fig.4 a-c). Focused reconstructed images of crystals scratches located on its different surfaces (marked by white rings) are presented on Fig. 4d,e. Defects which exceed the size of 20 microns (that corresponds to the scheme spatial resolution) are not found in the volume of a single crystal between these scratches. The detection of smaller defects requires the use of additional microscopic system.

Digital holograms of various samples are given as examples in Fig. 5. Dark areas near the boundaries of the single crystals are located in the pictures. One of the reasons of their appearance is related to nonflatness and nonparallelism of sample surfaces, and lead to vignetting (refractive index of $ZnGeP_2$ is more than 3.1).

Examples of internal defects detected for another single crystal are shown in Fig. 6a According to the number of detected defects, it is possible to reject it as defective and prevent further test.

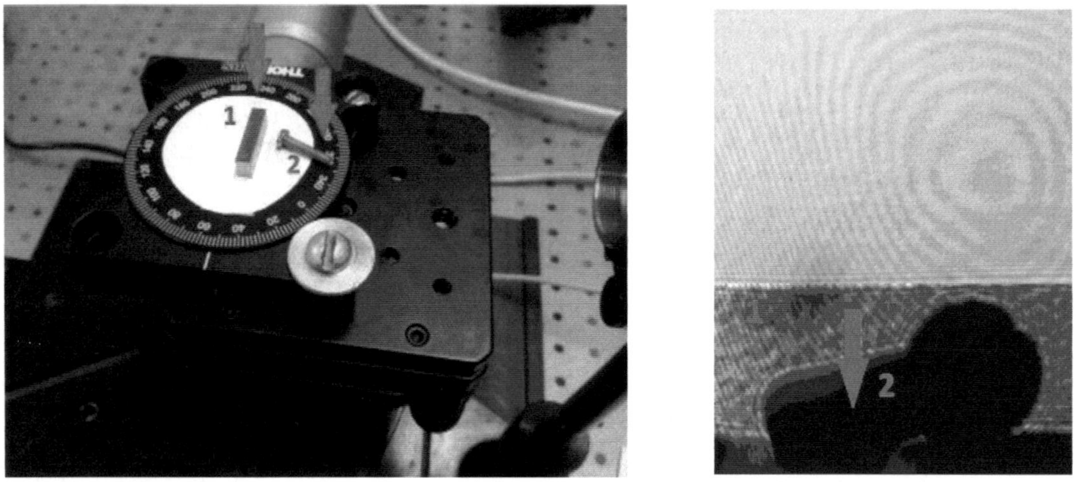

a b

Figure 3. Subject scene (a) and its digital hologram (b).

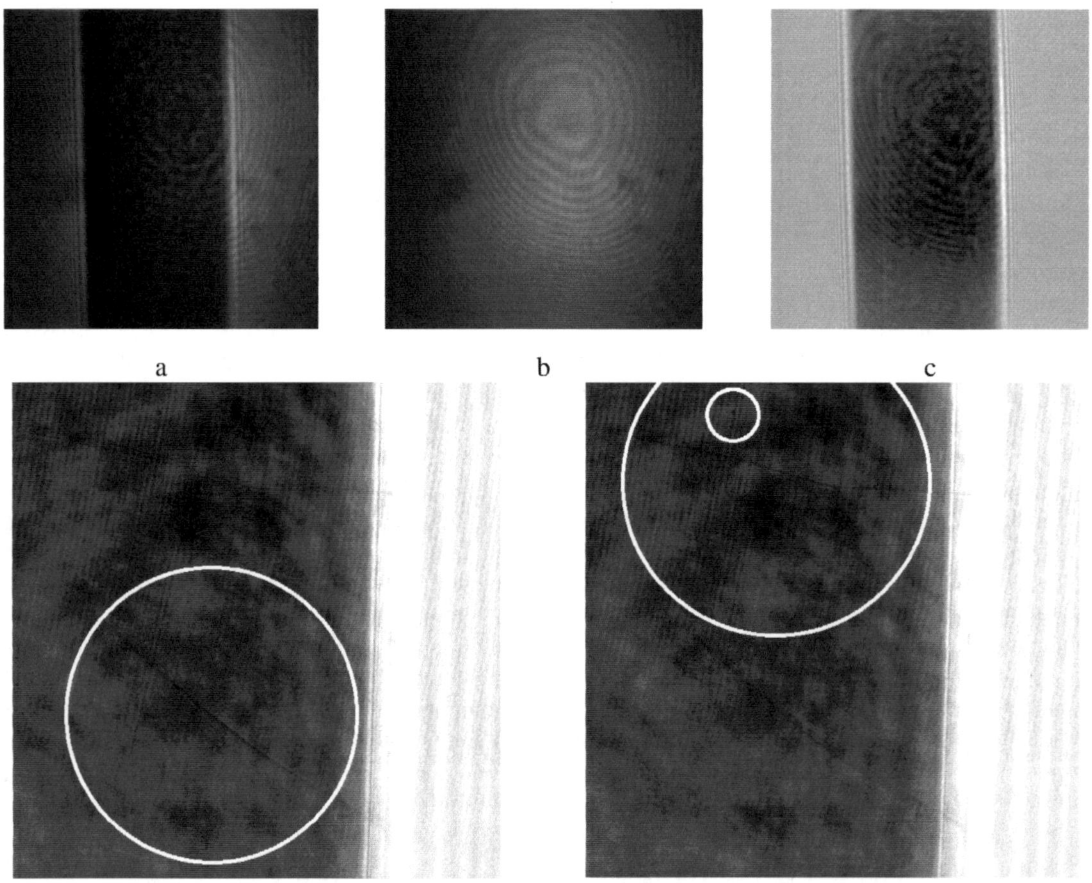

a b c

d e

Figure 4. Single crystal $ZnGeP_2$ defects investigation by digital holography
Note: a digital hologram of $ZnGeP_2$ sample, b – background, c - result of background subtraction, d, e - reconstructed images of single crystal scratches on different surfaces of the sample.

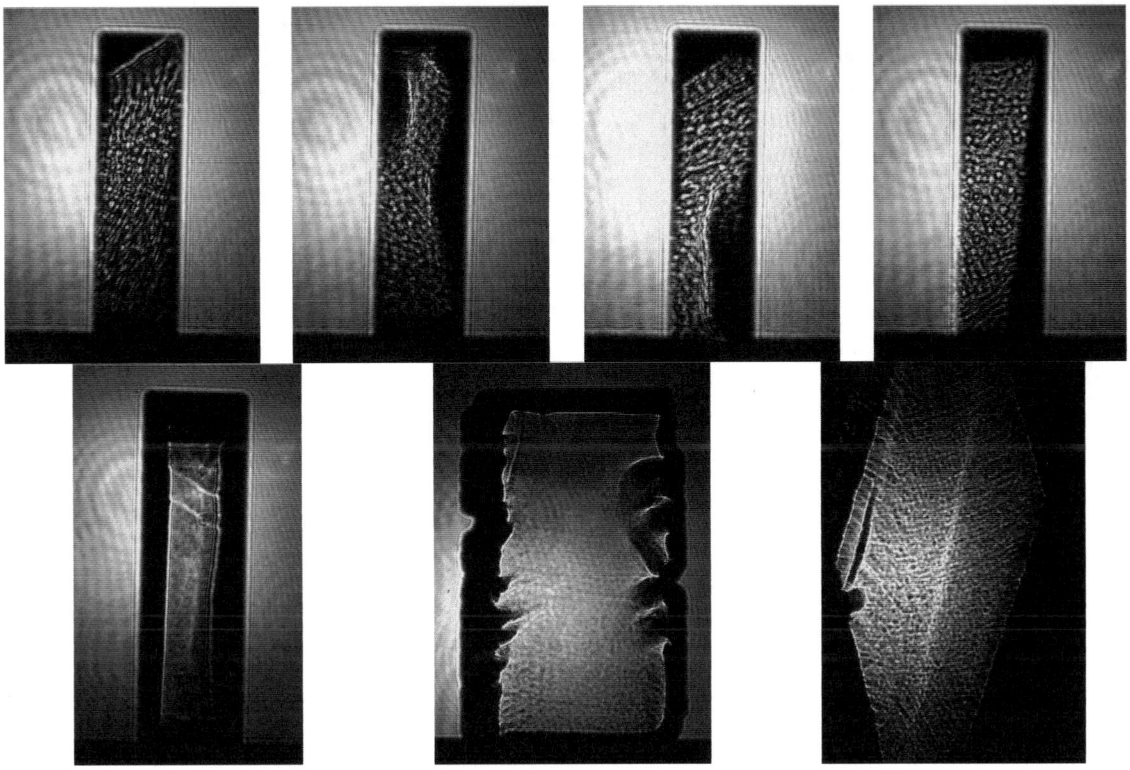

Figure 5. Digital holograms of several single crystals $ZnGeP_2$

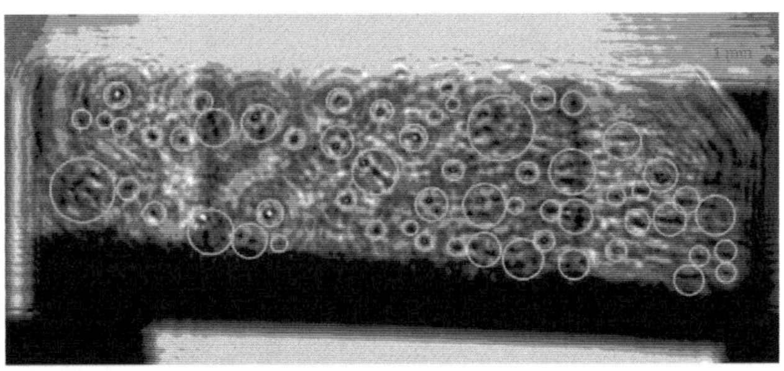

a

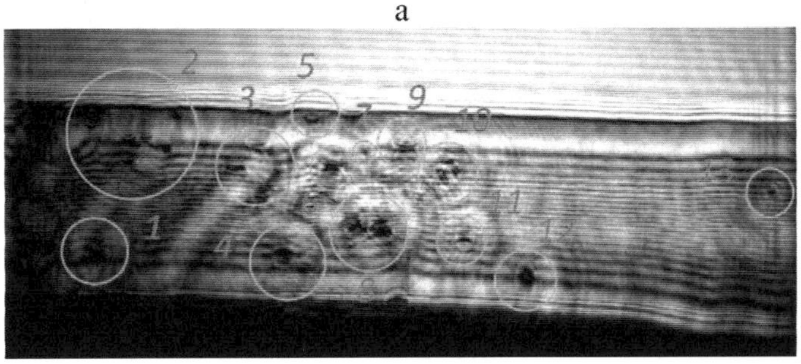

b

Figure 6. Defects in $ZnGeP_2$ single crystal visualized by holographic visor at 1.06 nm

Journal of Physics: Conference Series **737** (2016) 012072 doi:10.1088/1742-6596/737/1/012072

The images reconstructed from a digital hologram, allow us to determine the type of defect, to measure its dimensions and position (Fig. 6a). Such data can not be received by shadow projection image only. The results of digital hologram processing of another single crystal sample are presented below. For advanced visual representation of received data, 2D display of reconstructed volume holographic image of the sample according to the method described in [9] is used. Based on received data the size of defects are determined and presented in Table 1.

Table 1. Sizes of defects in $ZnGeP_2$ sample defined by digital holography

Defect №	1	2	3	4	5	6	7	8	9	10	11	12	13
Size, μm	150	100	100	150	120	300	50	400	150	100	100	350	80

4. Conclusion

Results of this paper show that in order to get successful implementation of the quality control method it is very important to match the crystal transmittance, the wavelength of the laser radiation and the spectral sensitivity of the CCD. Since the crystals may have very different spectral transmission (Fig. 7 shows the diagrams for some IRC), it is necessary to choose properly the receiver and the laser source in each case

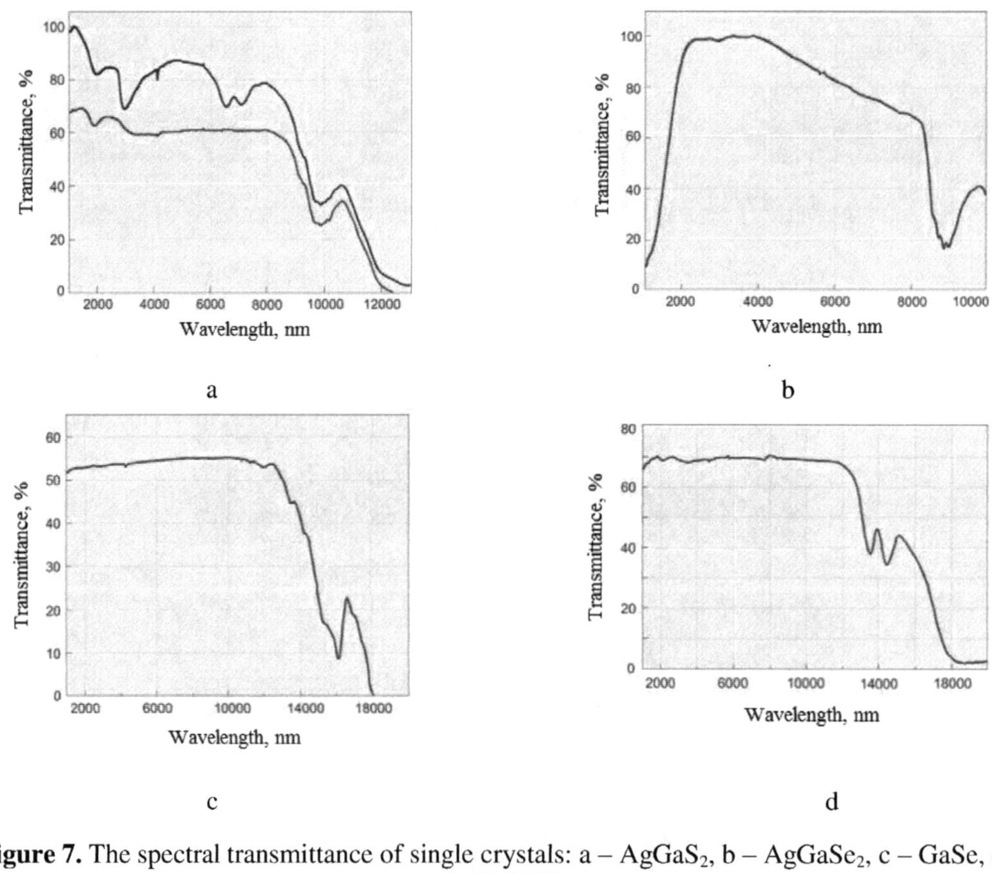

Figure 7. The spectral transmittance of single crystals: a – $AgGaS_2$, b – $AgGaSe_2$, c – GaSe, d – ZnGeP [5]

The method proposed in this paper is tested on 20 samples, defective blanks are identified. It was established experimentally that the method allows detecting and evaluating the location of internal $ZnGeP_2$ single crystals defects, the presence of chips, cracks, and the irregularities of refractive index.

Acknowledgments
We express our gratitude to LLC "Laboratory of Optical Crystals", Tomsk, in particular, Ph.D., Senior Scientist AI Grebenyukov for samples kindly provided for testing

This work was supported by project № 8.1.12.2015 within «D.I. Mendeleev research foundation of Tomsk State university» program in 2014-2016

References

[1]. Grishechkin M B, Denisov I A, Silina A A, Smirnova N A, Shmatov N I, Yakovenko A G 2014 *Applied Physics* **6** 9-15
[2]. Dyomin V V, Polovcev I G, Simonova G V 2014 *Optical Measurements* (Tomsk: Izdatelstvo Dom TGU) 580 p
[3]. Chuchupal S V, Komandin G A, Zhukova E S, et al 2015 *J. Phys. of the Solid State* **57** **(8)** 1607-1612
[4]. Zawilski K T, Setzler S D, Schunemann P G and Pollak T M 2006 *J. Opt. Soc. Am. B* **23** **(11)**
[5]. http://eksmaoptics.com *Spectral transmission of single crystals* (accessed 01.04.2016)
[6]. http://www.videoscan.ru/ *Specification of Video Scan CCD- cameras* (accessed 01.04.2016)
[7]. Dyomin V V, Kamenev D V, Olshukov A S 2014 *Proc. Ocean-Taipei*, 14767788
[8]. Dyomin V V, Kamenev D V 2016 *Proc. of SPIE* **9771** 97710
[9]. Dyomin V V, Kamenev D V 2013 *Journal of Optical Technology* **80** **(7)** 450-456

Simple method of modelling of digital holograms registering and their optical reconstruction

N N Evtikhiev, P A Cheremkhin, V V Krasnov, E A Kurbatova, D Yu Molodtsov, L A Porshneva, V G Rodin

National Research Nuclear University MEPhI (Moscow Engineering Physics Institute), Kashirskoe shosse 31, 115409 Moscow, Russia

E-mail: holo@pico.mephi.ru, CheremhinPavel@mail.ru

Abstract. The technique of modeling of digital hologram recording and image optical reconstruction from these holograms is described. The method takes into account characteristics of the object, digital camera's photosensor and spatial light modulator used for digital holograms displaying. Using the technique, equipment can be chosen for experiments for obtaining good reconstruction quality and/or holograms diffraction efficiency. Numerical experiments were conducted.

1. Introduction

Reconstruction quality of images and the diffraction efficiency (DE) are the major characteristics in classical (analog) holography [1-3]. Problem of estimation of DE and quality of reconstructed from analog holograms images were in detail described for various recording media. In digital holography [4-6], digital camera's photosensor can be considered as "recording medium".

There are two types of reconstruction of images from digital holograms: numerical [4-11] and optical [11-19] ones. In the case of numerical reconstruction, only reconstruction images quality is important and DE makes only speculative sense. However in case of optical reconstruction, DE has practical meaning. The most popular method of optical image reconstruction is displaying of the holograms on spatial light modulator (SLM) [11-19]. Optical reconstruction of images from holograms is used for practical applications in remote display of the static and dynamic 2D- and 3D-scenes including recorded in non-visible radiation [12], for projective systems [13-14], 3-dimensional vision, etc.

The problem of estimation of DE and quality of the reconstructed images can be solved by numerical modeling of process of recording of digital holograms and optical reconstructed of images from holograms. In this paper simple method of modeling of digital holograms registering and their optical reconstruction is described.

2. Description of the method of modeling of digital holograms registering and their optical reconstruction

The used scheme for numerical experiments on modeling of recording of digital holograms and optical reconstruction of images consists of 4 stages [20]:

1. creation of a model object,
2. modeling of hologram recording (modeling of wave propagation from the object plane to the hologram plane, synthesis of the hologram),
3. quantization of the hologram and modeling of displaying hologram to SLM,
4. modeling of optical reconstruction of the image (modeling of wave propagation from the SML plane to the plane of the focused reconstructed image)

The technique considers the following values:

- parameters of the used object:
- characteristics of the digital camera, used for displaying digital holograms:
    - resolution,
    - spatial and temporal noises,
    - radiometric parameters (dynamic range, maximum linear signal).
- characteristics of the SLM, used for displaying digital holograms:
    - resolution,
    - quantity of gradations,
    - SLM type (phase or amplitude).

Distribution of complex amplitudes of the object wave $B(u, v, z)$ in the plane of registration of the hologram by the photosensor of the digital camera located at z distance from the plane of object is calculated by direct calculation of Fresnel diffraction method [4]:

$$B(u,v,z) = \frac{\exp(ikz)}{i\lambda z} \cdot \exp\left(\frac{i\pi(u^2+v^2)}{\lambda z}\right) \cdot FFT\left\{A(x,y,0) \cdot \exp\left(\frac{i\pi(x^2+y^2)}{\lambda z}\right)\right\} \quad (1)$$

where

- FFT( ) — fast Fourier transform,
- $k = \frac{2\pi}{\lambda}$ — wave number,
- i — imaginary unit,
- λ — wavelength,
- (x,y) are discrete coordinates in the object plane.

Then intensity distribution of the registered hologram $H(u, v)$ is calculated:

$$H(u,v) = |B(u,v,z) + C|^2 \quad (2)$$

$$C = \sqrt{p \cdot \langle |B(u,v,z)|^2 \rangle} \quad (3)$$

where

- <*> – average value of the intensities of the object beam in hologram plane,
- C is amplitude value of each pixel of the reference beam in the hologram plane,
- p is ratio of average intensities of reference and object beams.

Journal of Physics: Conference Series **737** (2016) 012073 doi:10.1088/1742-6596/737/1/012073

Then the values of the hologram pixels are quantized:

$$H_Q(u,v) = \begin{cases} Q-1, \ if \ H(u,v) = max\{H(u,v)\} \\ fix\left\{\dfrac{H(u,v)}{max\{H(u,v)\}} * Q\right\}, \ in \ other \ cases \end{cases} \tag{4}$$

where

- $H_Q(u,v)$ – array of elements of the quantized hologram,
- $max\{*\}$ – maximum value of intensity of pixels of the hologram,
- Q – number of gradations of the hologram (2 or 256),
- $fix\{*\}$ – function of rejection of fractional part.

Then we modeled the displaying of the holograms on SLM. We should consider amplitude ($H^{AMP}{}_Q(u,v)$) and phase ($H^{PH}{}_Q(u,v)$) SLM separately:

$$H^{AMP}{}_Q(u,v) = H_Q(u,v) \cdot exp(i \cdot 0) \tag{5}$$

$$H^{PH}{}_Q(u,v) = 1 \cdot \exp\left(i \cdot \left[H_Q(u,v) \cdot 2\pi \cdot \left(\frac{Q-1}{Q}\right)\right]\right) \tag{6}$$

Final step is modeling of the optical reconstruction of the digital holograms displayed on the amplitude and phase SLM by the direct calculation of Fresnel diffraction method:

$$D^{AMP}(x',y',2z') = \frac{\exp(ikz)}{i\lambda z} \cdot \exp\left(\frac{i\pi(x'^2+y'^2)}{\lambda z}\right) \cdot FFT\left\{H^{AMP}{}_Q(u,v) \cdot \exp\left(\frac{i\pi(u^2+v^2)}{\lambda z}\right)\right\} \tag{7}$$

$$D^{PH}(x',y',2z) = \frac{\exp(ikz)}{i\lambda z} \cdot \exp\left(\frac{i\pi(x'^2+y'^2)}{\lambda z}\right) \cdot FFT\left\{H^{PH}{}_Q(u,v) \cdot \exp\left(\frac{i\pi(u^2+v^2)}{\lambda z}\right)\right\} \tag{8}$$

where

- (x',y') are discrete coordinates in the reconstruction plane.

3. Numerical application of the method of modeling of digital holograms registering and their optical reconstruction

Different amplitude-phase two-dimensional objects were as the test objects. Examples of intensity distribution of the initial objects are shown in Fig. 1:

- standard grayscale image with size of 256 × 256 pixels (Fig. 1a),
- binary contour image with size of 128 × 128 pixels (Fig. 1b).

Amplitude values of pixels of object were equal to the square root of the intensity. Phase values of the object pixels had uniform distribution between 0 and 2π.

Journal of Physics: Conference Series **737** (2016) 012073 doi:10.1088/1742-6596/737/1/012073

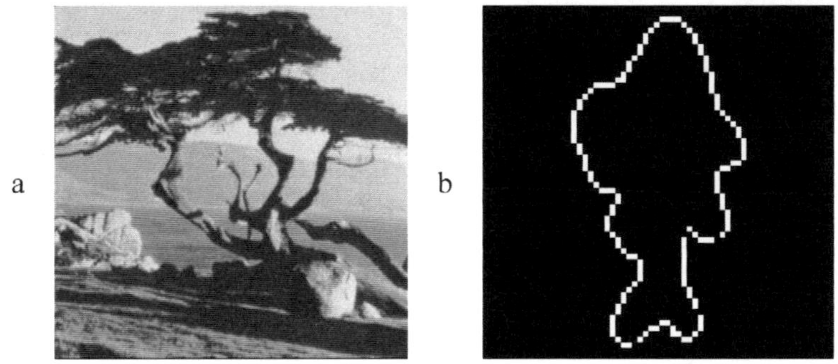

Figure 1. Examples of objects used in numerical experiments.

Fragments of synthesized digital hologram with 2 gradations (binary hologram) and with 256 gradations are shown in Fig. 2a,b correspondingly. Fragments have 128×128 pixels.

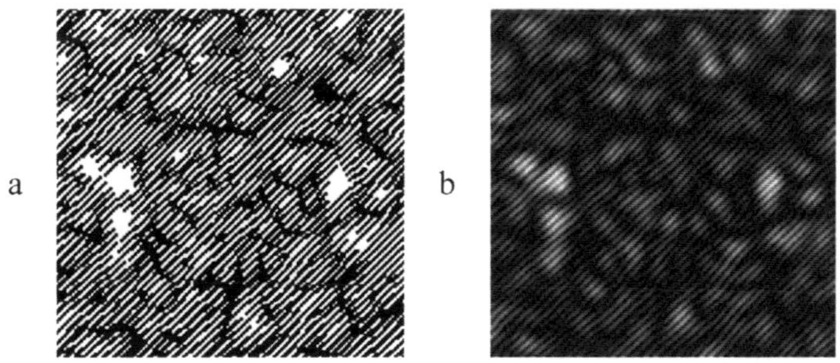

Figure 2. Fragments of synthesized digital hologram with 2 gradations (a) and with 256 gradations (b).

After digital holograms synthesis, we simulated optical reconstruction of images from the holograms using spatial light modulators. Two basic types of such modulators were considered: phase and amplitude SLMs [21-23]. Quantity of amplitude levels (in amplitude SLM) and phase levels (in phase SLM) are equal to 2 and 256.

Results of modeling of optical reconstruction of images object are shown in Fig. 3,4. Reconstruction of images using amplitude SLM are shown in Fig. 3a, 4a (binary digital hologram with 2 gray levels) and Fig. 3b, 4b (digital hologram with 256 gray levels). Reconstruction using phase SLM are shown in Fig. 3c, 4c (halftone image with uniform histogram) and Fig. 3d, 4d (binary contour image).

As can be seen, the reconstructed images from the amplitude SLM with 256 levels has a significantly better quality than that from the phase modulator. This is due to the fact that digital hologram is intensity that was registered by digital camera. In result amplitude SLM should provide better quality of reconstruction that phase one.

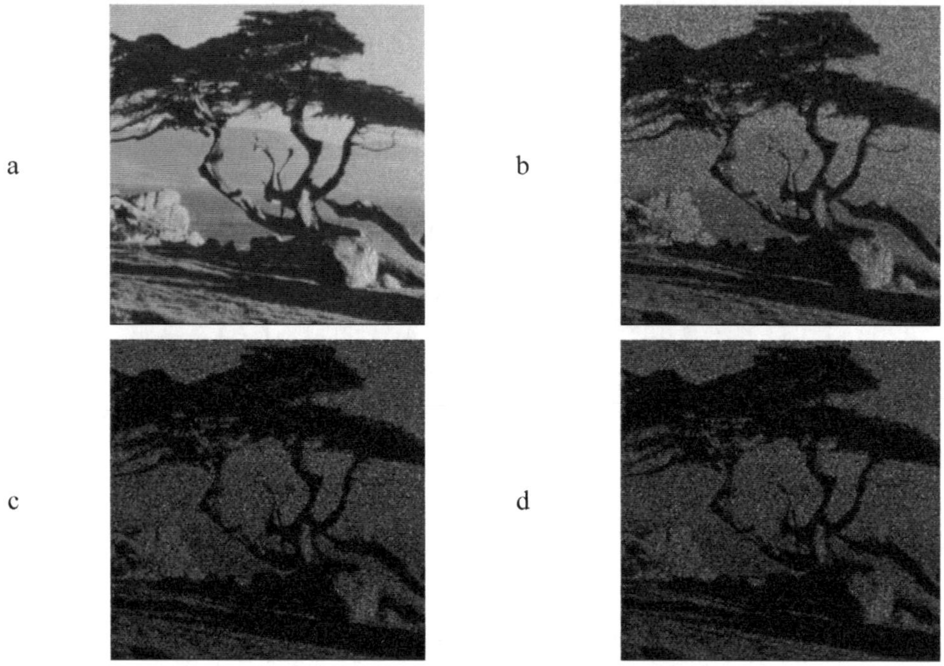

Figure 3. Reconstructed images of the halftone object in the cases of simulation of the amplitude (a, c) and phase (b, d) spatial light modulators. Modulators (holograms) with 256 (a and b) or 2 (c and d) levels are considered.

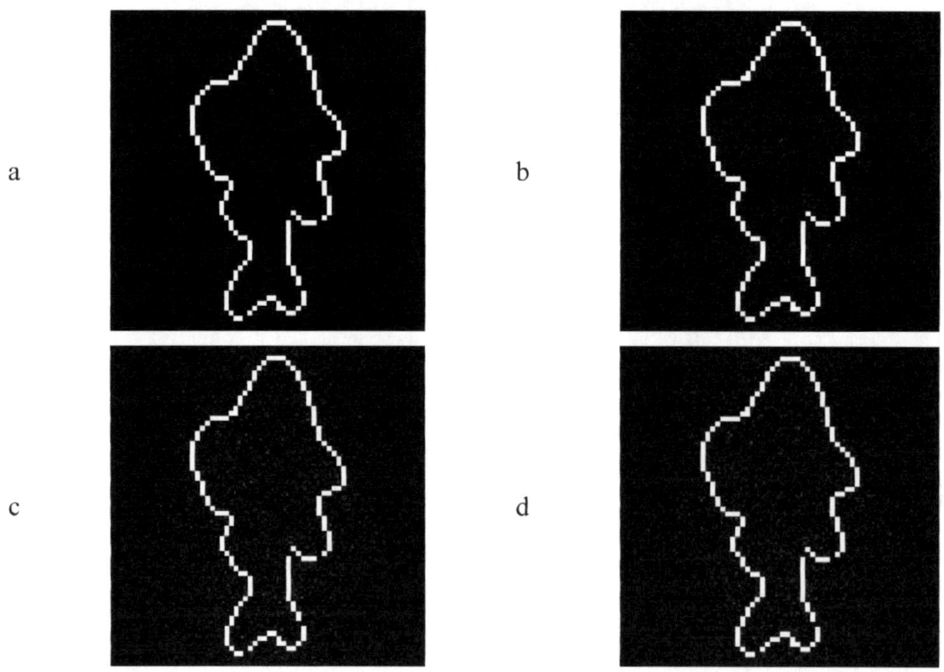

Figure 4. Reconstructed images of the binary object in the cases of simulation of the amplitude (a, c) and phase (b, d) spatial light modulators. Modulators (holograms) with 256 (a and b) or 2 (c and d) levels are considered.

4. Conclusion
A simple technique of digital holograms registering and their optical reconstruction was described. This technique makes possible to select possible conditions of experiments, parameters of the digital camera and the objects, the SLM type, and etc. This technique can be useful for optical information processing, for registration and displaying of volume scenes in real time.

Acknowledgments
This work was partially supported by Ministry of Education and Science of the Russian Federation, task 3.2066.2014/K.

References
[1] Collier R J, Burckhardt C B and Lin L H 1971 *Optical Holography* (Academic Press)
[2] Goodman J W 2004 *Introduction to Fourier Optics* (W. H. Freeman)
[3] Born M and Wolf E 1980 *Principles of Optics* (Pergamon, Oxford)
[4] Juptner W and Schnars U 2005 *Digital Holography: Digital Hologram Recording, Numerical Reconstruction, and Related Techniques* (Berlin-Heidelberg: Springer-Verlag)
[5] Ackermann G K and Eichler J 2007 *Holography: A Practical Approach* (Wiley)
[6] Schnars U, Falldorf C, Watson J and Jüptner W 2015 *Digital Holography and Wavefront Sensing* (Berlin-Heidelberg: Springer-Verlag)
[7] Memmolo P, Miccio L, Merola F, Gennari O, Netti P A and Ferraro P 2014 *Cytometry Part A* **85** 1030-1036
[8] Porshneva L A, Cheremkhin P A, Evtikhiev N N, Krasnov V V, Rodin V G and Starikov S N 2015 *Physics Procedia* **73** 333-337
[9] Pejchang D, Coëtmellec S, Gréhan G, Brunel M, Lebrun D, Chaari A, Grosges T and Barchiesi D 2015 *Optics Express* **23** 18351-18360
[10] Shechtman Y, Eldar Y C, Cohen O, Chapman H N, Miao J and Segev M 2015 *IEEE Signal Processing Magazine* **32** 87-109
[11] Osten W, Faridian A, Gao P, Körner K, Naik D, Pedrini G, Singh A K, Takeda M and Wilke M 2014 *Applied Optics* **53** G44-G63
[12] Locatelli M, Pugliese E, Paturzo M, Bianco V, Finizio A, Pelagotti A, Poggi P, Miccio L, Meucci R and Ferraro P 2013 *Optics Express* **21** 5379-5390
[13] Gopinathan U, Monaghan D, Hennelly B, McElhinney C P, Kelly D P, McDonald J B, Naughton T J, Sheridan J T 2008 *Journal of Display Technology* **4** 254-261
[14] Koreshev S N, Korepin I N and Vakhnin M S 2016 *Journal of Optical Technology* **83** 106-112
[15] Cheremkhin P A, Evtikhiev N N, Krasnov V V, Kulakov M N, Kurbatova E A, Molodtsov D Yu and Rodin V G 2016 *Proceedings of SPIE* **9889** 98891M
[16] Kujawinska M, Kozacki T, Falldorf C, Meeser T, Hennelly BM, Garbat P, Zaperty W, Niemelä M, Finke G, Kowiel M and Naughton T 2014 *Optics Express* **22** 2324-2336
[17] Zwick S, Haist T, Warber M and Osten W 2010 *Applied Optics* **49** F47-58
[18] Evtikhiev N N, Ivanov P A, Lyapin A S, Reyzin B M, Shevchuk A V, Sirotkin S I, Starikov R S, Zaharcev A V 2005 *Proceedings of SPIE* **5851** 242-245
[19] Li X, Chen C P, Li Y, Zhou P, Jiang X, Rong N, Liu S, He G, Lu J and Su Y 2016 *Journal of Display Technology* **12** 362-367
[20] Evtikhiev N N, Starikov S N, Cheremkhin P A and Kurbatova E A 2015 *Radiophysics and Quantum Electronics* **57** 635-649
[21] Wu S T and Yang D K 2005 *Reflective Liquid Crystal Displays* (Chichester: John Wiley)
[22] Bondareva A P, Cheremkhin P A, Evtikhiev N N, Krasnov V V, Starikov R S and Starikov S N 2014 *Journal of Physics: Conference Series* **536** 012011.
[23] Wang Z, Millet L, Mir M, Ding H, Unarunotai S, Rogers J, Gillette M U and Popescu G 2011 *Optics Express* **19** 1016-1026

Impact of DMD-SLMs errors on reconstructed Fourier holograms quality

D Yu Molodtsov, P A Cheremkhin, V V Krasnov, V G Rodin

National Research Nuclear University MEPhI (Moscow Engineering Physics Institute), Kashirskoe shosse, 31, 115409 Moscow, Russia

E-mail: holo@pico.mephi.ru, DYMolodtsov@mephi.ru

Abstract. The quality of reconstructed impulse response of holograms displayed on a three DMD modulators of different models was experimentally evaluated. The integral parameter of modulators` optical quality – the size of sufficiently flat area of matrix surface – was estimated according to the developed technique. The results of holograms reconstruction are analyzed.

1. Introduction

The DMD technology (Digital Micromirror Device) is widely applied in various areas of science and engineering: DLP projectors, holographic printers [1], 3D-displays [2, 3], etc.

In our researches possibility of using of DMD technology in optical holographic correlators is considered. Unlike LC SLM, this type of modulators possesses much higher speed. It is theoretically possible to realize the dispersive correlator for object recognizing not only on its spatial, but also on its spectral characteristics [4, 5]. In order to DMD modulator could be used as the device for displaying of holographic filters, it should satisfy several additional requirements. Existence of undesirable phase modulation, which is possible because of manufacture errors in a modulator design, can cause essential negative effects. Following characteristics are related to manufacture errors in the documentation for DMD and affect the optical quality:

- micromirror tilt angle variation,
- orientation of the micromirror axis-of-rotation variation,
- window flatness.

The values of these parameters for tested in this paper DMD are respectively ± 1°, ± 1° and 4 fringes. Flatness of a micromirrors matrix substrate is not specified in documentation, however it is probably indirectly considered in the first two characteristics.

Only amplitude modulation of radiation is important for many scopes of this kind of modulators, and the specified errors have no essential impact. However when using DMD as holograms displaying device, as it was shown in works [6, 7], there is a notable deterioration of the reconstructed impulse response because of the existing manufacture errors. It is limited considerably by possibilities of application of DMD for a conclusion of holograms: than big DMD matrix. Square with brought to it by the hologram is lit with the collimated laser bunch, that more strongly distorts the restored pulse response. This significantly limits the possibility of the application of DMD to holograms displaying: the bigger area of DMD-matrix is used to display a hologram and illuminated by a collimated laser beam, the more distorted impulse response is reconstructed.

Journal of Physics: Conference Series **737** (2016) 012074 doi:10.1088/1742-6596/737/1/012074

2. The setup and experiment description

The quality of the reconstructed impulse responses of the holograms was experimentally estimated in this work for three various DMD modulators produced by Texas Instruments [7]. All modulators belong to chips family 0.7 XGA 12° DDR, one has the model number s1076 7071b and two - s1076 7072 (these are used in DLP Discovery 4100 Development Kit). The technique proposed in work [7] was applied for estimation. The sufficiently flat area of DMD matrix L_f was defined as the characteristic of optical quality of the DMD modulator. The technique consists in reconstructing of several test holograms, which are displayed on the DMD modulator. Impulse responses of holograms contain one-dimensional gratings with the set period (see Fig. 1). Holograms differ from each other in resolution (number of pixels), and, respectively, an area which they will occupy on a matrix surface when being output on it. For estimation of the size of sufficiently flat area of a matrix surface it is enough to determine the maximum size of the hologram which impulse response contains any gratings with the demanded spatial frequency and which will be solvable in the reconstructed image.

The program was made for realization of this technique. It requires a file with the hologram and its coordinates on a matrix surface as input data, and then it outputs it to the DMD modulator at the specified coordinate. The program has function of automatic movement of the hologram on the set path for estimating the quality of various areas of a matrix.

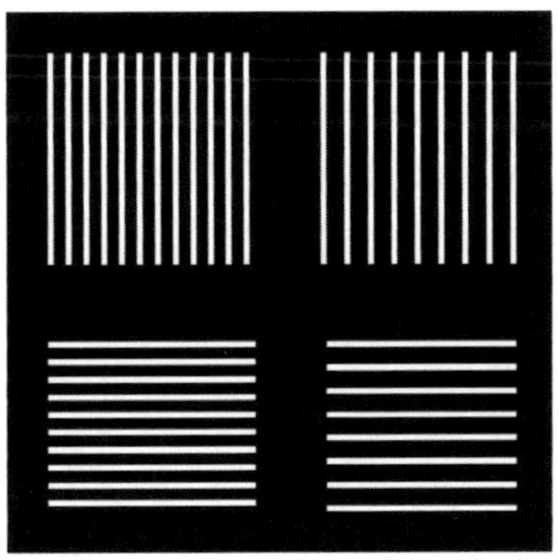

Figure 1. Impulse response for an estimating of resolution of the scheme (the size 200×200 pixels, contains gratings with the periods of 3 pixels and 4 pixels)

Laser radiation with the wavelength of 532 nanometers was used for holograms reconstruction, it was applied at illumination angle of 24°, so the reflected light was normal to the plane of the modulator. The focal length of a Fourier transforming lens is 500 mm. The scheme of setup is shown in Fig. 2.

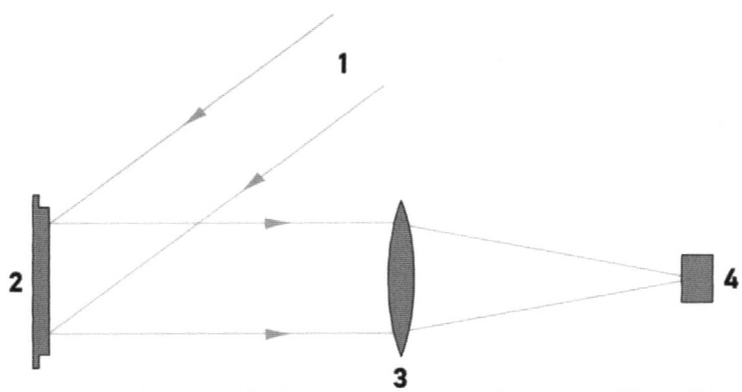

Figure 2. Scheme of setup of holograms reconstruction: 1 – collimated laser beam, 2 – DMD-chip, 3 – Fourier lens, 4 – digital camera for impulse response registration

Journal of Physics: Conference Series **737** (2016) 012074 doi:10.1088/1742-6596/737/1/012074

3. Experiment results
The influence of the overall curvature of DMD-chip has been shown in [7] for the s1076 7071b chip. The hologram placement in different areas of the matrix greatly affects the result of the reconstruction. It is an interesting fact that the central region of matrix on this chip is not the most qualitative for recovery of holograms.

Results of holograms reconstruction with use of newer chips (model s1076 7402) were significantly better. One of chips showed quality approximately by 1.5 times better, the second – by 3.8 times. Thus, results considerably differ even for chips of one model. The examples of the impulse responses registered from the chip with the worst and from the chip with the best optical qualities are shown in Fig. 3.

Zero diffraction order keeps its form much better when using DMD chip s1076 7402 (2), this is the good indicator of better optical quality. The result of "scanning" of this DMD chip is interesting also. It, as well as the second chip s1076 7402, provided the best quality of hologram reconstruction when using the central region of a matrix. The impulse responses registered for holograms outputted to edges of the best quality modulator are shown in Fig. 4. This kind of symmetric concerning the center distortions allows us to speak about some sphericity of this matrix.

For a numerical estimation of optical quality of DMD modulators the sizes of sufficiently flat area (L_f) for all three copies were obtained. The requirement to resolution was highest: in the reconstructed image the grating with the period of 2 pixels had to be resolved. The following results were obtained. Size of sufficiently flat area L_f of a matrix for DMD chip s1076 7071b was equal 1.4 mm, for DMD chip s1076 7402 (1) was equal 2.1 mm and for DMD chip s1076 7402 (2) was equal 5.3 mm. The matrix pixel pitch of all three DMD modulators is 13.68 μm.

The obtained results can be qualitatively compared to the result published in work [8]. In this work the Texas Instruments DLP Discovery 4100 Development Kit modulator is used. From this article it is possible to make a conclusion that the copies of DMD chips supplied with Discovery 4100 are also not deprived of the problems described above.

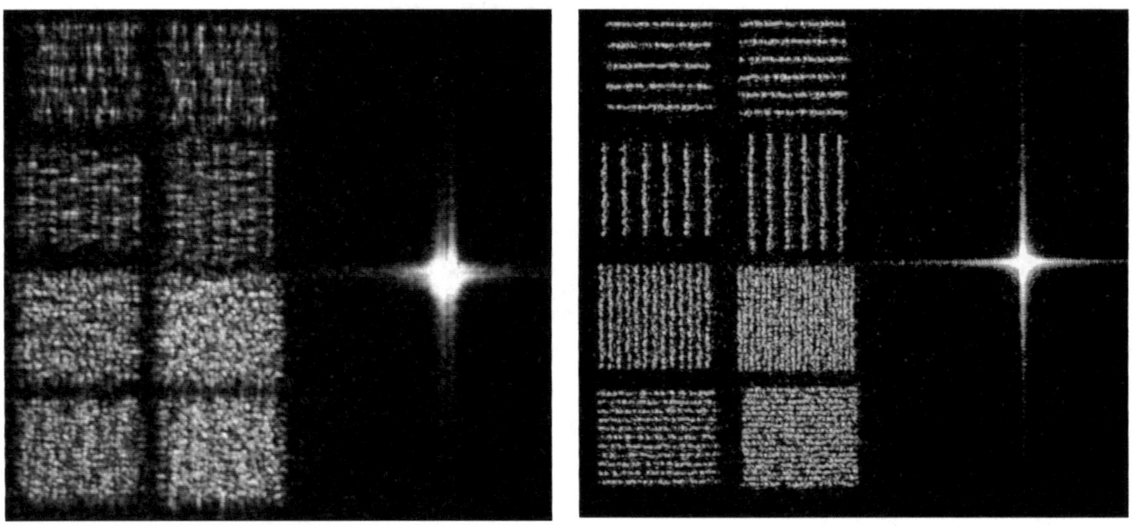

a b

Figure 3. Impulse response of the hologram (384×384 pixels) containing gratings with the periods of 2, 3, 7 and 8 pixels (the central arrangement of the hologram on a DMD-matrix of the chip s1076 7071b (a), of the best chip s1076 7402(b)

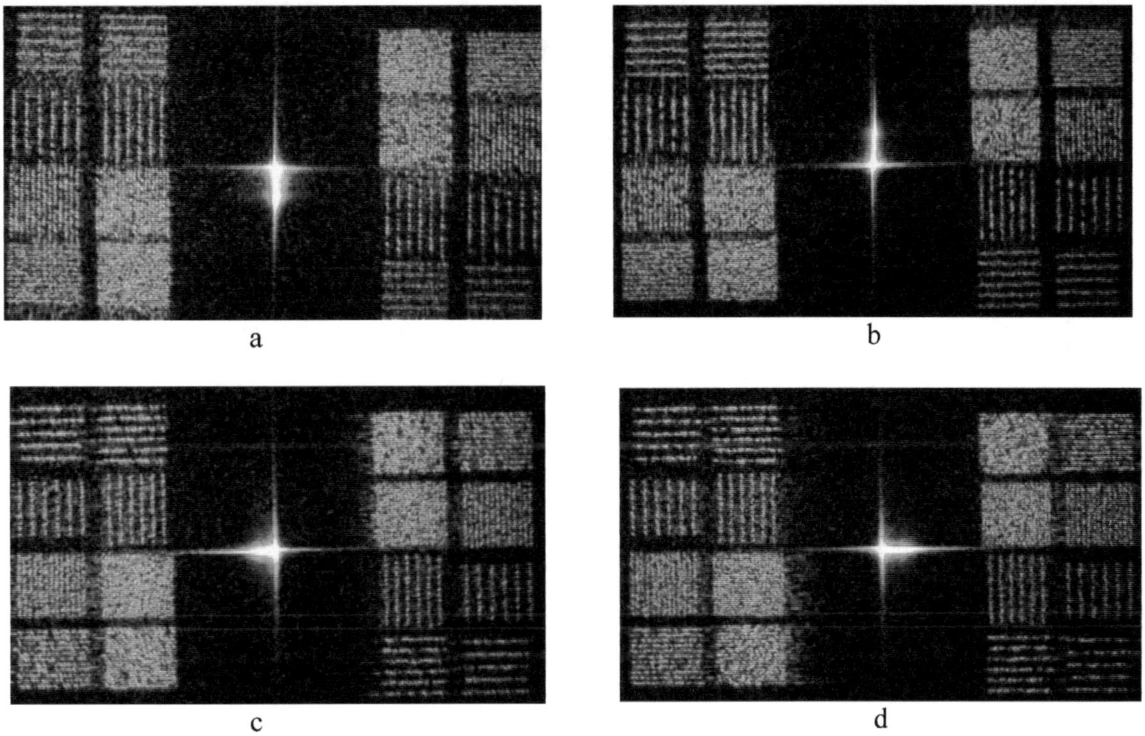

Figure 4. Impulse responses of the hologram used in Fig. 3b, outputted to the centres of various edges of matrix of the best chip s1076 7402: a) lower edge of a matrix; b) upper edge; c) left edge; d) right edge

4. Conclusion

Optical quality of three DMD-chips of two various models was experimentally estimated in this work. The obtained values showed a significant difference of devices quality, and even within one model. All DMD chips estimated in this work were taken from consumer DLP projectors, however one of models of the DMD chips (s1076 7402) is delivered as a part of Texas Instruments DLP Discovery 4100 Development Kit.

The modulator of model s1076 7402 appeared the best among the tested ones, it allows to obtain well solvable reconstructed image from holograms with the size up to 380×380 pixels (L_f = 5.3 mm). These values for two other DMD-chips have are equal to 150×150 (L_f = 2.1 mm) and 100×100 pixels (L_f = 1.4 mm).

Thus, it was established that according to modulators documentation it is impossible to find out, whether their optical quality is satisfactory for use as the holograms displaying device. It is necessary to analyze quality of an individual DMD-chip, because technological errors can have significant impact on resolution of system when reconstructing of the holograms outputted on it, and essential differences in quality are observed even among copies of one model.

Acknowledgments

This work was partially supported by Ministry of Education and Science of the Russian Federation, task 3.2066.2014/K.

Journal of Physics: Conference Series **737** (2016) 012074 doi:10.1088/1742-6596/737/1/012074

References

[1] Nesbitt R S, Smith S L, Molnar R A, Benton S A 1999 *Proceedings of SPIE* **3637** pp 12-20
[2] Huebschman M L, Munjuluri B, Garner H R 2003 *Optics Express* **11** pp 437-445
[3] Munjuluri B, Huebscham M L, Garner H R 2005 *Applied Optics* **44 (24)** pp 5076-5085
[4] Rodin V G and Starikov S N 2012 *Journal of Optical Technology* **79 (4)** pp 212-216
[5] Molodtsov D Yu, Rodin V G and Starikov S N 2014 *Journal of Physics: Conference Series* **536** p 012024
[6] Molodtsov D Yu, Rodin V G, Starikov S N 2015 *Physics Procedia* **73** pp 338–342
[7] Molodtsov D Yu, Cheremkhin P A , Krasnov V V, Rodin V G 2016 *Proceedings of SPIE* **9889** p 988926
[8] Chao T-H, Lu T T, Walker B, Reyes G *Proceedings of SPIE* **9094** p 909402